Research Anthology on Securing Mobile Technologies and Applications

Information Resources Management Association
USA

Volume I

IGI Global
PUBLISHER of TIMELY KNOWLEDGE

Published in the United States of America by
IGI Global
Information Science Reference (an imprint of IGI Global)
701 E. Chocolate Avenue
Hershey PA, USA 17033
Tel: 717-533-8845
Fax: 717-533-8661
E-mail: cust@igi-global.com
Web site: http://www.igi-global.com

Library of Congress Cataloging-in-Publication Data

Names: Information Resources Management Association, editor.
Title: Research anthology on securing mobile technologies and applications / Information Resource Management Association, editor.
Description: Hershey, PA : Information Science Reference, [2021] | Includes bibliographical references and index. | Summary: "This book includes a variety of contributions that seek to fill the void for an all-encompassing and comprehensive reference book covering the latest and most emerging research, concepts, and theories for preventing, identifying, and mitigating the spread of fake news both within a human and technological context"-- Provided by publisher.
Identifiers: LCCN 2021009745 (print) | LCCN 2021009746 (ebook) | ISBN 9781799885450 (hardcover) | ISBN 9781799886013 (ebook)
Subjects: LCSH: Mobile computing--Security measures. | Cell phone systems--Security measures. | Computer crimes--Prevention.
Classification: LCC QA76.59 .R48 2021 (print) | LCC QA76.59 (ebook) | DDC 005.8--dc23
LC record available at https://lccn.loc.gov/2021009745
LC ebook record available at https://lccn.loc.gov/2021009746

British Cataloguing in Publication Data
A Cataloguing in Publication record for this book is available from the British Library.

List of Contributors

Table of Contents

Preface

The widespread adoption and integration of mobile technologies into daily life has continued to spread and affect many aspects of society. With the usage of mobile devices and technologies being so widespread, from using mobile devices to access banking information, social media, and healthcare records to making actual payments, there are subsequently massive concerns regarding security and privacy of the information being stored and transferred across these technologies and applications. New methods and models for security and privacy have emerged to counteract and resolve these threats that can be highly detrimental and invasive. The need for accurate threat detection, risk analysis technologies, and mitigation efforts is more important now than ever. Along with technological solutions, users have proven to not be fully aware of potential threats or consequences of using mobile devices and especially downloading suspicious apps or clicking on links in SMS messaging, leading to a need for better user awareness in these areas. Only by advancing technology to become more secure and having users better understand methods for protection and risks will mobile technologies and applications become safer for the future. By presenting research in this area, the latest tools, technologies, and security methods and models can be better developed and disseminated to secure mobile technology.

Thus, the *Research Anthology on Securing Mobile Technologies and Applications* seeks to fill the void for an all-encompassing and comprehensive reference book covering the latest and most emerging research, concepts, and theories for preventing, identifying, and mitigating the spread of fake news both within a human and technological context. This two-volume reference collection of reprinted IGI Global book chapters and journal articles that have been handpicked by the editor and editorial team of this research anthology on this topic will support IT specialists, security analysts, technicians, engineers, computer scientists, technologists, professionals, students, researchers, and academicians with an advanced understanding of critical issues and developments in privacy, security, risk analysis, and threat detection of mobile technologies and applications.

The *Research Anthology on Securing Mobile Technologies and Applications* is organized into three sections that provide comprehensive coverage of important topics. The sections are:

1. Cyber Attacks, Threat Detection, and Security Frameworks;
2. Data Protection, Privacy, and User Risk Awareness;
3. Security Issues for Apps, Mobile Banking, and Mobile Payment.

The following paragraphs provide a summary of what to expect from this invaluable reference tool.

Section 1, "Cyber Attacks, Threat Detection, and Security Frameworks," opens this comprehensive reference work with research on the types of cyber-attacks taking place in mobile technologies, the security frameworks that can be put in place, and the development of systems that can detect potential threats. The opening chapter to this book, "Smartphone Security and Forensic Analysis," by Profs. Deepak Kumar Sharma, Kartik Kwatra, and Manan Manwa of Netaji Subhas University of Technology, India, discusses the need for mobile forensic analysis and basic smartphone vulnerabilities along with the types of attacks and their effects on smartphones followed by the security mechanisms employed to deal with the attacks and threats. The next chapter, "Mobile Agent Communication, Security Concerns, and Approaches: An Insight into Different Kinds of Vulnerabilities a Mobile Agent Could Be Subjected to and Measures to Control Them," by Profs. Kamat Pooja, Gite Shilpa, and Patil Shruti of Symbiosis Institute of Technology Pune, India, examines the various security issues in Mobile Agent systems and approaches used to overcome them. The following chapter, "Recent Advances Delivered in Mobile Cloud Computing's Security and Management Challenges," by Profs. Christos Stergiou and Kostas E. Psannis of the University of Macedonia, Greece, defines the most important issues and challenges in the field of mobile cloud computing (MCC) technology by illustrating the most significant works related to MCC during recent years. The chapter "Identification of Cryptographic Vulnerability and Malware Detection in Android" by Profs. Anjali Kumawat, Anil Kumar Sharma, and Sunita Kumawat of Government Women Engineering College, Ajmer, India, implements static and dynamic analysis of android application for vulnerable and malicious app detection. Next, "Android Botnets: A Proof-of-Concept Using Hybrid Analysis Approach" by Prof. Victor Chang of Teesside University, UK; Prof. Ahmad Karim of Bahauddin Zakariya University, Pakistan; and Prof. Ahmad Firdaus of the Faculty of Computer Systems and Software Engineering, Malaysia proposes a hybrid analysis framework combining static and dynamic analysis as a proof of concept to highlight and confirm botnet phenomena in Android-based mobile applications. Another chapter, "Preserving Security of Mobile Anchors Against Physical Layer Attacks: A Resilient Scheme for Wireless Node Localization," by Prof. Mrinal Kanti Naskar of Jadavpur University, India; Prof. Swarup Kumar Mitra of MCKV Institute of Engineering, India; and Prof. Rathindra Nath Biswas of A. J. C. Bose Polytechnic, India, introduces a new security scheme for mobile anchors avoiding the physical layer attacks towards localization in wireless sensor networks (WSNs). "A Framework for Various Attack Identification in MANET Using Multi-Granular Rough Set" by Profs. Debi Prasanna Acharjya and N. Syed Siraj Ahmed of VIT University, Vellore, India uses a multigranular rough set (MGRS) for the detection of intrusion and detection in MANET. The chapter "A Systematic Study and Analysis of Security Issues in Mobile Ad-hoc Networks" by Prof. Bibudhendu Pati of C. V. Raman College of Engineering, Bhubaneswar, India and Profs. Jhum Swain and Binod Kumar Pattanayak of Siksha 'O' Anusandhan University (Deemed), Bhubaneswar, India discusses the various protocols developed for security issues in MANET and IoT and presents a comparative analysis on their performance. In the following chapter, "A Survey of Efficient Trust Management Schemes in Mobile Ad-Hoc Network: Reliable Trust Management Framework of MANET," by Prof. S. Kaur of PURCITM, Punjab, India and Prof. J. Kaur of Punjabi University, Patiala, India, a survey is carried out to find some of the limitations behind the existing works which has been done by the researchers to implement various approaches thus to build the trust management framework. Next, "Communiqué Issues in MANET and VANET Protocols With Network Security Disquiet" by Prof. Mamata Rath of Birla Global University, India; Prof. Bibudhendu Pati of Rama Devi Women's University, India; and Prof. Jhum Swain of Centurion University of Technology and Management, India discusses one of the rule challenges in MANET which is to plot a ground-breaking security course of action that can shield

MANET from different routing assaults. Another chapter, "Optimized Three-Dimensional Security Framework to Mitigate Risks Arising From BYOD-Enabled Business Environment," by Profs. Prabhakar Rontala Subramaniam, Lizzy Oluwatoyin Ofusori, and Ncamiso Nkululeko Jahalenkhosi Dlamini of the University of KwaZulu-Natal, South Africa, addresses three categories of security threats in bring your own device (BYOD) and the existing security frameworks in place along with a potential framework as a future solution to protect BYOD-enabled business environments. "Simple Screen Locking Method Using Randomly Generated Number Grid on Image" by Profs. Hazleen Aris, Zailani Ibrahim, and Ariff Azman of the Universiti Tenaga Nasional, Kajang, Malaysia presents a new screen locking method (SyS) that is both simple, secure from shoulder surf and smudge attacks, and not expensive to implement. Next, "Phishing Attacks in Mobile Platforms" by Profs. Thangavel M., Yaamine A. M., and Nandhini J. T. of Thiagarajar College of Engineering, India discusses about the types of Phishing attacks in Mobile Platforms. The chapter "A Novel Approach to Detect Spam and Smishing SMS using Machine Learning Techniques" by Profs. Ankit Kumar Jain and Neelam Choudhary of the National Institute of Technology, Kurukshetra, India and Prof. Sumit Kumar Yadav of the Income Tax Department, Government of India, India presents a novel method to filter smishing message from spam message that is divided into two phases. The chapter "Feature Based Approach for Detection of Smishing Messages in the Mobile Environment" by Profs. Ankit Kumar Jain and B. B. Gupta offers 10 novel features that distinguish the fake messages from the ham messages and identifies 19 of the most suspicious keywords used by attackers to lure victims. This section concludes with "Securing Communication 2FA Using Post-Quantic Cryptosystem: Case of QC-MDPC- Mceliece Cryptosystem" by Profs. Kouraogo Yacouba, Orhanou Ghizlane, and Elhajji Said, LabMiA-SI, Faculty of Sciences, Mohammed V University in Rabat, Rabat proposes an SMS encryption mechanism using a post quantic cryptosystem quasi-cyclic MDPC and an electronic signature of the OTPs.

Section 2, "Data Protection, Privacy, and User Risk Awareness," presents extensive coverage and case studies on the latest findings on privacy issues with mobile devices, how data can be protected, and user awareness of threats. Beginning this section is the chapter "Smartphone Confrontational Applications and Security Issues" by Prof. Abhishek Kumar of Chitkara University, Punjab, India; Prof. Jyotir Moy Chatterjee of Lord Buddha Education Foundation (Asia Pacific University of Technology & Innovation), Kathmandu, Nepal; and Prof. Pramod Singh Rathore of ACERC, Delhi, India, which discusses applications and services that have been designed to track down and locate lost or stolen smartphones. The next chapter, "Theft Preventive Measures for Interconnected Personal Computer Devices as Proactive Physical Security of Data," by Profs. Ekaterina Pshehotskaya and Oleg Mikhalsky of Moscow Polytechnic University, Moscow, Russia, is concerned with the arising problems and implications of physical security and privacy of personal and control data on portable computer devices, especially smartphones. The following chapter, "Mobile Cyber Information Leakage: Users' Attitudes and Information Protection Strategies," by Profs. Hongbo Lv and Zhiying Zhou of Zhejiang Wanli University, China, investigates users' attitudes toward cyber information leakage and methods to protect personal information. Next, "Protection of the Communication and Digital Security of Mobile Phones as a Fundamental Right to Privacy: A Study of the Jurisprudence of the Brazilian Superior Court of Justice" by Prof. Nestor Eduardo Araruna Santiago and Ana Monica Anselmo Amorim of the Universidade de Fortaleza, Brazil discusses the notion of privacy arising from the use and exchange of messages through the use of mobile phones and whether police officers in charge of patrolling and public security may have access to the messages contained therein without proper judicial authorization. Another chapter, "Achieving Secure and Privacy-Preserving in Mobile Social Networks," by Prof. Mohamed Amine Ferrag of Guelma University, Algeria and Prof. Abdelaziz Amara Korba of Badji Mokhtar-Annaba University, Algeria, proposes a security framework for achieving secure and privacy-preserving mobile social networks named ASPP. "User Identity Hiding Method of Android" by Prof. Yi Zhang of the Institute of Information Engineering, Chinese Academy of Sciences, China proposes a user identity hiding method of Android and constructs a hidden user identity in Android based on the multi-user mechanism. The chapter "Biometric Systems in Mobile Devices: A Study" by Profs. Sukhdev Singh and Chander Kant of Kurukshetra University, Kurukshetra, India presents research about the impact that mobile devices have in people's lives and the presence of biometric systems in this kind of device. The following chapter, "Cell Phone Security: User Awareness of Security Issues and Mobile Device Management," by Profs. Scott E. Mensch and LeAnn Wilkie of Indiana University of Pennsylvania, USA, examines mobile security behaviors of college students and how students may not take appropriate measures to protect their personal data from hackers and thieves. Next, "Importance of Information Security and Strategies to Prevent Data Breaches in Mobile Devices" by Prof. Maulik Desai of Swinburne University of Technology, Australia and Prof. Swati Jaiswa of SKNSITS, India looks at the different ways in which their privacy can be invaded via a mobile phone and the mindset of a cell phone hacker and, potential damages, and strategies to prevent them. "Mobile Phone Usage Patterns, Security Concerns, and Security Practices of Digital Generation" by Profs. Sonya Zhang and Saree Costa of California State Polytechnic University, Pomona, CA, USA discusses the impact of recent mobile technology advancements on mobile threat environment and mobile security practices. The chapter "Expectations and Experiences of Implementing a Mobile Secure Communication Application" by Profs. Tuomas Kari, Markus Makkonen, Lauri Frank, and Eeva Kettunen of University of Jyvaskyla, Jyvaskyla, Finland examines user expectations before and use experiences during the implementation phase of a mobile secure communication application. The following chapter, "Smart Phone Security Practices: Item Analysis of Mobile Security Behaviors of College Students," by Profs. Scott E. Mensch and LeAnn Wilkie of Indiana University of Pennsylvania, USA, explores the widespread adoption of smartphones with a particular focus on their usage within college students. Next, "Awareness, Knowledge, and Ability of Mobile Security Among Young Mobile Phone Users" by Prof. River Yan of Noble and Greenough School, Dedham, MA, USA suggests that young mobile phone users are usually not aware of potential mobile security issues when using a mobile phone. This section ends with "Coping Strategies and Paradoxes Related to BYOD Information Security Threats in France" by Prof. Paméla Baillette of the University of Bordeaux, Bordeaux, France and Prof. Yves Barlette of Montpellier Business School, Montpellier, France, which investigates employees' perception of threats to their personal information security (ISS) when using a BYOD, especially in a professional context, with a focus on a survey of 233 employees.

Section 3, "Security Issues for Apps, Mobile Banking, and Mobile Payment," reveals the latest research on security and privacy issues and solutions with mobile banking. Opening the final section in this book is "Threats and Vulnerabilities of Mobile Applications" by Profs. Thangavel M., Divyaprabha M., and Abinaya C. of Thiagarajar College of Engineering, India, which explores the treats and vulnerabilities in mobile applications and how app developers, app market administrators, and users can prevent these threats. The next chapter, "Techniques for Analysis of Mobile Malware," by Profs. Gopinath Palaniappan and Balaji Rajendran of the Centre for Development of Advanced Computing (CDAC), India; Prof. S. Sangeetha of the National Institute of Technology Tiruchirappalli, India; and Prof. NeelaNarayanan V. of VIT University, India, captures the attack surface of mobile devices, lists the various mobile malware analysis techniques, and lays the ground for research on mobile malware by providing tools for detection and analysis. The following chapter, "Native vs. Hybrid Mobile Applications as Society Enters the

Internet of Things," by Prof. Irvin Renzell Heard of the University of the Cumberlands, Pembroke Pines, USA and Prof. Norman R. Ardila, an Independent Researcher in the USA, initiates a discussion on hybrid mobile app security, market, role in IoT, advantages, disadvantages, security challenges, and the best practices as society enters a new paradigm envisioned as a ubiquitous global network. Next, "Android Application Security" by Prof. Maurice Dawson of Illinois Institute of Technology, USA and Profs. Marwan Omar, Derek Mohammed, Van Nguyen, and Mubarak Banisakher of Saint Leo University, USA investigates how hackers use malicious apps as attack vectors to compromise Android smartphones and steal confidential data and that no security solutions exist to combat malicious apps. Another chapter, "A Hybrid Approach to Detect the Malicious Applications in Android-Based Smartphones Using Deep Learning," by Profs. Manokaran Newlin Rajkumar, Varadhan Venkatesa Kumar, and Ramachandhiran Vijayabhasker of Anna University Coimbatore, India, proposes a hybrid method based on deep learning methodology to identify the malicious applications in Android-based smart phones. The chapter "Towards Privacy Risk Analysis in Android Applications Using Machine Learning Approaches" by Profs. Kavita Sharma and B. B. Gupta of the National Institute of Technology, Kurukshetra, India presents an approach for privacy risk analysis in Android applications using machine learning that can analyze and identify the malware application permissions. "How Do Mobile Applications for Cancer Communicate About Their Privacy Practices? An Analysis of Privacy Policies" by Prof. Rukhsana Ahmed of University of Albany, SUNY, USA; Prof. Devjani Sen of Algonquin College, Canada; and Prof. Zerin Mahzabin Khan of Virginia Tech, USA, assesses the content of online privacy policies of selected cancer mobile apps through applying a checklist and performing an in-depth critical analysis to determine how the apps communicated their privacy practices to end users. The following chapter, "How Relevant Are Risk Perceptions, Effort, and Performance Expectancy in Mobile Banking Adoption?" by Profs. Aijaz A. Shaikh and Heikki Karjaluoto of the University of Jyväskylä, Jyväskylä, Finland and Prof. Richard Glavee-Geo of NTNU-Norwegian University of Science and Technology, provides a comprehensive overview of the adoption process using evidence from m-banking adoption in Pakistan. Next, "The Mediating Role of Perceived Value in the Effect of Multi-Dimensional Risk in Mobile Banking" by Profs. Stan Williamson and Long Pham of the University of Louisiana at Monroe; Prof. Cyrus Mohebbi of New York University, New York City, USA; Prof. Binh Nguyen of Military Commercial Joint Stock Bank, Hanoi, Vietnam; and Prof. Hien Nguyen of Foreign Trade University, Hanoi, Vietnam, examines which risk factors contribute most to perceived risk in Vietnam's mobile banking environment, explores the relationships among perceived risk, perceived value and intention to use mobile banking, and investigates any mediating role perceived value may have in the relationship between perceived risk and intention to use mobile banking. The chapter "An Investigation into Permissions Requested by Mobile Banking on Android Platform" by Profs. Latifa Er-Rajy and M. Ahmed El Kiram of Cadi Ayyad University, Marrakesh, Morocco investigates the danger of the permissions requested by mobile banking applications, their effects on sensitive user data, and their relationship with the attack called "man in the middle" and its different forms. Following is the chapter "Mobile Commerce Technologies and Management" by Prof. Kijpokin Kasemsap of Suan Sunandha Rajabhat University, Thailand, which reveals the prospect of mobile commerce (m-commerce); m-commerce and trust, privacy, and security issues; m-commerce adoption and technology acceptance model (TAM); and the significant perspectives on m-commerce. Next, "A Study of e-Commerce Transaction Risk Assessment Model in Mobile Internet" by Prof. Zhihong Huang of Shanxi Institute of Technology, Shanxi, China analyzes the risk of mobile e-commerce transactions and uses a reasonable evaluation system to build a mobile e-commerce transactions risk assessment model. The chapter "LD-Based Framework to Mitigate Threats in Mobile Based Payment System" by

Prof. Jitendra Singh of Dyal Singh Evening College, University of Delhi, New Delhi, India identifies threats applicable to smartphones and classifies them based on a broader category of threats. This book concludes with the chapter "Identity Authentication Security Management in Mobile Payment Systems" by Prof Xianrong Zheng of Old Dominion University, Norfolk, USA; Profs. Feng Weng, Sun Mingwei, and Ge Bao Shan of Jilin University, Changchun, China; Prof. Yong Chen of Texas A&M International University, Laredo, USA; Prof. Hong Wang of North Carolina A&T State University, Greensboro; and Prof. Li Haihua of Changchun Automobile Industry Institute, Changchun, which proposes an anonymous authentication and management flow for mobile payment to support secure transaction to prevent the disclosure of users' information and to reduce identity theft.

Although the primary organization of the contents in this work is based on its three sections, offering a progression of coverage of the important concepts, methodologies, technologies, applications, social issues, and emerging trends, the reader can also identify specific contents by utilizing the extensive indexing system listed at the end. As a comprehensive collection of research on the latest findings related to security and privacy issues in mobile devices and applications, the *Research Anthology on Securing Mobile Technologies and Applications* provides IT specialists, engineers, computer scientists, technologists, professionals, students, researchers, academicians, and all audiences with a complete understanding of how security issues in mobile technologies are being addressed with new models, methods for threat detection, user awareness, and security protocols. Given the issues that have risen since technology has become integrated in daily life, the security of these mobile devices and applications that are being utilized every day has become a critical exploration that this extensive book investigates and addresses with the most pertinent research on securing mobile technologies and applications.

Section 1

Cyber Attacks, Threat Detection, and Security Frameworks

Chapter 1
Smartphone Security and Forensic Analysis

Deepak Kumar Sharma
https://orcid.org/0000-0001-6117-3464
Netaji Subhas University of Technology, India

Kartik Kwatra
Netaji Subhas University of Technology, India

Manan Manwani
Netaji Subhas University of Technology, India

ABSTRACT

Modern day smartphones are capable of performing every single task that a desktop computer can do. Smartphones being a new technology with less than a decade in the communication industry are vulnerable to security attacks and data leakage. In the current scenario, Android devices are the majority are the prime target of hackers and criminal organisation. Thus, smartphone security is of utmost priority for software companies. In this chapter, the authors begin by discussing the need for mobile forensic analysis and discussing basic smartphone vulnerabilities. they then discuss the types of attacks and their effects on smartphones followed by the security mechanisms employed to deal with the attacks and threats. they also list and give a brief description of tools used commercially for forensic analysis of smartphones, and two experiment-based case studies are provided to allow the readers to get a better understanding of the merits of existing works and practical application of investigation process.

INTRODUCTION

In the past few years, technology has seen a huge boost. Advancements in electronics industry have helped software manufacturers in a huge way. With more processing power and memory, modern day computers have perform complex calculations in just a few seconds. Heavy computers have been replaced by small

DOI: 10.4018/978-1-7998-8545-0.ch001

portable yet powerful laptops and smartphones. These smartphones having such advanced technology are capable of doing online payments, e-banking, network data transfers etc.

Smartphones, being a new technology with less than a decade in the communication industry become vulnerable to security attacks and data leakage. To counter the problem of data theft and misuse of personal information, various security defects need to be analysed and hence the need of smartphone forensics arises.

Smartphone security has been a major topic of study in the recent years. Apart from implementing security features, the organisation also needs to gain customer trust on the security features that are present in the smartphone device.

In a research conducted by Matthews and Pierce (2009), they found out that while users preferred mobile devices for online searching and shopping, most of them defer to computer devices to make online payments.

Chin et al. (2012) expanded more on this by measuring user confidence on smartphones devices and concluded that smartphone security designs should be more user friendly and should offer better support for privacy of personal data.

Several studies have analysed the security aspect of smartphone based applications. Riadi et al. (2018) evaluated forensic tools on LINE messenger application in a smartphone running Android Operating System. Ntantogian (2014) evaluated the possibility of recovering authentication credentials of mobile applications from the volatile memory of Android mobile devices.

Furthermore, Ahvanooey (2017) surveyed smartphone security by evaluating security features of different smartphone operating systems and possible malwares, attacks and threats that can put user privacy to risk.

Key Concepts

Digital Forensics: Digital forensics is in simple words means using scientific and proven methods to present, validate, verify and analyse evidence derived from digital data. This evidence is used further to reconstruct malicious activities in order to gain information about any criminal activity related with the data.

Cyber-Attack: A cyber-attack is an illegal attempt by any person or an organisation to breach a software system and steal or destroy unauthorized information from the victim. Mobile cyber-attack are targeted attacks used by hackers to disrupt any working system by stealing private and confidential from smartphones devices.

Mobile Security: Mobile Security refers to the techniques used to protect portable devices like smartphones, laptops, tablets etc. from threats and attacks that jeopardize personal data security and private information of enterprises and individual users.

Some common types of cyber-attacks are:

- Phishing
- Denial of Service
- Malware
- SQL injection

DATA SECURITY AND LEAKAGE

Data Security is the process of securing digital data from any type of unauthorised access or a cyberattack. It involves encrypting data through cryptographic techniques and protecting private information through various authentication processes.

A system can never be perfectly secure. Hackers all over the world use different techniques to gain access to someone else's computer in order to steal/modify/erase information. The illegal transfer of classified information from one computer to another in known as Data Leakage. Data Leakage happens when systems are not secure enough and are breached easily. Hackers usually try to find the weakest link in the network and breach into the victim's computer through that link.

Nowadays, with rising popularity of smartphones and increasing ability of smartphones to perform almost every task that a computer can perform, smartphones have become most vulnerable to security attacks. Hence, there is an urgent need to improve the security of smartphones.

Data Leakage is a serious threat to individuals as well as organizations. It can lead to financial, reputational and other social and non-social losses. Data leakage can be caused by both internal and external factors. Most hackers and attackers have monetary gains as the objective of data theft but data leakage can occur due to variety of other reasons like destruction of evidence, defamation, evidence retrieval etc.

Some common causes of data leakage are:

1. **Weak Passwords:** A study by Verizon, US found out that close to 45% of data breaches involved stolen passwords. Some attackers use complex algorithms, while others rely on intelligent guessing. Smartphone applications employ various techniques like multi-factor authentication, restricted password input attempts etc to prevent password leaks.
2. **User Negligence:** Leaving the smartphone unlocked unattended, accidentally emailing sensitive information, entering passwords or other sensitive details on unverified apps and pages are some examples of user negligence.
3. **Malicious Attacks:** An attacker might inject malware into the smartphone through apps, links, email attachments etc. An attacker may also monitor the network that the smartphone uses and extract information through it. This is called network snooping.

Holding data hostage for a ransom is a new way of attack that has surfaced in the recent past and has gained a lot of popularity. An attacker may also exploit the inherent vulnerabilities of the smartphone and its operating system. Port scanning is one such method.

4. **Phishing:** It is a method of fraudulently acquiring sensitive information by impersonating a website or an app. The user may input information like password, bank details etc thinking that the website is in fact the original one. Now a days, mobile and desktop browsers come equipped with anti-phishing toolbars and other plugins that can prevent phishing attacks to some extent.

Smartphone forensics is one of the most popular topics in the software industry. With the ever increasing number of applications being developed daily and improving functionality of smartphones, it becomes essential to analyse and develop new security features for protection of digital information of individuals and organisations.

DATA STORAGE IN SMARTPHONES

Data is stored in 3 major areas as shown in Figure 1:-

1. **SIM Card:** A sim card stores most of the contact information and SMS texts. This includes One Time Passwords(OTP), Access tokens etc.
2. **Device Memory:** Device memory stores user created data like music, movies, photos and other personal media information. Application logs are also stored in the device memory which include past texts or internet searches.
3. **Cloud:** Cloud refers to online servers where data is stored so that it can be accessed through various devices and even at times when you don't have your personal device with you. This is the most vulnerable type of data storage system as the data is stored on some server and anyone with access to the server can steal your data.

Figure 1. Data storage in smartphones

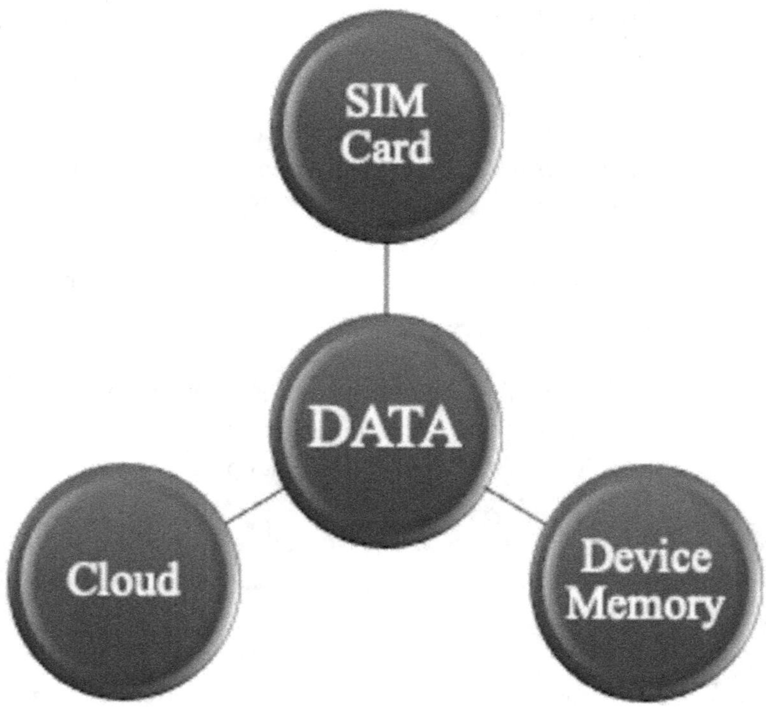

ANALYSIS OF SMARTPHONE SECURITY ANALYSIS

Security Objectives

The basic objective of smartphone security is to protect the assets of a smartphone against threats and attacks. The assets of a smartphone can be categorized into 3 categories as shown in Figure 2 (Woongryul Jeon, Jeeyeon Kim, Youngsook Lee, and Dongho Won, 2011):

Figure 2. Smartphone assets

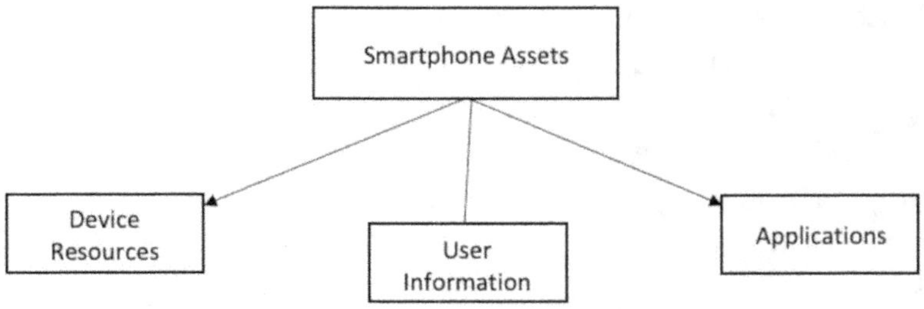

Device resources like battery, RAM, CPU are considered assets. If a malicious attacker gets access to the smartphone, he can harm these assets by overcharging, altering the circuit board with the help of tools, destroying it by force etc. Smartphone itself is also regarded as an asset since one can make cellular calls through it and connect to the internet.

User information needs to be guarded against attacks. It can be considered as the most important asset of a smartphone user. Information includes data such as contacts, messages, call logs, location information, saved documents, media files, e-mails, browser cache etc. It also includes all the data that is transmitted out of the smartphone. Access to information stored on the smartphone can cause serious damages to the user. For example, an attacker may extract saved passwords, documents etc.

Applications manage all the information and data on the smartphone, which is why security of applications is essential. For example, Smartphone browsers store user IDs and passwords for of the user since it is very inconvenient to type the password every time on a small screen. If the browser application is hacked, all this data will be compromised (Milad Taleby Ahvanooey, Qianmu Li, Mahdi Rabbani and Ahmed Raza Rajput, 2017). Hence applications are considered an asset that need to be secured.

Figure 3 represents the Security Objectives associated with a smartphone.

Figure 3. Security objectives

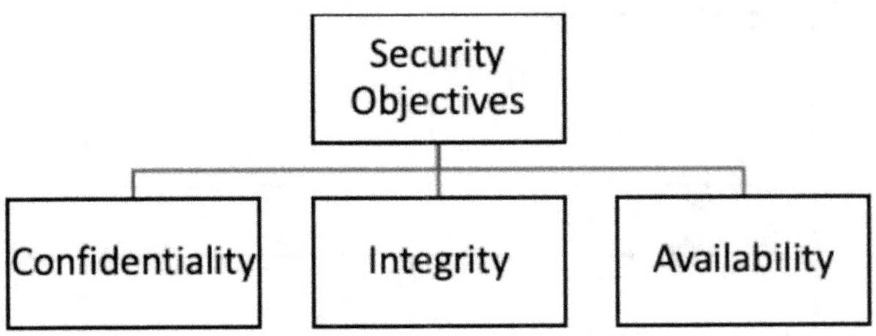

Confidentiality: There is need to protect the confidential information from unauthorized access. Confidentiality applies to both storage and transmission of information. For example: A user might store sensitive information like bank account details, account passwords, messages etc on their smartphone and such information should stay only with the owner pf the smartphone i.e. it should be confidential.

Integrity: Integrity means that the information can be modified only by authorized entities and through authorized mechanisms. Integrity may be compromised be factors other than malicious attacks like sudden power surge or abrupt crashing of smartphone software.

Availability: The information on the smartphone should be accessible and available to the user. Any information is useless if it is not available to the user. Attacks may delete or make data unavailable to the user. An attacker might also encrypt the data on the smartphone and render it useless for the user.

Vulnerabilities

Smartphone vulnerabilities as per Milad Taleby Ahvanooey, Qianmu Li, Mahdi Rabbani and Ahmed Raza Rajput (2017), can be categorised into two categories as shown in Figure 4:

1. Internal to smartphone
2. External to smartphone

Figure 4. Smartphone vulnerabilities

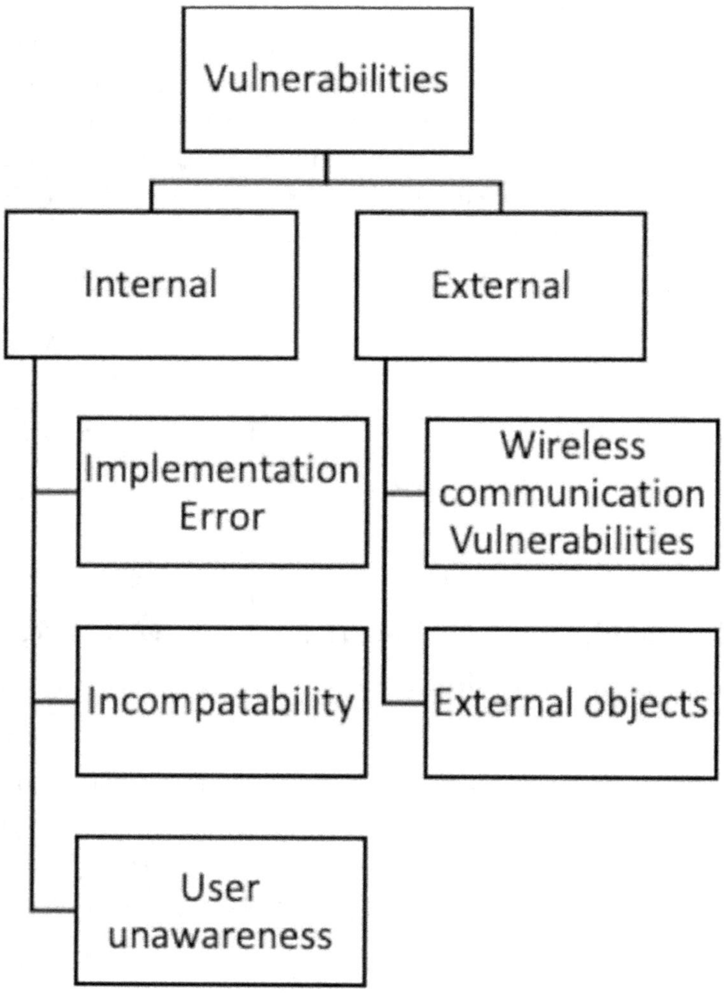

Internal to Smartphones

A smartphone is vulnerable to attacks when malfunctions are caused due to errors in implementation of internal code. Such errors may be present in kernel code. Malfunctions can also be caused due to presence of arbitrary code. A malicious attacker can easily exploit such vulnerabilities and harm the device.

Sometimes, an application installed on the smartphone may not be compatible with some already existing application or the platform itself and as a result it might get disabled and fail to perform the expected function. Such incompatibility vulnerabilities can be exploited by an attacker to disable some already existing app on a smartphone and extract information form the smartphone.

One of the very common vulnerabilities now a days is the lack of user awareness. A device owner might install application from untrusted sources, may connect to untrusted Wi-Fi or Bluetooth networks. An unaware user may visit harmful websites.

For example: Phishing websites are webpages that impersonate the content of some other site and have a similar looking URL. The user may enter sensitive information on these websites like passwords, credit card details etc.

EXTERNAL TO SMARTPONE

Wireless networks can be exploited to modify or block information by traffic snooping, sniffing or eavesdropping. External objects like PC, cellular base station can introduce vulnerabilities due to improper management.

THREATS

Threats give shape to smartphone attacks by exploiting the vulnerabilities (Sharma, K., & Gupta, B. B., 2018). The threats to smartphone security and the vulnerabilities exploited by them are shown in in Table 1:

1. **Malware**
   Malware is a software that is specifically written to harm, damage, or gain unauthorized access to a smartphone. Malware can leak private information, run arbitrary code on a smartphone to hog up its resources and even exploit services that incur cost to the user like SMS, MMS etc.

2. **Wireless network attacks**
   Wireless networks can be exploited to modify or block information by traffic snooping, sniffing or eavesdropping.

3. **Break in**
   An attacker can gain admin access of a smartphone by using logical flaws and loopholes in the code or through code injection.

4. **Phishing**
   Phishing sites, SMS phishing and messenger phishing are different types of phishing that exploit user unawareness to extract private data of a user

5. **Denial of service**
   A DOS attack is initiated by sending bogus requests to the base station, web servers etc. so that the smartphone device is unable to access these, hence rendering them unavailable for the smartphone user.

6. **Malfunction**
   This is a treat caused by user unawareness. A user may unknowingly disable some features or functionalities that may lead to device malfunction.

7. **Loss**
   User may lose his/her smartphone device hence threatening all the information present on it.

8. **Platform alteration**
   Rooting (android) and jail breaking(iPhone) are ways of altering a smartphone platform intentionally.

Table 1. Security threats

S.No.	Threats	Vulnerabilities
1.	Malware	Implementation error User unawareness External objects
2.	Wireless Network attacks	Wireless communication
3.	Denial of service	Wireless communication External objects
4.	Break-in	Implementation error
5.	Phishing	User unawareness
6.	Malfunction	Incompatibility User unawareness
7.	Loss	User unawareness
8.	Platform alteration	User unawareness

Effects of Attacks

Attacks can affect and harm the smartphone in a number of ways some of which are:-

- **Hardware Malfunction:** Attacks can cause the hardware to malfunction. For e.g. cause the speaker to give no output, result in hard bricking or blocking of smartphone motherboard
- **Privacy Concerns:** By exploiting the permission granted by the users, it can cause malicious activities such as exposing location data or transmitting sensitive and private data

- **Denial of Functionality:** Core functionalities of the smartphone can be blocked by the attacker
- **Unauthorized Access:** Attacker can misuse the functionalities that cost the user such as accessing the net, sending SMS/MMS.
- **Loss of User Experience:** Attacker can cause a loss of good experience by sending spam advertisements.
- **Permanent Data Loss:** All data may be edited or wiped from the device by an unauthorized person.

Security Mechanisms

Security mechanisms as cited by Sharma, K., & Gupta, B. B., 2018 are solutions to prevent the smartphone against threats and vulnerabilities. The need for security mechanisms arises from the fact that solutions like antivirus only prevent the device from outside attacks caused by malware etc. but do not prevent it from attacks caused from inside using implementation errors and or unawareness.

Security mechanisms can be classified into 3 categories:

1. **Add-On Applications:** It involves installing of suitable applications on the smartphone by the user. Examples of such applications are anti-virus, anti-malware, vpn, firewall etc.
2. **System Modification:** This involves modifying kernel code to make smartphone more resistant to attacks.
3. **System Add-Ons:** This refers to system updates that the manufacturer provides. Regularly updating a smartphone updates its security patches and makes it resistant to newly invented viruses

Figure 5. Security mechanisms

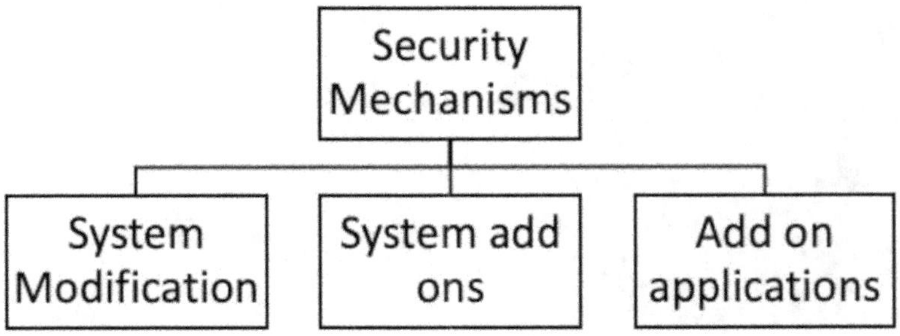

SMARTPHONE SECURITY INVESTIGATION

Various methods of investigating the security of a smartphone are discussed in this topic. We begin by looking at Android Debug Bridge(ADB), a command line tool used for monitoring android smartphones. It is made up of 3 components: a client, a daemon and a server.

Table 2. Security mechanism

Mechanism	Description	Type
Anti-Virus	Scans files on the memory, email and attachments, cookies from the web etc for virus and removes it.	Add-on application
Secure-API	It provides cryptographic functions for applications	System add-on
Firewall	A firewall protects the smartphone against network attacks	System modification
Spam filter	It blocks SMS, MMS, emails from unknown sources.	Application add-ons.
Authentication	Prevents unauthorized users from accessing the data on smartphone.	System Modification
Updating smartphone software	Regularly updating the smartphone software makes it immune against new viruses and threats	System Modification

Android Debug Bridge(ADB)

According to the google developer documentation, ADB is a "versatile command-line tool that lets you communicate with a device." It provides access to a Unix shell that can be used to run various commands on the device.

INSTALLING ADB (on Windows)

1. Head to android developers website and download the appropriate zip file.

Figure 6. Screenshot of download page

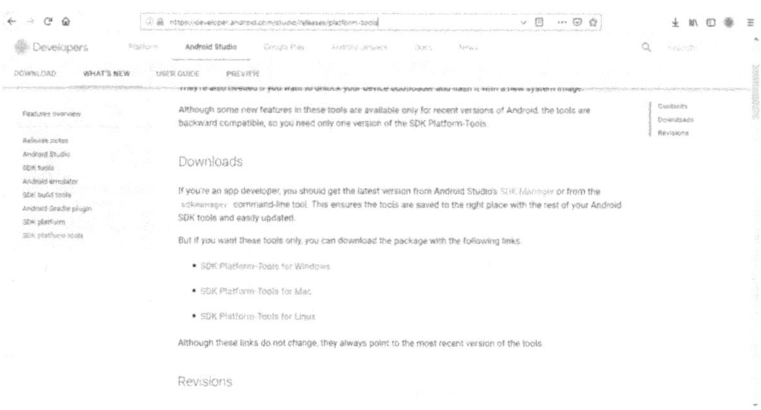

2. Extract the contents of the zip folder. Navigate to /platform-tools. Add the location of "adb.exe" to the PATH in the environment variables.
3. Open a command prompt window and run the command adb to check if it is correctly installed.

Figure 7. Screenshot of command prompt

Some Basic Commands

Here we list some basic commands used to perform basic tasks such as interacting with file system, monitoring the resources, installing apk's etc.

- **adb Devices**: Gives a list of devices currently attached through the bridge to the pc
- **adb Install**: This command is used to install an apk file on the smartphone
- **adb Push**: This command is used to send some data from the host pc to the connected device
- **adb Pull**: It is used to pull or fetch some data from the smartphone device to the pc
- **adb Shell**: Adb starts a UNIX shell that can be used to run a variety of commands on the connected device. It is a very powerful tool and can be used for penetration testing as well as launching various attacks.

Attacks Launched Using ADB

1. **Lock Screen Hack**: An android smartphone is unlocked using some kind of pin, password or gesture pattern. On android devices having android version less than or equal to 5.0, this information is loaded in a key located at data/system. If an attacker connects to an android smartphone through the adb, then by using the "adb shell" command the attacker can delete the gesture.key or password.key file and bypass the lock screen and hence have access to all the data stored on the smartphone.
2. **Physical Driveby**: This can be initiated by physically having he access to a device for a very short period of time hence the name "Physical driveby". An attacker could plug in to the smartphone device at coffee shop or some other place having public charging kiosk and enable adb with just one command. The attacker can then return to their system and access the device as long as it is connected to the same public Wi-Fi.

Port Scanning

Port scanning is a technique that allows an authorized admin or a malicious attacker identify what services are running and are publicly accessible. Applications offer network services through ports and hence are vulnerable to attacks (Sharma, K., & Gupta, B. B., 2016).

Before discussing various techniques for port scanning, we need to know how a TCP connection is established.

Three Way Handshake

A handshake as per Abdalazim Abdallah Mohammed Alamin, Dr.Amin Babiker A/Nabi Mustafa, 2015, is a process that must be completed to establish a reliable connection between two machines using TCP/IP. The process begins when the client sends a request packet with the SYN flag set to the server. The server, in response sends a SYN_ACK packet back to client. Finally, the client sends an ACK packet to the server. This completes the three way handshake process and connection is established.

Figure 8. Three way handshake

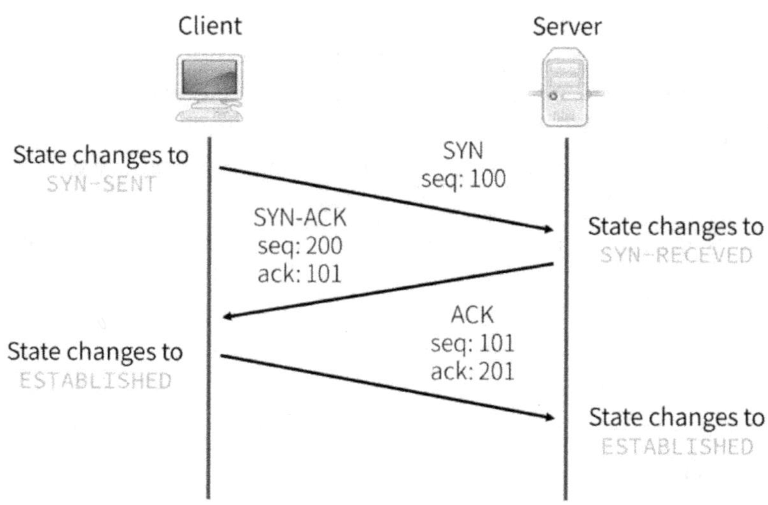

Port Scanning Techniques

Various port scanning techniques are discussed in this section.

- **Ping Sweep Scan**
  This is not a "true port scan", but it detects whether the device is active or not and if it is protected by a firewall.

- **Operating system fingerprinting**
  This is used to find which operating system is running on the device

- **SYN Scan**
  It is also called a 'half-open' scan since it initiates a 3-way handshake but never completes it. Instead of sending an acknowledgement back to the server, RST is sent. This way open ports can be detected without triggering Intrusion-Detection Systems.

- **Fin Scan**
  The attacker sends a packet with FIN flag set. If the port is open, this packet is ignored. If it is closed, a response packet having RST flag set is sent back. This way the attacker can check if the port is open or not.

- **UDP Scan**
  This is used for determining the status of UDP ports. It is different from other techniques since UDP does not use a three-way handshake like TCP.

- **ACK Scan**
  This type of scan is very useful and is used to determine whether a port is filtered or not. It is used with other scan techniques that determine the open/close status of the port.

Commercially Used Tools

Some of the common tools and application used for forensic analysis of smartphones are:

- NMAP(Network mapper)
- Autopsy
- Kiuwan
- Android Debug Bridge(adb)
- zANTI
- FaceNIFF

EXPERIMENT BASED CASE STUDY

Smartphone Forensics Investigation of Android Devices

Android is the most popular operating system for smartphones. As of 2018, android has a market share of about 85% which means that approximately every 9 out of 10 smartphone is running on the android operating system.

In this scenario, android devices being the majority are the prime target of hackers and criminal organisations. Leaks in data could prove devastating for countries as well as their citizens especially in the banking sector. Thus, smartphone security in Android Devices is of utmost priority for software companies.

This case study presents forensics techniques to gain administrative access to an android smartphone also known as the root access and the using this access to modify and extract information from the android device thus revealing the shortcomings in the security features of Android.

We also explore a cheaper method of doing forensic analysis on android smartphones as an alternative to the expensive methods currently available in the market.

This case study was demonstrated by Faheem, M., et al. (2014) where the smartphone device used was Samsung S3 phone running Android version 4.1.2.

AIM

The main aim of the case study is to explore Android forensics using the famous root access technique and then using the root access, we analyse a DD image using commercially available tools.

DD is an image creation tool which creates exact duplicates of data from hard drives. It is mainly used for forensic purposes and works using the command line or the terminal interface.

Hardware

- Smartphone Device running Android
- External Storage Device having memory larger than that of the Android device.

Software

- Android Development Tool (ADT)
- Rooting Software (in this case study, SRSRoot)
- UFED Physical Analyzer

Obtaining Root Access

To obtain root access we initially start by installing Android Development Tool (ADT) which is bundled along with Android Software Development Kit (SDK) obtained from the official website of Android Operating System.

Following steps are followed which grant us root access to our smartphone:-

1. Enable USB debugging on the android device which is present in the Developer options under the setting tab on the smartphone.
2. Install root software, SRSRoot on the computer device through which forensic analysis will be done.
3. Use the following list of commands

```
$ adb push /SRSRoot /data/local/tmp
$ adb shell
$ ls
$ chmod 777 <filename>
$ ./ <filename> (./ is used to execute a file)
$ su
# adb shell ($ replaced by # to indicate root)
```

NOTE-chmod 777 changes file permissions and grants everyone the access to read, write and execute

Creating DD Image of Partitions

Android file system contains partitions each of which has a specific task to perform. Partitions like cache, boot, recovery and data are present in all Android systems and our task is to create a DD image of all the data present in the partitions. This data is then used for forensic analysis and retrieval of vital information from the device.

DD image of 4 partitions data, cache, system and EFS are created using simple DD commands and is saved into any external storage device which acts as our forensic utility.

Forensic Analysis

The most important step is to analyse the DD image and extract data out of it. We use the standard UFED Physical Analyser software. Using the DD image we already created, we extract the data of the device. The software was able to extract critical information such as call logs and messages which are extremely useful in forensic analysis.

The experiment done using Samsung Galaxy S3 phone provided certain artefacts from the messaging app Viber. These artefacts were:

- SIM Number
- Message Time Stamps
- Viber Activation Code
- Country Code
- Media files shared via the app

RESULT AND CONCLUSION

Artefacts collected in the UFED Physical Analyser software are necessary data components of an application and are used for forensic analyses of security services provided by the smartphone manufactures as well as the application developers. A large amount of data was retrieved from the phone using root access and a simple forensic software which indicated that certain amount of data exchanged or present in the device is extractable and this forensic technique can be used to find important evidence in investigatory cases. Proper guidelines especially for evidence handling should be followed while carrying out investigatory processes using Forensic Analysis tools. (Mubarak Al-Hadadi and Ali AlShidhani)

HARMONISED DIGITAL FORENSIC INVESTIGATION PROCESS

Harmonised Digital Forensic Investigation Process is a standardised model of digital forensics proposed by Valjarevic and Venter and presents a novel approach towards the implementation of the various forensic principles through *parallel actions.*

Figure 9. Android forensic analysis

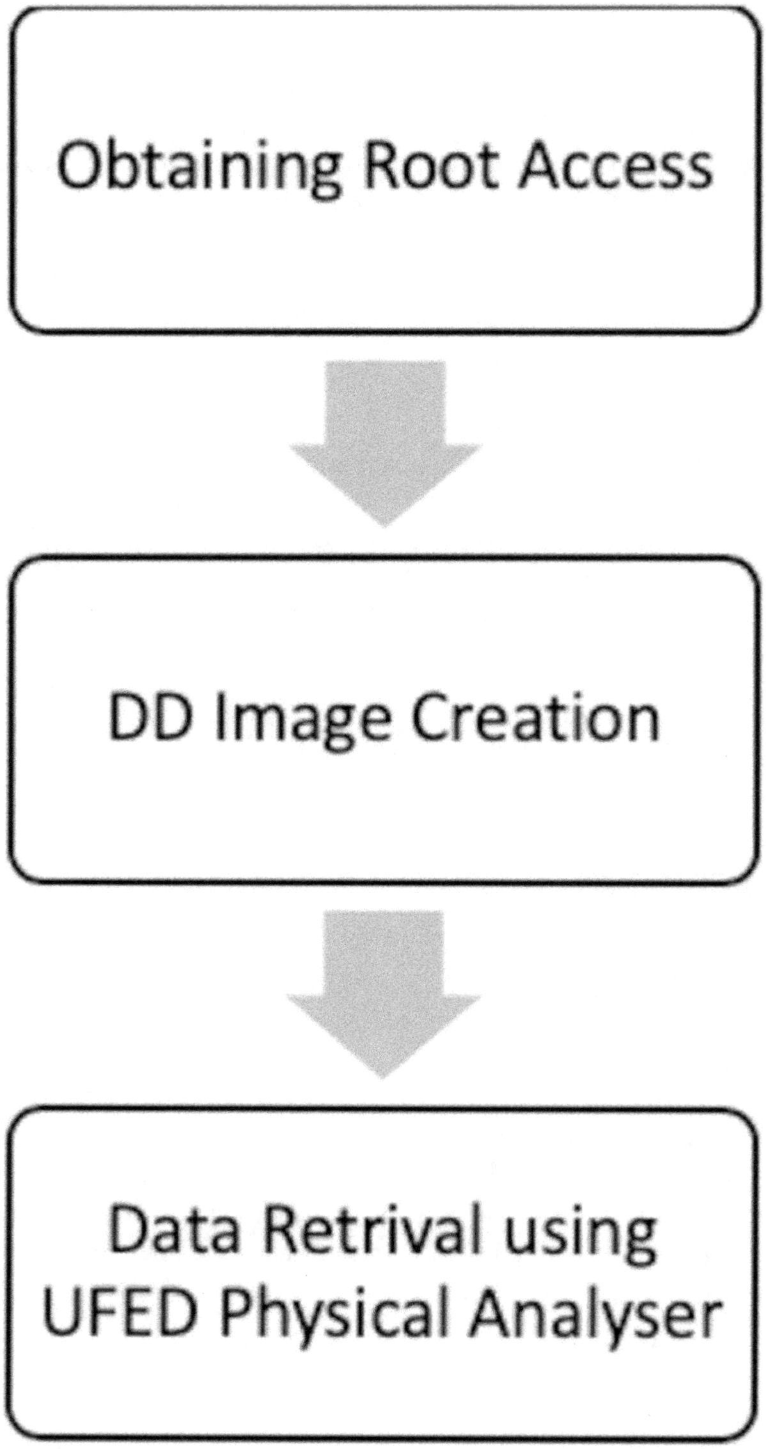

It consists of twelve phases and six actions that run parallel to the phases (parallel actions) and helps in removing possibility of errors and enhance the efficiency of mobile forensics.

This case study will present a method of using this process for mobile forensics which will help forensics investigators gain digital evidence present in mobile devices thus making smartphones more secure.

The main aim of HDFIP is to standardise the digital forensic investigation process and harmonises with all the previous digital forensic models. The HDFI Process is a standard (International Standard Organisation ISO/IEC 27043) comprehensive model and includes all the benefits of the earlier proposed models. This model introduces a novel approach to implement some of the common digital forensic principles using actionable items calls parallel actions. (Valjarevic, 2012)

HDFIP is divided into 5 classes. Each of these class comprises of a number of processes necessary for forensic investigation. These 5 classes ad their processes are as follows:-

THE READINESS CLASS

Forensic Readiness is the ability to maximise the potential of an organization to use digital evidence while minimizing the cost of forensic investigation. This is an optional step in the HDFIP model as it does not involve investigatory process and is mainly used for optimization.

THE INITIALISATION CLASS

The initialisation class is the beginning of the investigatory project. During this phase, the investigators become involved in the process. There are 4 sub sections in this class which are:

- Incident Detection Process is the identification of the type of incident and is categorised into types such as mobile forensic, cloud forensic, network forensic etc.
- First Response Process is the initial response taken by the investigators on the incident.
- Planning Process is the most important process as it involves development of proper procedure for investigation, defining methodologies and listing out various tools required in the procedure.
- Preparation process is the step where the investigator prepares the tools and equipment needed for investigation.

THE ACQUISITIVE CLASS

Acquisitive class procedures help in the acquisition of potential evidence required for the investigation. Evidence collection is a crucial step of investigation because evidence acts as a legal proof of the investigation and without official evidence, no investigatory process is considered valid by the judiciary. Acquisitive class includes 3 sub-processes:-

- Digital Evidence Identification Process is the initial step of evidence collection and is conducted at the incident scene. Potential evidence is identified and labelled during the step. The evidence collection is done in a forensically sound manner so as to safeguard the integrity of the potential evidence.

- Digital Evidence Acquisition Process is the step where all evidence follows proper legal procedure so that all evidence can be made admissible in court at all times.
- Digital Evidence Transportation Process involves transferring of the digital evidence to the place where forensic analysis would be done. Sometimes, digital evidence cannot be immediately used for forensic purposes and in such a case storage and preservation of the evidence is done.

THE INVESTIGATIVE CLASS

Investigative class consists of the data analysis part where the evidence is analysed using software tools and conclusion are derived based on the findings of the forensic analysis. It consists of the following sub-processes:-

- Digital Evidence Examination and Analysis Process involves examining and analysing digital evidence using several techniques available to the forensic experts. Analysed data is then used to obtain results and conclusions.
- Digital Evidence Interpretation Process takes the analysed data and uses the observations to derive conclusions of the investigation. This process uses scientifically proven techniques to analyse the facts derived from the investigation.
- Reporting and Presentation Process is the penultimate step in the investigation. The results from the forensic analysis are combined into a report which acts as official document and this document is then presented to the stakeholders.
- Investigation Closure Process is the conclusion of the investigatory project and proper outcome or findings are confirmed in this last step.

THE CONCURRENT CLASS

This class consists of those processes which run concurrently with all the other processes. There processes are also known as parallel actions which help in maintain integrity, availability and confidentiality of the digital evidence. The sub-processes under concurrent class are:-

- Obtaining Authorization Process ensures that investigators get proper authorization and all the legal provisions required to carry out the investigation process.
- Managing Information Flow Process makes sure that there is proper flow of information and findings between the investigators and the stakeholders.
- Interaction with Physical Investigation Process helps in the interaction between the digital investigation and physical investigation. It involves the interdependence of the two processes and handles their interconnection.

A brief overview of all the classes and processes is present in the Figure 10

Figure 10. HDFIP process

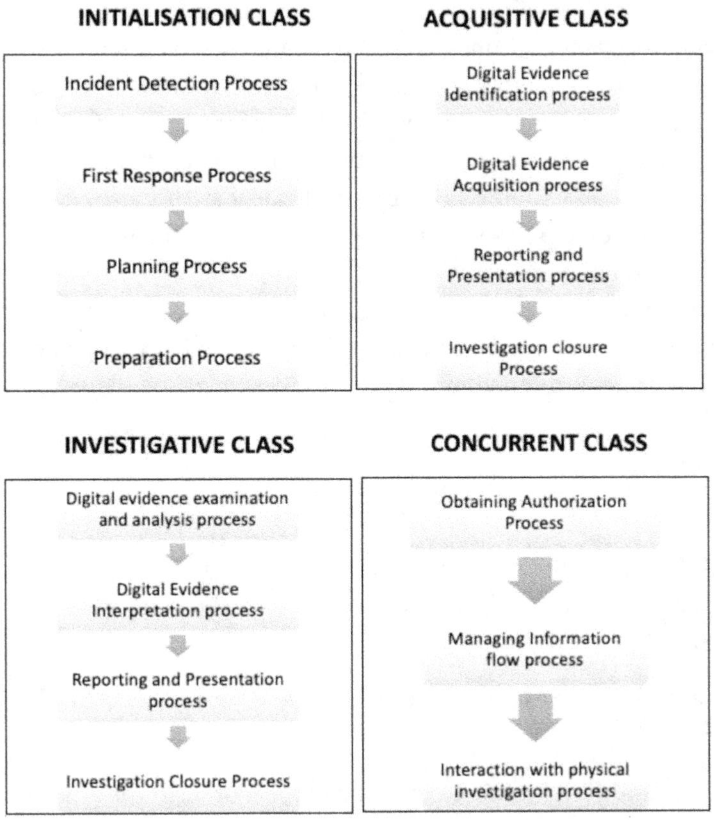

FUTURE RESEARCH AND DEVELOPMENT

The advancement of mobile technology has made cell phone the most popular electronic device in the world. Ease access to technology and new data exploration techniques has paved the way for criminal activities (Kausar, 2014). Not only have hackers started stealing crucial information from mobile devices, they have been using private information and analytic information to manipulate the users' decision making. This poses a huge threat to personal security and individual privacy.

With advancements happening in the digital world, we need continuous development of forensic techniques to counter any kind of security attack and to gather vital information in form of digital evidence. This makes mobile security an extremely important field of research and forensic analysis techniques need frequent revisions in its methods to keep up to date with the new security features that most of the mobile devices possess.

Future aspects of mobile security include:

- Developing anti-forensic security techniques to counter misuse of mobile forensics.
- Developing more secure cryptographic methods and improving authentication processes.
- Improving security features of cloud computing by exploring cloud forensic research techniques.

CONCLUSION

With millions of people using smartphones every day, data security becomes the topmost priority for smartphones manufactures. In order to improve security of data in smartphones, the devices need to be evaluated for risks and threats through forensic analysis of the security features these devices offer.

The chapter goes over various threats and vulnerabilities to smartphone security. It also discusses various security mechanisms and solutions employed to deal with these vulnerabilities. Smartphone forensics are analysed with the help of experimental case studies which provide a systematic and standardised model for mobile forensics research.

Despite of having newer and safer security features, smartphones can still be attacked, and data can be leaked through the smartphone devices. Security features need to be updated frequently and newer encryption techniques must be developed in order to keep smartphone data secure.

REFERENCES

Ahvanooey, M. T., Li, Q., Rabbani, M., & Rajput, A. R. (2017). A Survey on Smartphones Security: Software Vulnerabilities, Malware, and Attacks. *International Journal of Advanced Computer Science and Applications*, *8*(10), 2017.

Al-Hadadi & AlShidhani. (2013). Smartphone Forensics Analysis: A Case Study. *International Journal of Computer and Electrical Engineering, 5*(6).

Alamin, Babiker, & Mustafa. (2015). A Survey on Mobile Forensic for Android Smartphones. *IOSR Journal of Computer Engineering, 17*(2), 15-19.

Anjali, Geetanjali, Shivlila, Swati, & Kadu. (2013). Data leakage detection. *International Journal of Computer Science and Mobile Computing, 2*(5).

Chen, J. V., Yen, D. C., & Chen, K. (2009). The acceptance and diffusion of the innovative smart phone use: A case study of a delivery service company in logistics. *Information & Management*, *46*(4), 241–248. doi:10.1016/j.im.2009.03.001

Chin, Felt, Sekar, & Wagner. (2012). Measuring user confidence in smartphone security and privacy. In *Symp. on Usable Privacy and Security.* Washington, DC: Advancing Science Serving Society.

Damodaram. (2016). Study on phishing attacks and anti-phishing tools. *International Research Journal of Engineering and Technology, 3*(1).

Faheem, M., Le-Khac, N.-A., & Kechadi, T. (2014). Smartphone Forensic Analysis: A Case Study for Obtaining Root Access of an Android Samsung S3 Device and Analyse the Image without an Expensive Commercial Tool. *Journal of Information Security*, *5*(3), 83–90. doi:10.4236/jis.2014.53009

Habib, S. M., Jacob, C., & Olovsson, T. (n.d.). A Practical Analysis of the Robustness and Stability of the Network Stack in Smartphones. *11th IEEE International Conference on Computer Science & Information Technology*. 10.1109/ICCITECHN.2008.4803083

Jeon, W., Kim, J., Lee, Y., & Won, D. (2011). *A Practical analysis of Smartphone Security. Human Interface - Part I, HCII 2011, LNCS 677*. Springer-Verlag Berlin Heidelberg.

Kausar. (2014). New research directions in the area of smart phone forensic analysis. *International Journal of Computer Networks & Communications, 6*(4).

Kumar, N., & Muhammad, E. U. H. (2011). *Penetration Testing of Android-based Smartphones* (Master's thesis). Chalmers University of Technology.

Matthews, T., Pierce, J., & Tang, J. (2019). *No smart phone is an island: The impact of places, situations, and other devices on smart phone use.* Research Report RJ10452 IBM, 2009.

Maus, S., Hoefken, H., & Schuba, M. (2019). *Forensic Analysis of Geodata in Android Smartphones.* Academic Press.

Miglani, A., Bhatia, T., Sharma, G., & Shrivastava, G. (2017). An Energy Efficient and Trust Aware Framework for Secure Routing in LEACH for Wireless Sensor Networks. Scalable Computing. *Practice and Experience, 18*(3), 207–218.

Mumba, E. R., & Venter, H. S. (2014). Mobile forensics using the harmonised digital forensic investigation process. *Information Security for South Africa, Johannesburg, 2014*, 1–10. doi:10.1109/ISSA.2014.6950491

Ntantogian, C., Apostolopoulos, D., Marinakis, G., & Xenakis, C. (2014). Evaluating the privacy of Android mobile applications under forensic analysis. *Computers & Security, 42*, 66-76.

Riadi, I., Fadlil, A., & Fauzan, A. (2018). A Study of Mobile Forensic Tools Evaluation on Android-Based LINE Messenger. *International Journal of Advanced Computer Science and Applications.*, *9*(10), 201–206. doi:10.14569/IJACSA.2018.091024

Saxena, A., Shrivastava, G., & Sharma, K. (2012). Forensic investigation in cloud computing environment. *The International Journal of Forensic Computer Science, 2*, 64-74.

Sharma, K., & Gupta, B. B. (2016). Multi-layer defence against malware attacks on smartphone wi-fi access channel. *Procedia Computer Science, 78*, 19–25. doi:10.1016/j.procs.2016.02.005

Sharma, K., & Gupta, B. B. (2018). Attack in Smartphone Wi-Fi Access Channel: State of the Art, Current Issues, and Challenges. In *Next-Generation Networks* (pp. 555–561). Singapore: Springer. doi:10.1007/978-981-10-6005-2_56

Sharma, K., & Gupta, B. B. (2018). Mitigation and risk factor analysis of Android applications. *Computers & Electrical Engineering, 71*, 416–430. doi:10.1016/j.compeleceng.2018.08.003

Shrivastava, G. (2016, March). Network forensics: Methodical literature review. In *3rd International Conference on Computing for Sustainable Global Development (INDIACom)* (pp. 2203-2208). IEEE.

Shrivastava, G. (2018). Investigating New Evolutions and Research in Digital Forensic & Optimization. *Recent Patents on Engineering, 12*(1), 3–4. doi:10.2174/187221211201180308113130

Shrivastava, G., & Kumar, P. (2017). Privacy Analysis of Android Applications: State-of-art and Literary Assessment. Scalable Computing. *Practice and Experience, 18*(3), 243–252.

Shrivastava, G., Sharma, K., & Dwivedi, A. (2012). Forensic computing models: Technical overview. CCSEA, SEA, CLOUD, DKMP. *CS & IT*, *5*, 207–216.

Shrivastava, G., Sharma, K., Khari, M., & Zohora, S. E. (2018). *Role of Cyber Security and Cyber Forensics in India. In Handbook of Research on Network Forensics and Analysis Techniques* (pp. 143–161). IGI Global. doi:10.4018/978-1-5225-4100-4.ch009

Shrivastava, G., Sharma, K., & Kumari, R. (2016, March). Network forensics: Today and tomorrow. In *3rd International Conference on Computing for Sustainable Global Development (INDIACom)* (pp. 2234-2238). IEEE.

Singh, V., & Sharma, K. (2016, March). Smartphone Security: Review of Challenges and Solution. In *Proceedings of the Second International Conference on Information and Communication Technology for Competitive Strategies* (p. 8). ACM. 10.1145/2905055.2905214

Valjarevic & Venter. (2012). Harmonised digital forensic investigation process model. *Information Security for South Africa, Johannesburg, Gauteng, 2012*, 1–10. doi:10.1109/ISSA.2012.6320441

Waheed, Khan, & Vapen. (n.d.). *Attacks against Smartphones*. Project Report for Information Security Course Linköpings Universitet, Sweden.

This research was previously published in Forensic Investigations and Risk Management in Mobile and Wireless Communications; pages 26-50, copyright year 2020 by Information Science Reference (an imprint of IGI Global).

Chapter 2
Mobile Agent Communication, Security Concerns, and Approaches:
An Insight into Different Kinds of Vulnerabilities a Mobile Agent Could Be Subjected to and Measures to Control Them

Kamat Pooja
Symbiosis Institute of Technology Pune, India

Gite Shilpa
Symbiosis Institute of Technology Pune, India

Patil Shruti
Symbiosis Institute of Technology Pune, India

ABSTRACT

Mobile Agent Systems model has attracted attention of various researchers and scholars all over the world due to a wide array of features it offers. The capability of mobile agent to hop independently from one network to another, carrying out various computational processes on remote network, enables them to operate in fixed and mobile networks more efficiently and robustly than typical client-server systems. However little attention is paid to the security management of the mobile agents due to which it is still not widely used in the industry domain. . In this chapter, the authors examine the various security issues in Mobile Agent systems and approaches used to overcome them.

DOI: 10.4018/978-1-7998-8545-0.ch002

INTRODUCTION

A mobile agent can be termed as a unique type of mobile code. A mobile agent is a kind of a program that migrates from one host to another in a distributed network mobile agent. It has many advantages over existing distributed techniques like optimum resource utilization, minimize network traffic etc(Bhanot, R., & Hans, R.,2015). A Mobile Agent exhibits the following characteristics:

- **Mobility:** Mobility is a characteristic which allows a mobile agent to jump from one node to another. During this traversal, the mobile agent can perform any communication processes.
- **Independent:** An independent or autonomous Mobile Agent can take its own decision or act on behalf of another mobile agent.
- **Communication:** A mobile agent can talk to another mobile agent, fixed or mobile servers and other client systems.
- **Learning:** A mobile agent is known to be 'smart' that gains knowledge from its past experience and changes its behaviour accordingly.
- **Interoperability:** Mobile agents have property to execute on different platform or over different clients and adapt to changes in the environment.
- **Persistence:** Mobile agent has no need to establish continuous connection for execution of programs.

BACKGROUND

Due to large number of features supported by mobile agents, they have a higher probability of facing security threats. To get a better understanding of the security threats, the authors consider the two integral components of a mobile agent system - a mobile agent and an agent platform. An agent represents the mobile code and the state information which is needed to carry certain processing. An agent platform is a computational setting wherein the mobile agent 'operates'. A mobile agent can hop between agent platforms and also communicate with the platform. The platform where a mobile agent is created is called as a 'home platform' and it is often the safest environment for an agent. An agent platform can be termed as a communication place wherein mobile agents can speak to each other.

SECURITY THREATS IN A MOBILE AGENT SYSTEM

Belal Amro(2014) categorizes the security threats faced by a Mobile Agent System into four broad categories: threats due to a mobile agent harming an agent platform, a mobile agent attacking another mobile agent, an agent platform attacking a mobile agent and finally outside entities harming the mobile agent system. The fourth category includes threats when an agent harms a destination platform or one agent platform attacking another platform. The threats mentioned above have their analogues in a typical client-server system and have taken place in the past.

1. **Mobile Agent Harming Agent Platform:** In this category of threats, the mobile agent tries to identify the security loopholes in the agent platform and harms the platform accordingly. The type of threats which fall in this category are:
    a. **Masquerading:** Masquerading is a threat wherein a dubious agent poses to be a legitimate agent and tries to enjoy the services which it is not authorised to use. Also the dubious agent performs illegal operations for which the legitimate agent is blamed for. A dubious agent tries with all its ability to harm the reputation of the real agent and the platform on which it is currently operating making it a grave threat.
    b. **DOS Attack Against Platform:** A rogue agent harms the agent platform by maliciously consuming all its resources in a short frame of time making the platform undeliverable. This is termed as Denial of service(DOS) attack. It identifies the vulnerabilities of the platform and runs malicious scripts which further degrades the performance of the platform. The rogue agent tries to access data over which it has no authority. It has the capability to interrupt or completely terminate the platform processes.
    c. **Illegal Access:** A rogue agent tries to utilise the platform services which it is not entitled to. The organisations implementing mobile agent systems must strictly ensure that no unauthorised agent has any kind of view or edit access to the organization's critical data including cache data.
2. **Mobile Agent Attacking Another Mobile Agent:** This category of threats represents security weaknesses wherein a rogue agent which maliciously exploits the legitimate agent either by snooping on the communication between two legitimate agents, or masquerading as a legitimate agent. The examples of threats in this category are as follows:
    a. **Impersonation:** A very common and grave threat is when a fraud agent impersonates as a true agent and deceives the clients availing the services of the platform. For example, a fraud agent can pose as a bank employee and ask its customers for critical data such as account passwords, secure codes, etc. This kind of threat is grave as it not only harms the agent whose identity has been impersonated but also the client who has been unknowingly deceived by the fraud agent. Such kind of threat might disrupt the net banking services provided by reputed bank establishments.
    b. **Denial of Service Against Other Agents:** A rogue agent might launch DOS attack against another agent by bombarding it with an infinite number of messages which the true agent will not be able to handle. Also it can send many 'IP pings' to the true agent posing as if the pings are sent by other true agents on the platform. Due to this the poor agent will not be able to differentiate between real requests and fake requests.
    c. **Repudiation:** Another serious kind of a security threat can be when a mobile agent who was previously a part of a transaction or communication process later denies of such transaction being taken place. This is known as repudiation. On analysis of such kind of threats, often debate takes place whether this denial is purposeful or accidental. However, the harm caused is severe as it would lead to internal disputes amongst the organisation and hence it is extremely important that organisations and other involved parties must maintain proofs of the transactions taken place. Unless strong evidences are in proper place, such kind of threats are not easily resolved.

   d. **Illegal Access:** Every organisation must ensure strict access parameters for their mobile agents and agent platforms so that rogue agents are unable to invade the security. If the access mechanisms are weak, the rogue agent can easily attempt to access resources of the agent platform, reset the initial state of the platform and perform various such malicious actions. A worse case could be that the rogue agent gets access to modify the internal code of a true agent and in turn change its behaviour such as involving the true agent in some kind of malicious transactions. A rogue agent could also snoop on the communication taking place on the agent platform if proper countermeasures are not taken to block illegal access.

3. **Agent Platform Harming Mobile Agent:** This category of threats occurs when the agent platform itself turns out to be malicious and harms the mobile agents residing on its platform. Some of the common examples are Facade, DOS against Mobile Agent, Snooping and Alteration.
   a. **Facade:** Similar to a rogue mobile agent masquerading as a true agent, there is a threat of a rogue agent platform impersonating as the original platform. The rogue platform in this case deceives the agent in letting out sensitive information such as credit card passwords. The mobile agent is made to believe that the environment set by the rogue platform is secure and lure him into giving in critical information which can be later on used by the rogue platform for its own malicious purpose. In this way the rogue platform deceives both the mobile agent as well as the impersonated platform.
   b. **DOS Against Mobile Agent:** A Denial of Service attack can also be launched by a rogue agent platform against the agent. Consider a scenario wherein the agent visits the agent platform with some request to be fulfilled. The agent expects that the platform to service its requests and allocate it some resources of desirable quality without unacceptable delays. But in this case the rogue agent platform keeps the agent 'waiting' and does not fulfil its requests or executes its code. In some cases, the rogue platform also terminates the agent execution without prior notification.
   c. **Eavesdropping:** Eavesdropping involves threats stemming from interception of agent platform in the communication processes carried out by the mobile agent. The platform may eavesdrop and infer meaning from the communication and might try to sell this valuable information to third parties. Consider a case wherein a customer's agent is communicating with a bank agent about its account transaction records. The rogue agent platform can sell this information to fraudulent insurance and mutual funds companies which might pester the customer with advertisements and promotions. A worse situation can occur when some fraudulent attacker hacks the customer's account data.
   d. **Alteration:** An agent on reaching a platform provides information about its internal code, state and communication processes. A malicious platform must be blocked from accessing or modifying the same information. Stringent security measures such as use of digital signature must be taken against any kind of alteration attempt by a rogue platform on the agent's code information.

4. **External Threats Harming Agent Platform:** This category includes type of threats posed by external components including agents and agent platforms against agent platform:
   a. **Unauthorised Access:** This threat indicates any kind of unauthorised remote access to agent resources by remote terminals. Remote users and processes must be carefully analysed before granting access permissions as rogue users can make use of conventional internet attack schemes to gain direct access to resources.

b. **Copy and Replay:** From the time an agent's transaction process is initiated till it is completed, an agent hops through various agent platforms. During this course, the mobile agent exposed to various security threats. A typical security attack by malicious third party during the agent transition is the copy and replay attack. The third party might copy the agent's message and retransmit it again and again putting the entire communication in a loop. This could lead to severe errors and would lead to termination of agent's process.

SOLUTIONS AND RECOMMENDATIONS

Some of the most common security threats in MAS emerge from the fact that the entities of MAS i.e mobile agent and agent platform are unable to mutually authenticate each other or are oblivious to the malicious track record of the entity with which they are communicating. Also there is a general misconception that only hardware protection can secure a mobile agent system. However some countermeasures in form of algorithms can effectively tackle the security threats (Shen, Z. et al.2009). The authors discuss some of these solutions as follows:

1. **Host Revocation Authority:** Esparza et al. (2003) suggested a technique wherein a trusted third party called the Host Revocation Authority(HoRA) maintains a list of rogue hosts with previous malicious records. The sender's agent must consider the HoRA's decision before carrying out any communication process with the host. The HoRA also has the authority to declare a host as rogue if the sender's agent proves that the host did not act truthfully.
2. **Execution Tracking:** Vigna, G. (1998) proposed a technique to identify abnormal behaviour by using cryptography and checking in the agent directory to verify if the agent has completed all jobs dutifully. This require the mobile agent to preserve a huge log file.
3. **Peer-Learning Mobile Agents for Protection of Mobile Agents:** This method proposed by S. M. M. Ebrahimi et al suggested the use of an effective method of peer-learning for the purpose of protecting the security of mobile agents by the use of mobile agent characteristics such as cooperation, learning, mobility and presentation duplication.
4. **Distributed Lightweight Kerberos Protocol (DLKP):** H.M.N. Al- Hamadi et al. (2011) proposed DLKP algorithm which is based on enhancements modification of Kerberos and is suited to provide confidentiality, integrity, authentication and authorization (Yeun, C.Y, 2005). One of the biggest advantages of DLKP is that it ensures mutual authentication along with confidentiality for a dynamic structure like Mobile Agent System (MAS). Further, it also lessens the count of messages transferred between the various entities of MAS. In the next section, the authors discuss DLK protocol in more detail.

DISTRIBUTED LIGHTWEIGHT KERBEROS PROTOCOL (DLKP)

The DLKP will be explained using the case scenario of Airline Reservation system. Consider that a user wants to attend an important exhibition in London and he opens an application AirBook on his cellphone.

Design Algorithm of DLKP

The algorithm is explained using the case study of airline booking system. The user is prompted to enter his login details. After the login is successful, the user will choose to fill an inquiry form amongst the various services provided by the application. The inquiry form will capture details such as destination location, departure and return dates, etc. The Mobile Agent would carry this e-application details to Ticket Granting Server. The home platform will be asked to prove its authenticity first, and then it will generate a hashed secret key K_{HP}. The list of abbreviations used in the algorithm is as follows in Table 1:

Table 1. List of abbreviations in DLKP algorithm

HP	Home Platform
SPP_X	Service Provider Platform ID
SID	Service ID
MA_x	Mobile Agent with label X
R_X	MA's results and status from X
AS	Authentication Server
TGS	Ticket Granting Server
K_X	Private Key for X (Secret Key)
K_{XY}	Session Key between X and Y
Add_P	Platform's Network Address
VP_T	Ticket's Validation Period
t_X	Timestamp for X
$t_{X,Y}$	Timestamp for X updated by Y
$\{m\}K_X$	Message m encrypted by X's private key
A_X	Authenticator for X
T_X	Ticket to communicate with X
$\|\|$	Concatenation

The Authentication Server generates a session key $K_{HP,TGS}$ for the HP with a ticket T_{TGS} to further communicate with the TGS. The DLK protocol follows flow of Kerberos protocol very closely. It just improves the manner in which the TGS operates:

Step 1: The HP forwards an authenticator similar to the Kerberos protocol but includes the service ID and the mobile agent as seen below:

$HP \rightarrow AS: [HP]$

$$AS \rightarrow HP\left\{K_{HP,TGS}\right\}K_{HP}\left\|\left\{T_{TGS}\right\}K_{TGS}\right.$$

$$HP \rightarrow TGS : \{T_{TGS}\} K_{TGS} \| \{HP, t_{HP}, SID, MA_0\} K_{HP,TGS}$$

where:

$$T_{TGS} = [HP, Add_{HP}, K_{HP,TGS}, VP_T]$$

MA_0 = the original mobile agent

Step 2: Similar to the Kerberos protocol, the TGS decodes the ticket to obtain the session key $K_{HP,TGS}$. The HP authenticator is then decrypted, in order to match the Home Platform ID and the address network. The TGS receives the SID and searches in its database for SP platforms that provide this service.Consider the case that the TGS find several SP platforms which provide this service. The TGS will send the mobile agent to the same number of available SP platforms. A ticket T_{HP} will be created and encrypted by the TGS using the session key between itself and the home platform $K_{HP,TGS}$. An important thing to note here is that the TGS will create, on behalf of the HP, an authenticator which contains the HP ID, timestamp, SID, the session key $K_{SPPx, HP}$ and a clone of the original mobile agent. This authenticator is encoded using the SP platform secret key K_{SSPx}. Then the TGS sends both the ticket and the authenticator to the SP platform as seen below:

$$TGS \rightarrow SPP_X : \{HP, t_{HP}, SID, K_{SPPx,HP}, MA_C\} K_{SPPx} \| \{T_{HP}\} K_{HP,TGS}$$

where:

$$T_{HP} = [SPP_X, Add_{SPPx}, K_{SPPx,HP}, VP_T]$$

MA_C= A clone mobile agent

Step 3: Once the SP platform receives the mobile agent, it can decide on whether to serve it. If yes then, it will allow the mobile agent to execute itself. The SP platform will forward the ticket $T_{SPPx,HP}$ and sends an authenticator which includes its ID, the updated time stamp and the results as seen below:

$$SPP_X \rightarrow HP : \{SPP_X, t_{HP}, SPP_X, R_x\} K_{SPPX,HP} \| \{T_{HP}\} K_{HP,TGS}$$

Step 4: The Home Platform decodes the ticket using its session key $K_{HP,TGS}$. The ticket includes the session key which is then used to decode the authenticator. It is then compared with the SPP and Add_{SPP} between the ticket and the authenticator.

At the end the home platform will provide the results to the owner after the verification process.

Figure 1. Stages of DLK protocol

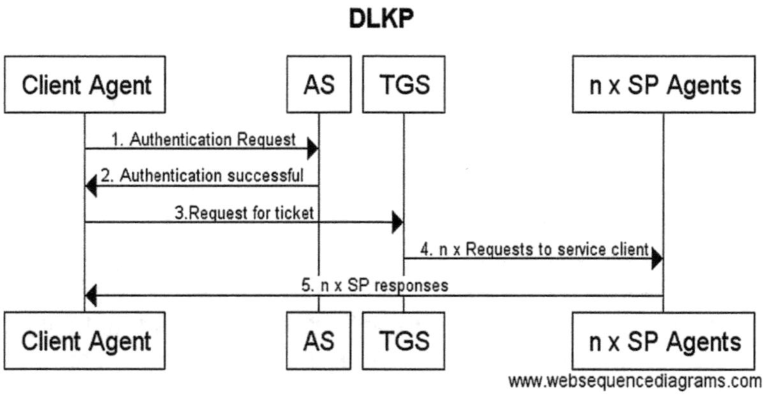

DLKP Design Architecture

Figure 1 depicts the various stages of DLK protocol. The DLK protocol is similar to Kerberos wherein the AS Agent represents Authentication Server and the TGS Agent represents Ticket Granting Server. The Service Provider is represented by the SP Agent and n is the number of service providers which can be contacted by the TGS agent with specific criteria. In the case of Kerberos, a client is a part of communication process involving six messages even if it only wants to access one service provider. This increases the overload on the client to a considerable extent. Hence, one of the key purposes of the DLK protocol is to lessen the count of messages and at the same time ensure the same level of security as that of Kerberos. Further, the client in this case can get to choose amongst several SP Agents based on its request. However, on the flipside, the DLK protocol is more complicated than the Kerberos as it requires more number of interactions. Instead of three interactions in Kerberos protocol, there are four in the DLK protocol. Al Hamadi et al. (2012) have compared the Kerberos and DLK protocol as shown in Table 2.

Table 2. Comparison between Kerberos and DLK protocols

Kerberos Protocol Stages	DLKP Stages
Client ← → AS	Client ← → AS
Client ← → TGS	Client ← → TGS
Client ← → SP	TGS ← → SP
	SP ← → Client

Security and Design Analysis of DLKP

1. **Security Analysis:**
   a. **Confidentiality:** Confidentiality expects that a mobile agent is not harmed by unauthorised access to its code and state information. Some of the common techniques are use of encryption algorithm to ensure confidentiality. The mobile agent's sensitive code and state information

are encrypted using session keys. The DLKP protocol makes use of session keys with a fixed validation period decided by the involved parties such as Home Platform and Service Provider Platform. Encryption mechanisms that can be used to encrypt session keys are as follows:

i. Symmetric Encryption.
ii. Java Security Key (It is a top-level interface for all keys which defines the functionality shared by all key objects).
iii. Javax Crypto Cipher (It is a class which provides the functionality of a cryptographic cipher for encryption and decryption).

b. **Integrity:** The security requirement of integrity verifies whether a particular piece of information is sent by a trusted sender. It ensures that the information is not modified in any way by a malicious third party. One of the popular ways to ensure this is by the use of one-way hash functions. In DLKP, the integrity of passwords is ensured using one-way hash function. Also every session is accompanied with a timestamp which is good way to identify if the message sent is fresh. This would further reduce playback attacks. Hash functions that can be used are as follows:

i. Password-Based Encryption(PBE) (which is hashing+ symmetric encryption)

c. **Authentication:** Authentication is one of the most popular and essential security requirement which verifies the authenticity of the identity of the parties involved in the communication. It's effective in preventing attacks such as facade and man-in-middle attack.DLK protocol provides mobile agent system the feature of mutual authentication wherein before the communication process is initiated, both the home platform and service provider platform authenticate each other. Mutual authentication is also helpful in combating eavesdropping attacks.

d. **Authorization:** Authorization involves specification and permission of rights of mobile agent to be carried out by the service provider platform.

In the DLK protocol, Ticket Granting Server(TGS) provides information to the Service Provider(SP) platform about the home platform and the type of task carried by mobile agent. The SP platform then allows the mobile agent to run its own code. Also the owner has the permission to select which platform to contact after receiving the results. In this system, the user password is encrypted by MD5 algorithm before sending it on the network.

2. **Design Analysis:** H.M.N. Al- Hamadi et al. (2011) performed design analysis by drawing comparisons between DLK protocol and other two famous protocols: Kerberos and Kryptoknight protocol (Janson, P et al. 1997). The comparison is done on basis of the count of messages via the count of servers the home platform needs to communicate with. See Table 3.

The equations used in this analysis are as follows:

Kerberos protocol: $4x + 2$.

Kryptoknight: $4x$.

DLK protocol: $2x + 3$.

where x is the number of servers.

Table 3. Tabular comparison between the Kerberos, Kryptoknight and DLK protocols with respect to count of messages passed

	1	2	3	4	5	6	7
Kerberos	6	10	14	18	22	26	30
Kryptoknight	4	8	12	16	20	24	28
DLK	5	7	9	11	13	15	17

FUTURE RESEARCH DIRECTIONS

Mobile agents have drawn much attention as a fundamental technology in next generation computing. Mobile Agents provide many advantages such as facilitating high quality, high performance, economical mobile applications, enabling use of portable, low-cost, personal communications devices, permitting secure Intranet-style communications on public networks, efficiently and economically using low bandwidth, high latency, error prone communications channels. However, other than the threats discussed in the chapter, there is a need to address many other security issues including: on running agent geo-localization, inter- mobile agent collaboration, real time attack detection, and mutual authentication between host and mobile agent.

CONCLUSION

The authors have described a number of techniques which can be used to make the mobile agent systems more secure. There can be scenarios wherein a particular technique is not suitable for a particular application or an organisation. In some other cases two techniques might face compatibility issues with each other. Some applications might demand that the security requirements are built rigidly within the agent framework and some applications can give the flexibility in the manner in which the security requirements operate. The organisations before implementing such kind of mobile agent systems must make a comprehensive study of all the countermeasures and trade-offs of the different techniques with respect to performance, compatibility, scalability and of course the cost to build such kind of a secure system.

REFERENCES

Al-Hamadi, H. M. N., Yeun, C. Y., Zemerly, M. J., & Al-Qutayri, M. (2011, February). Distributed lightweight kerberos protocol for mobile agent systems. In *GCC Conference and Exhibition (GCC), 2011 IEEE* (pp. 233-236). IEEE. doi:10.1109/IEEEGCC.2011.5752502

Al Hamadi, H. M. N., Yeun, C. Y., Zemerly, M. J., Al-Qutayri, M. A., & Gawanmeh, A. (2012). Verifying Mutual Authentication for the DLK Protocol using ProVerif tool. *International Journal for information Security Research, 2*, 256-265.

Amro, B. (2014, March). Mobile Agent Systems, Recent Security Threats and Counter Measures. *International Journal of Computer Science Issues*, *11*(2).

Bhanot, R., & Hans, R. (2015). A Secure and Fault Tolerant Platform for Mobile Agent Systems. *International Journal of Security and Its Applications*, *9*(5), 85–94.

Ebrahimi, S. M. M. (2016). Using of colleague learning mobile agents for protecting the confidentiality of mobile agents in a multi-agent environment. *Journal of Fundamental and Applied Sciences*, *8*(3), 579–598.

Esparza, O., Soriano, M., Muñoz, J. L., & Forné, J. (2003, July). Host revocation authority: A way of protecting mobile agents from malicious hosts. In *International Conference on Web Engineering* (pp. 289-292). Springer Berlin Heidelberg. 10.1007/3-540-45068-8_54

Janson, P., Tsudik, G., & Yung, M. (1997, April). Scalability and flexibility in authentication services: the KryptoKnight approach. *Proceedings of the IEEE*, *2*, 725–736.

Neuman, B. C., & Tso, T. (1994). Kerberos: An authentication service for computer networks. *IEEE Communications Magazine*, *32*(9), 33–38. doi:10.1109/35.312841

Shen, Z., & Tong, Q. (2009, August). A security technology for mobile agent system improved by trusted computing platform. In *Hybrid Intelligent Systems, 2009. HIS'09. Ninth International Conference on* (Vol. 3, pp. 46-50). IEEE. 10.1109/HIS.2009.222

Vigna, G. (1998). Cryptographic traces for mobile agents. In *Mobile agents and security* (pp. 137–153). Springer Berlin Heidelberg. doi:10.1007/3-540-68671-1_8

Yeun, C. Y. (2005). Security for emerging ubiquitous networks. *Networks*, *1*, 2.

ADDITIONAL READING:

Kamat, P., Gite, S., Kumar, M., & Patil, S. (2014). A Critical Analysis of P2P Communication, Security Concerns and Solutions. *International Journal of Applied Engineering Research*, *9*(24), 30899–30909.

Pai, P., Shinde, S. K., & Khachane, A. R. (2012). Security in Mobile Agent Communication'. *International Journal of Advanced Engineering Research and Studies*, *1*(4), 74–80.

KEY TERMS AND DEFINITIONS

Authentication Server: An authentication server (AS) component is used by the Kerberos protocol to authenticate the client and to further create a session key for communication.

Autonomy: A principle of executing independently and according to its own will.

Denial of Service: An act of refusing or ignoring service request with a purpose to introduce huge delays for critical tasks.

Kerberos: Kerberos is a trusted third party computer network authentication protocol that works on the basis of 'tickets' to allow nodes communicating over a non-secure network to prove their identity to one another in a secure manner.

Kryptoknight: Kryptoknight is an authentication and key distribution system that provides facilities for secure communication in any type of network environment.

Masquerading: An act of disguise by a malicious agent with a purpose to deceive another agent/platform.

Mobile Agent: A term which represents composition of computer software and data which is able to migrate (move) from one computer to another autonomously and continue its execution on the destination computer.

Platform: A platform provides computational environment where a mobile agent operates and carries out its functions.

Service Provider: A service provider fulfils client requests.

Ticket Granting Server: A ticket granting server (TGS) is a decision-making key distribution centre (KDC) of Kerberos which is used in the validation of a ticket used in communication processes.

This research was previously published in Detecting and Mitigating Robotic Cyber Security Risks; pages 300-311, copyright year 2017 by Information Science Reference (an imprint of IGI Global).

Chapter 3
Recent Advances Delivered in Mobile Cloud Computing's Security and Management Challenges

Christos Stergiou
https://orcid.org/0000-0002-7713-3667
University of Macedonia, Greece

Kostas E. Psannis
University of Macedonia, Greece

ABSTRACT

Mobile cloud computing provides an opportunity to restrict the usage of huge hardware infrastructure and to provide access to data, applications, and computational power from every place and in any time with the use of a mobile device. Furthermore, MCC offers a number of possibilities but additionally creates several challenges and issues that need to be addressed as well. Through this work, the authors try to define the most important issues and challenges in the field of MCC technology by illustrating the most significant works related to MCC during recent years. Regarding the huge benefits offered by the MCC technology, the authors try to achieve a more safe and trusted environment for MCC users in order to operate the functions and transfer, edit, and manage data and applications, proposing a new method based on the existing AES encryption algorithm, which is, according to the study, the most relevant encryption algorithm to a cloud environment. Concluding, the authors suggest as a future plan to focus on finding new ways to achieve a better integration MCC with other technologies.

DOI: 10.4018/978-1-7998-8545-0.ch003

INTRODUCTION

The need of "*cloud*" support has become inefficient due to the intensive computations, the mass storage, and the security issues. Some examples include limited storage capacity, communication capabilities, energy and processing. Inefficiencies like these have motivated us in order to find a model for the combination of CC and other technologies such as Internet of Things and Big Data. As a "*base*" technology, Cloud Computing consolidates various technologies and applications to get the maximum capacity and performance of the existing infrastructure (Kryftis et al., 2016; Stergiou & Psannis, 2016; Stergiou et al., 2018b).

Mobile Cloud Computing improved through the recent years by a new generation of services based on the concept of the "*cloud computing*", which aims to provide access to the information and the data from anywhere at any time, and simultaneously restrict or eliminate the need for hardware equipment (Rahimi et al., 2014; Stergiou & Psannis, 2016; Stergiou & Psannis, 2017b; Stergiou et al., 2018f). In particular, Mobile Cloud Computing (MCC) could be defined as an integration of Cloud Computing (CC) technology and mobile devices in order to make mobile devices resourceful in terms of computational power, memory, storage, energy, and context awareness (Fremdt et al., 2013; Keskin & Taskin, 2014; Haung, 2011; Stergiou et al., 2018c). As a result of the operations of Cloud Computing, it could be used as useful base for several technologies, such as Internet of Things (IoT), Big Data (BD) and Surveillance Systems, and could provide improvements on their functions.

Therefore, the term mobile cloud is generally referred to in two perspectives: (a) infrastructure based, and (b) ad-hoc mobile cloud. In infrastructure based mobile cloud, the hardware infrastructure remains static, and provides services to the mobile users. Although cloud is useful for computing and storage (Rahimi et al., 2014; Stergiou & Psannis, 2017a; Stergiou et al., 2018d), the traditional computation offloading techniques cannot be used directly for smartphones because these techniques are generally energy-unaware and "*bandwidthhungry*".

Furthermore, the term "*cloud computation*" is defined as "*the use of computing logistical resources by using services transported over the internet*" (Stergiou & Psannis, 2016; Stergiou & Psannis, 2017b). Nowadays, Cloud Computing services constitute one of the world's largest areas of competition between giant companies in the IT sector and software (Mell & Grance, 2011). However, Cloud Computing security is an evolving sub-domain of computer security, network security, and, more broadly, information security. It refers to a broad set of policies, technologies, and controls deployed to protect data, applications, and the associated infrastructure of cloud computing (Rahimi et al., 2014; Mell & Grance, 2011; Haghighat, 2015). Thus, as the MCC is the outcome of CC it faced the same security and privacy challenges and issues.

In addition to this, CC additionally used to be a base technology for other technologies due to its types of services (Stergiou & Psannis, 2016; Stergiou et al., 2018b; Plageraset al., 2017). One of those is the Big Data (BD). BD is a term used to describe the expected, due to the connected to the Internet devices, rapid increase in the volume of data production. Subsequently, these large amounts of data could be defined as "*a broad term for data sets so large or complex that traditional data processing applications are inadequate*" (Stergiou et al., 2018b). Furthermore, BD is often associated to the use of predictive analytics or certain advanced methods to extract knowledge from the data. Rarely, are also related to a particular size of set of data (Hilbert & López, 2011; Fu et al. 2015). Precision in BD could result in more confident decision making, and better decisions may drive in increased operational efficiency, reduced costs, and minimized risk (Hilbert & López, 2011). From this scope, it can be observed that BD is now equally important both for business and internet. This happens because more information drives to more accurate analysis (Stergiou & Psannis, 2016). The real problem is not that the large amounts of data have been obtained, but whether they have any value or not. Hopefully, by predicting that organizations would be able to acquire information from any source, harness the relevant data, and analyze them in a specific way in order to get quick answers, the following should be achieved: 1) reduce costs, 2) reduce time, 3) produce new items and optimize their offerings, and 4) take more ingenious decisions (Stergiou & Psannis, 2017b).

Regarding Security and Privacy issues and challenges in the field of CC and particularly of MCC, and in order to succeed a secure communication over the network, encryption algorithm plays an important role. It is a valuable and fundamental tool for the protection of the data. Encryption algorithm converts the data into scrambled form by using "*a key*" and only the user have the key in order to decrypt the data. Regarding the researches that have been made in the field, an important encryption technique is the Symmetric key Encryption. In Symmetric key encryption, only one key is used to encrypt and decrypt the data. In this encryption technique the most used algorithm is the AES algorithm (Stergiou & Psannis, 2017b; Kumar et al., 2011; Kaur & Kinger, 2014; Negi et al. 2013).

AES (Advanced Encryption Standard) is the high developed encryption standard recommended by NIST aiming to replace the older DES algorithm. Brute force attack is the only effective attack known against it, in which the attacker tries to test all the characters combinations to unlock the encryption. The AES algorithm block ciphers. Also, AES has been carefully tested for many security applications (Huang, 2011; Stergiou et al., 2018d; Singh & Kinger, 2013; Gupta et al. 2016).

The rest of the paper is organized in 6 sections as follows. In section 2 we present previous work that address challenges and solutions related to Mobile Cloud Computing. Section 3 provides a comparative analysis of the previous works which we have studied offering a brief analysis about evolution of the number of researches that have been made through the years. In section 4 we offer a brief literature analysis of CC and MCC. In section 5 we discuss the outcomes of the literature study and additionally we propose an algorithm method as an indicated solution for Security and Management challenges. Final conclusions and future research directions are given in section 6.

BACKGROUND

For the purpose of this paper we study and analyze previous literature which has been studying two aspects, Security and Privacy Management in MCC and Security and Privacy of MCC. In addition to this, we also present some former works of our research group which have been made in field of CC in general. All the papers presented with ascending form, from the older to the newest. The following paragraphs present the papers which contributed significantly in our study.

Security and Privacy Management in MCC

Initially, the papers that deal with the Security and Privacy issues of Management in MCC are illustrated [23-34]. As we can realize there are several works in this field. More particular, Angin et al. (2010) propose an entity-centric approach for an IDM model in Cloud environment. The proposed approach based on two aspects: a) active bundles, and b) anonymous identification. The active bundles include a payload of Personally Identifiable Information, privacy policies and a virtual machine that enforces the policies and additionally the active bundles use a set of protection mechanisms in order to protect themselves. As regard the anonymous identification, they use it with the aim to mediate interactions between the entity and the Cloud services using entity's privacy policies. Moreover, the authors present the main characteristics of the approach which are: a) independent of third party, b) provides minimum information to the Service Provider, and c) provides ability to use identity data on untrusted hosts. Then, Doukas et al. (2010) demonstrate the implementation of a mobile system that enables electronic healthcare data storage, update and retrieval using CC. The proposed mobile application based in Google's Android OS and offers management of patient health records and medical images. This system was evaluated with the use of Amazon's S3 cloud service. Finally, the authors summarize the details of the implementation and then present initial results of the system in practice. Moreover, Dinh et al. (2011) survey the MCC technology, which could help the general readers to have an overview of the MCC including the definition, the architecture, and the applications. Also, this work presents the issues, the existing solutions, and the recent approaches of the MCC technology. At the end, the authors discuss a number of future research directions of the MCC. Through Habib et al. (2011) propose a multi-faceted Trust Management system architecture for a cloud computing marketplace, with the aim to support the customers in reliably identifying trustworthy cloud providers. The proposed system offers means to identify the trustworthy cloud providers in term of different attributes that assessed by multiple sources and roots of trust information. Furthermore, Prasad et al. (2012) presents a sort survey of MCC evolution and additionally explains how CC and Mobile Devices could be combined with good terms for future opportunities, implications and legal issues for developing countries. In another research, Shiraz et al. (2012) try to review the existing Distributed Application Processing Frameworks, also known as DAPFs, for SMDs in MCC domain. The main objective of this work is to highlight issues and challenges to existing DAPFs in developing, implementing, and executing computational intensive mobile applications within MCC domain. Thus, through this work the authors propose a thematic taxonomy of the current DAPFs, and then they review current offloading frameworks by using thematic taxonomy, and analyze the implications and critical aspects of current offloading frameworks. Finally, this work puts forward open research issues in distributed application processing for MCC that remain to be addressed. Also, Kim & Park (2013) proposes a trust management approach by making an analysis of user behavioral patterns for a reliable MCC. So, the authors suggest a method in order to quantify a one-dimensional trusting relation count on the analysis of telephone call data from Mobile Cloud Environment. Subsequently, it is enhanced trustworthiness of data production, management, and overall application. Whaiduzzaman et al. (2014) present a state-of-the-art survey of vehicular Cloud Computing. More detailed, the authors present a taxonomy for vehicular Cloud in which special attention has been devoted to the extensive applications, Cloud formations, key management, inter-Cloud communication systems, and broad aspects of privacy and security problems. Additionally, this work illustrates the design of an architecture for Vehicular Cloud Computing, itemize the properties required in vehicular Cloud which support the proposed model. In order to achieve their goal, authors compare the proposed mechanism with normal Cloud Computing and then discuss about open research issues and the future directions. Additionally, Khalil et al. (2014) discuss the limitations of the state-of-the-art Cloud Identity Managements with respect to mobile clients. In particular, the authors demonstrate that the current IDMs are vulnerable to three attacks. As a result of their research, the authors propose and validate a new IDM architecture dubbed Consolidated IDM that countermeasures these attacks. Through their experimental results the authors illustrates that CIDM offers its clients with better security guarantees and that it has less energy and communication overhead compared to the current IDM systems. Furthermore, Ali et al. (2015) offer a detailed survey of the security issues that arise due to the very nature of Cloud Computing. Furthermore, this work presents a number of recent solutions offered in the literature with the aim to counter the security problems. In addition, there is a brief view of the highlighted security vulnerabilities in the Mobile Cloud Computing. Then, Rittinghouse & Ransome (2009) discuss about the evolution of the computing as regarding the historical perspective, with focus on advances which primarily led to the evolution of the Cloud Computing. Additionally, there is a survey of some of the critical components that are vital in order to make the Cloud Computing paradigm feasible. The authors by addressing a number of particular legal and philosophical problems try to conclude with a hard look at all the successful Cloud Computing vendors. Finally, Mollah et al. (2017) try to present the major security and privacy issues and challenges in the field which have grown much interest among the academia and research community. Also, they try to illustrate reports of recent works of literature. Moreover, they present a comparative analysis of the literature works based on different security and privacy requirements, and concluding they present the open issues in the field.

Security and Privacy of MCC

Subsequently, the papers that deal with the Security and Privacy issues of the MCC are illustrated. Starting from the oldest, a survey of the various security issues that pose a threat to the Cloud as presented by the work of Subashini & Kavitha (2011). The survey that illustrated in this work could be more specific to different security problems which have emanated due to the nature of the service delivery models of a Cloud Computing system. Continuously, Liang et al. (2011) propose a Security Service Admission Model based on Semi-Markov Decision Process in order to model the system reward for the Cloud provider. Initially, through this work try to define the system states by a tuple represented by the numbers of Cloud users and the associated to them security service categories, and the current event type. Then, the authors derive the system steady-state probability and service request blocking probability with the use of their proposed model. Then, Hada et al. (2011) propose a trust model for Cloud architecture which uses mobile agent as security agents to acquire useful information from the virtual machine that the user and the service provider could utilize with the aim to keep track of privacy of their data and virtual machines. In the proposed model the security agents can dynamically move through the network, replicate themselves according to the requirement, and perform the assigned tasks like accounting and monitoring of virtual machines. Also, Popa et al. (2013) propose a framework with the aim to secure the data transmitted between the components of the same Mobile Cloud Application, and with the aim to ensure the integrity of the applications at the installation on the mobile device and when being updated. Furthermore, the proposed framework of this work allows applying different security properties to various kinds of data and not the same properties to all the data processed by the application. In addition to this, the proposed approach takes into account the user's preferences and the mobile device performances. Moreover, Hashizume et al. (2013) discuss and identifies the main vulnerabilities in CC systems, and the most important threats that found in the literature related to CC technology and its environment, as well as to identify and relate vulnerabilities and threats with possible solutions. Suo et al. (2013) with the aim to facilitate the emerging domain of MCC security and privacy, in brief review the advantages and system model of MCC, and pay attention to the security and privacy in the MCC. At the end, the authors provide the current security and privacy approaches, by deeply analyzing the security and privacy problems from the aspects of mobile terminal, mobile network and Cloud. Additionally, Shahzad & Hussain (2013) present a comprehensive literature review of MCC and the security and privacy issues that MCC faced. Also, the authors present a complete understanding analysis of MCC, in where they explain its architecture, advantages and applications. At the end, the authors conclude that their research is significant useful for the mobile service providers, and thus they can improve the security technologies and mechanisms used for Cloud security in order to minimize the user's security concerns. Then, Khan et al. (2013) illustrate particular efforts that have been devoted in research organizations and academia in order to build secure MCC environments and infrastructures. Thus, in the spite of the efforts, there are a number of loopholes and challenges that still exist in the security policies of MCC. The authors through the literature review conclude in three things that discussed in this work. Firstly, they highlight the current state of the art work which proposed to secure MCC infrastructures. Secondly, they identify the potential problems. Finally, they provide taxonomy of the state of the art. Furthermore, Zhang et al. (2017) propose a novel technique that called "match-then-decrypt", in which a matching phase is also presented before the decryption phase. The proposed technique operates by computing special components in ciphertexts, which are used in order to perform the test that if the attribute private key matches the hidden access policy in ciphertexts without decryption. Moreover, in this work there is a proposal of a basic anonymous ABE construction, and then obtain a security-enhanced extension based on strongly existentially unforgeable one-time signatures. More specifically, the authors conclude that the formal security analysis and performance comparisons indicate that their proposed solutions simultaneously ensure attribute privacy and improve decryption efficiency for outsourced data storage in MCC. Finally, Jiang et al. (2018) initially identify that the scheme that proposed in former work of Tsai and Lo, which was privacy aware authentication scheme for distributed MCC services, fails to achieve mutual authentication. The authors of this work conclude to this regarding it is vulnerable to the service provider impersonation attack. In addition to this, the authors state that the former scheme also suffers from some minor design flaws, including the problem of biometric measure, wrong password, and fingerprint login, no user revocation facility when the smart card is lost or stolen. At the end, they offer some a number of suggestions in order to avoid the aforementioned design flaws in future design of authentication schemes.

Research Group's Previous Works in CC and MCC

At this point there will be present a number of former works which deal with problems and solutions in the field of Cloud Computing in general. More particular, some of them deal with problems and solutions in the field of Mobile Cloud Computing. Starting again from the oldest, Stergiou & Psannis (2016) try to combine the MCC and IoT with the Big Data with the aim to examine the common characteristics and in addition to discover which of MCC and IoT benefits improve the operation of BD applications. Also, the authors of this work present the contribution of MCC and IoT individually to Big Data. Moreover, Stergiou et al. (2018a) present a survey of Internet of Things and CC focusing on the security problems of both of them. More particular, the authors try to combine these technologies aiming to examine the common features, and also aiming to discover the benefits of their integration. At the end, there is a presentation of the contribution of Cloud Computing to the IoT. So, this work illustrates how the Cloud Computing improves the functionality of Internet of Things, and additionally, surveys the security challenges of the integration of Cloud Computing and Internet of Things. Continuously, Stergiou & Psannis (2017a) present a survey of Big Data and Cloud Computing, illustrating their basic characteristics, and focusing on the security and privacy problems of both of them. Regarding this, the authors try to combine the functionality of Big Data and Cloud Computing aiming to examine the frequent characteristics, and in addition to this to discover the benefits which are related in security problems of their integration. Furthermore, Stergiou & Psannis (2017b) survey Cloud Computing and Big Data technology, and their major characteristics, focusing on the security and privacy problems of both of them. Particularly, the authors combine the functionality of two technologies aiming to examine the common characteristics, and additionally to discover the benefits related in security issues of their integration. Then, there is a presentation of a novel method of an algorithm that can be used for the purpose of improving Cloud Computing's security through the use of algorithms that can offer more privacy in the data related to Big Data. At the end, there additionally a survey about the challenges of the integration of Cloud Computing and Big Data related to their security level. Also, Stergiou et al. (2018b) in order to achieve a type of network that will offer more intelligent media-data transfer new technologies were studied. Thus, the authors initially studied the use of various open source tools of Cloud Computing analyzers and simulators. So, the authors after they measure the simulated network performance with CloudSim simulator, they use the Cooja emulator of the Contiki OS aiming to confirm and access more metrics and options. In particular, in this work there is an implementation of a network topology from a small section of the script of CloudSim with Cooja, so that the authors could test a single network segment. The results that have been produced of the experimental procedure illustrate that there are not duplicated packets received during the whole procedure. Finally, Stergiou et al. (2018e) propose a novel system for Cloud Computing integrated with Internet of Things as a base scenario for Big Data. Moreover, the authors try to establish an architecture relaying on the security of the network with the aim to eliminate the security issues. The solution proposed in this work installs a security "*wall*" between the Cloud server and the outer Internet, aiming to eliminate the privacy and security problems. As regard the main goal of this paper a sort survey of IoT and Cloud Computing also presented, focusing on the security issues of both of them. Additionally, the authors state that through their study conclude that Cloud Computing could offer a more "*green*" and efficient "*fog*" environment for sustainable computing scenarios.

Literature Comparative Analysis

Taking into account the Related Research Review Section we realized that the study of MCC's Security and Privacy issues become more popular in the research and academic community over the recent years. We have come to the above conclusion counting on our study of several works in the field of Security and Privacy of Mobile Cloud Computing. The main balk of these works presented in Related Research Review Section. Consequently, there is a need for further research in this area as the growing numbers studied indicate.

As we can observe, the last four years the interest of the researchers has increased considerably compared to the previous decade. Figure 1 illustrates the growth of studies in Security and Privacy Management in Mobile Cloud Computing through the years. Equally important Figure 2 represents the growth of studies in Security and Privacy of Mobile Cloud Computing through the years.

The study of previous works motivates us to survey the Security and Privacy challenges and issues of the Mobile Cloud Computing technology. Thus, count on the major works related to Security and Privacy challenges and issues we have tried to figure out them in Table 1.

Table 1 lists the major challenges and issues which we have distinguished from the related works. Through Table 1 we can figure out which are the major issues and challenges of the Mobile Cloud Computing that have been addressed by the literature.

Figure 1. Growth of works made in security and privacy management in mobile cloud computing over the years

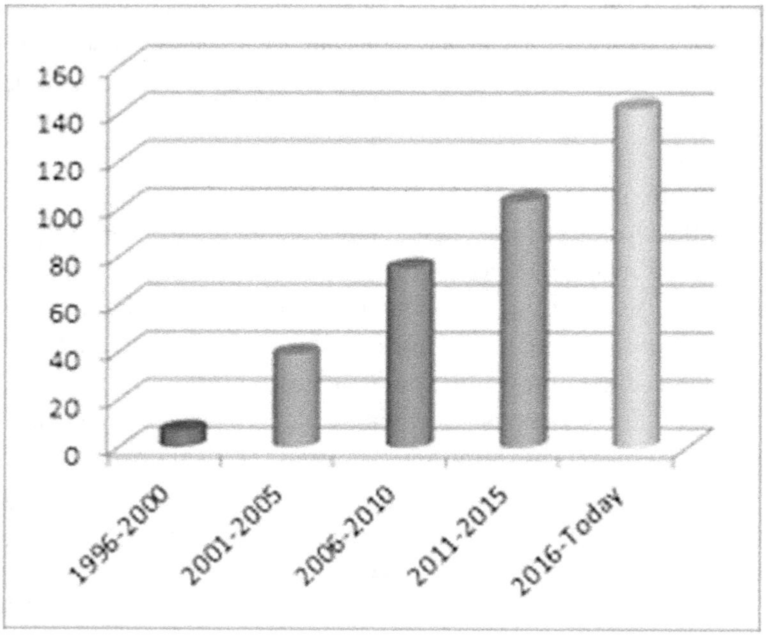

Table 1. Related research work's challenges and issues of the Mobile Cloud Computing technology

Year	Author	Challenge/Issue
2009-08	J. W. Rittinghouse & J. F. Ransome (2009)	• Cloud reliability to users • Users desired outcomes of the Cloud • Levels of trust in Cloud environment
2010-09	C. Doukas et al (2010)	• Sharing and management of medical information resources through mobile healthcare systems
2010-11	P. Angin et al (2010)	• Entities authentication to service providers • Entities multiple accounts associated with multiple service providers
2011-01	S. Subashini & V. Kavita (2011)	• Cloud environment safety • New Cloud model targeting to improve the existing one
2011-04	H. Liang et al (2011)	• Increased number of Critical and Normal Security service users • Allocate Cloud resource aiming to maximize the system rewards with the considerations of the Cloud resource consumption and incomes generated from Cloud users.
2011-10	H. T. Dinh et al (2011)	• Low bandwidth of Mobile Cloud Computing • Availability of Mobile Cloud Computing • Heterogeneity of Mobile Cloud Computing • Computing offloading of Mobile Cloud Computing • Security of Mobile Cloud Computing • Enhancing the efficiency of data access of Mobile Cloud Computing • Context-aware mobile Cloud services of Mobile Cloud Computing
2011-11	S. M. Habib et al (2011)	• Not consistent descriptions in Service Level Agreements among Cloud providers • Uncertain reliably identifying trustworthy Cloud providers for the customers
2011-12	P. S. Hada et al (2011)	• Major need of bringing reliability, transparency and security in Cloud model for client satisfaction

continues on following page

Table 1. Continued

Year	Author	Challenge/Issue
2012-10	M. R. Prasad et al (2012)	• Cloud Computing challenges: Performance, Security & Privacy, Control, Bandwidth Costs, Reliability • Mobile Cloud Computing legal issues
2012-11	M. Shiraz et al (2012)	• Users expectation to run computational intensive applications on Smart Mobile Devices in the same way as powerful stationary computers • Establishment of distributed application processing platform at runtime which requires additional computing resources on Smart Mobile Devices
2013-01	D. Popa et al (2013)	• Use of MCC increases security risks and privacy invasion • MCC application model not well-defined
2013-06	A. Shahzad & M. Hussain (2013)	• Security and privacy risks faced be MCC uses • Architecture and Cloud service delivery models issues • Mobile Cloud infrastructure issues • Mobile Cloud communication channel issues
2013-07	H. Suo et al (2013)	• Security and privacy issues and challenges in MCC
2013-07	A. N. Khan et al (2013)	• Security threats have become a hurdle in the rapid adaptability of the MCC paradigm
2013-12	M. Kim & S. O. Park (2013)	• MCC architecture, design and implementation need to be improved due to limited computing capability and storage issues • Trust management approach for reliable MCC
2013-12	K. Hashizume et al (2013)	• Risk of outsourcing data to third party Cloud providers • Cloud Computing inherits many technologies security issues
2014-02	M. Ali et al (2014)	• Services provided by third party Cloud providers entail additional security threats • Migration of user's assets outside the administrative control in the Cloud environment escalates the security concerns.
2014-03	I. Khalil et al (2014)	• Mobile devices are easy to be compromised • Mobile users store Personal Identifiable Information in unprotected text files, cookies and applications • Limitations of state-of-the-art Cloud Identity Management Systems with respect to mobile clients
2014-04	Md. Whaiduzzaman et al (2014)	• Traffic management and road safety by instantly using vehicular resources • Vehicular Cloud Computing security challenges: Authentication, Secure location & localization, Securing vehicular communication, Vehicular public key infrastructure, Data security, Network heterogeneity, Access control
2017-02	Y. Zhang et al (2017)	• Each decryption usually requires many pairings and the computation overhead grows with the complexity of the access formula • Existing schemes suffer a serve efficiency drawback and not suitable for MCC where users may be resource-constrained
2017-04	M. B. Mollah et all (2017)	• Data security challenges • Partitioning and offloading security challenges • Virtualization security challenges • Mobile Cloud applications security challenges • Mobile devices security challenges • Privacy challenges
2018-06	Q. Jiang et al (2018)	• A previous proposed scheme of Tsai & Lo fails to achieve mutual authentication and withstands all major security threats

Figure 2. Growth of works made in security and privacy of mobile cloud computing over the years

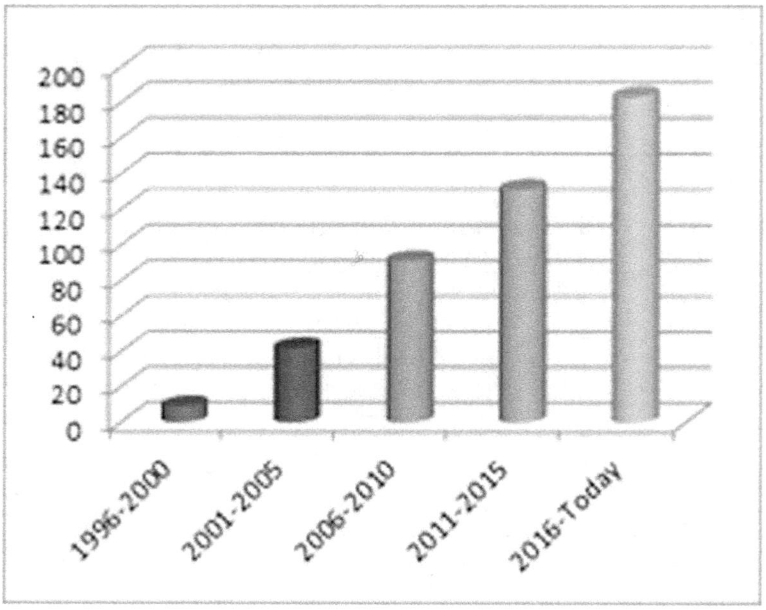

Introducing To Cloud Computing

CC offers abilities and functions such as computing, storage, services, and applications over the Internet. In general, to render smartphones energy efficient and computationally capable, major changes to the hardware and software levels are required. This causes the cooperation of developers and manufacturers (Rahimi et al., 2014).

Features

As all technologies, so the CC technology has a number of characteristics which determine its operation. These characteristics are represented and outlined below.

CC(a): Storage over Internet

Storage over Internet can be defined as "*a technology framework that uses Transmission Control Protocol/Internet Protocol (TCP/IP) networks to link servers and storage devices and to facilitate storage solution deployment*" (Whaiduzzaman et al., 2014; Garg et al., 2013).

CC(b): Service over Internet

The Service over Internet has as major objective is to "*help customers all over the world in order to transform aspirations into achievements by harnessing the Internet's efficiency, speed and ubiquity*" (Whaiduzzaman et al., 2014; Garg et al., 2013).

CC(c): Applications over Internet

Cloud Applications, or as scientific known as Applications over Internet, are the programs which have been written to do the job of a current manual task, or virtually anything, and which perform their job on the server through an internet connection (Whaiduzzaman et al., 2014; Garg et al., 2013).

CC(d): Energy Efficiency

Energy Efficiency could be defined as "*a way of managing and restraining the growth in energy consumption*" (Whaiduzzaman et al., 2014; Garg et al., 2013). By delivering more services for the same energy input or for the same services for less energy input may be something more energy efficient (Whaiduzzaman et al., 2014; Garg et al., 2013).

CC(e): Computationally Capable

The services of computational clouds are leveraging the computationally concentrated and ubiquitous mobile applications which have been enabled by the technology of MCC. Thus, a system can be considered as computationally capable when it meets the requirements to offer us the results we want, by making the right calculations (Whaiduzzaman et al., 2014; Garg et al., 2013).

Security on Cloud Computing

CC security is an evolving sub-domain of computer security, network security and information security. It makes an allusion to a broad set of policies, technologies, and controls deployed to protect data, applications, and the associated infrastructure of CC.

CC technology offers through its storage solutions to users and industries various capabilities with the aim to store and process their data in third-party data centers (Haghighat et al., 2015). Thus, by aiming to offer secure communication through the network, encryption algorithm plays a vital role. As regards the researches that have been made, an important encryption technique is the Symmetric Key Encryption. In Symmetric key encryption, only one key is used to encrypt and decrypt the data. In this encryption technique the most used algorithm is the AES (Kumar et al., 2011; Kaur & Kinger, 2011; Stergiou et al., 2018g; Gupta & Badve, 2017).

AES (Advanced Encryption Standard) is the newest encryption standard and the more reliable, recommended by NIST to replace DES algorithm. The only effective scenario of attacking in AES is the Brute force attack, in which the attacker tries to test all the characters combinations to unlock the encryption. AES encryption model is fast and flexible, and in addition, it can be implemented on different platforms (Singh & Kinger, 2013).

Mobile Cloud Computing Trade Offs

CC has some disadvantages-limitations which should be eliminated over the years in order to achieve a better and more ideal use. Some businesses and especially the smaller ones need to be aware of these limitations before going in for this technology.

CC(I-a): Security

One major issue of the MCC is the security issue. Before someone adopts this technology, they should know that all the company's sensitive information would be surrender to a third-party Cloud service provider. This could potentially put the company in great risk. Hence, someone must be absolutely sure that they would choose the most reliable service provider, who will keep the information completely safe (Stergiou & Psannis, 2016; Viswanathan, 2012; Pfarr et al., 2015; Stergiou et al.,2018g; Gupta et al., 2016).

CC(I-b): Connectivity

Internet connection is critical to CC. Thus, the user should be certain that there is a good result before opting for these services. Since someone owes a mobile device which is connected to the internet has become the norm in the wireless world of today, CC has a very large potential user base (Stergiou & Psannis, 2016; Almrot & Andersson, 2013).

CC(I-c): Performance

Another major concern of the CC pertains to its performance. Some users feel performance is not as good as in native applications. Thus, checking with one service provider and understanding their track record is advisable (Stergiou & Psannis, 2016; Fremdt et al., 2013; GetCloud Services, 2014).

CC(I-d): Latency (Delay)

In CC, latency (sometimes referred as turnaround time) is defined as the time involved in offloading the computation and getting back the results from the nearby infrastructure or cloud (Stergiou & Psannis, 2016; Li et al., 2017).

CC(I-e): Privacy

Data privacy is important and is one of the main bottlenecks that restrict consumers from adopting CC. Therefore, to gain consumers trust in the Cloud, the application models must support application development with privacy protection, and implicit authentication mechanisms (Stergiou & Psannis, 2016; Pfarr et al., 2014; Shi et al., 2010).

Challenges Outcomes by Combining Cloud COMPUTING with Other Technologies

When critical applications, such as the IoT applications, move towards the Cloud Computing technology, concerns arise due to the lack of trust in the service provider or the knowledge about service level agreements (SLAs) and knowledge about the physical location of data. Consequently, new challenges require specific attention as mentioned in surveys (Botta et al., 2016; Bhattasali et al., 2013; Simmhan et al., 2011). Multi-tenancy could also compromise security and lead to sensitive information leakage. Moreover, public key cryptography cannot be applied at all layers due to the computing power constraints imposed by the things. These are examples of topics that are currently under investigation in order to tackle the big challenge of security and privacy in integrating Cloud Computing with other technologies (Botta et al., 2016; Book, 2018; Stergiou et al., 2018g; Gou et al., 2017).

Subsequently, some challenges about the security issue in the integration of Cloud Computing with other technologies are listed below (Botta et al., 2016).

1. **Heterogeneity:** A big challenge in Cloud Computing integration with other technologies is related to the wide heterogeneity of devices, operating systems, platforms, and services available and possibly used for new or improved applications (Grozev & Buyya, 2014).
2. **Performance:** Often Cloud Computing integration with other technologies applications introduce specific performance and QoS requirements at several levels (i.e. for communication, computation, and storage aspects) and in some particular scenarios meeting requirements may not be easily achievable (Rao et al., 2012).
3. **Reliability:** When Cloud Computing integration with other technologies is adopted for mission-critical applications, reliability concerns typically arise. When applications are deployed in resource constrained environments a number of challenges related to device failure or not always reachable devices exists (He et al., 2014).
4. **Big Data:** With an estimated number of 50 billion devices that will be networked by 2020, specific attention must be paid to transportation, storage, access, and processing of the huge amount of data they will produce (Dobre & Xhafa, 2014).
5. **Monitoring:** As largely documented in the literature, monitoring is an essential activity in Cloud environments for capacity planning, for managing resources, SLAs, performance and security, and for troubleshooting (Aceto & Botta, 2013).

Cloud Computing as Base Technology for Big Data

Cloud Computing (CC) is a new generation of services which aims to provide accessibility to information and data from any place at any time. With this kind of novel technology there is no limitation and no further need for hardware equipment. The recent years, Cloud Computing services compose one of the major areas in the world of competition among the giant companies in the field of IT and software (Mell & Grance, 2011; Skourletopoulos et al., 2017; Dorgham et al., 2018).

More specifically, Cloud Computing is consisted of a technology of internet services providing remote use of hardware and software. As a result, the users of Cloud Computing could have access to information and data from any place at any time. In the concept of Cloud Computing there is another technology called Mobile Cloud Computing (MCC) that refers in general concept to in two prospects (Stergiou et al., 2018f): a) infrastructure based, and b) ad-hoc mobile cloud. In the prospect of infrastructure based mobile cloud, the infrastructure of the hardware is still static and delivers services to the mobile users (Stergiou et al., 2018c; Batalla et al. 2016). Particularly, MCC is defined as "*the integration of Cloud Computing and Mobile technology in order to make any type of mobile devices resourceful in terms such as computational power, memory, storage and energy*" (Stergiou & Psannis, 2017b). Regarding the usage of Cloud services in Mobile devices many types of services could be processed through it. Thus, high quality media could be transmitted through Cloud environment progressed in applications which were installed and operated in Cloud.

Considering this Cloud Computing could be settled as a base technology to operate other technologies such as Big Data and consequently to be accomplished an integration of Cloud and Big Data (Haghighat et al., 2015; Garg et al., 2013). In addition to this, Cloud Computing also used to be a base technology for others technologies due to its type of services (Stergiou & Psannis, 2016; Stergiou et al., 2018d).

As already mentioned, one of those is Big Data (BD). Big Data used to describe the surprisingly rapid increase in volume of data in structured and unstructured form. It is a broad term for data sets so large or complex that traditional data processing applications are inadequate. Furthermore, Big Data often refers to the use of predictive analytics or certain advanced methods to extract value from data. Rarely, it also refers to a particular size of data set (Hilbert & López, 2011; Fu et al., 2016). Precision in Big Data could result in more confident decision making, and better decisions may lead to an increased operational efficiency, reduced costs, and minimized risk [18]. From this scope we realize that the Big Data is now equally important both for business and internet. This happens because more information leads to more accurate analyses (Stergiou & Psannis, 2016). The real problem is not that you have acquired large quantities of data, but whether it has any value or not. Hopefully, by envisaging that the organizations would be able to obtain information from any source, harness the relevant data and analyze it with the aim to get quick answers, we will achieve the following: 1) to reduce costs, 2) to reduce time, 3) to produce new products and to optimize their offerings, 4) to make more intelligent decisions (Stergiou & Psannis, 2017a; Stergiou & Psannis, 2017b).

RESEARCH OUTCOMES & PROPOSED SOLUTIONS

Table 2 presents the challenges that have been addresses by the literature work which we have studied. As we could observe most of the works that we have studied focus on the "*Trusted environment*" which in our opinion is the major issue of the Mobile Cloud Computing, and as a result the "*Security*" issue is most popular in the literature work. Additionally, concerning the rest of the challenges we come to the conclusion that the "*Reliability*" and the "*Management*" are also basic issues that need to be addressed. More detailed, through the number of 22 works that we have stood out the statistic results are the following: Privacy 6 of 22, Security 15 of 22, Trusted environment 15 of 22, Bandwidth 5 of 22, Environment limitations 10 of 22, Management 11 of 22, Reliability 12 of 22, User authentication 9 of 22, Efficiency 5 of 22. Equally important, the less mentioned issues are the "*Bandwidth*" and the "*Efficiency*" which are really vital for the functionality of the Mobile Cloud Computing technology. Regarding this, we could realize that more researches need to be done in the field in order to find better solutions aiming to improve Bandwidth and Efficiency of MCC.

Thus, based on previous works (Stergiou et al., 2018b; Stergiou & Psannis, 2017a; Stergiou et al., 2018f; Stergiou et al., 2018c; Stergiou et al., 2018d; Stergiou & Psannis, 2017b; Plageras et al., 2017; Stergiou et al., 2018a; Stergiou et al., 2018e) the optimal solution in order to achieve a reliable and trusted environment with a more "*safe*" authentication and encryption for the users the MCC system have to adapt the AES encryption algorithm.

The AES algorithm has variable key length of 128, 192, or 256 bits, with default 256 bits. It encrypts data blocks of 128 bits in 10, 12 and 14 round depending on the key size. AES encryption is fast and flexible algorithm, and it can be implemented on various platforms especially in small devices, such as mobile devices. Also, AES has been carefully tested for many security applications (Singh & Kinger, 2013). As a result, the AES algorithm provides the ability to have speed key setup time a good key agil-

Table 2. Literature challenges in the field of Mobile Cloud Computing

Literature work	Challenges								
	Privacy	*Security*	*Trusted environment*	*Bandwidth*	*Environment limitations*	*Management*	*Reliability*	*User authentication*	*Efficiency*
Rittinghouse & Ransome, 2009			X			X	X	X	
Doukas et al., 2010			X			X			
Angin et al., 2010	X	X				X		X	
Subashini & Kavitha, 2011			X		X	X			
Liang et al., 2011		X	X			X	X	X	
Dinh et al., 2011		X	X	X			X		X
Habib et al., 2011		X	X				X	X	
Hada et al., 2011		X	X				X		
Prasad et al., 2012	X	X		X		X	X	X	
Shiraz et al., 2012					X	X			X
Popa et al., 2013	X	X			X	X			
Shahzad & Hussain, 2013	X	X	X		X		X		
Suo et al., 2013	X	X	X						
Khan et al., 2013		X			X				
Kim & Park, 2013			X	X	X		X		X
Hashizume et al., 2013		X	X				X	X	
Ali et al., 2015		X	X		X	X	X		
Khalil et al., 2014			X		X	X	X	X	
Whaiduzzaman et al., 2014		X				X	X	X	X
Zhang et al., 2017			X	X	X				X
Mollah et al., 2017	X	X		X					
Jiang et al., 2018		X	X		X			X	

ity. So, if we use this algorithm we could have a trusted relay method with an encryption of a speed key setup. Therefore, instead the trust relay use we can seize also there no serious weak keys in AES and so we could have a beneficial security use of the encryption in the integrated new model. Similarly, we can take advantage the less memory which AES needs for implementation that makes it for restricted-space environments. Thus, we can seize the transmit power that AES offers and as a result we can have a better and more trusted transmission in a MCC environment (Stergiou & Psannis, 2017b; Singh & Kinger, 2013; Stergiou et al., 2018e; Gupta, 2018).

Count our study and the reliability of AES encryption algorithm for a MCC environment we suggest the following part of pseudocode based on the AES encryption algorithm.

With this proposed method we can extend the advances of MCC, in particular when it integrates with other technologies like IoT and Big Data, by developing a highly innovative and scalable service platform to enable secure and privacy services. Through tour research we can propose the algorithm 1 which extends the security advances of MCC environment, especially when integrates with other technologies. As a proposal of this work could be this part of pseudocode algorithm which uses the original key consists of 128 bits/16 bytes which are represented as a 6x6 matrix.

Table 3. Suggested pseudocode based on the AES encryption algorithm

Algorithm 1

```
input -> byte[]
byte[] + R.Key -> state[]
for 6 to 66
W[i-1] -> T
if i mod 6 = 0
rotate T + 6
W[i-6] / T -> W[i]
R.Key+1
i+1 -> i
Row +1 -> Row
state[] -> output[]
```

Experimental Results

Compared our proposed method with the existing one we come to some measurements that have been through time and showing the benefits of our proposed method.

Figure 3. Security level of encryption algorithms of measurement used for the study of AES model algorithm

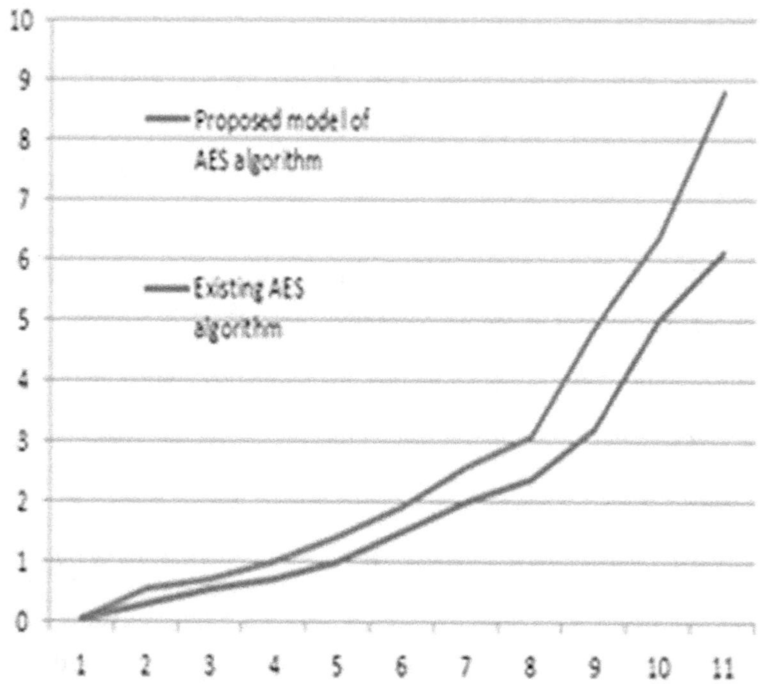

As we can observe by Figure 3 the more often is the combined use of the algorithms, the higher level of security of the data usage we get every time. The upper line represents our proposed model of AES algorithm and the other (down line) represents the existing AES algorithm. Based on Figure 3 we can also figure out that our proposed method is over from the existing model regarding the higher security level of encryption that it can achieve through the time.

CONCLUSION AND FUTURE DIRECTIONS

The MCC technology provides a number of possibilities, but additionally places several challenges and issues that need to be addressed as well. Mobile Cloud Computing refers to an infrastructure where data, applications and information could be processed through a mobile device, but simultaneously outside of the mobile device. The main objective of the use of MCC is to decrease the use of stronger hardware and to have the access to data and applications, and in many times to more computational power, from every place and in any time, through a mobile device.

With our study, in regard on the huge benefits of the Mobile Cloud Computing technology, we try to achieve a more safe and trusted environment for the MCC users in order to operate the functions, and transfer, edit and manage data and applications. This could be achieved proving a novel method count on the AES encryption algorithm, which is, according to our study, the most relevant encryption algorithm to a Cloud environment.

Furthermore, we try to define the most important issues and challenges in the field of Mobile Cloud Computing technology by presenting a number of the most significant works related to MCC through the last eight years.

As a future work, we could focus to find novel ways to achieve a better integration MCC with other technologies, focusing on security algorithms and all the challenges that the technologies faced on security level. Regarding the rapid development of Cloud technology the security issues of Mobile Cloud Computing must be solved or reduced to a minimum in order to have a better and safer model. The security challenges and issues that surveyed in this work could be the sector for further research as a case study, with the goal of minimizing them.

ACKNOWLEDGMENT

The authors would like to thank the anonymous reviewers for their valuable comments and feedback which was extremely helpful in improving the quality of the paper.

REFERENCES

Aceto, G., Botta, A., de Donato, W., & Pescape, A. (2013, June). Cloud monitoring: A survey. *Elsevier, Computer Networks*, *57*(9), 2093–2115. doi:10.1016/j.comnet.2013.04.001

Ali, M., Khan, S. U., & Vasilakos, A. V. (2015, February). Security in cloud computing: Opportunities and challenges. *Elsevier, Information Sciences*, *305*, 357–383. doi:10.1016/j.ins.2015.01.025

Almrot, E., & Andersson, S. (2013). *A study of the advantages & dis-advantages of mobile cloud computing versus native environment* (Bachelor Thesis). Blekinge Institute of Technology, Karlskrona.

Angin, P., Bhargava, B., Ranchal, R., Singh, N., & Linderman, M. (2010). An Entity-centric Approach for Privacy and Identity Management in Cloud Computing. *Proceedings of 29th IEEE International Symposium on Reliable Distributed Systems*. 10.1109/SRDS.2010.28

Batalla, J. M., Mastorakis, G., Mavromoustakis, C. X., & Zurek, J. (2016, October). On cohabitating networking technologies with common wireless access for Home Automation Systems purposes. *IEEE Wireless Communications, 23*(5), 76–83. doi:10.1109/MWC.2016.7721745

Bhattasali, T., Chaki, R., & Chaki, N. (2013). Secure and trusted cloud of things. *Proceedings of INDICON Annual IEEE India Conference.*

Blog: Follow what's happening at Get Cloud Services. (2014, December 23). Mobile Cloud Com-puting – Pros and Cons. *GetCloud Services*. Available: https://www.getcloudservices.com/blog/mobile-cloud-computing-pros-and-cons/

Botta, A., de Donato, W., Persico, V., & Pescape, A. (2016, March). Integration of Cloud Computing and Internet of Things: A Survey. *Journal of Future Generation Computer Systems, 56*, 1–54. doi:10.1016/j.future.2015.09.021

Dinh, H. T., Lee, C., Niyato, D., & Wang, P. (2011, October). A survey of mobile cloud computing: Architecture, applications, and approaches. *Wireless Communications and Mobile Computing, 13*(18), 1587–1611. doi:10.1002/wcm.1203

Dobre, C., & Xhafa, F. (2014, July). Intelligent services for big data science. *Future Generation Computer Systems, 37*, 267–281. doi:10.1016/j.future.2013.07.014

Dorgham, O., Almonani, A., Alauthman, M., Albalas, F., & Obeidat, A. (2018, April). An Online Intrusion Detection System to Cloud Computing Based on Neucube Algorithms. *International Journal of Cloud Applications and Computing, 8*(2), 96–112. doi:10.4018/IJCAC.2018040105

Doukas, C., Pliakas, T., & Maglogiannis, I. (2010). Mobile Healthcare Information Management utilizing Cloud Computing and Android OS. *Proceedings of 32nd Annual International Conference of the IEEE EMBS*. 10.1109/IEMBS.2010.5628061

Fremdt, S., Beck, R., & Weber, S. (2013). Does Cloud Computing Matter? An analysis of the Cloud Model software-as-a-service and its impact on operational agility. *Proceedings of 46th Hawaii International Conference on System Sciences*, 1025-1034. 10.1109/HICSS.2013.182

Fu, Z., Ren, K., Shu, J., Sun, X., & Huang, F. (2015, December). Enabling Personalized Search over Encrypted Out-sourced Data with Efficiency Improvement. *IEEE Transactions on Parallel and Distributed Systems.*

Fu, Z., Ren, K., Shu, J., Sun, X., & Huang, F. (2016, September). Enabling Personalized Search over Encrypted Outsourced Data with Efficiency Improvement. *IEEE Transactions on Parallel and Distributed Systems*, *27*(9), 2546–2559. doi:10.1109/TPDS.2015.2506573

Garg, S. K., Versteeg, S., & Buyya, R. (2013). A framework for ranking of cloud computing services. *Future Generation Computer Systems*, *29*(4), 1012–1023. doi:10.1016/j.future.2012.06.006

Gou, Z., Yamaguchi, S., & Gupta, B. B. (2017). Analysis of Various Security Issues and Challenges in Cloud Computing Environment: A Survey. In *Handbook of Research on Modern Cryptographic Solutions for Computer and Cyber Security* (pp. 221–247). IGI Global.

Grozev, N., & Buyya, R. (2014, March). Inter-cloud architectures and application brokering: Taxonomy and survey. *Software, Practice & Experience*, *44*(3), 369–390. doi:10.1002pe.2168

Gupta, B. B., Agrawal, D. P., & Yamaguchi, S. (2016). Handbook of Research on Modern Cryptographic Solutions for Computer and Cyber Security. IGI Global. doi:10.4018/978-1-5225-0105-3

Gupta, B. B., & Badve, O. P. (2017, December). Taxonomy of DoS and DDoS attacks and desirable defence mechanism in a Cloud computing environment. *Springer, Neural Computing and Applications*, *28*(12), 3665–3682.

Gupta Brij, B. (2018). *Computer and Cyber Security: Principles, Algorithm, Applications, and Perspectives*. CRC Press, Taylor & Francis.

Habib, S. M., Ries, S., & Muhlhauser, M. (2011). Towards a Trust Management System for Cloud Computing. *Proceedings of IEEE International Joint Conference TrustCom-11/IEEE ICESS-11/FCST-11*. 10.1109/TrustCom.2011.129

Hada, P. S., Singh, R., & Meghwal, M. M. (2011, December). Security Agents: A Mobile Agent based Trust Model for Cloud Computing. *International Journal of Computers and Applications*, *36*(12), 12–15.

Haghighat, M., Zonouz, S., & Abdel-Mottaleb, M. (2015, November). CloudID: Trustworthy cloud-based and cross-enterprise biometric identification. *Expert Systems with Applications*, *42*(21), 7905–7916. doi:10.1016/j.eswa.2015.06.025

Hashizume, K., Rosado, D. G., Fernandez-Medina, E., & Fernandez, E. B. (2013, December). An analysis of security issues for cloud computing. *Journal of Internet Services and Applications*, *4*(5), 1–13.

He, W., Yan, G., & Xu, L. D. (2014, May). Developing vehicular data cloud services in the IoT environment. *IEEE Transactions on Industrial Informatics*, *10*(2), 1587–1595. doi:10.1109/TII.2014.2299233

Hilbert, M., & López, P. (2011). The World's Technological Capacity to Store, Communicate, and Compute Information. *Science*, *332*(6025), 60–65. doi:10.1126cience.1200970 PMID:21310967

Huang, D. (2011, October). Mobile cloud computing. *IEEE COMSOC Multimedia Communications Technical Committee (MMTC). E-Letter*, *6*(10), 27–31.

Jiang, Q., Ma, J., & Wei, F. (2018, June). On the Security of a Privacy-Aware Authentication Scheme for Distributed Mobile Cloud Computing Services. *IEEE Systems Journal, 12*(2), 2039–2042. doi:10.1109/JSYST.2016.2574719

Kaur, R., & Kinger, S. (2014, March). Analysis of Security Algorithms in Cloud Computing. *International Journal of Application or Innovation in Engineering & Management, 3*(3), 171–176.

Keskin, T., & Taskin, N. (2014). A pricing model for cloud computing service. *Proceedings of 47th Hawaii International Conference on System Science*, 699-707. 10.1109/HICSS.2014.94

Khalil, I., Khreishah, A., & Azeem, M. (2014, March). Consolidated Identity Management System for secure mobile cloud computing. *Elsevier, Computer Networks, 99*, 99–110. doi:10.1016/j.comnet.2014.03.015

Khan, A. N., Mat Kiah, M. I., Khan, S. U., & Madani, S. A. (2013, July). Towards secure mobile cloud computing: A survey. *Elsevier, Future Generation Computer Systems, 29*(5), 1278–1299. doi:10.1016/j.future.2012.08.003

Kim, M., & Park, S. O. (2013, December). Trust management on user behavioral patterns for a mobile cloud computing. *Springer, Cluster Computing, 16*(4), 725–731. doi:10.100710586-013-0248-9

Kryftis, Y., Mastorakis, G., Mavromoustakis, C., Mongay Batalla, J., Pallis, E., & Kormentza G. (2016). Efficient Entertainment Services Provision over a Novel Network Architecture. *IEEE Wireless Communications Magazine.*

Kumar, Munjal, & Sharma. (2011). Comparison of Symmetric and Asymmetric Cryptography with Existing Vulnerabilities and Countermeasures. *International Journal of Computer Science and Management Studies, 11*(3).

Li, J., Huang, L., Zhou, Y., He, S., & Ming, Z. (2017, January). Computation partitioning for mobile cloud computing in big data environment. *IEEE Transactions on Industrial Informatics, 11.*

Liang, H., Huang, D., Cai, L. X., Shen, X. S., & Peng, D. (2011). Resource Allocation for Security Services in Mobile Cloud Computing. *Proceedings of IEEE Conference on Computer Communications Workshops INFOCOM WKSHPS 2011.* 10.1109/INFCOMW.2011.5928806

Mell, P., & Grance, T. (2011). NIST Definition of Cloud Computing - Recommendations of the National Institute of Standards and Technology. National Institute of Standards and Technology Special Publication.

Mollah, M. B., Azad, M., & Vasilakos, A. (2017, April). Security and privacy challenges in mobile cloud computing: Survey and way ahead. *Journal of Network and Computer Applications, 84*, 38–54. doi:10.1016/j.jnca.2017.02.001

Negi, P., Mishra, A., & Gupta, B. B. (2013). Enhanced CBF Packet Filtering Method to Detect DDoS Attack in Cloud Computing Environment. *Cryptography and Security.* arXiv preprint arXiv:1304.7073

Pfarr, F., & Buckel, T. (2014). Cloud Computing Data Protection – A Literature Review and Analysis. *Proceedings of 47th Hawaii International Conference on System Sciences*, 5018-5027. 10.1109/HICSS.2014.616

Plageras, A. P., Stergiou, C. L., & Psannis, K. E. (2017). Solutions for Inter-connectivity and Security in a Smart Hospital Building. *Proceedings of 15th IEEE International Conference on Industrial Informatics (INDIN 2017)*. 10.1109/INDIN.2017.8104766

Popa, D., Cremene, M., Borda, M., & Boudaoud, K. (2013). A Security Framework for Mobile Cloud Applications. *Proceedings of 11th RoEduNet International Conference*. 10.1109/RoEduNet.2013.6511724

Prasad, M. R., Gyani, J., & Murti, P. R. K. (2012, October). Mobile Cloud Computing: Implications and Challenges. *Journal of Information Engineering and Applications*, *2*(7), 7–15.

Rahimi, M. R., Ren, J., Liu, C. H., Vasilakos, A. V., & Venkatasubramanian, N. (2014, April). Mobile Cloud Computing: A survey, State of Art and Future Directions. *Mobile Networks and Applications*, *19*(2), 133–143. doi:10.100711036-013-0477-4

Rao, B. B. P., Saluia, P., Sharma, N., Mittal, A., & Sharma, S. V. (2012). Cloud computing for Internet of Things & sensing based applications. *Proceedings of IEEE 6th International Conference on Sensing Technology (ICST 2012)*, 374–380. 10.1109/ICSensT.2012.6461705

Rittinghouse, J. W., & Ransome, J. F. (2009). *Cloud Computing: Implementation, Management, and Security*. CRC Press.

Shahzad, A., & Hussain, M. (2013). Security Issues and Challenges of Mobile Cloud Computing. *International Journal of Grid and Distributed Computing*, *6*(6), 37–50. doi:10.14257/ijgdc.2013.6.6.04

Shi, E., Niu, Y., Jakobsoon, M., & Chow, R. (2010). Implicit Authentica-tion through Learning User Behavior. *Proceedings of ISC'10 13th International Conference on Information Security*, 99-113.

Shiraz, M., Gani, A., Khokhar, R. H., & Buyya, R. (2012, November). A Review on Distributed Application Processing Frameworks in Smart Mobile Devices for Mobile Cloud Computing. *IEEE Communications Surveys and Tutorials*, *15*(3), 1294–1313. doi:10.1109/SURV.2012.111412.00045

Simmhan, Y., Kumbhare, A. G., Cao, B., & Prasanna, V. (2011). An analysis of security and privacy issues in smart grid software architectures on clouds. *Proceedings of IEEE 4th International Conference on Cloud Computing*, 582–589. 10.1109/CLOUD.2011.107

Singh, G., & Kinger, S. (2013). Integrating AES, DES, and 3-DES Encryption Algorithms for Enhanced Data Security. *International Journal of Scientific & Engineering Research, 4*(7).

Skourletopoulos, G., Mavromoustakis, C. X., Mastorakis, G., Batalla, J. M., & Sahalos, J. N. (2017, August). An Evaluation of Cloud-Based Mobile Services with Limited Capacity: A Linear Approach. *Springer, Soft Computing*, *21*(16), 4523–4530. doi:10.100700500-016-2083-4

Stergiou, C., & Psannis, K. E. (2016, May). Recent advances delivered by Mobile Cloud Computing and Internet of Things for Big Data applications: A survey. *International Journal of Network Management*, 1–12.

Stergiou, C., & Psannis, K. E. (2017a). Algorithms for Big Data in Advanced Communication Systems and Cloud Computing. *Proceedings of 19th IEEE Conference on Business Informatics 2017 (CBI2017)*. 10.1109/CBI.2017.28

Stergiou, C., & Psannis, K. E. (2017b, November). Efficient and Secure Big Data delivery in Cloud Computing. *Springer, Multimedia Tools and Applications*, *76*(21), 22803–22822. doi:10.100711042-017-4590-4

Stergiou, C., Psannis, K. E., Gupta, B., & Ishibashi, Y. (2018e, June). Security, Privacy & Efficiency of Sustainable Cloud Computing for Big Data & IoT. *Elsevier, Sustainable Computing, Informatics and Systems*.

Stergiou, C., Psannis, K. E., Gupta, B., & Ishibashi, Y. (2018g, September). Security, Privacy & Efficiency of Sustainable Cloud Computing for Big Data & IoT. *Elsevier, Sustainable Computing, Informatics and Systems*, *19*, 174–184. doi:10.1016/j.suscom.2018.06.003

Stergiou, C., Psannis, K. E., & Gupta, B. B. (2018f). Advanced Media-based Smart Big Data on Intelligent Cloud Systems. *IEEE Transaction on Sustainable Computing*.

Stergiou, C., Psannis, K. E., Kim, B.-G., & Gupta, B. (2018a, January). Secure integration of IoT and Cloud Computing. *Elsevier, Future Generation Computer Systems*, *78*(3), 964–975. doi:10.1016/j.future.2016.11.031

Stergiou, C., Psannis, K. E., Plageras, A. P., Ishibashi, Y., & Kim, B.-G. (2018b, March). Algorithms for efficient digital media transmission over IoT and cloud networking. *Journal of Multimedia Information System*, *5*(1), 27–34.

Stergiou, C., Psannis, K. E., Plageras, A. P., Xifilidis, T., & Gupta, B. B. (2018c). Security and Privacy of Big Data for Social Networking Services in Cloud. *Proceedings of IEEE conference on Computer Communications (IEEE INFOCOM 2018)*. 10.1109/INFCOMW.2018.8406831

Stergiou, C. L., Psannis, K. E., Gupta, B. B., & Ishibashi, Y. (2018d). Security, Privacy & Efficiency of Sustainable Cloud Computing for Big Data & IoT. *Elsevier, Sustainable Computing, Informatics and Systems*.

Subashini, S., & Kavitha, V. (2011, January). A survey on security issues in service delivery models of cloud computing. *Journal of Network and Computer Applications*, *34*(1), 1–11. doi:10.1016/j.jnca.2010.07.006

Suo, H., Liu, Z., Wan, J., & Zhou, K. (2013). Security and Privacy in Mobile Cloud Computing. *Proceedings of 9th International Conference on Wireless Communications and Mobile Computing (IWCMC 2013)*. 10.1109/IWCMC.2013.6583635

Viswanathan, P. (2012). *Cloud Computing – Is it Really All That Beneficial?* Available: http://mobiledevices.about.com/od/additionalresources/a/Cloud-Computing-Is-It-Really-All-That-Beneficial.htm

Wang, C., Ren, K., Lou, W., & Li, J. (2010, July). Toward publicly auditable secure cloud data storage services. *IEEE Network*, *24*(4), 19–24. doi:10.1109/MNET.2010.5510914

Whaiduzzaman, Md., & Haque, M. N. (2014, June). A Study on Strategic Provision of Cloud Computing Services. *The Scientific World Journal*, 1–8.

Whaiduzzaman, M., Sookhak, M., Gani, A., & Buyya, R. (2014, April). A survey on vehicular cloud computing. *Journal of Network and Computer Applications*, *40*, 325–344. doi:10.1016/j.jnca.2013.08.004

Zhang, Y., Chen, X., Wong, D. S., Li, H., & You, I. (2017). *Ensuring attribute privacy protection and fast decryption for outsourced data security in mobile cloud computing. Elsevier, Information Sciences*, 379, 42–61.

Zhu, S., & Han, Y. (2018, January). Secure data outsourcing scheme in cloud computing with attribute-based encryption. *International Journal of High Performance Computing and Networking*, *12*(2), 128–136. doi:10.1504/IJHPCN.2018.094363

This research was previously published in Modern Principles, Practices, and Algorithms for Cloud Security; pages 21-43, copyright year 2020 by Information Science Reference (an imprint of IGI Global).

Chapter 4
Identification of Cryptographic Vulnerability and Malware Detection in Android

Anjali Kumawat
Government Women Engineering College, Ajmer, India

Anil Kumar Sharma
Government Women Engineering College, Ajmer, India

Sunita Kumawat
Government Women Engineering College, Ajmer, India

ABSTRACT

Android based Smartphones are nowadays getting more popular. While using Smartphone, user is always concerned about security and malicious attacks, cryptographic vulnerability of the applications. With increase in the number of Android mobiles, Android malwares are also increasing very rapidly. So the authors have proposed the "Identification of cryptographic vulnerability and malware detection in Android" system. They have designed a user friendly android application, through which user and developer can easily test the application whether it is benign or vulnerable. The application will be tested firstly using static analysis and then the dynamic analysis will be carried out. The authors have implemented static and dynamic analysis of android application for vulnerable and malicious app detection. They have also created a web page. User can either use the application or the web page.

INTRODUCTION

The Internet has turned out to be an essential part of the daily life of a large population. Increasingly people are utilizing the services which are provided by the Internet. The Internet has also developed from a basic communication system to an interconnected set of information sources enabling, among other things, new forms of interactions and market places for the sale of services and products. There

DOI: 10.4018/978-1-7998-8545-0.ch004

are people on the Internet with harmful intentions that struggle to endow themselves by taking benefits of legal users. Malware (i.e. malicious software) helps these people fulfilling their intents.

To protect justified users from these threats, there are various tools that have an objective to isolate malicious software components. Usually, these tools use signature matching to identify known threats. This technique needs a signature database. Then, these signatures are compared with possible threats. When a sample of a new possible threat is detected, it is analyzed to check whether this anonymous sample signifies a threat to users. If the sample poses a threat, a pattern or the signature is defined to identify this sample. The analysis of malware and the successive construction of signatures by humans are time consuming and error prone. An anti-virus vendor that receives thousands of unknown samples per day is not extraordinary nowadays. Symantec (4,300 per day) as well as McAfee (12,300 per day) report to have received over 1.6M new samples during the year 2008. This ample amount requires an automated approach to quickly differentiate between known and anonymous threats. This automatic analysis can be performed in two ways which are dynamic and static analysis. Dynamic analysis refers to techniques that uses a sample and verify its actions, while static analysis performs its task without actually executing the sample.

Malware

"Malware" is a term that can be used for any software that gets installed and activated on user's device and performs unwanted actions and tasks. It often does so for the benefit of the third party. Malware programs can range from being simple annoying programs like adwares (pop up advertisements) to cause serious damage or invasion on one's system (Rossow, Dietrich, Gier et al., 2012). The damage can be unmentionable as it can cause various threats like stealing passwords and data or infecting other machines that network with the infected device.

Majority of the malware programs are developed with an intention of data theft. They are designed to transmit information about the user for any third-party interest. They generally keep a track of the user's browsing habits for benefit of any third party.

Malwares are malignant software. They are designed such that to damage computer systems and affect their functioning without the knowledge of the owner of the system or the user of the system. Sometimes, software from reputed vendors also contains malwares. These malwares have malicious code that affects the system that leaks the essential information or steal confidential data to the remote servers. The point here is that the user is unaware of the data robbery. Malwares generally includes computer viruses, spywares, rootkits, adware, dialers, Trojans, etc.

There are various malware detection techniques available for successful detection of the malwares. With the increase in the use of mobiles, the mobile malwares are imposing a great threat to public as well as big organizations. Particularly in mobiles also, Android mobiles are in great demand.

Android, currently being the most popular operating system in term of mobile is also one of the most prone operating system to malwares. Day to day the number of malwares are increasing and approximately every day thousands of new malwares are introduced to the internet via various resources. Thus it becomes a need for the sake of one's safety that a system to detect and curd these malwares has to be developed.

Figure 1. Malicious Attacks

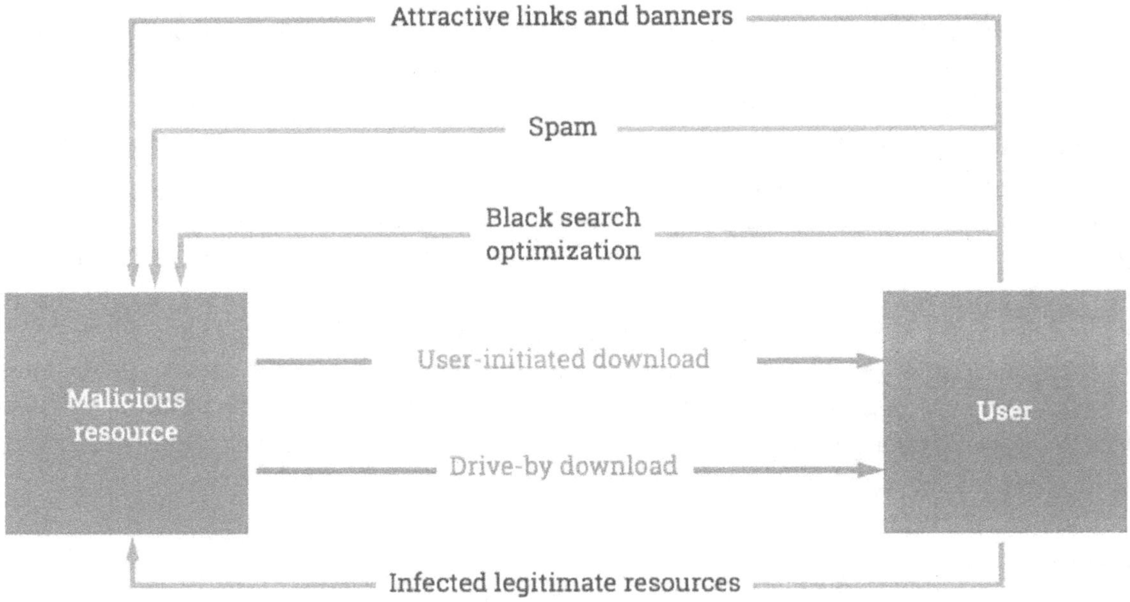

Malware Detection Techniques

Malware detection techniques can be generally classified into two categories: anomaly-based detection and signature-based detection. A special type of anomaly based detection is specification-based detection.

Signature Based Malware Detection Technique

Signature based detection uses the knowledge of malicious threats or actions to finds out the maliciousness of the program under inspection (Burguera, Zurutuza, & Nadjm-Tehrani, 2011). It uses a signature database to keep records of previous threats and use it to identify new ones. Commercial antivirus scanners look for signatures which are typically a sequence of bytes within the malware code to declare that the program scanned is malicious in nature.

Anomaly Based Malware Detection Technique

Anomaly based (Behavior-based) detection identifies the action performed malware rather than the binary pattern (Gascon, Yamaguchi, Arp et al., 2013). The program with different syntaxes and same behavior are collected. A baseline of behavior is defined and any action which deviated from this threshold behavior is detected as threat. This types of detection mechanisms helps in detecting the malwares which keeps on generating new mutants since they will always use the system resources and services in the similar manner. The behavior detector basically consists of following components which are data collection, interpretation and matching algorithm.

Figure 2. Types of Malware Detection Techniques

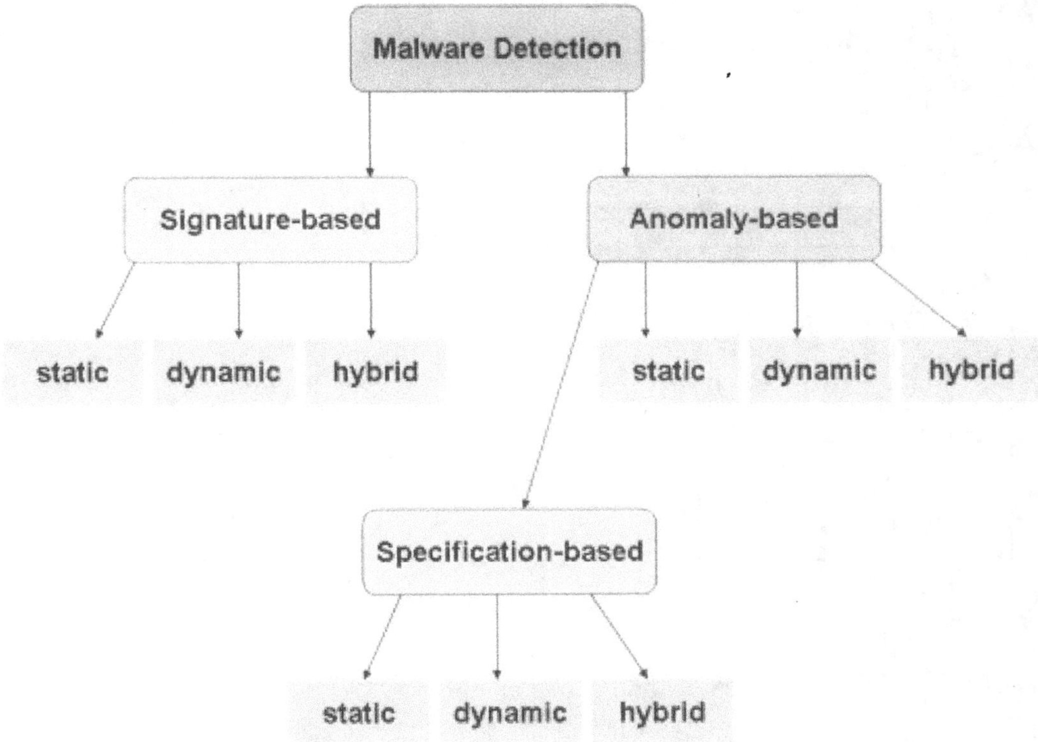

Figure 3. Basic Virus

Specification Based Malware Detection Technique

Specification based detection makes use of certain rule set of what is considered as normal in order to decide the maliciousness of the program violating the predefined rule set. Thus, programs violating the rule set are considered as malicious program. Instead of approximating the implementation of a system or application, specification-based detection approximates the requirements of application or system. In specification-based system, there exists a training phase which attempts to learn all the valid behaviour of a program or system which needs to inspect. The main limitation of specification based system is that it is very difficult to accurately specify the behavior the system or program.

Figure 4. Behavior (Anomaly) Detector

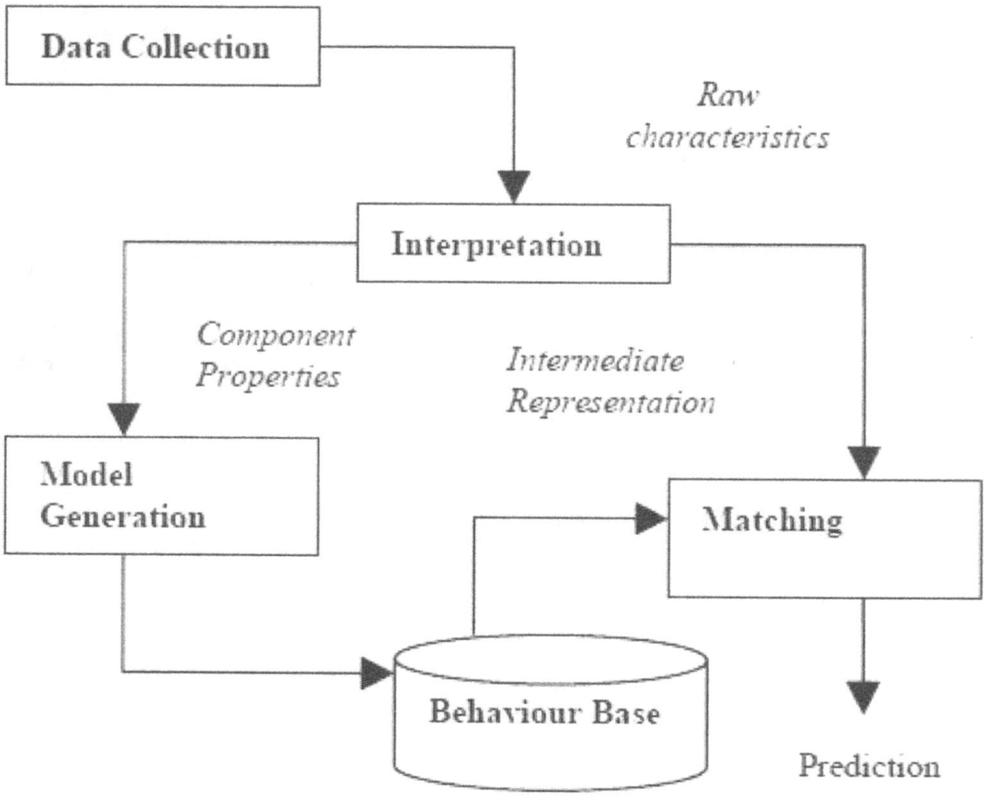

Cryptographic Vulnerabilities and Malware

Cryptographic Vulnerability attacks come into the picture when an app developer wants to take advantage of encryption in his application. This covers the possible ways where vulnerabilities associated with cryptography Vulnerabilities may be introduced in Android apps. There are some ways by which an attacker can exploit this kind of vulnerability.

Cryptography Vulnerabilities in Android apps can be introduced due to various reasons. The two main reasons as mentioned in OWASP Mobile Top 10 Projects are using a weak algorithm for encryption/decryption or a strong encryption algorithm but implementing it in an insecure way.

RELATED WORK

This section deals with the review of work done on various aspects with respect to malware detection in android.

Veen (2014) shows lightweight approaches to detect Android malware on the network level. The preliminary results show that the presented detection techniques are viable, but large-scale testing is required to determine real world performance. As Android malware evolves the effectiveness of these measures will decrease. However, these techniques can still be valuable as they raise the bar of entry for repackaged and newly created malware and come at low overhead.

Min and Cao (2012) have discussed critical elements which began with the key terms of malware analysis. After defining and identifying malware analysis, tools are used to perform the analysis. Detailed methodology is outlined so that the reader can perform a successful analysis on malware, as well as build defenses with information gathered in the analysis to protect their organization utilizing the defense-in-depth philosophy.

Zhou and Jiang (2012) show that before developing countermeasures against malicious software, it is important to understand how malware behaves and what evasion techniques it might implement. This article presented a comprehensive overview of the state of the art analysis techniques as well as the tools that aid an analyst to quickly and in detail gain the required knowledge of a malware instance's behavior

Iland, Pucher and Schäuble (2011) described some differences between common personal computer security and mobile device security. Based on these differences, the framework of a mobile dynamic malware analysis system was proposed as a reasonable solution for the different environment that mobile devices like smart phones operate in.

Christodorescu and Jha (2003) presented Semantic aware malware detection. It is based on semantic characteristics of program. Those properties are data flow and control flow. Control flow graphs are created for each function of program, and then IR (intermediate representation) is generated using instruction transformer. Templates are created for comparison with malicious program. If the template has multiplication operation in the instruction and program has equivalent left shift operation, then they are not matched. It enforces that the ordering of memory updates of program should be same as that of template.

Feng, Anand, Dillig and Aiken (2014) presented 'Apposcopy: Semantics-Based Detection of Android Malware through static Analysis'. It provides high accuracy and can detect obfuscations like: Change of component, method, field names. The method invocations to android classes are redirected using proxy methods. It cannot detect unknown malware family and detect obfuscations like dynamic code loading. It also cannot detect malware instantly.

Arp, Spreitzenbarth, Hubner, Gascon, Rieck, and Siemens (2014) gave method DREBIN which is effective and explainable detection of Android malware. This method enables identifying malware on smart phones directly. It combines static analysis of all the features of application and then apply machine learning to classify the applications as benign or malware. It also provides explanation for malware detection. It gives very few false alarms. The accuracy of method is high. It is good for detecting unknown malware. But it cannot detect transformations that are not detectable by static analysis. For example, Reflection and byte code encryption.

PROBLEM IDENTIFICATION AND DEFINITION

The purpose of this task is to present the specific Static and Dynamic Malware Detection system in Android that the authors are interested in solving, and identify the aspects of the problem.

Problem Definition

The Android platform developed by the 'Android Open Source Project' is one of the most popular systems for mobile devices in the recent years. Software can be installed on android devices from the official Android market, from third party app stores, or by direct downloading and installation of an APK file by users. The Android Marketplace has no review process and makes it easy to publish a malicious application with relative anonymity. This problem is become more severe by an increasing number of unofficial app stores leading users to avoid any existing review and reporting mechanisms. Third-party app stores are a feasible distribution method for Android Malware, often hosting infected copies of benign applications and malware targeted towards users in a particular country. It is probable for a malware developer to misuse existing applications by injection of malicious functionalities and resultant redistribution of the trojanized versions. Application developers are interested in protecting their applications. Protection in this case means that it should be difficult to know what an application is doing and how its functionalities are implemented. While anti-virus software is also there, and android limits applications to user approved permissions, many users remain unaware of the threat posed by malware and of actual infections on their devices.

The truth that Android permits application installation from third party dealers means that Google has no control over the quality or safety of the applications provided in these stores. Some cases were encountered where benign applications from the Google Play Store were modified to inject malicious code and the modified apps were distributed in these third-party app stores. It is difficult to determine whether the application is genuine or not. In these cases, the reliability of the application depends upon the security measures implemented by the application store. This makes it necessary to provide a reliable means to check the authenticity of the applications.

Proposed System

The purpose of this task is to create a new comprehensive model for improving Static and Dynamic Malware Detection in Android. The authors developed an automated system which will test android app without any human interference. User or developer can upload the application they want to test, or they can upload absolute URL, at the server APK will be downloaded, the status of downloading will be mailed to the user, with estimate time. Firstly, APK will be tested statically if cryptography uses found, the apk will be tested further dynamically in which, app will be explored dynamically and all GUI action will be performed and the logcat will be logged. Thus, once the testing is completed, the result along a report will be mailed to user that app is vulnerable or benign.

To identify the cryptographic misuse vulnerabilities in Android applications, the authors proposed a system that can test the application for any cryptographic vulnerabilities and malware in android app for app developers and naive user to test the Android app. A large-scale experiment is performed to measure cryptographic misuse in Android and made some rules. An Android app will take the App URL (to be tested) and the app will be analyzed Statically and dynamically at the server and will be tested using several rules.

Figure 5. Overview of System

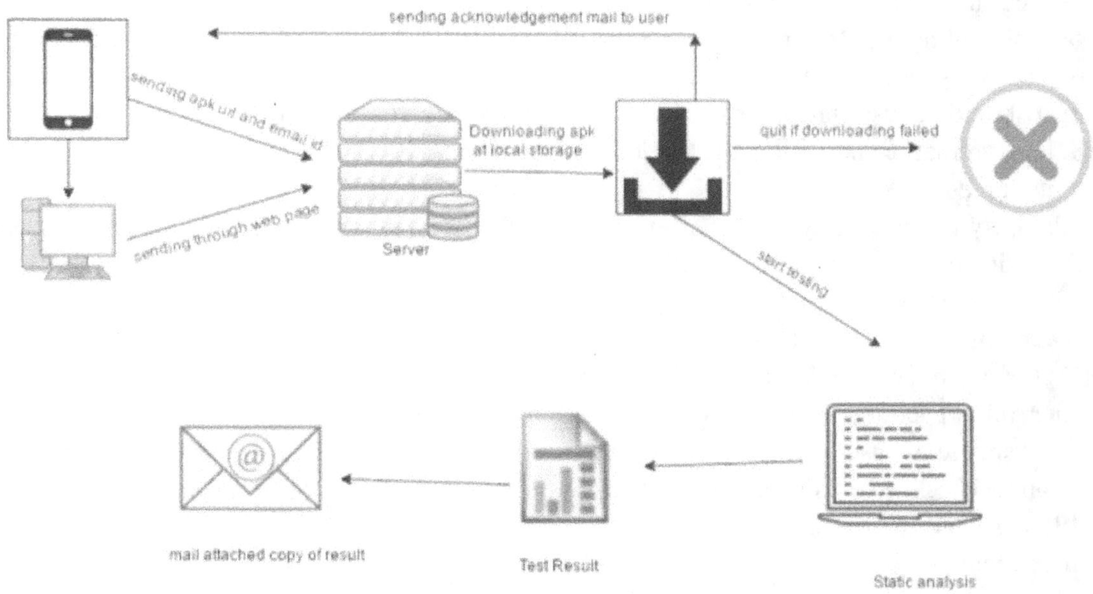

Advantages of Proposed System

- Application developers and app store maintainers can use the tools to identify likely misuses in cryptography before an end-user uses the application.
- End user can also use the tool before installing any app from third party by simply providing the APK URL.
- Android Application uses no bandwidth as only URL is sent to server not the APK file itself.
- If an app stores data on the SD card internal memory, the data is safe from any other app and attack.
- Android application size is less than 4 Mb can be easily downloaded.
- Easy and simple to use for naive user.
- Developer can easily remove cryptographic vulnerability, before publishing app in market.
- Third party domain can use to build an app market, where all the tested benign applications will be available.

Rules For Cryptography

Rule 1: It should not use DES, risk or broken algorithm for symmetric encryption because it is proved to be breakable.

Rule 2: For PBE, salt should not be static. It makes it vulnerable.

Rule 3: It should not use ECB mode for encryption because it is not IND-CPA secure and it is also not semantically secure.

Rule 4: It should not use non-random IV for CBC encryption because it allows for the possibility of a dictionary attack. [CWE-329]

Rule 5: For symmetric encryption, it should not use constant encryption keys or same cryptographic key multiple times because it is easily breakable.

Rule 6: It should not use fewer than 1000 iterations for PBE. This type of password based encryption is flagged as insecure. [RFC 2898]

Rule 7: It should not use static seeds to seed SecureRandom(). It is not safe.

Rule 8: For symmetric encryption, the key length should be no less than 128 bits otherwise it will cause brute force attack.

Rule 9: For asymmetric encryption, the keylength should not be inadequate i.e. it should be no less than 1028 bits to avoid brute force attack.

Rule 10: In case of hash algorithms, it should not use one-way hash because it is insecure (mainly two algorithms: MD4 and MD5).

Rule 11: For asymmetric encryption, it should not use RSA algorithm with OAEP (Optical Asymmetric Encryption Padding). It weakens the encryption. [CWE-780]

Rule 12: It should not use improper certificate validation. For android (It is due to SSL/TLS weakness. It can be checked by using a tool created and released by google known as nogotofail).

Rule 13: It should not exchange the key without entity authentication. This authentication makes it more secure.

Rule 14: It should not use a signing key past after it expiration date.

Rule 15: For secure cryptography, the default behavior in cryptographic libraries is often not recommended (change default behavior to more secure variant) otherwise it can be easily cracked. (DRD-18)

Rule 16: User should not broadcast sensitive information using implicit intents (It is vulnerable to passive eavesdropping or active denial of service attacks).

Key Differences between Static and Dynamic Analysis

Presented in Table 1.

Table 1. Key Differences between Static and Dynamic Analysis

S. No.	Static analysis	Dynamic Analysis
1.	Fast & safe	Time Consuming & vulnerable
2.	Analyze malicious software without executing	Analyzing the behavior of a malicious code (interaction with the system) while it is being executed
3.	Good in analyzing the multipath malware (Global view)	Difficult to analyze the multipath malware
4.	Can't analyze the obfuscated & polymorphic malware	Can analyze the obfuscated and Polymorphic malware
5.	Can't detect new, unknown malware	Detect known as well as unknown malware
6.	Low level of false positive (accuracy is high)	High level of false positive (accuracy is low)

IMPLEMENTATION AND WORKING OF IDENTIFICATION OF CRYPTO VULNERABILITIES AND MALWARE DETECTION

Algorithm of Proposed System

Step 1: Install xampp, PHPMYADMIN and Android SDK
Step 2: Open xampp control panel and organize all the modules
Step 3: Write Rules according the user requirements
Step 4: Create database using mySQL containing previously detected malicious application
Step 5: When the new application testing request is made, send acknowledgement mail to the user
Step 6: Start testing
Step 7: Apply static analysis
If new application is detected in the database
Then go to step 8
Else
Apply dynamic analysis
Step 8: Generate test results
Step 9: Mail the result file to the user and update the database
Step 10: Exit

Android Application

The application consists of three tabs which are Help, Upload URL and Statistics. Help tab consists of basic steps to be followed, if the user is using application for the first time. He can follow the steps easily as given in Help tab. Upload URL tab consists of two input text box with one submit button. One is File URL, where the file is available for download. Email -id field, where user will enter his email id. The app is validating URL, because if the URL is not a valid URL then it will cause the server performance and might crash the server by bulk invalid URL submission. If the user enters invalid URL, he will be notified by displaying toast, that URL is invalid, and the information will not post to server.

Web Page

There is also a web page, through which user can upload apk file if user doesn't have URL of the app and if he have URL then he can, there is also option of URL so that he doesn't need any bandwidth (Wifi, Mobile data) to upload apk.

Server Side Processing

Once the URL has sent to sever side using post method, the data will be sent in post data request form. The data will be stored in database and the user will be identified by device id of the user that is taken when he is requesting for any URL. Now if the URL is submitted, server will start downloading the apk at server side.

Figure 6. Uploading tab with input's

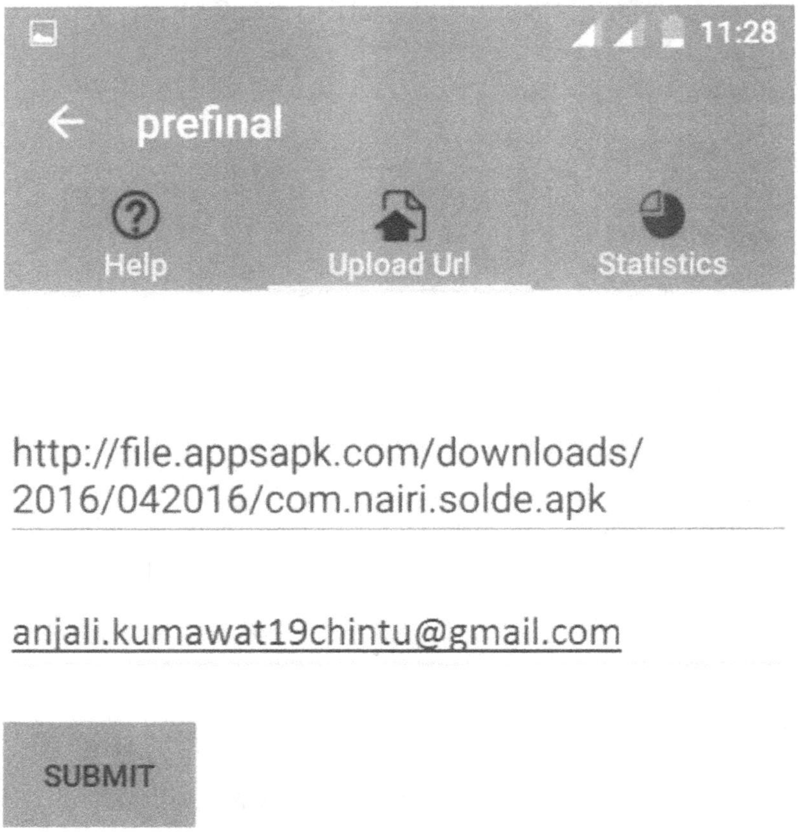

Figure 7. Web page for uploading apk, URL

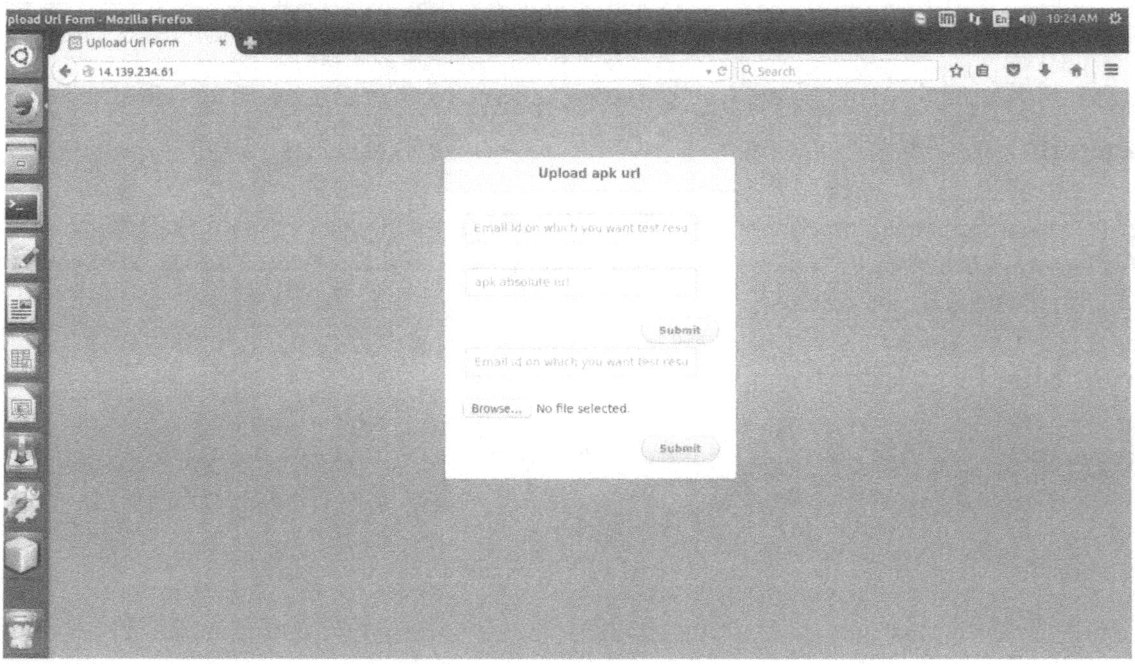

Figure 8. Final result report received via mail

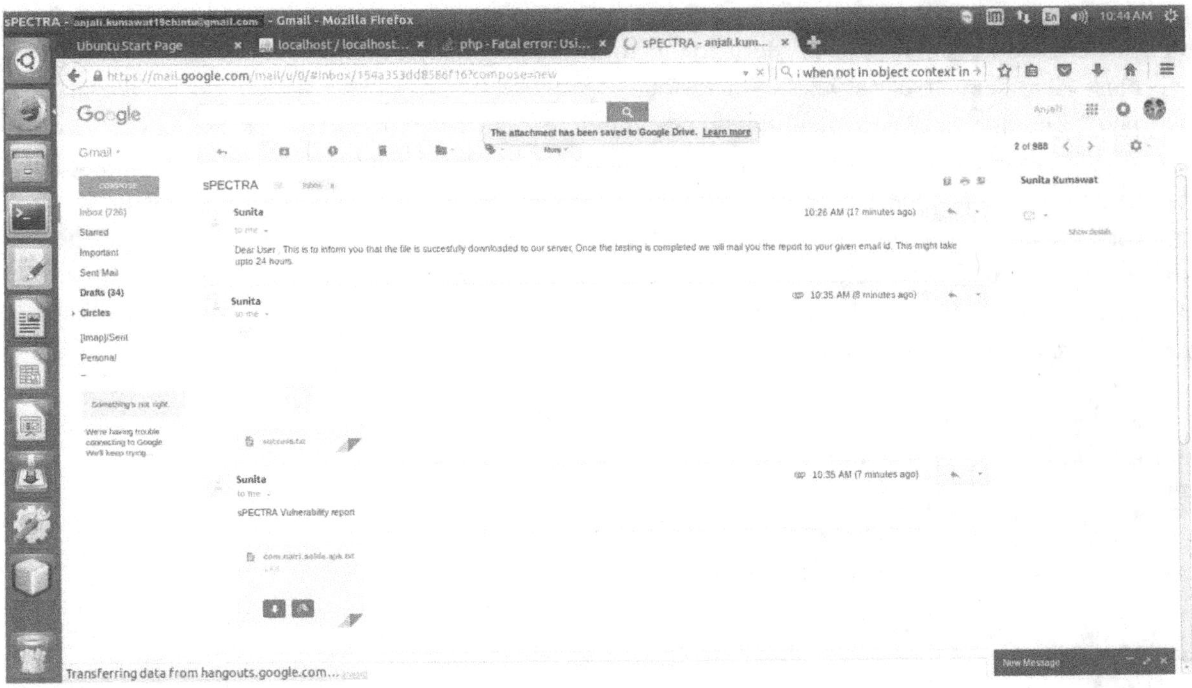

If the app is downloaded successfully then the status will be mailed to user and stored in database or if the downloading is failed, user will be notified regarding this failure. If the file is uploaded using web page then it is stored at server side, firstly it checks whether there is any name collision, if there is a same name file then the uploaded file name will be modified and stored at local storage. The status of uploaded file will be send to user via mail. Then the package name will be extracted from app and will be search if the app is already analysed then the result will be sent to user's email, else it will start static analysis. If in static analysis, encryption uses are found then the dynamic analysis for the app dynamic analysis will started by Robotium tool. It will explore the app dynamically and the logs will be captured in background. Then after analysing the logs it will check for the vulnerability or any malware in app.

Result

This following is a Pie chart of tested applications from different third party domain that was vulnerable.

The authors have tested various applications from different app stores and got these results from the analysis. The table shows the total number of vulnerable and infected application after the testing.

CONCLUSION AND FUTURE SCOPE

The need of malware detection in Android operating system is of a great concern. The authors started the work with the problem of static and dynamic analysis of android applications for their protection from viruses, worms. Now days, in many applications of android there are many ways to introduce malicious content. The authors tried to identify those malicious applications and for this, the authors collected the malicious samples as much as possible and took the permission combinations out of them. The authors added a class variable in logs which defined that the following sample or application is malicious or not. This class label helps in classification of various applications.

Once the authors have enough samples and enough permission combinations the authors made a rule database. When the authors have the signature database of rules, the authors took files as input and it in-turn first reverse engineer the application that is given as input. It matches the exploited permission set of the input application with the already existing permissions in the signature database. If the signature matches, the input application is malicious in nature. And if not, then there are two cases; first, that the application is not malicious, and second, the application is malicious but its signature is not in the signature database.

Figure 9. Pie chart of tested apps

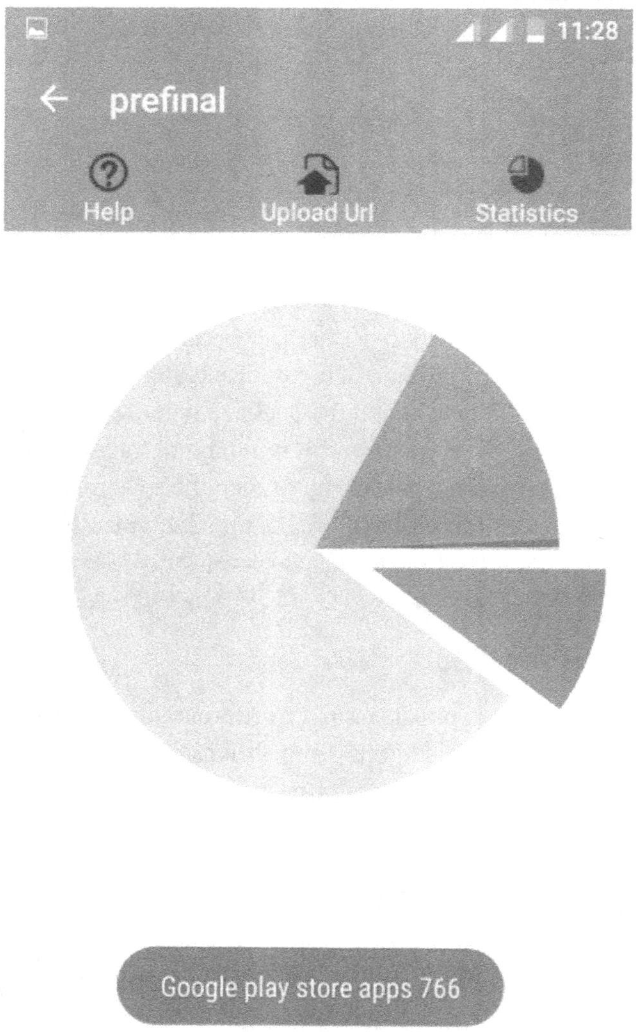

Table 2. Sample's Statistics

S.NO.	Domain/Market-Name	#Apps
1	Google Playstore	766
2	Gfan	5069
3	Nduoa	698
4	Androidpur	336
5	Mobomarket	48
6	Appsapk	72
7	Apkfun	11

The first result is obvious and doesn't need any verification. But the second result is absurd. To make sure the second situation never reaches in a scan, the authors collected as many malicious samples as the authors can and add their signature to the already existing signature database.

The goal behind developing the system is as the uses of android devices are increasing rapidly through the application development of android app is also increasing. There are many applications developed by developers which are free but they are not secure. Even Google restricts to download application from unknown source but that option can be allowed by accepting the risk.

So, the authors designed an application that will tell you whether the application is vulnerable or malicious or benign app. It is very easy to use the application. You just need to give the URL of the app and the test result will be sent to your email address. Even developer can also use this before publishing app for any cryptographic vulnerability. The dynamic analysis of app makes it different from other existing system. This can be integrating to any app store for acceptance of android app uploaded by developer. By using this, the authors can create our own app store that will be having all benign app. This can be added as an in-built in app in android device which will test every app before installing it to device. In future the app will also tell the correct use for the misuse vulnerability.

REFERENCES

Aafer, Y., Du, W., & Yin, H. (2013). DroidAPIMiner: Mining API-level features for robust malware detection in android. *Proceedings of the International Conference on Security and Privacy in Communication Networks (SecureComm).* 10.1007/978-3-319-04283-1_6

Arp, D., Spreitzenbarth, M., Hubner, M., Gascon, H., Rieck, K., & Siemens C. (2014, February). Drebin: Effective and explainable detection of android malware in your pocket. USA: NDSS Internet Society.

Barrera, D., Kayacik, H. G., Van Oorschot, P. C., & Somayaji, A. (2010). A methodology for empirical analysis of permission-based security models and its application to android. *Proceedings of the ACM Conference on Computer and Communications Security (CCS)* (pp. 73–84). 10.1145/1866307.1866317

Burguera, I., Zurutuza, U., & Nadjm-Tehrani, S. (2011). Crowdroid: behavior-based malware detection system for android. *Proceedings of the ACM Workshop on Security and Privacy in Smartphones and Mobile Devices (SPSM)* (pp. 15–26). 10.1145/2046614.2046619

Chakradeo, S., Reaves, B., Traynor, P., & Enck, W. (2013). Mast: Triage for market-scale mobile malware analysis. *Proceedings of the ACM Conference on Security and Privacy in Wireless and Mobile Networks (WISEC)*. 10.1145/2462096.2462100

Christodorescu, M., & Jha, S. (2003). Static analysis of executables to detect malicious patterns. *Proceedings of the 12th conference on USENIX Security Symposium SSYM'03* (Vol. 12, pp.12–12). Berkeley, CA, USA: USENIX Association.

Crussell, J., Stevens, R., & Chen, H. (2014, June). *MAdFraud: Investigating ad fraud in android applications. Proceedings of the International conference on Mobile systems, applications, and services (MobiSys'14)*, Bretton Woods, NH, USA. ACM. 10.1145/2594368.2594391

Enck, W., Octeau, D., McDaniel, P., & Chaudhuri, S. (2011). A study of Android application security. *Proceedings of the 20th USENIX conference on security*, San Francisco, CA, USA.

Enck, W., Ongtang, M., & McDaniel, P. (2009). Understanding android security. *Security & Privacy*, *7*(1), 50–57. doi:10.1109/MSP.2009.26

Felt, A. P., Finifter, M., Chin, E., Hanna, S., & Wagner, D. (2011). A survey of mobile malware in the wild. *Proceedings of the ACM Workshop on Security and Privacy in Smartphones and Mobile Devices (SPSM)* (pp. 3–14). 10.1145/2046614.2046618

Felt, A. P., Ha, E., Egelman, S., Haney, A., Chin, E., & Wagner, D. (2012). Android permissions: user attention, comprehension, and behavior. *Proceedings of the 8th symposium on usable privacy and security SOUPS '12*, Washington, DC, USA. 10.1145/2335356.2335360

Feng, Y., Anand, S., Dillig, I., & Aiken, A. (2014, November). Apposcopy: Semantics-Based Detection of Android Malware through Static Analysis. *Proceedings of the ACM SIGSOFT Int. Symp.* 10.1145/2635868.2635869

Gascon, H., Yamaguchi, F., Arp, D., & Rieck, K. (2013, November). Structural detection of android malware using embedded call 11 graphs. *Proceedings of the ACM Workshop on Artificial Intelligence and Security (AISEC)* (pp. 45–54).

Grace, M., Zhou, Y., Zhang, Q., Zou, S., & Jiang, X. (2012). Riskranker: scalable and accurate zero-day android malware detection. *Proceedings of the International Conference on Mobile Systems, Applications, and Services (MOBISYS)* (pp. 281–294). 10.1145/2307636.2307663

Iland, Danny, Pucher, Alexander, & Schäuble, Timm (2011, December 5). *Detecting Android Malware on Network Level.*

King, J., Lampinen, A., & Smolen, A. (2011). Privacy: is there an app for that? *Proceedings of the 7th symposium on usable privacy and security SOUPS '11*. New York, NY, USA: ACM Press. 10.1145/2078827.2078843

Liang, S., & Du, X. (2014). *Permission-combination-based scheme for Android mobile malware detection. 2014 IEEE international conference on communications (ICC)*. Sydney, Australia: IEEE.

Liu, B., Nath, S., Govindan, R., & Liu, J. (2014, April). DECAF: Detecting and characterizing ad fraud in mobile apps. *Proceedings of the USENIX Symposium on Networked Systems Design and Implementation (NSDI'14)* (pp. 57–70). Seattle, WA, USA: USENIX Association.

Lu, L., Li, Z., Wu, Z., Lee, W., & Jiang, G. (2012). CHEX: statically vetting Android apps for component hijacking vulnerabilities. *Proceedings of the ACM Conference on Computer and Communications Security (CCS)*. 10.1145/2382196.2382223

Min, L., & Cao, Q. (2012). Runtime-based Behavior Dynamic Analysis System for Android Malware Detection. *Proceedings of the 2012 2nd International Conference on Computer and Information Application (ICCIA '12)*. 10.2991/iccia.2012.57

Ongtang, M., McLaughlin, S., Enck, W., & McDaniel, P. (2009). Semantically rich application-centric security in android. *Proceedings of the Annual Computer Security Applications Conference*. (pp. 340–349). IEEE.

Portokalidis, G., Homburg, P., Anagnostakis, K., & Bos, H. (2010). Paranoid Android: versatile protection for smartphones. *Proceedings of the Annual computer security applications conference (ACSAC)*, Austin, Texas, USA. 10.1145/1920261.1920313

Rassameeroj, I., & Tanahashi, Y. (2011, May 15). Various approaches in analyzing Android applications with its permission-based security models. *Proceedings of the IEEE International Conference on Electro/Information Technology (EIT)*. 10.1109/EIT.2011.5978583

Rastogi, V., Chen, Y., & Jiang, X. (2013). DroidChameleon: evaluating android anti-malware against transformation attacks. *Proceedings of the ACM Symposium on Information, Computer and Communications Security (ASIACCS)*. 10.1145/2484313.2484355

Roesner, F., Kohno, T., Moshchuk, A., Parno, B., Wang, H. J., & Cowan, C. (2012). User-driven access control: rethinking permission granting in modern operating systems. *Proceedings of the IEEE Symposium on Security and Privacy*. 10.1109/SP.2012.24

Rossow, C., Dietrich, C., Gier, C., Kreibich, C., Paxson, V., Pohlmann, N., ... Van Steen, M. (2012). Prudent practices for designing malware experiments: Status quo and outlook. *Proceedings of the IEEE Symposium on Security and Privacy*. 10.1109/SP.2012.14

Veen, V. van der (2014). ANDRUBIS - 1,000,000 Apps Later: A View on Current Android Malware Behaviors.

Vidas, T., & Christin, N. (2013). Sweetening Android lemon markets: measuring and combating malware in application marketplaces. *Proceedings of the 3rd ACM conference on data and application security and privacy CODASPY '13*. New York, USA: ACM Press. 10.1145/2435349.2435378

Wu, D.-J., Mao, C.-H., Wei, T.-E., Lee, H.-M., & Wu, K. P. (2012). Droidmat: Android malware detection through manifest and API calls tracing. *Proceedings of the Asia Joint Conference on Information Security (Asia JCIS)* (pp. 62–69). 10.1109/AsiaJCIS.2012.18

Zhou, Y., & Jiang, X. (2012). *Dissecting android malware: Characterization and evolution. Proceedings of the IEEE Symposium on Security and Privacy* (pp. 95–109).

This research was previously published in the International Journal of Information Security and Privacy (IJISP), 11(3); pages 15-28, copyright year 2017 by IGI Publishing (an imprint of IGI Global).

Chapter 5
Android Botnets:
A Proof-of-Concept Using Hybrid Analysis Approach

Ahmad Karim
Bahauddin Zakariya University, Pakistan

Victor Chang
https://orcid.org/0000-0002-8012-5852
Teesside University, UK

Ahmad Firdaus
Faculty of Computer Systems and Software Engineering, Malaysia

ABSTRACT

Mobile botnets are gaining popularity with the expressive demand of smartphone technologies. Similarly, the majority of mobile botnets are built on a popular open source OS, e.g., Android. A mobile botnet is a network of interconnected smartphone devices intended to expand malicious activities, for example; spam generation, remote access, information theft, etc., on a wide scale. To avoid this growing hazard, various approaches are proposed to detect, highlight and mark mobile malware applications using either static or dynamic analysis. However, few approaches in the literature are discussing mobile botnet in particular. In this article, the authors have proposed a hybrid analysis framework combining static and dynamic analysis as a proof of concept, to highlight and confirm botnet phenomena in Android-based mobile applications. The validation results affirm that machine learning approaches can classify the hybrid analysis model with high accuracy rate (98%) than classifying static or dynamic individually.

INTRODUCTION

Although Android OS being an open source has promoted mobile applications developers, yet malware programmers have also contributed to exploit its open source nature to carry out malicious acts. MacAfee, an antimalware platform, has diagnosed more than 700K mobile malware in the second quarter

DOI: 10.4018/978-1-7998-8545-0.ch005

of 2014 (Weafer, 2014). Another report (Chebyshev, 2016) published in 2016 discovered that Internet access on smartphone devices had exceeded 61% in the first quarter of 2015. This study also revealed that 60.85% of Android users had started Internet access on their smartphone devices. Consequently, the similar growth is observed in malware program construction, i.e., 40,267 new mobile malware variants were analyzed and diagnosed by the security agencies at the end of 2015(Millman, 2015). In Q2 2016, it was observed that Android was used by 86.2% of smartphone users (Paul, 2017). Similarly, its wide-spread adaptation to other platforms such as televisions, tablets, wearable, and vehicles opened the new dimensions for multi-platform attacks. In the similar pace, IoT (Internet of Things) would be the next target for malware programmers. A more recent report (BILIĆ, 2017) states that the growth of mobile malware is constantly increasing since 2013. On average 200 new malicious code variants have been discovered per month during 2015, this number rose to 300 per month by the end of 2016. As a result, on average 400 new malicious code variants are expected to evolve by the end of 2017 (Weafer, 2016).

Mobile botnet or SMARTbot (Karim, Salleh, & Khan, 2016) is a malevolent action which is inspired from traditional botnets (PC based). The basic motive behind botnet is to gain illegitimate access to someone's personal device (smartphone, tablets, etc.) and makes this device compromised by a bot binary (app). After becoming the part of a bot network, another pivotal role (botmaster) is responsible for controlling this device remotely and to initiate various attacks using some command and control (C&C) channel. Consequently, these devices are then participating in numerous malicious activities including DDoS, ransom, making premium calls, sending text messages and emails without user's consent.

There are two most common analysis strategies exist, static and dynamic analysis. In static analysis, structural properties of program code are observed including permission usage, CFGs, function call graphs and API calls, etc. For static analysis, reverse engineering tools (Lukan, 2012) are deployed to disassemble program code (Schmidt et al., 2009) or directly fetching parameters from executable binaries (Petsas, Voyatzis, Athanasopoulos, Polychronakis, & Ioannidis, 2014; Yousafzai et al., 2016). In contrast, dynamic analysis requires execution of malware binaries in a secure environment (called sandbox) to extract runtime behavior of these applications. Following are some of the parameters which are of interest during dynamic analysis: (a) file operations (b) network traces (c) initiated services (d) HTTP and DNS traffic etc. Currently, some mobile malware detection approaches (Arp, Spreitzenbarth, Hubner, Gascon, & Rieck, 2013; Chen, Rong-Cai, ZHENG, Jia, & Li-Jing, 2016; Fereidooni, Conti, Yao, & Sperduti, 2016; Yang, Wang, Ling, Liu, & Ni, 2017) are introduced targeting either program code or runtime execution traces. However, at a higher level of abstraction, these approaches are targeting mobile malware detection rather than mobile botnet. This is the extension of our previously proposed approaches (Ahmad Karim & Shah, 2015; Karim, Salleh, Khan, Siddiqa, & Choo, 2016) in a way that it can highlight the need for a hybrid analysis framework for the detection of botnet mobile binaries.

In this paper, we will investigate and highlight the problem of the mobile botnet by comparing botnet properties of known botnet applications with existing malware families and benign samples.

Thus, the basic motive of this research is to confirm the existence of mobile botnet phenomenon in Android-based system which is continuously progressing with the technological advancements. Moreover, we argued that hybrid analysis systems can detect mobile botnet binaries more accurately. Consequently, mobile botnet problem may supersede traditional PC based botnets if the precautionary measures have not been devised timely. Overall the paper objectives are summarized as follows:

- To strengthen the existing efforts (mobile malware detection in general) towards mobile botnet (C&C) detection by identifying, highlighting and comparing the properties of known mobile botnets with existing malware and benign applications;
- We have divided our hybrid analysis framework into three different categories for mobile applications which includes botnet, malware and benign set of applications. Through this setup, we need to observe the most inherent properties and ongoing trends of mobile applications having botnet capabilities;
- Our claim of a hybrid analysis is more efficient to detect mobile botnet malware is verified by machine learning approaches;
- Conclusively, we have provided some guideline steps to avoid this growing hazard.

The rest of the paper is structured as follows: section 2 discusses related work, section 3 describes our proposed hybrid analysis framework, section 4 present evaluation results generated from code and runtime analysis, section 5 discuss some guidelines to avoid this growing hazard (mobile botnet), and conclude with section 6.

RELATED WORK

Mobile intrusion detection systems fall into the categories of detection and prevention as depicted in Figure 1. Overall mobile malware detection is a broader term used to identify the anomalous behavior of apps with respect to benign apps. However, the motivation behind malware programmers is diverse, i.e., to gain illegitimate access, hinder the resource utilization, ransom, root exploit, spam dissemination and botnet. Therefore, in this article, we are specifically discussing mobile botnet phenomenon which is getting popularity in recent times for intrusive personals. By making a single device as a part of a bot network, the controller (C&C) can initiate various other types of attacks as well. As the mobile botnet phenomena are still not as popular as is predecessor (PC-based botnet), therefore the major work is carried out to detect mobile malware in general. However, recent reports state that the mobile botnet is also growing at the same pace as technology evolves.

Android permissions are considered as one the major entry point to breach smartphone security. Malware writers are often used dangerous permission set to exploit device by getting the benefit of users unawareness about the complexities associated with such dangerous permissions. In (Aswini & Vinod, 2014) authors, presented a static analysis model to mine prominent permissions that are of interest for malware writers. After extracting a feature vector from AndroidManifest.xml of 436 Android applications the future pruning method is applied to validate the accuracy of features. However, the approach is unable to extract detailed code features which result in greater false positives.

A mobile botnet detection approach using VPN (virtual private networks) is presented in (Choi, Choi, & Cho, 2013). The study revealed communication flow characteristics by investigating C&C communication flow over a VPN. The characteristics observed are a total number of bytes and the total number of packets. The basic theme of this research is to gain insight of a particular C&C traffic flow by comparing it with existing abnormal models, predefined signatures or whitelists. Whereas, DeDroid (Karim, Salleh, Khan, et al., 2016) can detect zero-day attacks through static analysis of Android applications.

Figure 1. Mobile malware detection approaches

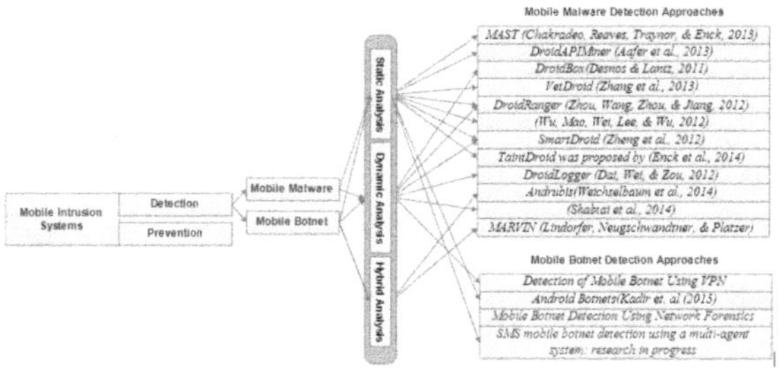

DroidRanger (Liang & Du, 2014), a hybrid analysis approach in which the app binaries are shortlisted based on the usage of dangerous permission set. Further, the malicious behavior is compared with existing malware's behavior by looking into the manifest, packages, function call graphs (FCG) and code sequence. Moreover, applications without a trusted signature are treated as zero-day attacks and selected for subsequent analysis. Nevertheless, the approach does not support to analyze network specific communication parameters. VetDroid (Pravin, 2015) is treated as dynamic analysis approach in which applications are scrutinized based on harmful permissions usage. The permission analysis component of VetDroid extracts all permissions an application is using and sketch a relationship among them. Consequently, system generates FCG to classify malicious applications. Further, each application is executed for a certain amount of time in the sandbox. The authors used monkeyrunner (Android, 2012) for triggering UI events. A sandbox known as DroidBox is proposed by Lantz (Desnos & Lantz, 2014) for behavioral analysis of applications for effective runtime analysis of applications. Similarly, it provides rich parameter support for dynamic analysis including network traces. One shortcoming of DroidBox is that it is not compatible with Android applications before version 4.2. We are also using DroidBox in our work to generate runtime analysis reports.

Another approach (Su & Fung, 2016) proposed a detection framework based on static analysis to detect malicious Android binaries. The features selected for evaluation are permissions, vulnerable functions applied by applications, intents, and native permissions. Moreover, the system validates the intrusive nature of applications by observing dynamic behavior through running and capturing runtime traces of applications. Further, the study highlighted the importance of short message (SM) in the detection of malicious intent. A more recent work on the detection of the mobile botnet is presented in (da Costa, Barbon, Miani, Rodrigues, & Zarpelão, 2017). The authors proposed an anomaly-based approach working on the host machine. They used machine learning classification algorithms to identify anomalous behaviors in Android applications using system calls as a feature vector. The approach uses dynamic properties of known malware (self-generated) and machine learning to detect botnet behavior. The said approach can detect malicious binaries with 92% of accuracy while relying only on dynamic features. However, we are analyzing both static and dynamic properties of botnet binaries. Moreover, in our hybrid analysis framework, Random Forest algorithm can detect more accurate results by achieving 98% of accuracy.

In a recent study (Zhi et al., 2017) the authors studied network traffic flow for botnets using Venn-abers predictor, K-nearest neighbor (KNN) and Kernel Density Estimation (KDE). The authors focused HTTP, IRC and P2P based botnet families for evaluation. Botrevealer (Khoshhalpour & Shahriari, 2018) is a behavior-based botnet detection framework based on botnet life cycle that analyzes network activities. Whereas, our proposed framework is based on hybrid analysis of android applications. A more recent work (Kirubavathi & Anitha, 2018) proposed structural analysis framework that can identify botnet applications from benign applications using machine learning classifiers. Contrary to this approach we are dealing with hybrid features of android applications using machine learning algorithms to detect C&C based Android applications.

PROPOSED FRAMEWORK

As described above that currently Android is the most prevalent operating system for malware program authors. Therefore, we have focused Android platform in our analysis experimentation. To carry out this analysis task, we have taken (McIlroy, Ali, & Hassan, 2016; Parkour, 2011) ten samples of mobile applications from each of the categories (benign, malware, botnet). We have performed analysis using hybrid analysis approach, i.e., by combining static/code and dynamic/runtime analysis techniques. From code analysis, we will highlight various properties related to static code observation, which includes Permissions, API calls, Intents and hash values. Whereas, through run-time analysis, we will run all thirty applications in a secure sandbox and log information during their execution. The properties we will observe during execution of applications are: opened connections, DNS queries, started services, HTTP request/response queries. After storing all feature values in comma separated files, we will then analyze and discuss the results of these properties. Finally, we will conclude the paper with our findings that applications having botnet intension has certain features which can be distinguishable. As a result, we should devise some proactive mechanisms to avoid such harmful threat.

We have carried out our experimentations on SANTOKU (a Linux distribution) which is specifically designed for task related to mobile malware analysis (NowSecure, 2015). For evaluation, Intel Xeon ® server (with 3.50GHz processing speed and 16GB of RAM) is deployed. The basic architecture of our analysis setup is depicted in Figure 2.

Figure 2. Proposed framework

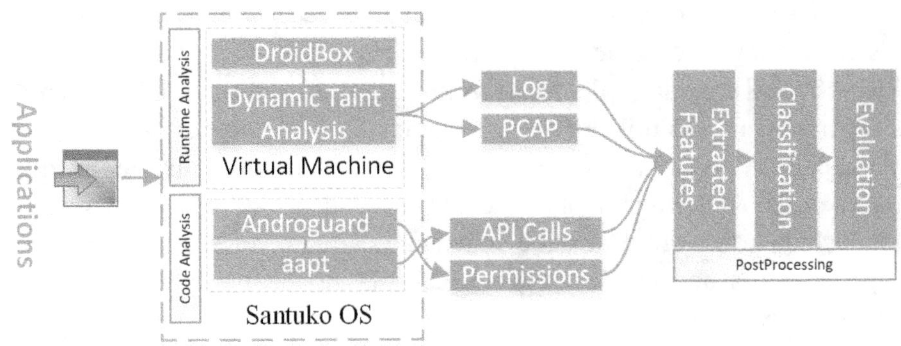

Initially, applications are passed through static and dynamic analysis and stores respective log and trace files. For static analysis, Androguard (Desnos, 2011) tool is used, whereas dynamic analysis is performed with open source DroidBox (Desnos & Lantz, 2014) tool. Next, the static features (Ahmad Karim & Shah, 2015) and dynamic features (Karim, Salleh, & Khan, 2016) are extracted and labeled according to input class. Lastly, the results obtained from the hybrid analysis are evaluated and compared.

DATA ACQUISITION

As discussed above we have chosen ten mobile application samples from each category. To carry out our analysis task, we have downloaded those samples from online repositories including google play-store (Viennot, Garcia, & Nieh, 2014), contagion (Parkour, 2011) malware repository and well-known Drebin(Arp et al., 2013) dataset.

Table 1. Sample dataset used

Sample	Apps	Repository	Observation	Category	Family
Botnet	10	Contagio	Code/ Runtime	HTTP/ SMS-based	NotCompatible.C, Obad, SpamSold, Plangton, Geinime, DroidDream
Malware	10	Drebin/Contagio	Code/ Runtime	Banking Trojan/ Information stealing / Premium SMS/self-installed binaries etc.	.Foncy, Opfake, FakeNotify, Hippo, SMSreg
Benign	10	Google Play store	Code/ Runtime	Games, Entertainment, Web Browser, Wall- papers, GPS Tracking	-

At the time of writing this article, it was the largest dataset publicly available through educational credentials. The dataset used for evaluation is described in Table 1.

PROGRAM CODE ANALYSIS

Usually, there are two approaches used in malware analysis: one is static/code analysis and the other is dynamic/behavior analysis. Through code analysis, an application is investigated through inspecting the downloaded mobile app's program code only. Signature-based systems mainly adopted by antivirus companies' falls in the category of static analysis. However, malware writers use obfuscation techniques to hinder the inspection of program code. Apart from various off-the-shelf obfuscation methods available, malware programmers also utilize Native API calls by hiding system activities and calling the functions from outside the Java/Runtime library. As a result, our aim in this step is to apply different forensic techniques on a binary program of application and gather results for comparative analysis. The popular forensic techniques involved in static code analysis are reverse engineering, de-compilation, pattern matching, decryption and analysis of system calls. Using these techniques, the common thing is that the program is not executed at all.

The sequence diagram in Figure 3 represents the working flow of our code analysis method to extract and compare interesting features with malware, botnet, and benign binaries.

The features selected for code analysis includes permission calls, and API calls which are already proven as interesting features of mobile botnet applications (Karim, Salleh, Khan, et al., 2016).

Therefore, we have selected permission calls, and functions call extracted from Manifest and .dex class respectively with the help of Androguard (Desnos) utility. To extract features automatically, we have executed a python script on all applications and stored all features on a CSV files. The contents of

Figure 3. Code analysis workflow

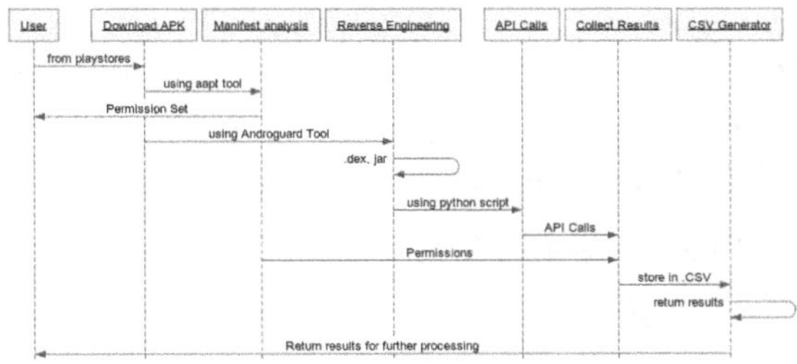

CSV files are binary numbers, i.e., stored "1" if the application has the particular feature enabled and "0" otherwise. SLet x and y be the number of applications and the set of features (included permissions and API calls, respectively. The feature vector for application *i* is $\left(a_{i,1}, a_{i,2}, a_{i,3}, \ldots, a_{i,y}\right)$ where:

$$a_{i,j} = \begin{cases} 1, & if\, application\, x\, uses\, feature\, k \\ 0, & otherwise \end{cases} \tag{1}$$

Similarly, we suppose class of an instance in the generated dataset is $c_i \in$ {*benign, malicious, botnot*} which shows the class of an application *i*. Then:

$$P\left(c_i \mid a_i\right) = P\left(c_i \mid a_{i,1}, a_{i,2}, a_{i,3}, \ldots, a_{i,y}\right) = \prod_{k=1}^{y} P\left(c_i \mid a_{i,k}\right) \tag{2}$$

Selected Feature Vector in Code Analysis

We can classify botnet working states into different phases depending upon their action type. These states are classified as (a) connection phase (b) communication phase and (c) status information phase which are shown in Table 2.

During connection phase, botnet applications try to establish a connection with the remote host. The permission and API calls for this phase include INTERNET, openConnection(), execute(), openStream(), etc. The bulk of negotiation and interactions happened once the connection is established. These interactions can be classified as (a) Direction of the Information (b) Communication Protocol. According to (Gu, Zhang, & Lee, 2008), the direction of C&C communication messages can be characterized as pull based or push based. In pull-based C&C mechanism, the bots periodically request for information in passive mode. In Contrast, for push-style C&C mode, bots explicitly send request messages to the server. Another important feature to recognize C&C communication patters is communication protocol. The most commonly used protocols in botnet communication are HTTP, IRC, and P2P. Moreover, communication phase causes initiation of malicious activity as well. Therefore, we have highlighted only those features which can cause communication and attack commencement. For this purpose the permissions include: READ_CONTACTS, READ_LOGS and SEND_SMS whereas the API calls includes: openInputStream, openOutputStream, openFileDescriptor,exec, getDefault and sendTextMessage. To retain and expand attack vector, the attacker should constantly observe the active status of the device and changing network conditions. For this purpose, the following features are of interest for malware writers to actively monitor device status and network state, READ_PHONE_STATE, ACCESS_NETWORK_STATE, ACCESS_WIFI_STATE, ACCESS_COARSE_LOCATION, ACCESS_FINE_LOCATION, getDeviceld, getLine1Number, getSubscriberld, getSimSerialNumber, getDeviceSoftwareVersion, getLastKnownLocation, requestLocationUpdates, isProviderEnabled, getActiveNetworkInfo, getNetworkInfo, getConnectionInfo, getWifiState, getCellLocation.

BEHAVIOR/RUNTIME ANALYSIS

Behavior analysis requires the application to be processed at run-time. By this analysis, we have to detect and capture the system's execution states including communication patterns, SMS messages, HTTP traffic on network interfaces, network access to outside world and DNS requests. Behavior or dynamic analysis overcomes all possible problems that can incur during code analysis such as obfuscation, encrypted or confused communication and native API calls. This kind of analysis framework requires secure operational environment by sandboxing, virtual devices, or Cloud-based models (Chang, 2015) to gather the runtime execution traces by simulating and executing the application binaries. While certain types of malware are recognized using code analysis, many other kinds of malware can reliably be detected at looking the runtime execution traces. For instance, newly deployed malware may not have signatures introduced while they may be widely separated.

One of the major problems with the dynamic analysis is scalable solutions. For large datasets, it considered a most time consuming and difficult task to gather traces from run-time analysis due to various resource constraints. Consequently, to overcome this problem cloud-based systems (Chang, 2015; VirusTotal; Weichselbaum et al., 2014) are available for the research community to execute their apps in sandboxes (controlled environment) with maximum available resources which are sometimes difficult for an individual researcher to manage. However, for our behavioral analysis task, as we are not dealing with huge dataset, therefore, we have chosen the well-known sandbox known as DroidBox (Desnos & Lantz, 2011) which provides enough logging capabilities useful for identification of malicious intention. DroidBox also provides a comprehensive picture of the applications runtime activities provided by logging all the transmitted/accessed data of a particular application. The activities include, data read from or written to files, SMS messages sent and received, HTTP traffic initiated to/from, DNS queries, dataset or received over the network and much more.

The steps we have taken to perform behavioral analysis are shown in Figure 4. Suppose, A is an instance of app comprising of, n number of applications and m number of features, then the runtime analysis collects all application with the following formula:

$$A = \left\{ app \mid app_i \ni \left| f_j \right| \ \forall \ \begin{matrix} i = 1,2,3,\ldots,n \\ j = 1,2,3,\ldots,m \end{matrix} \right\} \tag{1}$$

where:

n = total number of applications
m = total number of features

Table 2. Feature vector (permissions and API calls)

Phase	Permissions	API Calls
Connection Phase	INTERNET	openConnection(), execute(), connect(), openStream(), getInputStream(), Socket(), getContent()
Status Information Phase	READ PHONE STATE	getDeviceId(), getSimSerialNumber(), getSubscriberId(), getLine1Number()
	ACCESS NETWORK STATE	getActiveNetworkInfo(), getWifiState(),
	ACCESS WIFI STATE	getConnectionInfo(), getNetworkInfo()
	ACCESS COARSE LOCATION	getCellLocation()
	ACCESS FINE LOCATION	getLastKnownLocation(), isProviderEnabled(), requestLocationUpdates()
	READ CONTACTS	openOutputStream(), openInputStream(), openFileDescriptor()
	READ LOGS	exec()
Communication Phase	SEND SMS	getDefault(), sendTextMessage()

Figure 4. Behavior analysis workflow

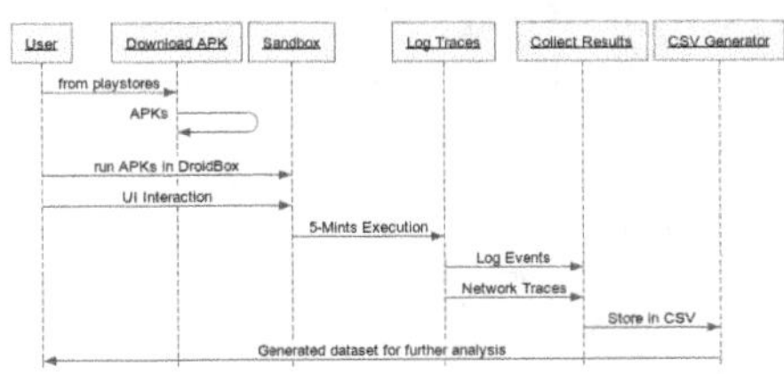

Feature Vector for Behavioral Analysis

The features we have selected for behavioral analysis can be characterized as file activities, network operations, Information Leaks, Services, SMS operations, Cryptographic Operations, DNS Traffic, HTTP Traffic, unknown Conversations, which are listed in Table 3.

Table 3. Feature vector (runtime-analysis)

Features	Parameters
File Activity	Read/Write
Network Operations	Opened connections, Network Reads, Network Writes
Information Leaks	File Leaks, Data Leaks
Services	Started Services
SMS	Sent SMS
DNS Traffic	DNS Requests
HTTP Traffic	HTTP Conversations, HTTP Connection attempts

EVALUATION

This section will evaluate static and dynamic analysis features among existing botnet, malware, and benign applications to highlight interesting trends in botnet applications.

Observations From Code Analysis

Malware families and benign application have different tendencies for requesting permissions. Malware generally requests more permissions than benign applications or even can request that permission which has risk associated with them. The risk factor provided with dangerous permissions is usually not descriptive to users at the time of installing an application. Therefore, user unknowingly installs such applications which have high risk related to them.

Figure 5. Permission frequency analysis

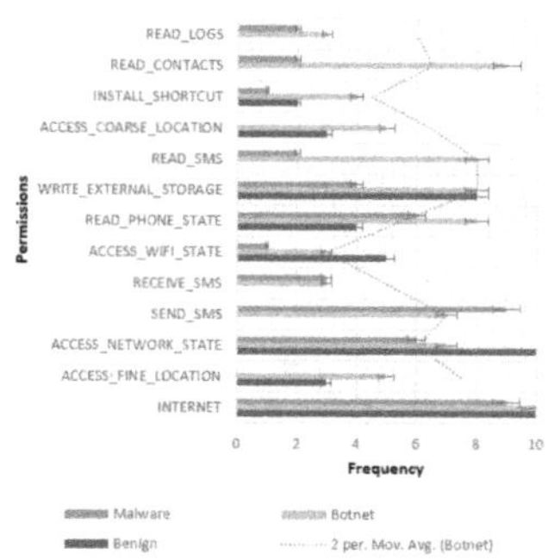

Figure 5 shows the average permissions used by benign, botnet and malware applications. The figure indicates that botnet applications are more frequently used feature vector specified in Table 2 than benign and malware applications.

The most prominent permissions used by botnet applications are INTERNET, ACCESS NETWORK STATE, READ PHONE STATE, READ SMS, READ CONTACTS, INSTALL SHORTCUT, and ACCESS COARSE LOCATION. These permissions are used by botnet applications to establish a remote connection with C&C by observing current status of device and network condition. Moreover, we have observed that botnet applications are using HTTP protocol as a medium of communication. Therefore, it is evident from the figure that on average SEND SMS permission is crucial for malware applications. Another interesting factor can be drawn from the figure is that on average benign application uses ACCESS WIFI STATE higher than botnet applications. Recently, WIFI has become the most widely used network access technology. Therefore, recent applications use this service more frequently. The Similar, trend is shown in a comparison of ACCESS WIFI STATE permission because our benign dataset is recently downloaded from official Google play store, whereas most botnet applications (60%) in our dataset are introduced during 2011-2017.

Another most valuable property which is exploited by boot master is to identify the active state of the mobile phone. Through observation of current status, the intruders can negotiate with bot clients and can instruct them accordingly. The same trend is observed that on average 80% of botnet applications utilize READ PHONE STATE permission than malware (60%) or benign (40%) applications.

The basic aim of a botnet initiator is not only to launch a botnet attack but also to expand its bot network. For this reason, botmaster tries to capture a broad audience through reading the contact list of affected mobile device and propagating the malware code to those contact persons. Consequently, on average 90% of botnet applications uses READ CONTACTS permission. In contrast, on average 20% malware and 0% benign applications use READ CONTACTS permission. Similarly, 80% of applications exploit READ SMS in comparison with 20% and 0% of malware and benign applications respectively.

Figure 6. API call frequency analysis

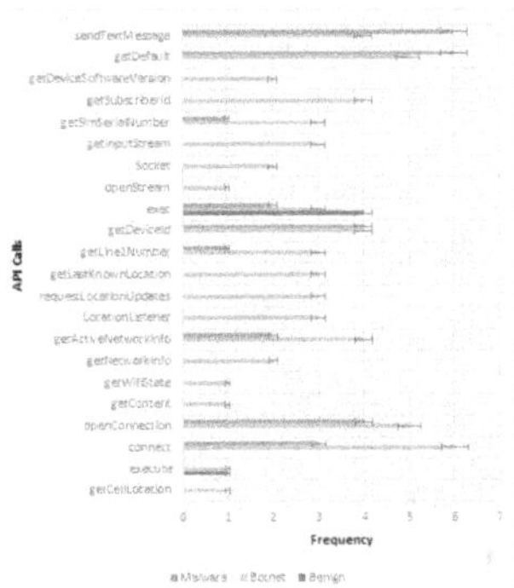

We have also examined API calls in order to analyze runtime execution traces. The usage frequency of API calls on botnet, benign and malware applications is depicted in Figure 6. The figure clearly indicates that botnet applications have frequent access to botnet specific API calls such as for connection establishment, connect () and openConnection() functions are called. Whereas, for continuous connection status and to gather network information, botnet apps use getNetworkInfo(), getConnectionInfo(), getConnectionInfo(), getNetworkInfor(), locationListener(), getLastKnownLocation(), getActiveNetworkInfo(), locationListener(), requestLocationUpdates(), getLine1Number() API calls.

To send/receive malicious commands to/from a particular bot, botmaster requires the unique identification of a device. For this purpose botnet application uses getDeviceID() and getSubscriberID() functions. We can observe the similar frequency of access for getDeviceID() function in both botnet and malware samples, i.e., on average 57% usage for each category. Whereas getSubscriberID() function is utilized on average 57% of botnet applications. In addition to that, as we have discussed above that mostly malware samples (60%) belongs to SMS Trojan. Therefore, we can observe the higher usage pattern (on average 90%) of sendTextMessage() function than in benign or botnet samples.

Observations From Run-Time Analysis

In this subsection, we will discuss the results obtained from runtime analysis of botnet, malware, and benign applications. Figure 7 shows the results generated from runtime analysis of botnet, benign and malware applications. The results clearly indicate the high upward trend in botnet applications with the selected feature vector.

Android applications work under various service models to start background services. Malware programmers deployed this model to establish communication pathway among C&C and mobile device. Further, the personal information is acquired and sends this information to an adversary. From the figure, it can be depicted that, on average botnet, each application has initiated 43 background services, whereas malware and benign application initiated at an average of 15 and 18 services.

Figure 7. Features call frequency during runtime analysis

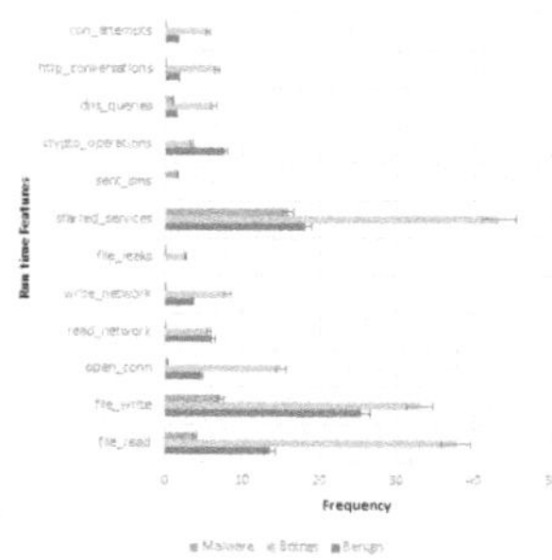

Android applications access internal and external storage of mobile devices to store and retrieve installation contents. Botnet applications access external storage to store execution code and initiate it on demand. According to the research conducted by (Weichselbaum et al., 2014) that 96% and 95% of malicious application access files for reading and writing respectively. In over findings, an overall upward trend is noticed for file reading and writing by botnet applications as well. On average each botnet application calls file_read 38 times and file_write 33 times.

Similarly, from opened connections, we can observe what communication channels are being used by botnet applications. For this purpose, we noticed that on average botnet applications opened 15 network connections mostly belongs to HTTP or HTTPS protocols, which clearly indicates malicious intention. In contrast, on average malware and benign applications opened one and four connections, respectively. In addition to that, the downward trend for opening a network connection in malware samples confirms our dataset choice as mostly malware samples belong to SMS Trojans that requires SMS as a communication channel rather than HTTP/s protocol to coordinate with C&C.

As we are targeting HTTP based mobile botnets, therefore, we also need to identify HTTP conversations. Through HTTP conversation we can notice, (a) actually what information is passing through the channel (b) what are the source and destination addresses (c) time stamp of each transaction. From the above figure, on average seven requests are made by botnet applications, in contrast, benign application on average call two times this feature.

One of the most important protocols used by botnet applications is DNS protocol. From the previous generation (PC Based) of botnets, it is proved that there are various parameters in DNS queries through which we can predict any botnet activity. Some of them includes: (a) failed DNS queries reveal botnet structure (b) DNS requests to backlist IP addresses (c) various suspicious DNS requests produced as round-robin DNS (c) DNS requests having short TTL for C&C domains (d) similar communication patterns, i.e., ratio of numerical characters in the DNS names. Based on these properties we can depict from the figure that, on average botnet, applications initiate 7 DNS requests. Whereas, a benign application on average calls one DNS request to resolve. The largest number of DNS queries that are resolved by the most recent mobile botnet NotCompatible.C which is 23 and among 4 of them are unsuccessful DNS queries. Moreover, we observed the similarity pattern in DNS names that are requested to resolve. Some of the queries are vhost240.181-4-8.telecom.net.ar, host26.186-126-172.telecom.net.ar, host105.181-4-113.telecom.net.ar, host30.186-127-146.telecom.net.ar, which clearly indicates a similar DNS query pattern.

Validation Through Machine Learning

In this section, we present validation of our proposed analysis platform to strengthen our claim that hybrid analysis is more effective to diagnose mobile malware having C&C features. The validation is performed through different machine learning classifier algorithms including J48, Naïve Bayes, Random Forest, and Logistic Regression. The j48 algorithm is typically applied on datasets to generate a pruned or unpruned classifier model which is based on decision trees algorithm logic. Similarly, another model which generates random trees to build a classifier model is called a random forest. Whereas Naïve Bayes algorithm derives prediction models based on independent assumptions. Moreover, logistic regression technique stems from the logistic model tree. In this paper, the motive behind the selection of diverse classifiers is to measure a higher level of accuracy. We have used 10-fold cross-validation on two different datasets. One of them is the dataset used in evaluation purpose (discussed above) and the second one comprises of 1371 botnet samples collected from (Kadir, Stakhanova, & Ghorbani, 2015) and 500 benign datasets collected from (Kang, Jang, Mohaisen, & Kim, 2014).

Table 4 presents the output of machine learning classifiers' performance on sample dataset which consists of 30 instances. The generated output comprises of correctly classified instances, incorrectly classified instances and accuracy in percentage against each classifier algorithm. The experiments involve one dataset of thirty botnet applications. Static-only classification experiment uses static features (as discussed above) of botnet applications, whereas dynamic-only experimentation takes dynamic features to pass through classifier models. Finally, hybrid experimentation involves the union of both static and dynamic features. The results indicate that the outcome of hybrid experimentation produced relatively good results other than static-only and dynamic-only experimentation. Similarly, based on accuracy results, random forest outperforms other classifier models with a classification accuracy of 90%, which is more than any other experimentation/classifier algorithm.

To validate our results on large-scale implementation we have applied the same classifier algorithm on Android botnet dataset. The results depicted from Table 5 confirm the viability of our hybrid analysis platform to detect botnet binaries on a large scale. Although all ML classifiers produced relatively good accuracy rates, however, simple Random Forest outperforms the other classifier algorithms. It correctly classifies 98% of Android botnet dataset using hybrid feature vector space to distinguish botnet applications. In difference, Naive Bayes, J48, and Logistic regression achieve an accuracy rate of 87%, 97%, and 95% respectively. However, the static-only experiment also produces almost the same accuracy result (fractionally lower).

Table 4. Classification on sample dataset

		Classifier Algorithms											
		J48			Random Forest			Logistic Regression			Naïve Bayes		
		Correctly Classified	Incorrectly Classified	Accuracy (%)	Correctly Classified	Incorrectly Classified	Accuracy (%)	Correctly Classified	Incorrectly Classified	Accuracy (%)	Correctly Classified	Incorrectly Classified	Accuracy (%)
Static Only	30	21	9	70	26	4	87	25	5	83	23	7	77
Dynamic Only	30	17	13	57	21	9	70	16	14	53	18	12	60
Hybrid	30	20	10	67	27	3	90	23	7	77	25	5	83

Table 5. Classification on android botnet dataset

		Classifier Algorithms											
		J48			Random Forest			Logistic Regression			Naïve Bayes		
		Correctly Classified	Incorrectly Classified	Accuracy (%)	Correctly Classified	Incorrectly Classified	Accuracy (%)	Correctly Classified	Incorrectly Classified	Accuracy (%)	Correctly Classified	Incorrectly Classified	Accuracy (%)
Static Only	1871	1803	68	96	1825	46	98	1777	94	95	1622	249	87
Dynamic Only	1871	1685	186	90	1742	129	93	1562	309	84	1038	833	56
Hybrid	1871	1810	61	97	1827	44	98	1785	86	95	1621	250	87

GUIDELINES TO AVOID MOBILE BOTNET

The following guidelines are suggested to avoid such harmful hazard growing in mobile phone technologies:

- Google Play store provides advanced security mechanisms to keep app repository intact from malicious attempts. Therefore, it is recommended for users not to download apps from third-party application stores;
- Applications' permissions enable gateway to access smartphone resources. Therefore, permissions are considered as one of the interesting attributes for malware writers. The recommendation is to carefully look carefully requesting permissions by the app, for instance, if a weather application requests for contact list permission, then it seems to be fishy. Thus, the permissions should be skimmed before app installation. After installation, keep paying attention to the behavior of the app. Most often malicious application requests more permissions after their installation. Reputable security solutions for your mobile device will certainly protect your mobile phone from active threats;
- Similarly, make sure that your security software is up to date by installing latest patches and for software and firmware;
- The worldwide botnet risk is best defined by close national and international collaboration amongst governments and in fact with legislative and technically oriented organizations. For an efficient detection and mitigation mechanism to work, liaison between stakeholders must be reinforced and strengthened through political support, will, and negotiations.

CONCLUSION

In this paper, we have analyzed and identified the existence of botnet phenomenon in Android-based mobile applications. We have proposed a hybrid analysis framework which is divided into two different steps. During first step (code-based analysis), initially, we have highlighted all those permissions and API calls that can lead to a botnet activity. Further, we have extracted this permission from a dataset comprising of ten applications from each category: known-botnets, benign and malwares. Finally, we compare and evaluate the results and conclude that botnet applications have frequent access to botnet specific permissions and API calls in comparison with benign and malware applications.

Furthermore, as the second step of our analysis task, we have executed our selected dataset onto a sandbox (DroidBox) and collected the dynamic features from the log files. Among the various runtime analysis parameters, we have chosen and compared the most interesting parameters which can cause a botnet attack to expand exponentially or can persist for a long time. The results are then compared by applying machine learning classification models on both sample dataset and an Android Botnet dataset. The machine learning algorithms also affirm the viability of our framework by achieving 98% of accuracy in case of hybrid feature vector space.

Conclusively, botnet phenomenon is gaining popularity in the computationally intensive mobile environment. So, there is a need to devise some proactive mechanisms through which user should be aware of the consequences by unknowingly installing an application which has botnet capabilities. Ultimately, this can not only affect the overall performance of user's device, but also a user can naively become the part of a distributed attack.

REFERENCES

Android, S. D. K. (2012). *Tools: monkeyrunner*. Retrieved from https://www.android.com/

Arp, D., Spreitzenbarth, M., Hubner, M., Gascon, H., & Rieck, K. (2013). Drebin: Efficient and explainable detection of android malware in your pocket. *Paper presented at the Proc. of 17th Network and Distributed System Security Symposium, NDSS*.

Aswini, A. M., & Vinod, P. (2014). Droid permission miner: Mining prominent permissions for Android malware analysis. *Paper presented at the 2014 Fifth International Conference on Applications of Digital Information and Web Technologies (ICADIWT)*. Academic Press. 10.1109/ICADIWT.2014.6814679

Bilić, D.G. (2017). *Mobile security: The reality of malware ... augmented.* We Live Security. Retrieved from https://www.welivesecurity.com/2017/03/10/mobile-security-the-reality-of-malware-augmented/

Chang, V. (2015). A cybernetics social cloud. *Journal of Systems and Software*.

Choi, B., Choi, S.-K., & Cho, K. (2013). Detection of mobile botnet using VPN. *Paper presented at the 2013 Seventh International Conference on Innovative Mobile and Internet Services in Ubiquitous Computing (IMIS)*. Academic Press. 10.1109/IMIS.2013.32

da Costa, V. G., Barbon, S., Miani, R. S., Rodrigues, J. J., & Zarpelão, B. B. (2017, May). Detecting mobile botnets through machine learning and system calls analysis. *Proceedings of the 2017 IEEE International Conference on Communications (ICC)* (pp. 1-6). IEEE.

Desnos, A. (2011). Androguard: Reverse engineering, malware and goodware analysis of android applications... and more (ninja!).

Desnos, A. & Lantz, P. (2011). Droidbox: An android application sandbox for dynamic analysis.

Fereidooni, H., Conti, M., Yao, D., & Sperduti, A. (2016). ANASTASIA: ANdroid mAlware detection using STatic analySIs of Applications. *Paper presented at the 2016 8th IFIP International Conference on New Technologies, Mobility and Security (NTMS)*. Academic Press.

Gu, G., Zhang, J., & Lee, W. (2008). BotSniffer: Detecting botnet command and control channels in network traffic.

Kadir, A. F. A., Stakhanova, N., & Ghorbani, A. A. (2015). Android botnets: What urls are telling us. *Paper presented at the International Conference on Network and System Security*. Academic Press. 10.1007/978-3-319-25645-0_6

Kang, H. J., Jang, J. W., Mohaisen, A., & Kim, H. K. (2014). Androtracker: Creator information based android malware classification system. *Paper presented at the Information Security Applications-15th International Workshop, WISA*. Academic Press.

Karim, A., Salleh, R., & Khan, M. K. (2016). SMARTbot: A Behavioral Analysis Framework Augmented with Machine Learning to Identify Mobile Botnet Applications. *PLoS One*, *11*(3). doi:10.1371/journal.pone.0150077 PMID:26978523

Karim, A., Salleh, R., Khan, M. K., Siddiqa, A., & Choo, K.-K. R. (2016). On the analysis and detection of mobile botnet applications. *Journal of Universal Computer Science*, *22*(4), 567–588.

Karim, A., Salleh, R., & Shah, S. A. A. (2015, August). DeDroid: a mobile botnet detection approach based on static analysis. *Proceedings of the 2015 IEEE 12th Intl Conf on Ubiquitous Intelligence and Computing and 2015 IEEE 12th Intl Conf on Autonomic and Trusted Computing and 2015 IEEE 15th Intl Conf on Scalable Computing and Communications and Its Associated Workshops (UIC-ATC-ScalCom)* (pp. 1327-1332). IEEE.

Khoshhalpour, E., & Shahriari, H. R. (2018). BotRevealer: Behavioral Detection of Botnets based on Botnet Life-cycle. *The ISC International Journal of Information Security*, *10*(1), 55–61.

Kirubavathi, G., & Anitha, R. (2018). Structural analysis and detection of android botnets using machine learning techniques. *International Journal of Information Security*, *17*(2), 153–167. doi:10.100710207-017-0363-3

Liang, S., & Du, X. (2014). Permission-combination-based scheme for android mobile malware detection. *Paper presented at the 2014 IEEE International Conference on Communications (ICC).* IEEE Press. 10.1109/ICC.2014.6883666

Lukan, D. (2012). *Reverse Engineering Tools*, Retrieved 8-9-2016, from http://resources.infosecinstitute.com/reverse-engineering-tools/

McIlroy, S., Ali, N., & Hassan, A. E. (2016). Fresh apps: An empirical study of frequently-updated mobile apps in the Google play store. *Empirical Software Engineering*, *21*(3), 1346–1370. doi:10.100710664-015-9388-2

Millman, R. (2015). Updated: 97% of malicious mobile malware targets Android. *SC Magazine UK*. Retrieved from https://www.scmagazineuk.com/updated-97-of-malicious-mobile-malware-targets-android/article/535410/

NowSecure. (2015). *SANTOKU Operating System*. Retrieved from https://santoku-linux.com/

Parkour, M. (2011). *Contagio malware dump. blog sobre compartición de malware*. Contagiodump. Retrieved from http://contagiodump.blogspot.com/

Paul. (2017). Android Malware Doubled in 2016, Adding to Mobile Malware Problem. Security Ledger. Retrieved from https://securityledger.com/2017/03/android-malware-doubled-in-2016-adding-to-mobile-malware-problem/

Peng, C., Zhao, R.-C., Zheng, S., Xun, J., & Yan, L.-J. (2016). *Android Malware of Static Analysis Technology Based on Data Mining. DEStech Transactions on Computer Science and Engineering*. AICE-NCS.

Petsas, T., Voyatzis, G., Athanasopoulos, E., Polychronakis, M., & Ioannidis, S. (2014). Rage against the virtual machine: hindering dynamic analysis of android malware. *Paper presented at the Seventh European Workshop on System Security*. Academic Press. 10.1145/2592791.2592796

Pravin, M. N. P. (2015). VetDroid: Analysis using permission for vetting undesirable behaviours in Android applications. *International Journal of Innovative and Emerging Research in Engineering*, *2*(3), 131–136.

Schmidt, A.-D. Bye, Rainer, Schmidt, H-G, Clausen, Jan, Kiraz, Osman, Yuksel, Kamer A, … Albayrak, Sahin. (2009). Static analysis of executables for collaborative malware detection on android. *Paper presented at the IEEE International Conference on* Communications *ICC'09*. IEEE Press.

Su, M.-Y., & Fung, K.-T. (2016). Detection of android malware by static analysis on permissions and sensitive functions. *Paper presented at the 2016 Eighth International Conference on Ubiquitous and Future Networks (ICUFN)*. Academic Press.

Unuchek, R., & Chebyshev, V. (2018). *Mobile malware evolution 2015*. AO Kapersky Lab.

Viennot, N., Garcia, E., & Nieh, J. (2014). A measurement study of Google Play. *Paper presented at the 2014 ACM international conference on Measurement and modeling of computer systems.*

VirusTotal. (2017) *VirusTotal*, Retrieved from https://www.virustotal.com/

Weafer, V. (2014). Threats Report. *McAfee Labs.*

Weafer, V. (2016). 2017 Threats Predictions. *McAfee Labs.*

Weichselbaum, L., Neugschwandtner, M., Lindorfer, M., Fratantonio, Y., van der Veen, V., & Platzer, C. (2014). Andrubis: Android malware under the magnifying glass. Vienna University of Technology.

Yang, M., Wang, S., Ling, Z., Liu, Y., & Ni, Z. (2017). Detection of malicious behavior in android apps through API calls and permission uses analysis. *Concurrency and Computation*, *29*(19), e4172. doi:10.1002/cpe.4172

Yousafzai, A., Gani, A., Noor, R. M., Naveed, A., Ahmad, R. W., & Chang, V. (2016). Computational offloading mechanism for native and android runtime based mobile applications. *Journal of Systems and Software*, *121*, 28–39. doi:10.1016/j.jss.2016.07.043

Wang, Z., Gao, H.-Z., Zhang, Y.-M., Hu, Yu-C., Qiu, K.-F., Cheng, X., & Jia, C.-F. (2017). Fortifying Botnet Classification based on Venn-abers Prediction. DEStech Transactions on Computer Science and Engineering Teaching and Research

This research was previously published in the Journal of Organizational and End User Computing (JOEUC), 32(3); pages 50-67, copyright year 2020 by IGI Publishing (an imprint of IGI Global).

Chapter 6
Preserving Security of Mobile Anchors Against Physical Layer Attacks:
A Resilient Scheme for Wireless Node Localization

Rathindra Nath Biswas
A. J. C. Bose Polytechnic, India

Swarup Kumar Mitra
MCKV Institute of Engineering, India

Mrinal Kanti Naskar
Jadavpur University, India

ABSTRACT

This chapter introduces a new security scheme for mobile anchors avoiding the physical layer attacks towards localization in wireless sensor networks (WSNs). In a network, anchors are made location-aware equipping them with GPS (global positioning system) receivers. Direction finding capabilities are also incorporated with smart antennas. The proposed algorithm is based on adaptive beamforming of smart array that always minimizes the probabilities of successful attacks, keeping the adversaries beyond its beam coverage. Particle swarm optimization (PSO) technique is used to compute array excitation coefficients, generating the desired pattern. Thus, anchors remain secured through pattern irregularities, deteriorating the information retrieval process even though chances of occurring adequate RSS (received signal strength)/AoA (angle of arrival) measurements may exist. Moreover, anchors are assumed to send pseudo references towards stationary nodes over private links, preserving data integrity for localization. Simulation results validate its effectiveness over the existing methods.

DOI: 10.4018/978-1-7998-8545-0.ch006

INTRODUCTION

During the last decades, continuous efforts were made to bring the diverse technologies in a common platform. The outcome can be viewed as an Internet of Things (IoT), an amalgamation of the Internet and smart sensors to ensure connectivity, computation and communications among the heterogeneous entities (Al-Gburi et al., 2018; Misra et al., 2017). Wireless sensor networks might be regarded as an example of such infrastructures. It inherently possesses numerous attractive features such as scalability, fault-tolerance capability, low power requirements, and less establishment cost, etc. Hence, such network architectures, comprising a considerable number of battery-driven tiny sensors, have now become much famous for data gathering applications under the hostile environments (Akyildiz et al., 2002; Biswas et al., 2014). For example, several essential services, both in civil and military sectors, often deploy such kind of network structures for various purposes like continuous monitoring of the environments, security surveillance, target detection, and tracking, etc. in the areas of interest. However, such applications frequently need location-based data to be relayed at the sink or base station (BS) for the realization of any event occurring within the networks. Usually, sensor devices are randomly deployed over the harsh fields from aircraft/ space vehicles to collect raw data from their surroundings. Accordingly, they remain scattered, unattended and unaware of their locations unless they are made location-aware by some system supports or being localized with the help of a localization system (Ou, 2011; Akter et al., 2018). In practice, a variety of network architectures are commonly existing. They are classified as (i) static networks consisting of stationary nodes only (ii) mobile networks comprised of mobile nodes only and (iii) quasi-static networks involving few mobile nodes along with the plentiful stationary nodes. However, random deployment causes some coverage problem in the first configurations because there is no provision of node relocations. Instead, the second arrangements could provide adequate coverage, but they seem to be much more energy inefficient. In this context, a more feasible solution might be obtained with the third configurations, and hence, it has now become the basis for the development of the state-of-the-art WSN architectures (Halder and Ghosal, 2016; Lin et al., 2017). Throughout the entire networks, mobile nodes can move freely, and they act as mobile data collectors (MDC) (Pazzi and Boukerche, 2008). For several location-based services (LBS), they also need to be location aware along their trajectories. Hence, they usually are made as embedded systems with GPS (Global Positioning System) receivers. Such nodes are referred to as the mobile anchors or beacons in WSNs literature. By employing suitable localization systems, on the other hand, stationary nodes are made capable of relaying location-based data to the base station in a cooperative manner. In most of the cases, anchors are also used as reference nodes for the localization process (Ssu et al., 2005; Naraghi-Pour and Rojas, 2014). Thus, mobile anchor based WSN architectures frequently possess several useful attributes like portability, self-configuration capability, self-localization capability, ease of installation and maintenance, etc. Therefore, they are frequently found to have their extensive use in various application domains.

Currently, the versatile applications of such wireless networks also make them tempting targets of several malicious attacks (Zou et al., 2016; Shiu et al., 2011). There are two types of attacks: internal and external, usually present in WSN environments. In case of internal attacks, the adversaries capture few static and/or mobile benevolent nodes and make them compromised either by directly accessing their control systems or tampering the radio environments of those particular areas over the networks. They usually aim to mislead the decision to be made upon analyzing all the collected data at the base station and hence, provide erroneous information. Sybil, replay and wormhole (Boukerche et al., 2008), etc. are typical examples of such attacks. On the contrary, adversaries enter into the networks from outside as different entities in external type attacks. They also make efforts to route spoofed information over the networks, suppressing their actual identities to the others. Although cryptographic solutions are found to be very effective against the external attacks, they often show vulnerability against the internal attacks (Chen et al., 2010). Moreover, such attacks would be more severe in the physical layer because the entire node set must share a shared wireless medium for relaying their information over the networks (Zhou et al., 2014). Hence, adversaries eavesdropping at some specific region or few compromised nodes might be able to capture the vital information during data transmission among the anchors and the stationary nodes (Kang, 2009). Anchors may also have the possibility of being detected and tracked while attackers can make sufficient measurements of the received signal strength (RSS) or the angle of arrival (AoA) information from their broadcast messages. Thus, they could corrupt the reference data easily by estimating the corresponding locations using trilateration or triangulation (Gezici, 2008; Wong et al., 2007) principles. They would also be able to disrupt the localization process by relaying these data throughout the entire networks. Moreover, the attackers can produce jamming signals or noise, introducing errors in the GPS receiver outputs. Consequently, preserving the security of mobile anchors over the sensor fields has now become a challenging task for many critical applications.

Several approaches have been adopted so far towards protecting the privacy of mobile users/ anchors at the upper layers (Sweeney, 2002; Meyerowitz and Roy Choudhury, 2009). These schemes are usually based on the principles of obscuring their actual location information or changing pseudonyms at appropriate intervals. Such techniques may not be suitable for preventing the attacks occurred in the physical layer. Because the anchors remain insecure and may be captured by the illegal hunters whenever they emit signals into air/freespace. Hence, the attackers need to be kept beyond their radio coverage. This important criterion was overlooked in most of the existing methods making them fragile in the presence of attackers.

Design of a security model for the mobile anchors defending the adversarial attacks at the physical layer and keeping the reference data intact towards the localization process is the primary concern of this chapter. The proposed algorithm is based on the adaptive beamforming principles of smart antennas. Particle Swarm Optimization (PSO) technique is used here to calculate the array element excitation coefficients producing the optimal pattern. Mobile anchors are considered to be equipped with smart antenna arrays. The desired pattern must consist of the main lobe focused at the desired node, and also several deep nulls steered towards the neighboring nodes during each time of data collection. It always ensures security in the networks reducing the probability of successful attacks and keeping the adversaries beyond the exposure of such patterns. Since the precise measurement of RSS/AoA data is dependent on the radio characteristics and the physical environments of the networks, the pattern irregularities each time enforce erroneous data estimation. Moreover, the algorithm provides an extra layer of protection for data integrity allowing pseudo anchor references along the trajectories. Thus, this scheme leads to a robust localization in wireless sensor networks.

RELATED WORKS

The research area of preserving security in WSNs has been much enriched with various pioneering works during past decades. Most of these methods are usually based on the k-anonymity, dummy/ fake ID transfer, and obfuscation principles, etc. Such schemes could guarantee the protection of nodes against the adversarial attacks acquiring accumulated data in the upper/application layer (Huang et al., 2005;

Myles et al., 2003). However, they fail to perform well in the physical layer because the nodes might be exposed to the adversaries each time during signal transmission over the radio (Rios & Lopez, 2011). The design of security model for this layer is slightly more complicated than that of earlier cases, and thus, very little research work was found satisfying all the necessary conditions (Bloch & Barros, 2011). In WSN scenarios, the attackers could make a threat to the privacy of sensor nodes by getting access to their RSS, AoA, ToA (time of arrival) and TDoA (time difference of arrival) information. Accordingly, they could break down the localization process corrupting this data and conveying such distorted information throughout the networks. Several innovative works to mitigate malicious attacks on the localization systems are discussed over here. Research work (Oh et al., 2011) proposed a novel jamming technique that mixes jamming signal from the neighboring nodes keeping similar link throughputs. Moreover, it introduced a multi cooperator power control (MCPC) algorithm to control the jamming noise power. Thus, it claims to ensure an additional increase of privacy through an active control on the strength of jamming noise. Another work (Wong et al., 2007) also described a privacy protection scheme under AoA based localization systems. It usually aims at decreasing the number of possible AoA measurements by using adaptive beamforming of smart antennas. Performance of this method is also dependent on the security margin defined with a pre-specified signal to noise ratio (SNR). Again, it has been shown that an acceptable performance could be obtained against the adversarial attacks on RSS based localization systems with a directional antenna made of tin cans (Bauer et al., 2009). However, this is an unrealistic approach because integrating directional antennas to each sensor makes the system bulky and expensive. A new method has also been introduced by altering anchor transmit power dynamically and adding a significant amount of noise with its transmitted signal (El-badry et al., 2010). Thus, it guarantees to avoid the precise RSS measurement of the radiated signal from an adversary in its vicinity. However, this approach makes a significant reduction in link throughput. Similarly, a security scheme has been described for mobile users against RSS based attacks on localization systems in wireless local area networks (WLAN) (Wang & Yang, 2011). An intelligent antenna pattern synthesis method was used to protect the location information of mobile users in the physical layer. However, this is also a simple scheme to perform well within network regions with a lower density of nodes and in lower communication ranges. Some researchers (Jiang et al., 2007) have also proposed a simple technique using intelligent transmit power control (TPC) in wireless nodes. It decreases the number of adversaries in range, thus, reducing the possibility of being captured. A physical layer spoofing detection scheme has also been proposed (Wang et al., 2017). Utilizing signal processing and feature recognition, improves the detection performance to some extent. Further, few (Soosahabi & Naraghi-Pour, 2012) described a scalable security method randomizing the key distributions to defeat any keyspace exploration attack. An analysis of outage probability for secrecy rate in multiple input multiple output (MIMO) wireless systems in the presence of eavesdroppers and jammers, has also been presented (Rawat et al., 2017). Further, research works (Mehta et al., 2012) proposed two techniques in each of the cases providing privacy both at the source and sink locations, making trade-offs between privacy, communication cost, and latency. However, the effectiveness of this method is also restricted to lower node density areas in the WSNs.

Therefore, new security model should be developed that must not only preserve the robustness of localization process but also ensures an uninterrupted data communication in location-based services, even over the higher node density regions of the WSNs. The proposed security scheme considers both of these aspects and becomes more resilient compared to other existing methods.

SECURITY ISSUES IN WIRELESS SENSOR NETWORKS

As mentioned in the earlier sections, wireless sensor networks constitute plentiful smart battery-driven tiny sensor devices. In most of the cases, after a random deployment in remote and hostile environments, they remain scattered, unattended and unaware of their locations. So, several issues regarding energy efficiency, coverage and security, etc. are pervasive to exist there. Towards fulfilling these conditions, network architectures and protocols should be appropriately framed. For example, energy efficiency can be enhanced by selecting appropriate operating modes (e.g., active, route, sleep, etc.) as per the requirements, to reduce the power consumption on each node in the networks. Likewise, an improvement in coverage can be made by introducing some mobility to facilitate the node relocation process. Preserving security in WSNs, on the other hand, appears to be a rather great challenging task and now attracts more research attention. Both in civil and military sectors, there is an increased demand for location-based services today. So, all sensor nodes need to be incorporated with GPS receivers, or there must have a suitable localization system enabling them to determine their locations autonomously. In this regard, the installation of a GPS to each node might be an unrealistic approach because it makes sensors bulky and expensive. In contrast, localization systems could play a crucial role through observation and detection of an event conveniently occurring within the networks. As a result, they are now becoming the tempting targets to the attackers. Attacks may occur in different ways such as direct (or external) and indirect (or internal) to the localization systems. However, localization systems are generally classified into two categories namely range-free and range-based methods depending on the requirement of range information (e.g., distance/angle) in position computation (Yao et al., 2016; Patwari et al., 2005). Although range-free schemes are simple enough, range-based systems are often used due to their higher accuracy. All range-based localization systems normally function with three successive parts which are range estimation, position computation, and localization algorithm. Thus, attacks may occurr on one or more segments to hamper their normal operations causing the overall breakdown of the localization systems (Boukerche et al., 2008; Chen et al., 2006). Moreover, GPS data could be corrupted with jamming signals or noise in WSNs.

Wireless Sensor Networks Preliminaries

The topic of preserving security in WSNs has become an emerging issue at present. From these perspectives, development of a useful security model to ensure continuous data gathering and routing through the nodes is urgently required. As the adversaries are supposed to attack the localization systems, the key feature of such schemes is to preserve the necessary data intact at each part keeping the attackers outside (Srinivasan and Wu, 2008). For this purpose, the reference data of all anchors along their trajectories also need to be protected from the adversarial attacks. All necessary assumptions on the network configurations, radio characteristics, and attack scenarios are introduced here to expedite the description and analysis of the proposed scheme in the following sections.

Network Topologies and Protocols

To implement the proposed security model, several assumptions are made at network architectures and protocols (as shown in Figure 1). These are furnished as follows:

1. The wireless networks are constructed with a random deployment of few mobile anchors and adequate stationary sensors over the areas of interest. The sensor fields must have many RF (radio frequency) reflectors making the adversaries challenging to find the direct/ dominant signal path.
2. Anchors are equipped with both the GPS receivers and smart antennas. The mobility features are also incorporated by mounting them on wheels/vehicles. They have an extra capacity for memory and refilling options to a power source.
3. Each sensor is configured with a unique ID and activated to broadcast message over the networks in a synchronous manner and with a periodic interval of time (τ). Also, they must have a localization system and some specific operating modes.
4. Anchors are capable of estimating the range information while the RSS exceeds the prescribed threshold limit on signal to noise ratio (SNR_{th}) using path loss model and ESPRIT (estimation of signal parameters via rotational invariance technique) algorithm.
5. Anchors must have the ability to establish secure communication links with the desired nodes, suppressing the effect of interferences from their surroundings by using an adaptive beamforming mechanism.
6. Anchors move in random trajectories within the networks and run with a scheduling technique.
7. The base station (BS) is fixed and located at a highly protected area in WSNs.

Figure 1. Wireless sensor network architectures

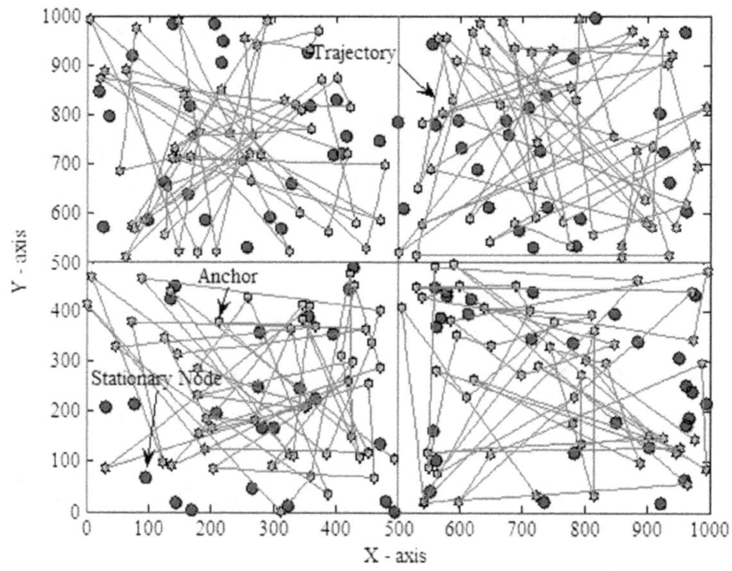

Radio Propagation Characteristics

The physical environments always cause various propagation features on the transmitted signal such as reflection, refraction, diffraction, and dispersion, etc. over the networks. Accordingly, the mean power of the received signal could be estimated with an exponentially decaying function of distance traveled. This phenomenon is known as path loss and is mostly used for the measurement of distance information in many RSS based localization systems (Wang and Yang, 2011; Patwari et al., 2005). However, selection of a proper signal propagation model, defining the path loss more realistically, is another important aspect for developing the security model in WSNs. The assumptions made in this regard are described as below:

1. Signals flowing through the radio environments are usually a composite form of large-scale path loss component, medium scale slow varying component, and small-scale fast varying component. The medium scale slow varying component uses log-normal distributions, and small-scale fast varying component uses Rician/ Rayleigh distributions in LoS (line of sight)/ NLoS (non-line of sight) data communications respectively (Liberti & Rappaport, 1999).
2. Path loss is determined using small-scale propagation characteristics with fast signal variations due to multipath fading over short distances (order of few wavelengths)/ short duration of time (order of few seconds).
3. The link budget equations are formulated assuming equal energy dissipation per bit in the electronic circuits of wireless sensor transceivers (transmitters/ receivers). However, extra energy consumption of the power amplifier circuit is considered in the transmission of each data bit at a unit distance over the free space (Biswas et al., 2014).

Attack Scenarios in Physical Layer

In general, the objective of attacks could be viewed as to mislead important plans/decisions taken over the accumulated data in WSNs. As mentioned earlier, localization systems take a significant role of detecting/tracking the events occurred in various location-based services. Hence, this chapter is focused on the description of possible attack scenarios for localization systems in the physical layer and effective countermeasures against them. Attacks may occur in direct/indirect ways to malfunction any segment of the localization systems, leading to overall systems break down (Bettini et al., 2005).

Indirect Attacks

Adversaries could establish passive links to control few benevolent nodes getting access to their stored data. Otherwise, they could compromise the radio environments of specific areas in WSNs deteriorating the propagation characteristics of some benevolent nodes over there. Such types of attacks are also termed as internal attacks, and it causes the localization process very tricky.

1. **Through Compromised Nodes:** Distance estimations could be made erroneous by varying the transmission power or delaying the data packets transmission time of compromised nodes towards RSS based or ToA/TDoA based measurement techniques respectively (Li and Ren, 2010). To make tampering of angle estimations through AoA based measurement methods, compromised nodes could send signals with reduced SNR. Besides, position computations could be made incorrect, corrupting the anchor references.
2. **Through Compromised Environments:** By introducing obstacles, smoke or noise, etc. to change the physical medium over the WSNs, both RSS based or ToA/TDoA based measurement methods could be made inaccurate. However, AoA based measurement schemes could be compromised deploying magnets over the sensor fields. Also, position computations could be hampered, jamming the GPS signals to make erroneous anchor references (Myles et al., 2003).

Direct Attacks

Adversaries could install large advanced electronic devices and circuits to eavesdrop at few nodes and capture MAC IDs from broadcast messages over the networks. Then they might appear with such IDs and communicate with the benevolent nodes. During communications, they could inject garbage/deceived information to these nodes, causing erroneous data routing throughout the entire networks in a convenient manner. Such attacks are most often termed as external attacks (Yilmaz and Arslan 2015).

PROPOSED SECURITY MODEL PERSPECTIVES

The IoT systems include all significant types of wireless network architectures. The entities of such systems usually interconnect themselves and operate through the exploitation of capabilities like data sensing, communication, and actuation, etc. So, they are also prone to be under security threats and vulnerabilities at various functional segments (Jeyanthi, 2016). Cryptographic approaches, checking authentication and data integrity of broadcast messages, well suited to provide security against external attacks. However, they become fragile to internal attacks because compromising few nodes; adversaries may get access to their stored keys and passwords. So, some methods based on the principles of encountering the compromised nodes and blocking them from data communications exist in the literature. Moreover, the performance of some schemes is validated with their reliability on position computation in the presence of compromised nodes. The proposed security model tries to trade off these two objectives as blocking the adversaries using optimal patterns and computing positions generating pseudo anchor references (as shown in Figure 2).

Adaptive Beamforming of Smart Antennas

In WSNs, most of the power is wasted while sensor nodes set up communication links producing an omnidirectional pattern with their conventional dipole antennas. The unused power may create interferences to the neighboring nodes deteriorating their quality of service (QoS) in the networks. Conversely, smart antennas can enhance the signal-to-interference-plus-noise ratio (SINR) in the data links mitigating multipath fading through spatial isolation (Winters, 2006). Therefore, adaptive beamforming means shaping of an optimal pattern by steering main-lobe and nulls at appropriate angles as per the direction-of-arrival (DoA) information (Godara, 1997). This improves channel capacity in the wireless links making them stable and secure to guarantee higher throughputs in data transmission. Towards localization, the precision of information regarding distance/angle and anchor references, etc. is vital. Hence, such a method may be advantageous to protect them against the adversarial attacks, establishing private and point-to-point links. In this chapter, the beamforming mechanism is realized using particle swarm optimization (PSO) to determine the element excitation coefficients of a linear array with uniform spacing. A brief overview of smart antennas and PSO algorithm is presented here to expedite the description and analysis of adaptive beamforming in smart antennas.

Figure 2. Block diagram representation of the proposed security model

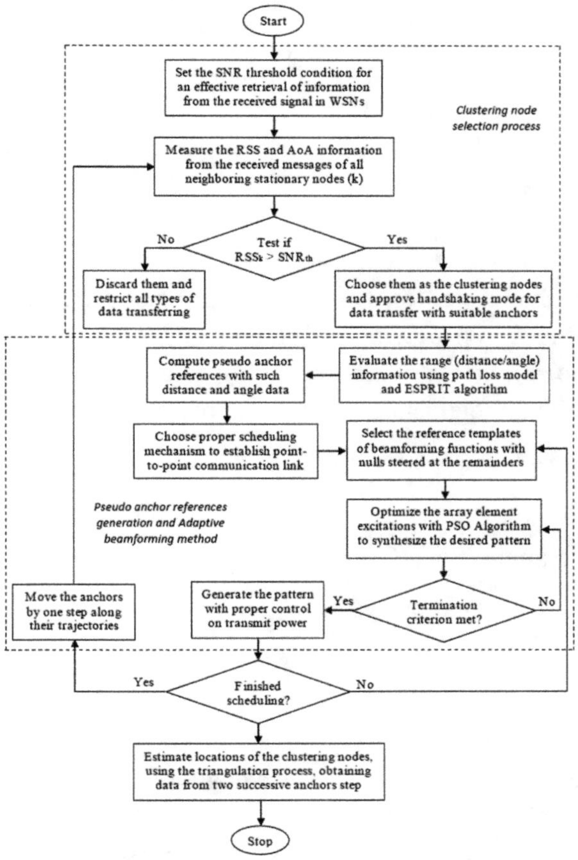

Basics of Smart Antennas

A smart antenna system (as shown in Figure 3) is comprised of three major components as (i) an antenna unit (ii) a signal processing unit and (iii) a beamforming network (Godara, 2004). Here, the antenna unit involves an array of 2N spatially separate antenna elements, providing space division multiple access (SDMA). Signal processing unit usually estimates DoA from the intercepted signals, x(t). Also, beamforming network updates the weight vectors, W, in an iterative process producing the output signal, y(t) by a reference template, r(t). Therefore, the output is obtained while the error function, e(t) = r(t)-y(t) reaches within an allowable limit. This can be expressed as follows:

$$y(t) = W^T x(t) \tag{1}$$

where,

$$x(t) = a(\theta_d)s(t) + \begin{bmatrix} a(\theta_1) & a(\theta_2) & \cdot & \cdot & a(\theta_{D-1}) \end{bmatrix} . \begin{bmatrix} i_1(t) \\ i_2(t) \\ \cdot \\ \cdot \\ i_{D-1}(t) \end{bmatrix} + n(t) \tag{2}$$

or, $x(t) = x_s(t) + x_i(t) + n(t)$

Thus, it signifies a composite vector form of the desired signal, $x_s(t)$ arriving at an angle, θ_d plus (D-1) number of interfering signals, $x_i(t)$ arriving at different angles from θ_1 to θ_{D-1} and zero mean AWGN (additive white Gaussian noise), n(t) for each channel. Again, W^T means weight vector of excitation coefficients. For an array with 2N-elements, it can be expressed as follows:

$$W^T = [W_1 W_2 \cdots W_{2N}] \tag{3}$$

For a particular scanning angle (θ_i), $a(\theta_i)$ represents an array steering vector that takes the form as follows:

$$a(\theta_i) = \begin{bmatrix} 1 & e^{j\beta\Delta\sin\theta_i} & e^{j2\beta\Delta\sin\theta_i} & \cdot & \cdot & e^{j(2N-1)\beta\Delta\sin\theta_i} \end{bmatrix}^T \tag{4}$$

Where $\beta = 2\pi/\lambda$ is the wave number, and λ is the wavelength of the transmitted/received signal. Again, Δ indicates inter-element spacing of the array.

Although two types of smart antennas such as (i) switched-beam and (ii) adaptive arrays often exist in practice (Gross, 2005), the latter can estimate signals more intelligently incorporating channel propagation characteristics under fluctuating radio environments. The state-of-the-art DoA estimation methods are based on time series analysis, spectrum analysis, Eigenstructure methods, parametric methods, and linear prediction methods, etc. Among some popular DoA estimation methods, Bartlett, Capon, ESPRIT (estimation of signal parameters via rotational invariance techniques), MUSIC (multiple signal classification) and Root-MUSIC, etc. are ubiquitous. On the other hand, beamforming algorithms usually work on the principles of maximizing the signal to interference ratio (SIR), minimizing the variance and minimizing the mean square error (MSE), etc. For example, least mean squares (LMS), sample matrix inversion (SMI), recursive least squares (RLS), constant modulus (CM) and conjugate gradient (CG) methods, etc. are usually used in various beamforming applications.

Overview of PSO Algorithm

Particle swarm optimization (PSO) is a population-based, stochastic evolutionary technique (Clerc and Kennedy, 2002). It inherently possesses useful features as a simple structure and higher convergence speed. Hence, it is now widely used to solve many multi-objective and complex optimization problems in various fields of applications. It's performance is further improved, avoiding local optima trapping, by introducing an inertia weight (w) or constriction factor to the velocity update equation. In this algorithm, each potential solution is termed as a 'particle' in the search space and must have a fitness value

Figure 3. Smart antennas configurations

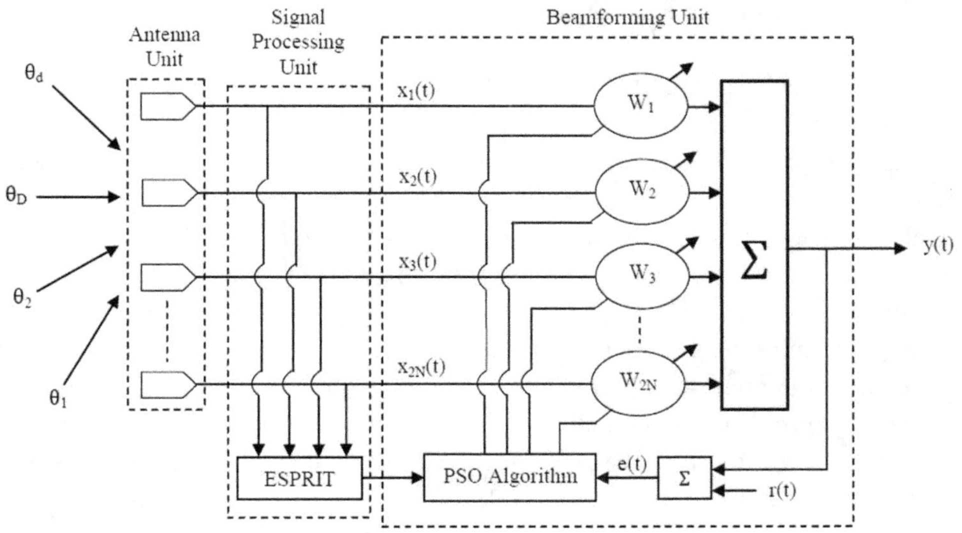

obtained by evaluating the objective function (also termed as cost/ fitness/ error function). Each particle begins the movement, in fact, with random position (X) and random velocity (V) over the search space. It is an iterative process, and each particle tries to find out the position with possibly higher fitness value at its every movement. Using previous knowledge of personal best position (P_{best}) and the global best position (G_{best}) attained ever in the solution space, each particle must update its velocity and position as per the following relations:

$$V = w.V + 2.rand().(P_{best} - X) + 2.rand().\ (G_{best} - X) \tag{5}$$

$$X = X + V \tag{6}$$

Here, t is taken as a unit time step. Also, P_{best} and G_{best} are chosen by particles fitness values during the search process.

The second and third terms of equation (5) are known as a 'cognitive' and a 'social' component respectively. Cognitive component encourages each particle to move toward its own best position, and the social component explores the optimal global solution, exploiting swarm behaviors. The constants represent the relative weights on stochastic acceleration, pulling each particle towards P_{best} and G_{best} positions. The 'Rand ()' is a function of uniformly distributed random numbers in the range [0, 1] that incorporate some randomness to mimic the analogy of a real scenario. The inertia weight is usually introduced to keep a balance between global exploration and local exploitation abilities in the search. It is found empirically that the process converges faster while inertia weight is linearly decreased from 0.9 to 0.4. The termination criterion for such methods is usually set by choosing a maximum number of iterations or allowing the fitness value within a predefined limit. Also, boundary conditions are applied to reinforce the particles movements inside the desired domain of interest. The dynamic range of the search space is defined with the maximum and minimum value in velocity and position respectively. Whenever any particle exceeds this limit, i.e. (V_{max},V_{min}) and (X_{max},X_{min}), its velocity and position are to be reset forcefully between the upper and lower boundaries towards an effective controlling on the convergence speed (Robinson and Rahmat-Samii, 2004).

PSO-Based Adaptive Beamforming Method

Patterns with higher directivity, narrower beam width, and lower side-lobes level are highly desirable for secured data transfer through long-haul point-to-point wireless communication links. As the single antenna is unable to generate such patterns, an array configuration is frequently used (Balanis, 2005). The optimum tradeoff among these beamforming attributes can be achieved by optimizing the electrical/ physical parameters of the antenna array (Zurita, 2014). Although such an optimum pattern alone cannot always offer security against several malicious attacks in WSNs, hence, the flexibility of placing deep nulls in the direction of interferences is crucial for providing a second layer of protection (Zhu, 2016). This makes the design problem very complicated, and there is always a probability of deteriorating the trade-off issues. However, the desired beam pattern with steered deep nulls towards the interferers can be synthesized by the PSO algorithm, evaluating the element excitation coefficients (W) conveniently. Assuming symmetry in amplitudes of the array element excitation coefficients about the origin, the normalized form of array factor (AF) for a linear array (as shown in Figure 3) with uniform inter-element spacing (Δ) can be written as follows:

$$AF_p(\theta) = \sum_{n=1}^{N} W_n \cos\left[\left(n - \frac{1}{2}\right)\beta.\Delta(\sin\theta - \sin\theta_d)\right] \tag{7}$$

Therefore, the optimization problem reduces to only half (N) dimensions for symmetrical array structures. Again, minimum and maximum boundary value in each dimension is set as [0, 1] giving a dynamic range of unity in the optimization process. Now, all the necessary beamforming attributes of desired patterns are stipulated on the reference templates of a simple time-scaled cosine function as follows:

$$AF_d(\theta) = \begin{cases} \cos\left[(\theta - \theta_d)\frac{\pi}{FNBW_d}\right] & if \quad |\theta - \theta_d| \le \frac{FNBW_d}{2} \\ \eta & if \quad \theta = \theta_n \\ SLL_d & elsewhere \end{cases} \tag{8}$$

Here, desired beam width ($FNBW_d$) is adjusted by choosing the appropriate value of phase angle for the cosine function. Also, desired side-lobes level (SLL_d) and depth of nulls (η) at interfering directions (θ_n) can be set as per the needs of the design.

Again, considering the perspectives of lower computational complexity in the optimization process, a simple formula is adopted here for calculation of the objective function (F) as:

$$F = \min\left\{ \sum_{\theta=-90^0}^{90^0} \delta(\theta) \right\} \tag{9}$$

where,

$$\delta(\theta) = \begin{cases} 1 & if \quad AF_d(\theta) - AF_p(\theta) \le 0 \\ 0 & otherwise \end{cases} \tag{10}$$

Here, $AF_d(\theta)$ and $AF_p(\theta)$ are taken as the desired and produced pattern components respectively. Also, $\delta(\theta)$ represents a weight assigned for each deviation obtained at any sample angle (θ) over the range [-90^0, 90^0].

Since each particle has (N) dimensions within the solution space, and now considering (M) particles, the algorithm would yield both position (X) vector and velocity (V) vector as (M × N) dimensional matrices. They are also randomly initialized towards faster convergence. The objective function is evaluated for each particle to find its P_{best} value as $P_{best} = [P_{best1}, P_{best2}.....P_{bestM}]^T$ and G_{best} value as $G_{best} = \min\{P_{best}\}$ in an iterative way. The P_{best} and G_{best} value along with their respective positions are recorded and used to update velocity and position according to equation (5) and (6) respectively. This algorithm is set to be terminated on attaining a maximum iteration of 1000.

Generation of Pseudo Anchor References

As mentioned in the earlier sections, localization systems involve anchor references in position computation of sensor nodes. However, attacks may occur to distort anchor references causing erroneous position estimation. Hence, such reference data needs to be kept secure and intact for the localization process. This is ensured in the proposed scheme considering anchors to send pseudo references along with their trajectories rather than the actual one. Thus, sensor nodes can autonomously estimate their locations having two such pseudo anchor references, spoofing the attackers in WSNs. The steps for generating pseudo anchor references are discussed as follows:

Calculation of Distance Information

Formulating an appropriate radio propagation model, distance information can be estimated with path loss component. In long-distance distributions (Liberti and Rappaport, 1999), large-scale path loss component (L_p) can be defined by a function of distance (d) and usually expressed (in dB) as:

$$L_p(d) = L_p(d_0) + 10\alpha \log\left(\frac{d}{d_0}\right) \tag{11}$$

Where α is known as path loss exponent. It indicates the rate of increase in path loss with distance. Also, d_0 denotes a small reference distance from the transmitter.

Now, considering variable cluttering effects of environments, the expression of path loss component is modified and defined with log-normal shadowing model (in dB) as:

$$L_{px}(d) = L_p(d) + X_\sigma \tag{12}$$

where X_σ is a random variable of zero mean Gaussian distributions with standard deviation σ.

The link budget of LoS propagation can be described with Frii's free space equation as:

$$P_r = \frac{P_t G_t G_r \lambda^2}{(4\pi d)^2} \tag{13}$$

Where P_t and G_t signify power output and gain of the transmitting antenna. P_r and G_r imply power intercepted and gain of the receiving antenna.

Again, the equation (13) can also be expressed (in dB) as:

$$P_r = P_t + G_t + G_r - 32.45 - 20\log_{10}(d/KM) - 20\log_{10}(f/MHz) \tag{14}$$

where L_P=32.45+20log$_{10}$(d/KM)+20log$_{10}$(f/MHz) is the path loss component.

To estimate distance (d_k) for the k-th node using received signal power (P_k), log-normal shadowing condition is to be applied in equation (14) and can be written as:

$$P_k = P_t + G_t + G_r - L_p(d_0) - 10\alpha \log(\frac{d_k}{d_0}) - X_\sigma \tag{15}$$

P_t and P_r are normally represented in dB/dBm. Also, G_t and G_r are represented in dBi. Parametric settings are made as per the Crossbow MICAz motes keeping compatibility with FCC limits (MICAz 2018). For low power tiny sensors, they are chosen as P_t = 30 dBm, G_t = 10 dBi, G_r = 0 dBi, d_0 = 1 m, SNR_{th} = -40 dBm, X_σ = 0 dB and α = 2.

Estimation of Angle Information

The angle information of all received signals can also be determined using one common high precision direction of arrival (DoA) technique known as ESPRIT in smart antennas literature (Roy and Kailath, 1989). ESPRIT algorithm exploits the rotational invariance property in the signal subspace. It usually utilizes a translational invariance structure that consists of two identical sub-arrays with a finite separation Δ (also known as doublets). Hence, assuming the number of signal sources (D) is less than that of array elements (2N), the signals induced on each of the sub-arrays can be written as:

$$x_1(t) = A_1.s(t) + n_1(t) \tag{16}$$

and

$$x_2(t) = A_2.s(t) + n_2(t) = A_1.\Phi.s(t) + n_2(t) \tag{17}$$

where,

$$\Phi = diag\{e^{j\beta.\Delta \sin\theta_1}, e^{j\beta.\Delta \sin\theta_2}, ..., e^{j\beta.\Delta \sin\theta_D}\}$$

is a diagonal unitary matrix with progressive phase shifts. Also, A_1 and A_2 are called Vandermonde matrix of steering vectors for two sub-arrays.

Now, considering the contributions of both sub-arrays, the total received signal and correlation matrix for the complete array can be expressed as:

$$x(t) = \begin{bmatrix} A_1 \\ A_1.\Phi \end{bmatrix} .s(t) + \begin{bmatrix} n_1(t) \\ n_2(t) \end{bmatrix} \tag{18}$$

$$R_{xx} = E[x.x^H] = A_1 R_{ss} A_1^H + \sigma_n^2 I \tag{19}$$

Again, correlation matrices for two sub-arrays are usually represented as:

$$R_{11} = E[x_1.x_1^H] = A_1 R_{ss} A_1^H + \sigma_n^2 I \tag{20}$$

and

$$R_{22} = E[x_2.x_2^H] = A_1 R_{ss} A_1^H + \sigma_n^2 I \tag{21}$$

Due to the invariance array structure, signal subspace (E_x) can be decomposed into two subspaces: E_1 and E_2 whose columns include the D eigenvectors corresponding to the largest eigenvalues of R_{11} and R_{22}. Since these arrays are related with translational invariance features, E_1 and E_2 would be related by a unique non-singular transformation matrix Ψ such that $E_1\Psi = E_2$.

Similarly, there must also be a unique non-singular transformation matrix Γ such that $E_1 = A_1\Gamma$ and $E_2 = A_1\Phi\Gamma$.

From the above relationships, it can be derived that $\Gamma\Psi\Gamma^{-1} = \Phi$.

Thus, the eigenvalues of Ψ must be equal to the diagonal elements of Φ such that

$$\lambda_1 = e^{j\beta.\Delta \sin\theta_1}, \lambda_2 = e^{j\beta.\Delta \sin\theta_2}, ..., \lambda_D = e^{j\beta.\Delta \sin\theta_D}$$

Also, the columns of Γ must be the eigenvectors of Ψ.

Now, the angle information of signals arriving at k-th node (θ_k) can be estimated as:

$$\theta_k = \sin^{-1}\left(\frac{\arg(\lambda_k)}{\beta.\Delta}\right) \tag{22}$$

where, k = 1,2,….,D

Formation of Anchor References

In WSNs, anchors are aware of their own locations through GPS receivers. As discussed earlier, they estimate distance and angle information from broadcast messages having SNR greater than a prescribed value. Such neighboring nodes (also termed as clustering nodes) are selected for serial data transfer from present anchor position. This anchor reference is essential for estimating the position of each clustering node. Hence, it needs to be kept confidential so that data integrity is retained in relayed data packets. Accordingly, pseudo forms of such references are prepared along the corresponding directions of each clustering node. For k-th clustering node, pseudo anchor reference (x_{pk},y_{pk}) is generated as follows:

$$\left.\begin{aligned} x_{pk} &= x_a + rand().d_k \cos\theta_k \\ y_{pk} &= y_a + rand().d_k \sin\theta_k \end{aligned}\right\} \tag{23}$$

Where, (x_a,y_a) denotes actual position in present anchor step.

PERFORMANCE EVALUATION FRAMEWORK

As discussed earlier, security of anchors is preserved through adaptive beamforming and pseudo references. Beamforming aims to provide a stable link with the desired node keeping all neighboring nodes away from its radio coverage. In contrast, the objective of generating pseudo references is to keep anchors' trajectories secured, reducing the chances of being detected/ tracked directly. This section deals with the formulation of security conditions, selection of benchmarks for evaluating performance and simulation results.

Security Criterions

All neighboring nodes are assumed to be suspicious because some of them might be compromised or externally implanted. So, anchors must control their transmission power in beamforming as per the least requirement of signal to noise ratio (SNR_{th}), maintaining reliable data transfer with the desired node. This would also cause deterioration of SNR at all adjacent nodes keeping them always below the threshold limit. Moreover, the second layer of protection is provided by steering deep nulls towards them. As the retrieval of contents is quite impossible at such a lower SNR level, the integrity of the relayed data packets is ensured. Hence, the conditions for preserving anchors security can be expressed (in dBm) as below:

For the desired node

$$P_0 | AF_d(\theta_d) | 2 \leq SNRt_h \tag{24}$$

And in the case of all other neighboring nodes,

$$P_0 | AF_d(\theta_k) |2 < SNRt_h \tag{25}$$

where P_0 is the effective isotropic radiated power (EIRP) at a reference distance from anchors.

However, SNR gives a quantitative measure of signal power to noise power. In the case of AWGN (additive white Gaussian noise) distribution, it usually varies with distance from the transmitter in free space (X_σ = 0 dB). Thus, SNR level for a k-th node with distance d_k can also be defined as:

$$SNR_k = P_k(d_k) \tag{26}$$

On the other hand, transmitting data packets along with pseudo anchor references may be helpful in camouflaging their actual positions to the attackers. However, the desired node can estimate its position, as an intersection point of lines originating from two distinct anchor positions with such reference data. In contrast, security conditions for anchors may also be violated if two or more neighboring nodes remain on and above SNR_{th} within patterns.

Performance Metrics

Performance of the proposed security algorithm depends typically on the accuracy of the beamforming process, and hence, it is verified under two metrics as follows:

- **Beamforming efficiency (γ):** It is the capability of reproducing the patterns as per the attributes defined on the desired beamforming functions. It measures all the deviations of the optimal pattern from the desired specifications and expressed (in percentage) as:

$$\gamma = (1 - \varepsilon) \times 100\% \tag{27}$$

Where ε denotes the error associated with the beamforming process. It is here defined as a ratio of the aggregated value of deviation weights to the total number of sample points (Ω) and expressed as:

$$\varepsilon = \frac{\sum_{i=1}^{\Omega} \delta(\theta_i)}{\Omega} \tag{28}$$

Typically, θ_i is the i-th sample angle over the range [-90°,90°].

- **Success rate (μ):** It measures the degree of effective attempts preserving security on particular anchor movements in WSNs. Therefore, it can also be defined as the ratio of the number of successful attempts (Λ_s) to total attempts (Λ) made by anchors over a specific time duration. It is expressed (in percentage) as:

$$\mu = \frac{\Lambda_s}{\Lambda} \times 100\% \qquad (29)$$

where successful attempt means permitting not more than one neighboring node (in the worst case) within patterns at a time.

Simulation Environments

Simulation environments are implemented on MATLAB software (version 7) platform. Several off-line PC (personal computer) generated data, making an analogy with the real-time WSNs scenarios are assumed here. A random deployment of 100, 200 and 300 stationary sensor nodes (making different node density) over a two-dimensional (2-D) field of 1000m×1000m area are considered as network architectures. Sensors are supposed to be with a reasonable far space for increasing network coverage avoiding interferences among them. Anchors are also considered to move along random trajectories over the sensor field and relay necessary data packets via private links for localization of the nodes. For connecting nodes, scheduling mechanism is adopted by their corresponding distances. Adversaries are assumed to either capture few nodes making them compromised or participate directly in data communications over the networks. As radio propagation over free space, log-normal shadowing model is assumed.

Simulation Parameters

Simulation parameters are set in keeping similarity with common wireless narrowband transmission systems such as WLAN IEEE 802.11 (IEEE 2010) and are given in Table 1. Antenna array design parameters and specifications of the desired patterns for producing optimal patterns are chosen as per Table 2. In the PSO algorithm, a maximum of 1000 iterations are set as termination condition, and boundary limits for optimization variables are kept in the range of {0,1}.

Table 1. Simulation parameters

Parameters	Values
Network size	1000m x 1000 m
Number of nodes	100, 200, 300
Number of anchors	2,4,6
Number of compromised nodes	10,20,30
Maximum transmission range (d_{max})	100 m
Transmit power of an anchor (P_t)	30 dBm
Threshold level in SNR (SNR_{th})	−40 dBm
Path loss exponent (α)	2
Shadowing noise variance (σ^2)	1

Table 2. Antenna design parameters and pattern attributes

Parameters	Values
Array size (2N)	20,30,40
Frequency (f)	2.4 GHz
Inter-element spacing (Δ)	0.5λ
Phase shift (ζ)	0
SLL_d	0.01 (-40 dB)
$FNBW_d$	20^0
η	0.00001 (-100 dB)
θ_d	30^0
θ_n	-20^0, -5^0, 10^0, 45^0, 60^0

Simulation Results

The proposed security model is tested under extensive simulations on its performance parameters as mentioned above. Such parameters are usually varied with array length and node density in WSNs. However, empirical cumulative distribution functions (ECDF) of the results are made, by taking 30 runs of the program, towards proper explanations in all the cases. Also, convergence curves and corresponding optimal patterns are shown in Figure 4 and Figure 5 respectively. The optimized values of array element excitation coefficients are given in Table 3. Beamforming efficiency is illustrated in Figure 6. Success rate (considering only 25 anchor steps in each run) under variable array lengths and node density is also shown in Figure 7 and Figure 8 respectively.

Figure 4. Convergence curve for optimization process (Node = 100)

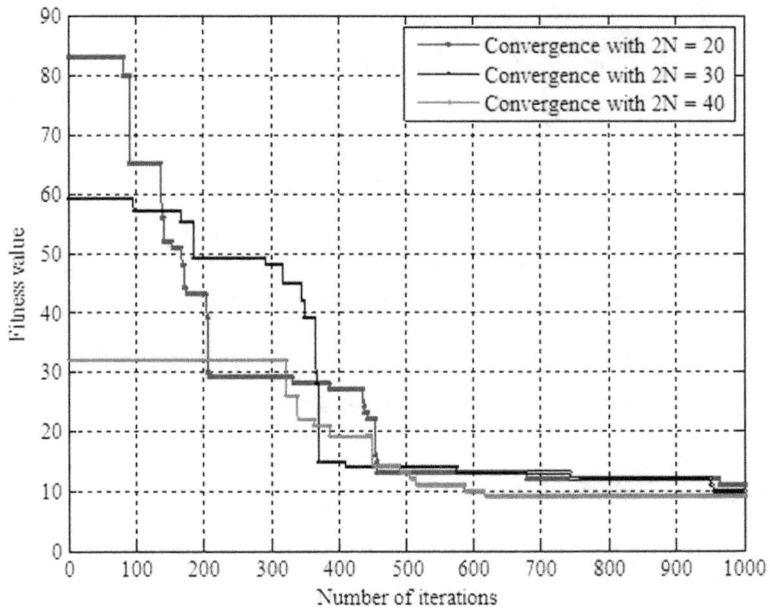

Performance Analysis

Performance of any security scheme can be explained under two key factors as (i) node deployment scenario and (ii) attack model in WSNs. Thus, attacks might be more severe in case of the networks with higher node density rather than its counterpart. Also, the complexity of the attack model might have a significant impact on the performance of a security algorithm. The ordinary security models often ignore to take up necessary preventive measures against such conditions and hence, fail to perform well in most of the cases. The proposed security model becomes fragile against replay type attacks to some extent at higher node/ compromised node density regions. Although combining cryptographic approaches, using pairwise key distribution at the time of data transfer with anchors might be a viable solution, it also becomes vulnerable. Because having no previous knowledge of compromised nodes' behaviors over the networks, anchors could share their secret keys with compromised nodes via direct links. Otherwise, a compromised node prevailing on the patterns of another link could capture anchor generated data packets. In both cases, they might be able to replay the data packets, tampering its contents towards adjacent nodes. Obviously, it would mislead the localization process on those particular benevolent nodes. Hence, some alternative measures should also be taken to identify the compromised nodes and eliminate their effects on the localization process.

Figure 5. Optimal patterns with various array lengths (Node = 100)

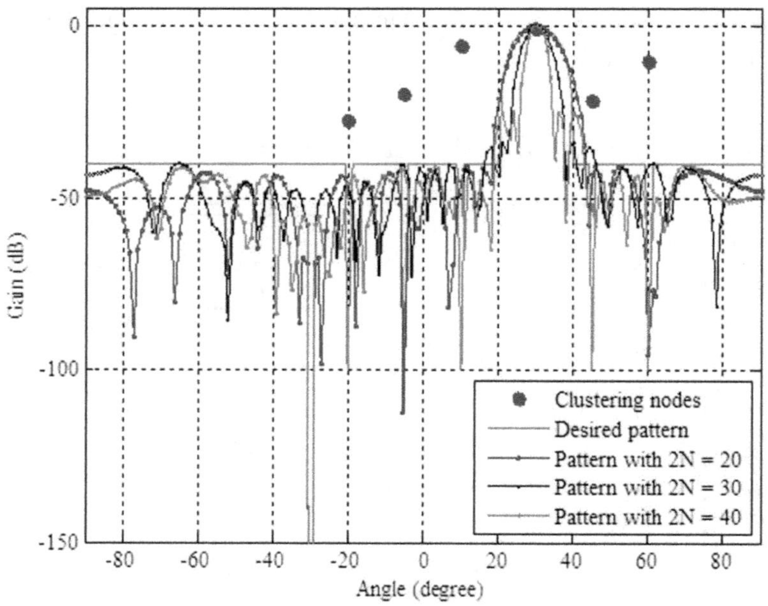

Figure 6. Beamforming efficiency with various array lengths (Node = 100)

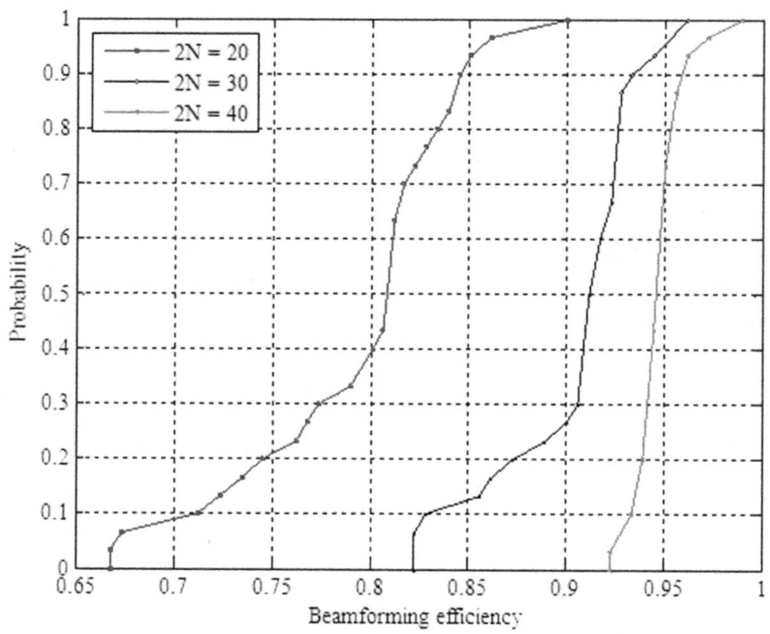

Figure 7. Success rate attained with various array lengths (Node = 100)

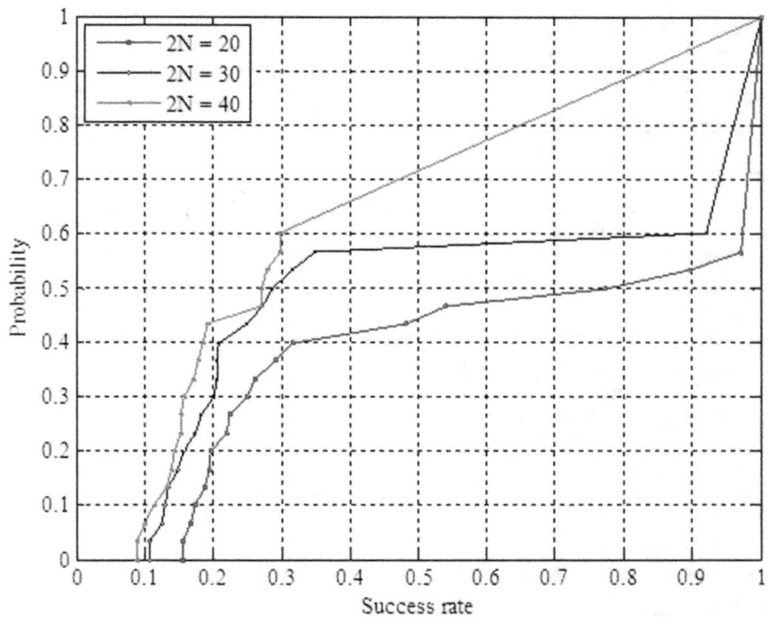

Figure 8. Success rate attained with various node density (2N = 20)

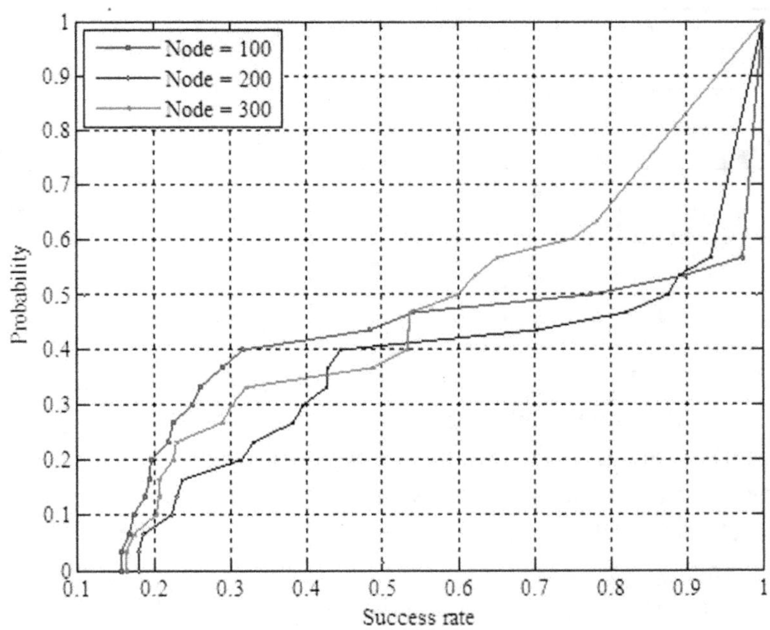

Table 3. Optimized element excitation coefficients

Parameters	Values		
Index (n)	Number of Array Elements (2N)		
	N = 10	N = 15	N = 20
1	1.0000	1.0000	1.0000
2	0.9340	0.9693	0.9873
3	0.8505	0.9307	0.9297
4	0.7358	0.8813	0.8912
5	0.6098	0.7999	0.8971
6	0.4618	0.7324	0.8106
7	0.3389	0.6745	0.7440
8	0.2083	0.5832	0.7555
9	0.1252	0.4907	0.7060
10	0.0636	0.4307	0.6783
11	--	0.3050	0.6354
12	--	0.2282	0.6038
13	--	0.1689	0.5556
14	--	0.1298	0.4760
15	--	0.0615	0.4152
16	--	--	0.3317
17	--	--	0.2565
18	--	--	0.1964
19	--	--	0.1335
20	--	--	0.0661

CONCLUSION

A new security model using adaptive beamforming of smart antennas is presented in this chapter. It can ensure privacy protection of mobile anchors against the attacks at the physical layer. As anchors are used as reference nodes in the localization process, preserving their location privacy over networks is always essential. In the proposed method, data transmission is secured via establishing more stable links with the optimal patterns. In most cases, such patterns reduce the chances of attacks keeping the adversaries beyond the radio coverage of anchors. Moreover, pattern irregularities cause data retrieval process trickier even if some attackers remain within the links. The relaying of pseudo anchor references also helps to keep data confidentiality against direct attacks. Simulation results validate its competency with higher success rates and accuracy in the localization process. However, this scheme is developed with several assumptions on attack models. Adversaries could also install large, sophisticated equipment/code to crack the relayed data packets accessing its contents. So, the feasibility of this scheme should be verified under several complex attack scenarios in the future works.

REFERENCES

Akter, M., Rahman, M. O., Islam, M. N., Hassan, M. M., Alsanad, A., & Sangaiah, A. K. (2018). Energy-efficient tracking and localization of objects in wireless sensor networks. *IEEE Access: Practical Innovations, Open Solutions*, *6*, 17165–17177. doi:10.1109/ACCESS.2018.2809692

Akyildiz, I. F., Su, W., Sankarasubramaniam, Y., & Cayirci, E. (2002). A survey on sensor networks. *IEEE Communications Magazine*, *40*(8), 102–114. doi:10.1109/MCOM.2002.1024422

Al-Gburi, A., Al-Hasnawi, A., & Lilien, L. (2018). Differentiating security from privacy in internet of things: A survey of selected threats and controls. In K. Daimi (Ed.), *Computer and network security essentials* (pp. 153–172). Springer International Publishing. doi:10.1007/978-3-319-58424-9_9

Balanis, C. A. (2005). *Antenna theory: Analysis and design*. New York: John Wiley and Sons Inc.

Bauer, K., McCoy, D., Anderson, E., Breitenbach, M., Grudic, G., Grunwald, D., & Sicker, D. (2009). The directional attack on wireless localization-or-How to spoof your location with a tin can. In *Proceedings of Global Telecommunications Conference* (pp. 1-6). Honolulu, HI: IEEE. 10.1109/GLOCOM.2009.5425737

Bettini, C., Wang, X. S., & Jojodia, S. (2005). Protecting privacy against location-based personal identification. In *Proceedings of the Second VDLB International Conference on Secure Data Management* (vol. 1, pp. 185-199). Trondheim, Norway: Springer. 10.1007/11552338_13

Biswas, R. N., Mitra, S. K. & Naskar, M. K. (2014). A robust mobile anchor based localization technique for wireless sensor networks using smart antenna. *International Journal of Ad-Hoc and Ubiquitous Computing*, *15*(1/2/3), 23-37.

Bloch, M., & Barros, J. (2011). *Physical-layer security: From information theory to security engineering*. New York: Cambridge University Press. doi:10.1017/CBO9780511977985

Boukerche, A., Oliveira, H. A. B. F., Nakamura, E. F., & Loureiro, A. A. F. (2008). Secure localization algorithms for wireless sensor networks. *IEEE Communications Magazine*, *46*(4), 96–101. doi:10.1109/MCOM.2008.4481347

Chen, Y., Kleisouris, K., Li, X., Trappe, W., & Martin, R. P. (2006). The robustness of localization algorithms to signal strength attacks: A comparative study. In *Proceedings of International Conference on Distributed Computing in Sensor Systems* (vol. 1, pp. 546-563). San Francisco, CA: Springer. 10.1007/11776178_33

Chen, Y., Yang, J., Trappe, W., & Martin, R. P. (2010). Detecting and localizing identity-based attacks in wireless and sensor networks. *IEEE Transactions on Vehicular Technology*, *59*(5), 2418–2434. doi:10.1109/TVT.2010.2044904

Clerc, M., & Kennedy, J. (2002). The particle swarm: Explosion, stability, and convergence in a multidimensional complex space. *IEEE Transactions on Evolutionary Computation*, *6*(1), 58–73. doi:10.1109/4235.985692

Data sheet for Crossbow MICAz mote. (n.d.). Retrieved on August 14, 2018, from http://edge.rit.edu/edge/P08208/public/Controls_Files/MICaZ-DataSheet.pdf

El-badry, R., Sultan, A., & Youssef, M. (2010). HyberLoc: Providing physical layer location privacy in hybrid sensor networks. In *Proceedings of International Conference on Communications* (pp. 1-5). Cape Town, South Africa: IEEE. 10.1109/ICC.2010.5502104

Gezici, S. (2008). A survey on wireless position estimation. *Wireless Personal Communications*, *44*(3), 263–282. doi:10.100711277-007-9375-z

Godara, L. C. (1997). Application of antenna arrays to mobile communications, part II: Beam-forming and direction-of-arrival considerations. *Proceedings of the IEEE*, *85*(8), 1195–1245. doi:10.1109/5.622504

Godara, L. C. (2004). *Smart antennas*. Boca Raton, FL: CRC Press LLC. doi:10.1201/9780203496770

Gross, F. B. (2005). *Smart antennas for wireless communications*. New York: McGraw-Hill Companies, Inc.

Halder, S., & Ghosal, A. (2016). A survey on mobile anchor assisted localization techniques in wireless sensor networks. *Wireless Networks*, *22*(7), 2317–2336. doi:10.100711276-015-1101-2

Huang, L., Matsuura, K., Yamane, H., & Sezaki, K. (2005). Enhancing wireless location privacy using silent period. In *Proceedings of Wireless Communications and Networking Conference* (vol. 2, pp. 1187–1192). New Orleans, LA: IEEE.

Jeyanthi, N. (2016). Internet of things (IoT) as interconnection of threats (IoT). In F. Hu (Ed.), *Security and privacy in internet of things (IoTs): Models, algorithms, and implementations* (pp. 1–19). Boca Raton, FL: Taylor & Francis Group, LLC. doi:10.1201/b19516-3

Jiang, T., Wang, H. J., & Hu, Y.-C. (2007). Preserving location privacy in wireless LANs. In *Proceedings of the 5th International Conference on Mobile Systems, Applications and Services* (pp. 246-257). San Juan, Puerto Rico: ACM.

Kang, L. (2009). Protecting location privacy in large-scale wireless sensor networks. In *Proceedings of the International Conference on Communications* (pp. 603-608). Dresden, Germany: IEEE. 10.1109/ICC.2009.5199372

Li, Y., & Ren, J. (2010). Source-location privacy through dynamic routing in wireless sensor networks. In *Proceedings of 29th IEEE Conference on Computer Communications* (pp. 1-9). San Diego, CA: IEEE. 10.1109/INFCOM.2010.5462096

Liberti, J. C., & Rappaport, T. S. (1999). *Smart antennas for wireless communications: IS-95 and third generation CDMA applications*. Englewood Cliffs, NJ: Prentice-Hall.

Lin, C.-F., Chi, K.-H., Hsu, Y.-Y., & Liu, C.-Y. (2017). Mobile anchor-assisted localization over android. *Wireless Networks*, *23*(8), 2379–2394. doi:10.100711276-016-1295-y

Mehta, K., Liu, D., & Wright, M. (2012). Protecting location privacy in sensor networks against a global eavesdropper. *IEEE Transactions on Mobile Computing*, *11*(2), 320–336. doi:10.1109/TMC.2011.32

Meyerowitz, J., & Roy Choudhury, R. (2009). Hiding stars with fireworks: location privacy through camouflage. In *Proceedings of the 15th Annual International Conference on Mobile Computing and Networking* (pp. 345–356). Beijing, China: ACM. 10.1145/1614320.1614358

Misra, S., Maheswaran, M., & Hashmi, S. (2017). *Security challenges and approaches in internet of things*. Springer International Publishing. doi:10.1007/978-3-319-44230-3

Myles, G., Friday, A., & Davies, N. (2003). Preserving privacy in environments with location-based applications. *IEEE Pervasive Computing*, *2*(1), 56–64. doi:10.1109/MPRV.2003.1186726

Naraghi-Pour, M., & Rojas, G. C. (2014). A novel algorithm for distributed localization in wireless sensor networks. *ACM Transactions on Sensor Networks*, *11*(1), 1–25. doi:10.1145/2632150

Oh, S., & Gruteser, M. (2011). Multi-node coordinated jamming for location privacy protection. In *Proceedings of the Military Communications Conference* (pp.1243-1249). Baltimore, MD: IEEE.

Ou, C.-H. (2011). A localization scheme for wireless sensor networks using mobile anchors with directional antennas. *IEEE Sensors Journal*, *11*(7), 1607–1616. doi:10.1109/JSEN.2010.2102748

Part 11: Wireless LAN medium access control (MAC) and physical layer (PHY) specifications. (n.d.). Retrieved on August 14, 2018, from http://www.di-srv.unisa.it/~vitsca/RC-0809I/IEEE-802-11.pdf

Patwari, N., Ash, J. N., Kyperountas, S., Hero, A. O., Moses, R. L., & Correal, N. S. (2005). Locating the nodes: Cooperative localization in wireless sensor networks. *IEEE Signal Processing Magazine*, *22*(4), 54–69. doi:10.1109/MSP.2005.1458287

Pazzi, R. W. N., & Boukerche, A. (2008). Mobile data collector strategy for delay-sensitive applications over wireless sensor network. *Computer Communications*, *31*(5), 1028–1039. doi:10.1016/j.comcom.2007.12.024

Rawat, D. B., White, T., Parwez, M. S., Bajracharya, C., & Song, M. (2017). Evaluating secrecy outage of physical layer security in large-scale MIMO wireless communications for cyber-physical systems. *IEEE Internet of Things Journal*, *4*(6), 1987–1993. doi:10.1109/JIOT.2017.2691352

Rios, R., & Lopez, J. (2011). Analysis of location privacy solutions in wireless sensor networks. *IET Communications*, *5*(17), 2518–2532. doi:10.1049/iet-com.2010.0825

Robinson, J., & Rahmat-Samii, Y. (2004). Particle swarm optimization in electromagnetics. *IEEE Transactions on Antennas and Propagation*, *52*(2), 397–402. doi:10.1109/TAP.2004.823969

Roy, R., & Kailath, T. (1989). Esprit - estimation of signal parameters via rotational invariance techniques. *IEEE Transactions on Acoustics, Speech, and Signal Processing*, *37*(7), 984–995. doi:10.1109/29.32276

Shiu, Y.-S., Chang, S. Y., Wu, H.-C., Huang, S. C.-H., & Chen, H.-H. (2011). Physical layer security in wireless networks: A tutorial. *IEEE Wireless Communications*, *18*(2), 66–74. doi:10.1109/MWC.2011.5751298

Soosahabi, R., & Naraghi-Pour, M. (2012). Scalable PHY-Layer security for distributed detection in wireless sensor networks. *IEEE Transactions on Information Forensics and Security*, *7*(4), 1118–1126. doi:10.1109/TIFS.2012.2194704

Srinivasan, A., & Wu, J. (2008). *A survey on secure localization in wireless sensor networks. In Encyclopedia of Wireless and Mobile Communications*. Taylor and Francis Group.

Ssu, K.-F., Ou, C.-H., & Jiau, H. C. (2005). Localization with mobile anchor points in wireless sensor networks. *IEEE Transactions on Vehicular Technology, 54*(3), 1187–1197. doi:10.1109/TVT.2005.844642

Sweeney, L. (2002). k- Anonymity: A model for protecting privacy. *International Journal of Uncertainty, Fuzziness and Knowledge-based Systems, 10*(5), 557–570. doi:10.1142/S0218488502001648

Wang, N., Li, W., Jiang, T., & Lv, S. (2017). Physical layer spoofing detection based on sparse signal processing and fuzzy recognition. *IET Signal Processing, 11*(5), 640–646. doi:10.1049/iet-spr.2016.0378

Wang, T., & Yang, Y. (2011). Location privacy protection from RSS localization system using antenna pattern synthesis. In *Proceedings of International Conference on Computer Communications* (pp. 2408-2416). Shanghai, China: IEEE. 10.1109/INFCOM.2011.5935061

Winters, J. H. (2006). Smart antenna techniques and their application to wireless ad hoc networks. *IEEE Wireless Communications, 13*(4), 77–83. doi:10.1109/MWC.2006.1678168

Wong, F. L., Lin, M., Nagaraja, S., Wassell, I., & Stajano, F. (2007). Evaluation framework of location privacy of wireless mobile systems with arbitrary beam pattern. In *Proceedings of Fifth Annual Conference on Communication Networks and Services Research* (vol. 2, pp. 157-165). Fredericton, Canada: IEEE. 10.1109/CNSR.2007.30

Yao, Y., Zou, K., Chen, X., & Xu, X. (2016). A distributed range-free correction vector based localization refinement algorithm. *Wireless Networks, 22*(8), 2667–2680. doi:10.100711276-015-1129-3

Yilmaz, M. H., & Arslan, H. (2015). A survey: Spoofing attacks in physical layer security. In *Proceedings of 40th Annual IEEE Conference on Local Computer Networks* (pp. 812-817). Clearwater Beach, FL: IEEE. 10.1109/LCNW.2015.7365932

Zhou, X., Song, L., & Zhang, Y. (Eds.). (2014). *Physical layer security in wireless communications.* Boca Raton, FL: Taylor & Francis Group, LLC.

Zhu, J. (2016). *Physical layer security in massive MIMO systems* (Unpublished doctoral dissertation). The University of British Columbia, Vancouver, Canada.

Zou, Y., Zhu, J., Wang, X., & Hanzo, L. (2016). A survey on wireless security: Technical challenges, recent advances, and future trends. *Proceedings of the IEEE, 104*(9), 1727–1765. doi:10.1109/JPROC.2016.2558521

Zurita, L. N. R. (2014). *Optimising multiple antenna techniques for physical layer security* (Unpublished doctoral dissertation). The University of Leeds, Leeds, UK.

This research was previously published in Cryptographic Security Solutions for the Internet of Things; pages 211-243, copyright year 2019 by Information Science Reference (an imprint of IGI Global).

Chapter 7
A Framework for Various Attack Identification in MANET Using Multi-Granular Rough Set

N. Syed Siraj Ahmed
https://orcid.org/0000-0003-3943-604X
VIT University, Vellore, India

Debi Prasanna Acharjya
https://orcid.org/0000-0003-3828-2050
School of Computing Sciences and Engineering, VIT University, Vellore, India

ABSTRACT

The topology changes randomly and dynamically in a mobile adhoc network (MANET). The composite characteristics of MANETs makes it exposed to interior and exterior attacks. Avoidance support techniques like authentication and encryption are appropriate to prevent attacks in MANETs. Thus, an authoritative intrusion detection model is required to prevent from attacks. These attacks can be at either the layers present in the network or can be of a general attack. Many models have been developed for the detection of intrusion and detection. These models aim at any one of the layer present in the network. Therefore, effort has been made to consider either the layers for the detection of intrusion and detection. This article uses a multigranular rough set (MGRS) for the detection of intrusion and detection in MANET. The advantage of MGRS is that it can aim at either the layers present in the network simultaneously by using multiple equivalence relations on the universe. The proposed model is compared with many traditional models and attained higher accuracy.

1. INTRODUCTION

Increasing popularity of wireless equipment and devices has inclined to study self-configuring and self-healing networks with exclusion of centralized or pre-established environment. Such a network having no pre-established or centralized environment is called ad hoc network in which nodes employ multi hop

DOI: 10.4018/978-1-7998-8545-0.ch007

links to communicate information between each other. The communication between nodes is carried out through intermediate nodes. These intermediate nodes act as a router in a static way whereas nodes of mobile ad hoc network, operates independently and connected through wireless associations in a dynamic way. Nodes in a mobile ad hoc network create lived short term or temporary network in which it can move freely and randomly. At the same time nodes can act as a router participating in maintaining and determining the paths of other nodes. But, the wireless links between the nodes are deeply vulnerable to attack with a chance of losing associations promptly due to mobility. Thus, in mobile ad hoc network, nodes behave like self-managing routers connected through wireless associations. Each router in mobile ad hoc network is selfish, free to decision making, and move arbitrarily. Therefore, such type of network able to operate separately as well as associated with superior network such as the internet. Additionally, such kind of network is very simple to connect, configure, and quick to use abruptly such as ad hoc meeting, ad hoc classroom, etc. The rise in use of this network needs novel protection system and strategies to assure accessibility, reliability, and attack free environment for effective communication. To maintain attack free and user-friendly environment in mobile ad hoc network, novel detection and attack prevention system unlike the intrusion detection system of wired network is essential for further research.

Over the preceding few years various attacks in mobile ad hoc network discussed are denial of service, jamming, sybil, flooding, black hole, worm hole, jelly fish, link spoofing, malicious message injecting, eaves dropping, etc. (Shrivastava & Jain, 2013; Wazid, Singh & Goudar, 2011; Ahmed & Acharjya, 2015a, 2015b). A scheme to control fake flooding route requests in ad hoc network is also discussed (Kataria, Dhekne &. Sanyal, 2006). Discovering of anomalous such as black hole, resource consumption, and packet dropping in mobile ad hoc network was addressed by Fattah, Dahalin & Jusoh (2010) using nearest neighbour and distance outlier. Similarly, identification of denial of service attack in MANET using principal component analysis with features reduction was also proposed. This model uses dynamic source routing protocol to recognize denial of service attacks on different wireless network traffics (Kabiri & Aghaei, 2011).

Later, Chouhan & Yadav (2011) discussed about ad hoc flooding attack detection and prevention in MANET by using ad hoc on demand distance vector (AODV) protocol and dynamic source routing (DSR). The major limitation of this model is that it takes extra bit of time for the detection of misbehaving nodes. Similarly, Pastrana, Mitrokotsa, Orfila and Lopez (2012) portrayed an assessment to detect intrusion in MANET by using various classifiers such as support vector machine, genetic programming, multi-layer perceptron, linear classifier, naive Bayes, and Gaussian mixture model. The efficiency assessment identified that genetic programming model was more suitable to detect intruder, whereas support vector machine classifier model was suitable for identification of distinct attacks such as packet dropping, flooding, black hole and forging. Mitrokotsa, Tsagkaris and Douligeris (2008) projected supervised learning classification algorithms such as support vector machine, multi- layer perceptron, Gaussian mixture model, and naive Bayes classification for detection of various attacks like packet dropping, black hole, flooding, and forging. It is observed that classification using support vector machine generates maximum accuracy (92%) for all attacks.

Garg & Sharma (2014) described about sybil attack identification in MANET. In this attack, an adversary uses different identities to take away the identity of various reliable nodes within the network. This attack is extremely destructive particularly for MANET where the network is infrastructure less and all the nodes are reliable to each other. To overcome this attack, they used three constraint rates such as frequency, speed, and energy of each node enters into the network. The node is treated as valid if their constraint rates are less than the predefined threshold values. Further, a technique to recognize jamming attack in MANET using clear to send (CTS) and ready to send (RTS) signals was also discussed in the literature. The major drawback of the proposed method is that it requires more time to detect jamming (Sharma & Singla, 2015). Likewise, a model to detect packet dropping attack in MANET using soft computing techniques such as neuro fuzzy classifier was also discussed (Chaudhary, Tiwari & Kumar, 2015). Similarly, rough set is also used for various attack identification in computer network. Detection of denial of service attack and jamming attack in wireless network using dominance based rough set is carried out (Ahmed & Acharjya, 2015a, 2015b). Further for phishing attack identification a hybridized model using rough set and formal concept analysis is developed (Ahmed, Acharjya & Sanyal, 2017). Rough set has been extended to fuzzy rough set (Dubois & Prade, 1990), variable precession rough set (Ziarko, 1993), rough set on fuzzy approximation space (Acharjya, 2013), rough set on two universal sets (Liu, 2010; Geetha, Acharjya, & Iyengar, 2014) to name a few. However, these theories are extensively applied to various real-life problems (Das, Acharjya, & Patra, 2015; Acharjya, & Bhattacharjee, 2013; Acharjya, 2013). It is observed that these theories are not applied to various attack identification in computer network.

Additionally, it is observed that jamming attack generally occurs at physical layer or data link layer; jellyfish, worm hole, black hole attacks occur at network layer; repudiation attack occurs at application layer, etc. It is clear that various attacks occur at different layers. Additionally, some attacks may occur at different layers. Considering all layers at a time and analyzing attacks in MANET requires multi-granulation rather than single granulation. Also, the features or attributes that are associated to identify these attacks contain uncertainties. Therefore, an effort has been made in this paper to consider both uncertainties and multi-granulation while detection of various attacks in MANET. The mathematical tool multi-granular rough set (Qian & Liang, 2006) that handles both uncertainties and multi-granulation is taken into consideration for identifying and recognizing various attacks in MANET. The prime objective of introducing multi-granular rough set is that, it generates 'OR' rules and it helps to detect an attack which may occur at different layers of a computer network. The proposed approach is validated through an empirical study and compared with existing techniques. The comparative study and validation show the viability of the proposed approach.

The organization of the paper is as follows. In order to unfold the article, foundations of rough set are discussed in section 2. Section 3 briefly discusses fundamentals of multi-granular rough set followed by various measures of multigranular rough set in section 4. Proposed model to identify the distinct categories of attacks in MANET using multigranular rough set is discussed in section 5. Section 6 presents an empirical study. Additionally, we present a comparative study in section 5. The article is concluded in section 7.

2. FOUNDATIONS OF ROUGH SET

Crisp rough set was introduced by Pawlak (1982) to handle uncertainties with the help of an indiscernibility relation. The indiscernibility relation is an equivalence relation employed over an information system. An information system is defined as $K = (U, A)$, where $U= \{x_1,x_2,\ldots,x_n\}$ is the finite set of objects, and $A= \{a_1,a_2,\ldots,a_m\}$ be the finite non-empty set of attributes. Let $P \subseteq A$ such that IND(P) is an equivalence relation defined over U. The equivalence relation IND(P) is defined as:

IND(P) = $\{(x_i, x_j) \in (U \times U) \mid a(x_i) = a(x_j)\ \forall\ a \in P\}$

The equivalence relation leads to equivalence classes $U/$ IND(P) or U/P. Given a target set $X \subseteq U$, the lower approximation of X, $(\underline{P}X)$, and the upper approximation of X, $(\overline{P}X)$, are defined in Equations (1) and (2):

$$\underline{P}X = \cup\{Y \in U / P\colon Y \subseteq X\} \tag{1}$$

$$\overline{P}X = \cup\{Y \in U / P\colon (Y \cap X) \neq \phi\} \tag{2}$$

If $\underline{P}X = \overline{P}X$, then the target set X is known as the crisp set. Equivalently, the target set X is said to be a rough set if $\underline{P}X \neq \overline{P}X$. From the lower and upper approximations, it is clear that the classification takes all the attributes of P and generates the class and hence in rule mining it generates 'and' rules.

3. FUNDAMENTALS OF MULTIGRANULAR ROUGH SET

Granulation can be considered as an equivalence relation defined over a universe of objects and the granulation space can be considered as a partition over the universe. Many real-life applications spread over more than one equivalence relation and the concept of multi-granular rough set was introduced to handle such problems (Qian, Liang, Yao, & Dang, 2009). The data analysis of multi-granular rough set is completely distinct from Pawlak's rough set model (Pawlak, 1982) since it is created on behalf of multi indiscernible relations rather than single indiscernible relation. The multi-granular rough set (MGRS) can be defined over a knowledge base $K = (U, \Re)$, where $\Re$ is a family of equivalence relations over U.

Let us consider a relational system $K = (U, \Re)$ be a knowledge base. Let $X \subseteq U$ and P,Q in $\Re$ generates two different partitions of the universe U. The lower and upper approximation of X on U based on two equivalence relations are defined as below, where $[x]_p$ is the partition generated by the equivalence relation P and X^c represents the complement of X:

$$\underline{P+Q}X = \cup\{\mathrm{x} \in \mathrm{U} : [\mathrm{x}]_P \subseteq X\, or\, [\mathrm{x}]_Q \subseteq X\} \tag{3}$$

$$\overline{P+Q}X = (\underline{P+Q}(X^c))^c \tag{4}$$

The boundary line objects $BN_{P+Q}(X)$ is defined as $BN_{P+Q}(X) = \overline{P+Q}X - \underline{P+Q}X$. The objects of the target set X is said to be (P+Q) certain if $BN_{P+Q}(X) = \phi$ and (P+Q) rough if $BN_{P+Q}(X) \neq \phi$.

A diagrammatic representation of Pawlak's single granulation with respect to relations P and Q is presented in Figure 1 and Figure 2 respectively. The multi-granulation is presented in Figure 3. From both the figures it is clear that multi-granular rough set provides better knowledge than single granulation and covers more element of the target set. In addition, more number of elements of the target set can be classified under multi-granulation. Hence, it is a better model as com-pared to single granulation.

Figure 1. Pawlak's single granulation with respect to relation P

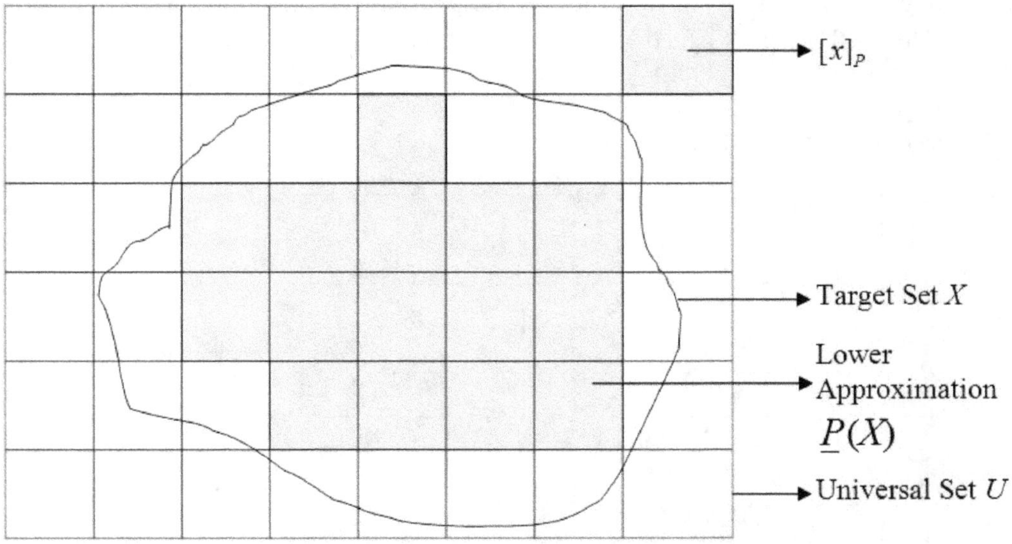

Figure 2. Pawlak's single granulation with respect to relation Q

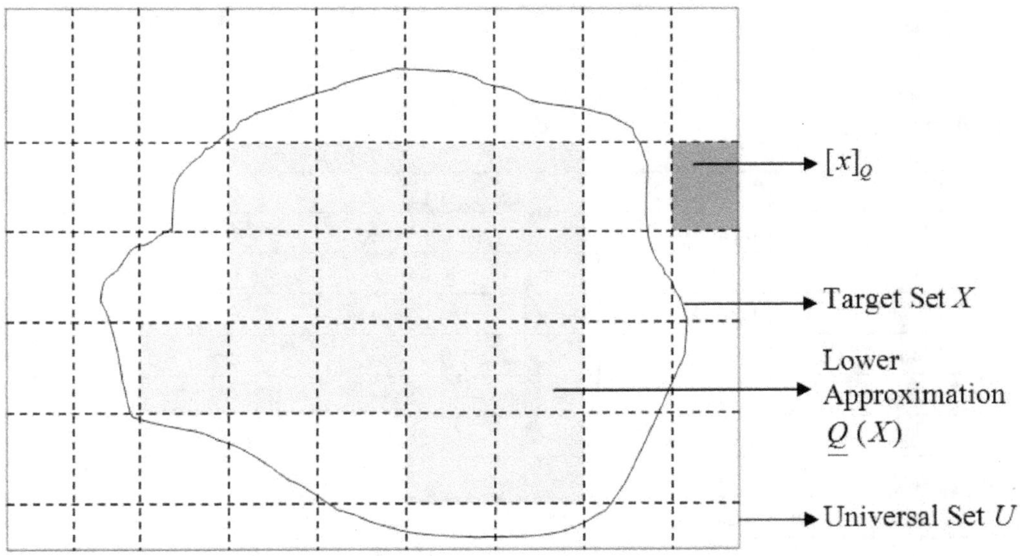

Example 1: Let us consider an example to explain the notions and computations involved in multi-granular rough set. Table 1 gives a view of an information system of attacks in mobile ad hoc network. Where U is a complete set of objects The set of objects U is characterised by a set of relations $\Re = \{P_1, P_2, P_3, P_4, P_5\}$, where the equivalence relation P_1 represents receiving rate of packets (Mbps), P_2 represents sending rate of packets (Mbps), P_3 represents packets dropping rate (Mbps), P_4 represents forwarding delay in time and mobility, and P_5 represents the number of hops. A decision is used to represent different types of attacks or benign is denoted as D. The domain of

the relations P_1, P_2 and P_3 = {Low, Moderate (Mod), High}, whereas for P_4 = {Short, Medium, (Med), Long}. The decision can be either benign, jelly fish or black hole. Thus, the domain of D = {Black Hole (BH), Jelly Fish (JF), Benign (BN)}.

Figure 3. A view of multigranular rough set

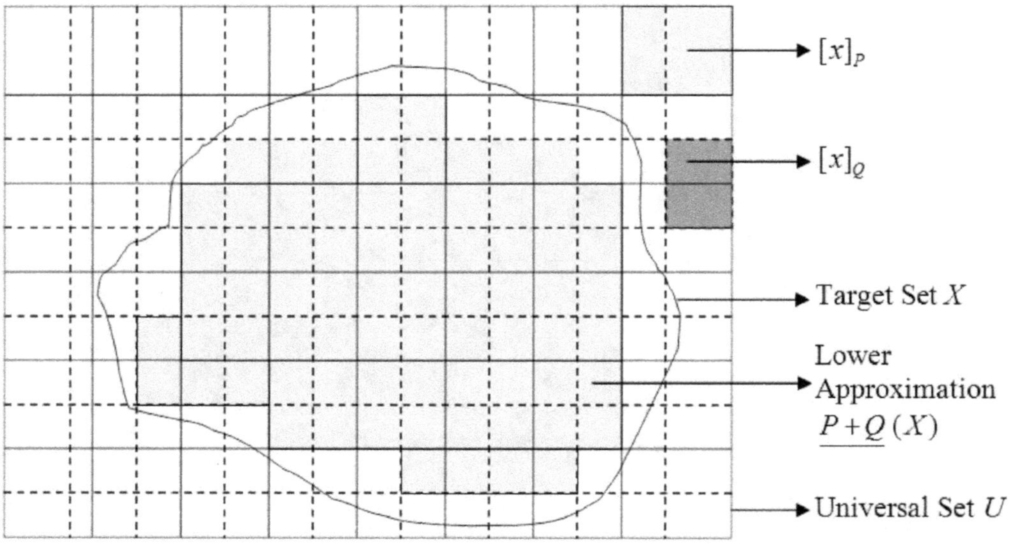

Table 1. An information system of intrusion detection

U	P_1	P_2	P_3	P_4	P_5	D
x_1	Low	Mod	Low	Long	4	BN
x_2	High	Low	High	Short	8	BH
x_3	High	Low	High	Short	5	JF
x_4	Mod	Mod	Low	Short	6	BN
x_5	Mod	Mod	Low	Short	6	BH
x_6	High	Low	Low	Short	6	BN
x_7	High	Mod	Low	Long	8	JF
x_8	Mod	High	Mod	Short	6	BH
x_9	Low	Low	Low	Med	6	JF

Let us consider a target set $X = \{x_1, x_4, x_6\}$. Our objective is to show how multi-granulation is different from single granulation. In order to show the difference, we compute the lower and upper approximation of multi-granular rough set as follows:

$$U / P_1 = \{\{x_4, x_5, x_8\}, \{x_2, x_3, x_6, x_7\}, \{x_1, x_9\}\}$$

$$U / P_2 = \{\{x_1, x_4, x_5, x_7\}, \{x_2, x_3, x_6, x_9\}, \{x_8\}\}$$

$$U / P_3 = \{\{x_1, x_4, x_5, x_6, x_7, x_9\}, \{x_2, x_3\}, \{x_8\}\}$$

$$U / P_4 = \{\{x_1, x_7\}, \{x_2, x_3, x_4, x_5, x_6, x_8\}, \{x_9\}\}$$

$$U / P_5 = \{\{x_1\}, \{x_2, x_7\}, \{x_3\}, \{x_4, x_5, x_6, x_8, x_9\}\}$$

$$U / \Re = \{\{x_1\}, \{x_2\}, \{x_3\}, \{x_4, x_5\}, \{x_6\}, \{x_7\}, \{x_8\}, \{x_9\}\}$$

Therefore, we get $\underline{\sum P_i}X = \{x_1\}$:

$$\overline{\sum P_i}X = (\underline{\sum P_i}X^c)^c = \{x_1, x_2, x_4, x_5, x_6, x_7\}$$

where i= 1,2,3,4 and 5.

The lower and upper approximation in accordance with Pawlak's rough set is given as:

$$\underline{\Re}X = \{x_1, x_6\}$$

$$\overline{\Re}X = \{x_1, x_4, x_5, x_6\}$$

From the above analysis, it is clear that $\underline{\sum P_i}X \subseteq \underline{\Re}X$ and. $\overline{\sum P_i}X \supseteq \overline{\Re}X$. Therefore, multi-granular rough set provides better information about objects while considering lower approximation. Additionally, it includes more objects of the universe while considering upper approximation.

4. NUMEROUS MEASURES OF MULTIGRANULAR ROUGH SET

In the section, we define numerous measures of multigranular rough set. It includes accuracy measure, degree of dependency, simple statistic for the precision of approximation, and significance of equivalence relations in multigranular rough set.

4.1. Accuracy Measure of Multigranular Rough Set

In the previous section, we have analyzed a sample information system of intrusion detection. From the analysis it is clear that the existence of boundary line region of a target set X leads to the uncertainty. The presence of larger boundary line region leads to less accuracy and vice versa. We define the accuracy measure as follows.

Let $K = (U, \Re)$, where $\Re = \{P_1, P_2, \cdots, P_m\}$ is a family of equivalence relations over U. Let $X \subseteq U$ be the target set and $[x]_{P_1}$, $[x]_{P_2}$, $\cdots$, and $[x]_{P_m}$ be the partitions obtained by the family of equivalence relations $P_1, P_2, \ldots$, and P_m, respectively. The accuracy measure of the target set X is defined as $\alpha_{\Sigma P_i}(X)$, where $X^1\phi$ and |X| represents the cardinality of X:

$$\alpha_{\Sigma P_i}(X) = \frac{\left|\underline{\Sigma P_i} X\right|}{\left|\overline{\Sigma P_i} X\right|} \tag{5}$$

It captures the degree of completeness of our knowledge about the target set. It is clear that the value of accuracy lies between 0 and 1. If $\alpha_{\Sigma P_i}(X) = 1$, then there will be no boundary line region and all the objects of the target set are certainly defined. If $\alpha_{\Sigma P_i}(X) = 0$, then the boundary line region is non empty and none of the objects of the target set are certainly defined.

4.2. Degree of Dependency

Let $K = (U, \Re)$, where $\Re = \{P_1, P_2, \cdots, P_m\}$ is a family of equivalence relations over U. Let $X \subseteq U$ be the target set and $[x]_{P_1}$, $[x]_{P_2}$, $\cdots$, and $[x]_{P_m}$ be the partitions obtained by the family of equivalence relationsP$_1$, $P_2, \ldots$, and P_m respectively. Let $Q \subseteq \Re$ and the partition induced by Q be $[x]_Q$. The degree of dependency of Q by $\Re$ is used to determine the deterministic part of the rough set description of Q. It counts those objects that can be reclassified to the blocks of $[x]_Q$ with the given knowledge ΣP_i. The degree of dependency of Q is defined as $\beta_{\Sigma P_i}(Q)$, where |Y| represents the cardinality of Y:

$$\beta_{\Sigma P_i}(Q) = \frac{\sum \left|\underline{\Sigma P_i}(Y) : Y \in [x]_Q\right|}{|U|} \tag{6}$$

It is clear that, the value of $\beta_{\Sigma P_i}(Q)$ lies in between 0 and 1. If $\beta_{\Sigma P_i}(Q) = 0$, then Q is totally independent of $\Re$. If $\beta_{\Sigma P_i}(Q) = 1$, then Q is totally dependent of $\Re$. In this case all objects of the universe can be classified to elementary categories of $[x]_Q$ by using knowledge of $\Re$. Similarly, Q is partially dependent on $\Re$ if $0 < \beta_{\Sigma P_i}(Q) < 1$.

4.3. Simple Statistic for the Precision of Approximation

Simple statistic for the precision of approximation was introduced by Gediga & Duntsch (2001). It is the relative number of objects in the target set X that can be approximated by $[x]_{P_i}$ for $i = 1,2,\ldots,n$. Let $K = (U, \Re)$, where $\Re = \{P_1, P_2, \cdots, P_m\}$ is a family of equivalence relations over U. Let $X \subseteq U$ be the target set and $[x]_{P_1}$, $[x]_{P_2}$, $\cdots$, and $[x]_{P_m}$ be the partitions obtained by the family of equivalence relations $P_1, P_2, \ldots$, and P_m respectively. The simple statistic for the precision of approximation, $\pi_{\Sigma P_i}(X)$, in the context of multigranular rough set can be defined as follows:

$$\pi_{\Sigma P_i}(X) = \frac{|\underline{\Sigma P_i}(X)|}{|X|} \tag{7}$$

4.4. Significance of Equivalence Relations in Multigranular Rough Set

Let $K = (U, \Re, D)$ be a complete decision system, where $\Re = \{P_1, P_2, \cdots, P_m\}$ is a family of equivalence relations over *U*, and *D* is the decision. Let $X \subseteq U$ be the target set. We say that an equivalence relation $P_i \in \Re$ is lower approximation significant in $\Re$ with respect to the target set *X* if $\underline{\Sigma \Re}(X) \supset \underline{\Sigma(\Re - P_i)}(X)$. Similarly, an equivalence relation $P_i \in \Re$ is upper approximation significant in $\Re$ with respect to the target set *X* if $\overline{\Sigma \Re}(X) \subset \overline{\Sigma(\Re - P_i)}(X)$. To measure the significance of equivalence relations, we introduce importance measure due to lower approximation and upper approximation with respect to the decision *D*. Let $[x]_D$ be the partition induced by the decision attribute *D*.

Let $\wp \subseteq \Re$ be a subset of equivalence relations defined over the decision system. The importance measure of the lower approximation of $\wp$ with respect to the decision *D* in multigranular rough set is defined as $S_\wp(D)$, where:

$$S_\wp(D) = \frac{\Sigma\,|\underline{\Sigma \Re}(X) : \underline{\Sigma(\Re - \wp)}(X)|}{|U|}\text{; for } X\hat{I}[x]_D \tag{8}$$

Likewise, the importance measure of the upper approximation of $\wp$ with respect to the decision *D* in multigranular rough set is defined as $S^\wp(D)$, where:

$$S^\wp(D) = \frac{\Sigma\,|\overline{\Sigma(\Re - \wp)}(X) : \overline{\Sigma \Re}(X)|}{|U|}\text{; for } X\hat{I}[x]_D \tag{9}$$

From the above definition it is clear that equivalence relations $\wp$ with respect to the decision *D* is not lower approximation significant if and only if $S_\wp(D) = 0$. Similarly, the equivalence relations $\wp$ with respect to the decision *D* is not upper approximation significant if and only if $S^\wp(D) = 0$.

Example 2: In this example, we show how the quantitative measure is being computed. We continue the discussion from Example 1. The decision *D* induces the decision classes as $\{x_1, x_4, x_6\}$, $\{x_2, x_5, x_8\}$, and $\{x_3, x_7, x_9\}$. The lower approximation significance of each equivalence relation with respect to the decision *D* is given below:

$$S_{P_1}(D) = \frac{\left|\{x_1, x_3\} : \{x_1, x_3\}\right| + \left|\{x_8\} : \{x_8\}\right| + \left|\{x_9\} : \{x_9\}\right|}{9} = 0$$

$$S_{P_2}(D) = \frac{\left|\{x_1, x_3\} : \{x_1, x_3\}\right| + \left|\{x_8\} : \{x_8\}\right| + \left|\{x_9\} : \{x_9\}\right|}{9} = 0$$

$$S_{P_3}(D) = \frac{\left|\{x_1, x_3\} : \{x_1, x_3\}\right| + \left|\{x_8\} : \{x_8\}\right| + \left|\{x_9\} : \{x_9\}\right|}{9} = 0$$

$$S_{P_4}(D) = \frac{\left|\{x_1, x_3\} : \{x_1, x_3\}\right| + \left|\{x_8\} : \{x_8\}\right| + \left|\{x_9\} : \{\phi\}\right|}{9} = \frac{1}{9}$$

$$S_{P_5}(D) = \frac{\left|\{x_1, x_3\} : \{\phi\}\right| + \left|\{x_8\} : \{x_8\}\right| + \left|\{x_9\} : \{x_9\}\right|}{9} = \frac{2}{9}$$

Similarly, the upper approximation significance of each equivalence relation with respect to the decision D is given below:

$$S^{P_1}(D) = \frac{\begin{array}{l}\left|\{x_1, x_2, x_4, x_5, x_6, x_7\} : \{x_1, x_2, x_4, x_5, x_6, x_7\}\right| + \\ \left|\{x_2, x_4, x_5, x_6, x_7, x_8\} : \{x_2, x_4, x_5, x_6, x_7, x_8\}\right| + \\ \left|\{x_3, x_4, x_5, x_6, x_7, x_9\} : \{x_3, x_4, x_5, x_6, x_7, x_9\}\right|\end{array}}{9} = 0$$

$$S^{P_2}(D) = \frac{\begin{array}{l}\left|\{x_1, x_2, x_4, x_5, x_6, x_7\} : \{x_1, x_2, x_4, x_5, x_6, x_7\}\right| + \\ \left|\{x_2, x_4, x_5, x_6, x_7, x_8\} : \{x_2, x_4, x_5, x_6, x_7, x_8\}\right| + \\ \left|\{x_3, x_4, x_5, x_6, x_7, x_9\} : \{x_3, x_4, x_5, x_6, x_7, x_9\}\right|\end{array}}{9} = 0$$

$$S^{P_3}(D) = \frac{\begin{array}{l}\left|\{x_1, x_2, x_4, x_5, x_6, x_7\} : \{x_1, x_2, x_4, x_5, x_6, x_7\}\right| + \\ \left|\{x_2, x_4, x_5, x_6, x_7, x_8\} : \{x_2, x_4, x_5, x_6, x_7, x_8\}\right| + \\ \left|\{x_3, x_4, x_5, x_6, x_7, x_9\} : \{x_3, x_4, x_5, x_6, x_7, x_9\}\right|\end{array}}{9} = 0$$

$$S^{P_4}(D) = \frac{\begin{array}{l}\left|\{x_1, x_2, x_4, x_5, x_6, x_7, x_9\} : \{x_1, x_2, x_4, x_5, x_6, x_7\}\right| + \\ \left|\{x_2, x_4, x_5, x_6, x_7, x_8, x_9\} : \{x_2, x_4, x_5, x_6, x_7, x_8\}\right| + \\ \left|\{x_2, x_3, x_4, x_5, x_6, x_7, x_9\} : \{x_2, x_3, x_4, x_5, x_6, x_7, x_9\}\right|\end{array}}{9} = \frac{2}{9}$$

$$S^{P_5}(D) = \frac{\begin{array}{l}\left|\{x_1, x_2, x_3, x_4, x_5, x_6, x_7\} : \{x_1, x_2, x_4, x_5, x_6, x_7\}\right| + \\ \left|\{x_1, x_2, x_3, x_4, x_5, x_6, x_7, x_8\} : \{x_2, x_4, x_5, x_6, x_7, x_8\}\right| + \\ \left|\{x_1, x_2, x_3, x_4, x_5, x_6, x_7, x_9\} : \{x_2, x_3, x_4, x_5, x_6, x_7, x_9\}\right|\end{array}}{9} = \frac{4}{9}$$

From the above analysis it is clear that:

$$S_{P_1}(D) = S_{P_2}(D) = S_{P_3}(D) = 0 = S^{P_1}(D) = S^{P_2}(D) = S^{P_3}(D)$$

and hence the equivalence relations P_1, P_2, and P_3 are neither lower approximation significant nor upper approximation significant. It can also be concluded that the equivalence relations for which $S_{\wp}(D)$ is maximum is most significant. Similar argument also holds for upper approximation measure $S^{\wp}(D)$.

5. PROPOSED MODEL TO IDENTIFY NUMEROUS ATTACKS OF MANET

A well-known distinctive category of attacks at MANETs are jamming, jelly fish, denial of service etc. These attacks can be found at several layers of TCP/IP. Therefore, identifying and recognizing such categories of attacks are of massive challenge. To overcome this, in this section, we propose a research model for identifying numerous attacks in MANET that occurs at various layers. A conceptual outlook of the proposed research model is depicted in Figure 4. To recognize the problem description, we require basic knowledge of the problem. Initially the information collected is pre-processed. The key notion is to translate the natural input information into appropriate design for succeeding analysis. The various steps that are carried out are integration of information from information repositories; cleaning of noise and duplicate observations; and then picking associated observations as per the requirement of the problem undertaken. The observations are picked in support of the analysis of the decision. Before we divide the data into training and testing we use multigranular rough set data reduction algorithm as discussed below. This generally removes the equivalence relations that are having no influences on the decisions. Further, the processed information system is divided into two categories such as training data (55%) and testing data (45%). The training information is analyzed through multi-granular rough set to detect the decision class that originates the decision. We use multigranular rough set based decision algorithm to obtain decision rules.

5.1. MGRS Data Reduction Algorithm

Let $K = (U, \Re, D)$ be a complete decision system, where $\Re = \{P_1, P_2, \cdots, P_m\}$ is a family of equivalence relations over U, and D is the decision. Let $\wp \subseteq \Re$ be a subset of equivalence relations defined over the decision system. Let $[x]_D = \{X_1, X_2, \ldots, X_r\}$ be the partition induced by the decision attribute D. Therefore, the lower approximation $\left(\underline{\wp} D\right)$ and upper approximation $\left(\overline{\wp} D\right)$ due to multigranular rough set is defined as below:

$$\underline{\wp} D = \left(\underline{\Sigma P_i} X_1, \underline{\Sigma P_i} X_2, \underline{\Sigma P_i} X_3, \cdots, \underline{\Sigma P_i} X_r\right) \text{ where } P_i \in \wp \tag{10}$$

and:

$$\overline{\wp} D = \left(\overline{\Sigma P_i} X_1, \overline{\Sigma P_i} X_2, \overline{\Sigma P_i} X_3, \cdots, \overline{\Sigma P_i} X_r\right) \text{ where } P_i \in \wp \tag{11}$$

Figure 4. A conceptual outlook of the proposed model

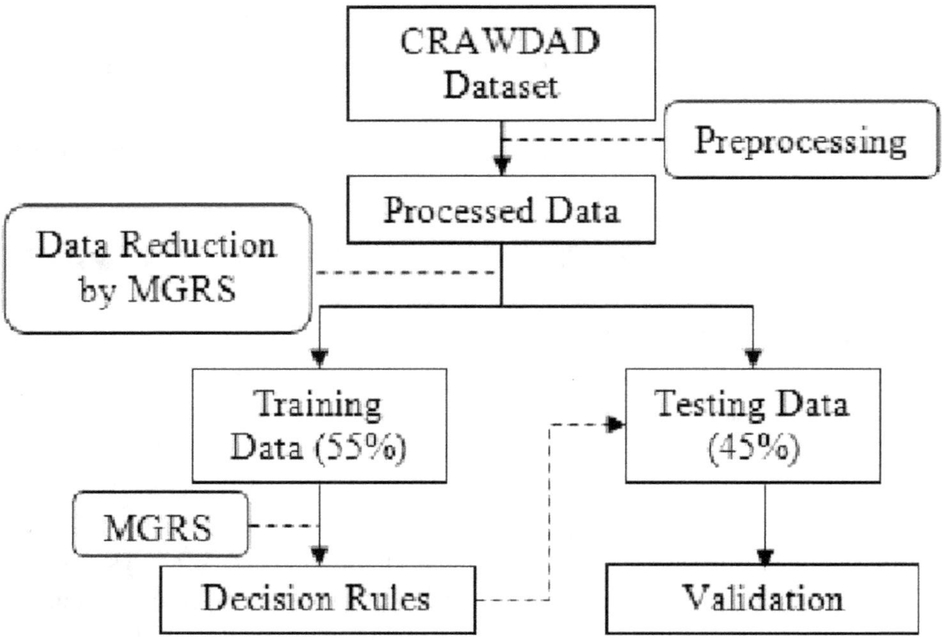

Instantly, $\wp \subseteq \Re$ is said to be a lower approximation consistent set of *K*, if $\underline{\wp}D = \underline{\Re}D$. We exclaim that, $\wp \subseteq \Re$ is said to be a lower approximation reduct of *K*, if $\wp$ is a lower approximation consistent set and no proper subset of $\wp$ is a lower approximation consistent set. Similarly, the argument also holds for upper approximation. Immediately, $\wp \subseteq \Re$ is said to be an upper approximation consistent set of *K*, if $\overline{\wp}D = \overline{\Re}D$. We exclaim that, $\wp \subseteq \Re$ is said to be an upper approximation reduct of *K*, if $\wp$ is an upper approximation consistent set and no proper subset of $\wp$ is an upper approximation consistent set.

We define the lower approximation reduct algorithm as below. The algorithm searches the subsets of $\Re$ starting from singleton set. We define the subsets of $\Re$ with the help of binomial coefficient, ${}^{|\Re|}C_z$, as below:

$${}^{|\Re|}C_z = \frac{|\Re|!}{z!\left(|\Re| - z\right)!}$$

Therefore, we define with the help of binomial coefficient, the *z*- equivalence relation subset, $\wp_z$, of $\Re$ as follows:

1. The 1- equivalence relation subsets, $\wp_1$ of $\Re$ as:

$$\wp_{11} = \{P_1\}, \wp_{12} = \{P_2\}, \wp_{13} = \{P_3\}, \cdots, \wp_{1\,{}^{|\Re|}C_1} = \{P_m\} \text{ where } |\Re| = m$$

2. The 2- equivalence relation subsets, $\wp_2$ of $\Re$ as:

$$\wp_{21} = \{P_1, P_2\}, \wp_{22} = \{P_1, P_3\}, \cdots, \wp_{2j} = \{P_1, P_{j+1}\}, \cdots, \wp_{2|\Re|C_2} = \{P_{m-1}, P_m\}$$

3. The 3- equivalence relation subsets, $\wp_3$ of $\Re$ as:

$$\wp_{31} = \{P_1, P_2, P_3\}, \wp_{32} = \{P_1, P_2, P_4\}, \cdots, \wp_{3j} = \{P_1, P_2, P_{j+2}\}, \cdots, \wp_{3|\Re|C_3} = \{P_{m-2}, P_{m-1}, P_m\}$$

4. The z- equivalence relation subsets, $\wp_z$ of $\Re$ as:

$$\wp_{z1} = \{P_1, P_2, P_3, \cdots, P_z\}, \cdots, \wp_{z|\Re|C_z} = \{P_{m-z+1}, \cdots, P_{m-1}, P_m\}$$

5. The *m*- equivalence relation subsets, $\wp_m$ of $\Re$ as $\wp_{m|\Re|C_m} = \{P_1, P_2, \cdots, P_{m-1}, P_m\}$.

Now, we define the various notions and symbols used in finding lower approximation reduct algorithm are stated below:

- *w*: Number of lower approximation reducts; $\wp_0 = \{\wp_{01}, \wp_{02}, \cdots, \wp_{0w}\}$;
- $\wp_{zy}$: *z*-equivalence relation subset of $\wp \subseteq \Re$, where *z* varies from 1 to $|\Re| = m$;
- *y*: Indexing variable varies from 1 to ${}^{|\Re|}C_z$ for each value of *z*.

Algorithm 1. Lower Approximation Reduct
Input: Decision system
Output: Minimal reduct $\wp_0 = \{\wp_{01}, \wp_{02}, \cdots, \wp_{0w}\}$

```
Set  y = 1,  z = 1,  w = 1,  ℘01 = ℘02 = ··· = ℘0w = ϕ
while  z ≤ |ℜ|  do
   y = 1
   while  y ≤ |ℜ|C_z  do
      for  t = 1  to  w  do
         if  ℘0t ≠ ϕ  then
            if ℘0t ⊂ ℘zy then
              exit
            Endif
         Endif
      Endfor
      if (ΣP_i D; P_i ∈ ℘zy) = (ΣP_i D; P_i ∈ ℜ) then
         w = w + 1, ℘0w = ℘zy
      Endif
      y = y + 1
   Endwhile
      z = z + 1
Endwhile
```

Example 3: In this example, we show how the lower approximation reduct is computed from the decision system. We continue the discussion from Example 1. The decision *D* induces the decision classes $X_1 = \{x_1, x_4, x_6\}$, $X_2 = \{x_2, x_5, x_8\}$, and $X_3 = \{x_3, x_7, x_9\}$. Consider a 3-equivalence relation subset of $\Re$ as $\wp_{3\ 10} = \{P_3, P_4, P_5\}$. Now we compute $\underline{\Sigma_3^5 P_i} D$ and $\underline{\Sigma_1^5 P_i} D$ from the decision system presented in Table 1. It is clear that $\underline{\Sigma_3^5 P_i} X_1 = \{x_1\}$; $\underline{\Sigma_3^5 P_i} X_2 = \{x_8\}$; $\underline{\Sigma_3^5 P_i} X_3 = \{x_3, x_9\}$. Similarly, $\underline{\Sigma_1^5 P_i} X_1 = \{x_1\}$; $\underline{\Sigma_1^5 P_i} X_2 = \{x_8\}$; $\underline{\Sigma_1^5 P_i} X_3 = \{x_3, x_9\}$. Therefore, we have:

$$\underline{\Sigma_3^5 P_i} D = \{\{x_1\}, \{x_8\}, \{x_3, x_9\}\} = \underline{\Sigma_1^5 P_i} D$$

It indicates that the reduct is $\wp_{01} = \{P_3, P_4, P_5\}$. Likewise, another minimal reduct $\wp_{02} = \{P_2, P_4, P_5\}$ can also be obtained.

5.2. Rule Extraction Using MGRS

In this section, we propose how decision rules ("OR" rules) due to multigranular rough set can be induced from lower approximation reduct. Pawlak's rough set rules (Pawlak, 2002) are "AND" rules whereas MGRS decision rules are "OR" rules. It indicates that, MGRS decision rules are weaker than rough set decision rules as MGRS decision rules can be decomposed into several decision rules. The following Algorithm 2 (LARMGRS) provides clear overview of obtaining decision rules from lower approximation reduct (LAR) using MGRS. Algorithm 3 extracts decision rules using MGRS without computing lower approximation reduct.

Let $K = (U, \Re, D)$ be a complete decision system, where $\Re = \{P_1, P_2, \cdots, P_m\}$ is a family of equivalence relations over *U*, and *D* is the decision. Let $[x]_D = \{X_1, X_2, \cdots, X_r\}$ be the partition induced by the decision attribute *D*. The input to the decision rule generation algorithm is the lower approximation reduct $\wp_0 = \{\wp_{01}, \wp_{02}, \cdots, \wp_{0w}\}$, where each element $\wp_{0w}$ contains some equivalence relations. Rename the equivalence relations present in the lower approximation reduct $\wp_{0w}$ as $P_1, P_2, P_3, \cdots$. The notions and symbols used in the algorithm are furnished below:

- i: Indexing variable varies from 1 to $|\wp_0|$;
- j: Indexing variable varies from 1 to number of equivalence relations present in $|\wp_{0i}|$;
- k: Indexing variable that varies from 1 to number of decisions;
- z: Indexing variable that varies from 1 to number of equivalence classes present in $[x]_{P_j}$;
- $[x]_{P_j}^z$: z^{th}- equivalence class induced by the equivalence relation P_j.

Ruleset: Set of all rules pertaining to a decision class X_r.

Algorithm 2. LARMGRS: Decision Rule using MGRS from LAR
Input: Lower approximation reduct $\wp_0 = \{\wp_{01}, \wp_{02}, \cdots, \wp_{0w}\}$
Output: "OR" decision rules

Set $i = 1, \quad j = 1, \quad k = 1, \quad \text{Rule} = \phi, \quad \text{Ruleset} = \phi$

while $k \leq r$ do

  while $i \leq |\wp_0|$ do

    while $j \leq |\wp_{0i}|$ do

      while $z \leq |[x]_{P_j}|$ do

        if $[x]^z_{P_j} \subseteq X_k$ then

          $\text{Rule} = \text{Rule} \cup \{[x]^z_{P_j}\}$

        Endif

        $z = z + 1$

      Endwhile

      $j = j + 1$

    Endwhile

    $i = i + 1$

  Endwhile

Ruleset = Rule

$k = k + 1, \quad i = 1, \quad j = 1$

Endwhile

Example 4: In this example, we show how the "OR" decision rules are generated from lower approximation reduct. We continue our discussion from Example 3. The decision D induces the decision classes $X_1 = \{x_1, x_4, x_6\}$, $X_2 = \{x_2, x_5, x_8\}$, and $X_3 = \{x_3, x_7, x_9\}$. The lower approximation reducts obtained are $\wp_{01} = \{P_3, P_4, P_5\}$ and $\wp_{02} = \{P_2, P_4, P_5\}$. Let us consider lower approximation reduct $\wp_{01} = \{P_3, P_4, P_5\}$. Rename the equivalence relations P_3, P_4, and P_5 as P_1, P_2, and P_3 respectively. Here, k varies from 1 to r=3. Let us consider k=3 and $X_3 = \{x_3, x_7, x_9\}$. On employing Algorithm 2 we get the following:

$i = 1, j = 1, z = 1 \quad [x]^1_{P_1} = \{x_1, x_4, x_5, x_6, x_7, x_9\} \not\subset X_3 = \{x_3, x_7, x_9\}$

$z = 2 \quad [x]^2_{P_1} = \{x_2, x_3\} \not\subset X_3 = \{x_3, x_7, x_9\}$

$z = 3 \quad [x]^3_{P_1} = \{x_8\} \not\subset X_3 = \{x_3, x_7, x_9\}$

$i = 1, j = 2, z = 1 \quad [x]^1_{P_2} = \{x_1, x_7\} \not\subset X_3 = \{x_3, x_7, x_9\}$

$z = 2 \quad [x]^2_{P_2} = \{x_2, x_3, x_4, x_5, x_6, x_8\} \not\subset X_3 = \{x_3, x_7, x_9\}$

$z = 3 \quad [x]^3_{P_2} = \{x_9\} \subseteq X_3 = \{x_3, x_7, x_9\}$

$\text{Rule} = \text{Rule} \cup \{[x]^3_{P_2}\}$

$i = 1, j = 3, z = 1 \quad [x]^1_{P_3} = \{x_1\} \not\subset X_3 = \{x_3, x_7, x_9\}$

$z = 2 \quad [x]^2_{P_3} = \{x_2, x_7\} \not\subset X_3 = \{x_3, x_7, x_9\}$

$z = 3 \quad [x]^3_{P_3} = \{x_3\} \subseteq X_3 = \{x_3, x_7, x_9\}$

$\text{Rule} = \text{Rule} \cup \{[x]^3_{P_2}, [x]^3_{P_3}\}$

$z = 4,\ [x]^4_{P_3} = \{x_4, x_5, x_6, x_8, x_9\} \not\subset X_3 = \{x_3, x_7, x_9\}$

$\text{Ruleset} = \{[x]^3_{P_2}, [x]^3_{P_3}\}$

Similarly, considering $\wp_{02} = \{P_2, P_4, P_5\}$ and on employing Algorithm 2 we get Ruleset= $\{[x]^3_{P_2}, [x]^3_{P_3}\}$. Thus, the corresponding rule according to Example 1 is given as below:

If P_4 is Med or P_5 is 5, then the decision D is JF.

Similarly, considering X_1= $\{x_1, x_4, x_6\}$ and on employing Algorithm 2 we get Ruleset= $\{[x]^1_{P_3}\}$. Additionally, on taking X_2= $\{x_2, x_5, x_8\}$ and employing Algorithm 2 we obtain $\text{Ruleset} = \{[x]^3_{P_1}\}$ in accordance with $\wp_{01}$ or $\{[x]^3_{P_1}\}$ in accordance with $\wp_{02}$. Collectively according to Example 1, Algorithm 2 generates the following rules:

If P_4 is Med or P_5 is 5, then the decision D is JF.
If P_5 is 4 then D= BN
If P_2 is High or P_3 is Mod then D= BH

Algorithm 3. MGRS: Decision Rule using MGRS
Input: Set of all conditional equivalence relations $\Re = \{P_1, P_2, \cdots, P_m\}$
Output: "OR" decision rules

```
Set  i = 1,  j = 1,  k = 1,  Rule = φ,  Ruleset = φ
while  k ≤ r  do
   while  i ≤ |ℜ|  do
      while  j ≤ |[x]_{P_i}|  do
         if  [x]^j_{P_i} ⊆ X_k  then
            Rule = Rule ∪ {[x]^j_{P_i}}
         Endif
      Endwhile
         j = j + 1
      Endwhile
      i = i + 1
   Endwhile
Ruleset = Rule
k = k + 1,  i = 1,  j = 1
Endwhile
```

Example 5: In this example, we show how the "OR" decision rules are generated from the set of conditional equivalence relations $\Re = \{P_1, P_2, \cdots, P_5\}$. We continue our discussion from Example 3. The decision D induces the decision classes $X_1= \{x_1,x_4,x_6\}$, $X_2= \{x_2,x_5,x_8\}$, and $X_3= \{x_3,x_7,x_9\}$. Here, k varies from 1 to r=3. Let us consider k=3 and $X_3= \{x_3,x_7,x_9\}$. On employing Algorithm 3 we get the following:

$$
\begin{aligned}
i = 1, j = 1 \quad & [x]^1_{P_1} = \{x_4, x_5, x_8\} \not\subset X_3 = \{x_3, x_7, x_9\} \\
j = 2 \quad & [x]^2_{P_1} = \{x_2, x_3, x_6, x_7\} \not\subset X_3 = \{x_3, x_7, x_9\} \\
j = 3 \quad & [x]^3_{P_1} = \{x_1, x_9\} \not\subset X_3 = \{x_3, x_7, x_9\} \\
i = 2, j = 1 \quad & [x]^1_{P_2} = \{x_1, x_4, x_5, x_7\} \not\subset X_3 = \{x_3, x_7, x_9\} \\
j = 2 \quad & [x]^2_{P_2} = \{x_2, x_3, x_6, x_9\} \not\subset X_3 = \{x_3, x_7, x_9\} \\
j = 3 \quad & [x]^3_{P_2} = \{x_8\} \not\subset X_3 = \{x_3, x_7, x_9\} \\
i = 3, j = 1 \quad & [x]^1_{P_3} = \{x_1, x_4, x_5, x_6, x_7, x_9\} \not\subset X_3 = \{x_3, x_7, x_9\} \\
j = 2 \quad & [x]^2_{P_3} = \{x_2, x_3\} \not\subset X_3 = \{x_3, x_7, x_9\} \\
j = 3 \quad & [x]^3_{P_3} = \{x_8\} \not\subset X_3 = \{x_3, x_7, x_9\} \\
i = 4, j = 1 \quad & [x]^1_{P_4} = \{x_1, x_7\} \not\subset X_3 = \{x_3, x_7, x_9\} \\
j = 2 \quad & [x]^2_{P_4} = \{x_2, x_3, x_4, x_5, x_6, x_8\} \not\subset X_3 = \{x_3, x_7, x_9\}
\end{aligned}
$$

$$
\begin{aligned}
j = 3 \quad & [x]^3_{P_4} = \{x_9\} \subseteq X_3 = \{x_3, x_7, x_9\} \\
\text{Rule} = & \text{ Rule} \cup \{[x]^3_{P_4}\} \\
i = 5, j = 1 \quad & [x]^1_{P_5} = \{x_1\} \not\subset X_3 = \{x_3, x_7, x_9\} \\
j = 2 \quad & [x]^2_{P_5} = \{x_2, x_7\} \not\subset X_3 = \{x_3, x_7, x_9\} \\
j = 3 \quad & [x]^3_{P_5} = \{x_3\} \subseteq X_3 = \{x_3, x_7, x_9\} \\
\text{Rule} = & \{[x]^3_{P_4}\} \cup \{[x]^3_{P_5}\} = \{[x]^3_{P_4}, [x]^3_{P_5}\} \\
j = 4 \quad & [x]^3_{P_5} = \{x_4, x_5, x_6, x_8, x_9\} \not\subset X_3 = \{x_3, x_7, x_9\} \\
\text{Ruleset} = & \{[x]^3_{P_4}, [x]^3_{P_5}\}
\end{aligned}
$$

Therefore, the rule according to MGRS for the decision class X_3 can be expressed as below.

If P_4 is Med or P_5 is 5, then the decision D is JF.

Similarly, considering $X_1= \{x_1,x_4,x_6\}$ and on employing Algorithm 3 we get Ruleset= $\{[x]^1_{P_3}\}$. Additionally, on taking $X_2= \{x_2,x_5,x_8\}$ and employing Algorithm 3 we obtain $\text{Ruleset} = \{[x]^3_{P_1}\}$. Collectively, Algorithm 3 generates the following rules:

If P_4 is Med or P_5 is 5, then the decision D is JF.
If P_5 is 4 then D= BN
If P_2 is High or P_3 is Mod then D = BH

6. EMPIRICAL STUDY ON IDENTIFICATION OF NUMEROUS ATTACKS IN MANET

This section illustrates how the proposed system is applied to a dataset. The dataset is pre-processed and became input to the proposed system. Data accumulation is a vital problem in research. There are many different ways to accumulate data. But complexities exist through these different ways. Further to model various categories of attacks in MANET, distinct types of data are required. To stay away from these complexities, we considered the dataset from CRAWDAD (Community Resource for Archiving Wireless Data at Dartmouth) repository for our experimental study.

The dataset includes 13,742 objects in which decisions for 3,311 objects falls under category of normal is discarded from the dataset keeping in mind the objective of our study. The remaining 10,431 objects are having numerous attacks such as jamming, jelly fish, packet dropping, black hole, worm hole, resource consumption, repudiation, message fabrication and denial-of-service in mobile adhoc network. These attacks are found at different layers present in the network. The different layers of TCP/IP model are partitioned into six relations such as physical layer, mac layer, network layer, transport layer, application layer, and general relation. The physical layer is used to verify signal to noise (SNIO) rate; number of neighbor (NNEG) nodes present, added, removed; carrier sensing (CSEN) time; signal strength (SSTH) received; and number of broken (NBRK) links during communication from one node to another. The mac layer is used to recognize the rate of packets (PRTE) send, received; packets retransmit (PRET); energy consumption (ECON); and packet error (PERR) while sending and receiving. Similarly the network layer is used to clarify the rate of packets forwarded (PFWR); rate of packets initiated (PINT); route request (RREQ) send, received; route reply (RREP) send, received; route error (RERR) send, received; rate of packets dropped (PDRP); percentage changes in route entries (PRCR); and percentage changes in number (PRCN) of hops during communication between source and sink. Likewise, the transport layer is employed to get the transmission delay (TDLY); route added to cache; difference in mobility (DIMB) speed; packets send through source (PSSR); and packets received through destination (PRDT) while in communication. The application layer is used to check the changes in number (CANO) of flag status during transmission. In addition, there are some general conditions such as type of protocols, traffic status and type of service (SERV) in mobile adhoc network during communication. The various attributes that act differently in various layers and general conditions is presented in Table 2.

The various categories of attacks found in mobile adhoc network are considered. The dataset taken into consideration includes nine decision classes such as jamming, jelly fish, packet dropping, black hole, worm hole, resource consumption, repudiation, message fabrication and denial-of-service. The sample dataset of 15 records is presented in the following Table 3.

6.1. Experimental Analysis

In this section we analyze identification of distinct categories of attacks in mobile adhoc network using C programming language and carry out the experimental study in a system with 2.67 GHz Intel core *i*3 processor and 2 GB RAM. Total 10,431 records are partitioned into two categories like training dataset of 5,737 (55%) and testing dataset of 4,694 (45%) records. Different classification of training and testing dataset in accordance with decisions is depicted in Table 4.

Table 2. Indiscernibility relation across various layers

<table>
<tr><th>Various Layers and General Conditions</th><th>Indiscernibility Relation</th><th>Attributes</th><th>Classification</th></tr>
<tr><td rowspan="5">Physical Layer</td><td rowspan="5">P_1</td><td>SNIO</td><td rowspan="5">Poor (P)
Low (L)
Average (A)
High (H)
Very High (VH)</td></tr>
<tr><td>NNEG</td></tr>
<tr><td>CSEN</td></tr>
<tr><td>SSTH</td></tr>
<tr><td>NBRK</td></tr>
<tr><td rowspan="4">Mac Layer</td><td rowspan="4">P_2</td><td>PRTE</td><td rowspan="4">Poor (P)
Low (L)
High (H)
Very High (VH)</td></tr>
<tr><td>PRET</td></tr>
<tr><td>ECON</td></tr>
<tr><td>PERR</td></tr>
<tr><td rowspan="8">Network Layer</td><td rowspan="8">P_3</td><td>PFWR</td><td rowspan="8">Poor (P)
Low (L)
Average (A)
High (H)
Very High (VH)</td></tr>
<tr><td>PINT</td></tr>
<tr><td>RREQ</td></tr>
<tr><td>RREP</td></tr>
<tr><td>RERR</td></tr>
<tr><td>PDRP</td></tr>
<tr><td>PRCR</td></tr>
<tr><td>PRCN</td></tr>
<tr><td rowspan="4">Transport Layer</td><td rowspan="4">P_4</td><td>TDLY</td><td rowspan="4">Poor (P)
Low (L)
High (H)
Very High (VH)</td></tr>
<tr><td>DIMB</td></tr>
<tr><td>PSSR</td></tr>
<tr><td>PRDT</td></tr>
<tr><td>Application Layer</td><td>P_5</td><td>CANO</td><td>Number of flags 1 to 8</td></tr>
<tr><td>General</td><td>P_6</td><td>SERV</td><td>Finger (F)
SMTP (S)
FTP (FT)
HTTP (HT)
Other (O)</td></tr>
</table>

From Table 4, it is clear that the decision D divides the dataset into 9 classes such as $[x]_D = \{X_1, X_2, \ldots, X_9\}$ where X_1 represents the decision class jamming, X_2 represents the decision class jelly fish etc. On considering the decision class X_1, decision rules are generated on employing LARMGRS and MGRS algorithm. The rules generated are again validated with testing dataset with predefined threshold of 70%.

Similarly, on considering decision class jelly fish, packet dropping, black hole, worm hole, resource consumption, repudiation, message fabrication, and denial-of-service, decision rules are generated from training dataset using LARMGRS. Further, the rules generated are validated with testing dataset. The rules generated for various categories of attacks and its accuracy is presented in the following Table 5.

Table 3. Sample dataset of empirical study

U	P_1	P_2	P_3	P_4	P_5	P_6	D
x_1	$SNIO_P$	$PRTE_{VH}$	$RREQ_{VH}$	$DIMB_H$	$CHNO_8$	$SERV_F$	Jamming
x_2	$SSTH_H$	$PRTE_H$	$PRCR_L$	$DIMB_H$	$CHNO_8$	$SERV_F$	Jellyfish
x_3	$SSTH_H$	$PRTE_{VH}$	$PINT_L$	$DIMB_H$	$CHNO_8$	$SERV_F$	Packet dropping
x_4	$SSTH_H$	$PRTE_H$	$PRCR_P$	$DIMB_H$	$CHNO_1$	$SERV_F$	Repudiation
x_5	$SSTH_H$	$PRTE_{VH}$	$PRCR_P$	$DIMB_H$	$CHNO_7$	$SERV_F$	Resource consumption
x_6	$CSEN_L$	$PRTE_H$	$PRCR_P$	$DIMB_P$	$CHNO_0$	$SERV_F$	Jellyfish
x_7	$SSTH_H$	$PRTE_H$	$PRCR_{VH}$	$DIMB_{VH}$	$CHNO_0$	$SERV_F$	Worm hole
x_8	$SSTH_H$	$PRTE_H$	$RREQ_{VH}$	$DIMB_{VH}$	$CHNO_8$	$SERV_{FT}$	Resource consumption
x_9	$SSTH_H$	$PRTE_H$	$PDRP_H$	$DIMB_{VH}$	$CHNO_0$	$SERV_F$	Black hole
x_{10}	$SSTH_H$	$PRTE_H$	$PRCR_P$	$DIMB_{VH}$	$CHNO_8$	$SERV_O$	DoS
x_{11}	$SSTH_L$	$PRTE_H$	$PRCR_P$	$DIMB_{VH}$	$CHNO_0$	$SERV_O$	Jamming
x_{12}	$CSEN_L$	$PRTE_H$	$PINT_H$	$DIMB_{VH}$	$CHNO_0$	$SERV_O$	DoS
x_{13}	$CSEN_L$	$PRTE_H$	$RREP_H$	$DIMB_{VH}$	$CHNO_0$	$SERV_O$	Message fabrication
x_{14}	$CSEN_L$	$PRTE_H$	$RREQ_{VH}$	$PRDT_H$	$CHNO_0$	$SERV_O$	DoS
x_{15}	$CSEN_L$	$PRTE_{VH}$	$PRCR_P$	$DIMB_{VH}$	$CHNO_6$	$SERV_O$	DoS

Table 4. Training, testing classification of datasets

Serial Number	Attack Name	Training Set	Testing Set
1	Jamming	971	594
2	Jelly Fish	177	345
3	Packet Dropping	521	608
4	Black Hole	677	373
5	Warm Hole	482	468
6	Resource Consumption	625	418
7	Repudiation	365	260
8	Message Fabrication	152	161
9	Denial-of-Service	1,767	1,467
	Total	5737 (55%)	4694 (45%)

The accuracy presented in Table 5 is computed by using precision, and recall. The precision, recall and accuracy is defined below. The precision, recall, and accuracy are calculated by using the Formulae (12), (13), and (14). The symbol T_P is used to represent right classification of objects to decisions while F_P is used to represent wrong classification of objects to decisions. The symbol T_N is used for number of objects which are rightly classified as negative whereas F_N is the number of objects wrongly classified as positive. The decision rules obtained are validated through testing dataset to determine the accuracy. The following Table 6 presents the computation of precision, recall and accuracy of various attacks of MANET:

$$\text{Precision} = \frac{|T_p|}{|T_p + F_P|} \tag{12}$$

$$\text{Recall} = \frac{|T_p|}{|T_p + F_N|} \tag{13}$$

$$\text{Accuracy} = \frac{|T_p + \mathrm{T}_N|}{|T_p + F_P + \mathrm{T}_N + \mathrm{F}_N|} \tag{14}$$

Table 5. LARMGRS decision rules for various attacks in MANET

i	Attacks	Rule Description	Accuracy (%)
1	Jamming	If P_1 is $SNIO_L$ or $SNIO_P$ or $SSTH_L$ or $SSTH_P$ received or $NNEG_H$ or $NNEG_A$ added or $NBRK_H$ or P_2 is $PERR_H$ or $PERR_{VH}$ received then D = Jamming	96.93
2	Denial-of-Service	If P_2 is $PRTE_H$ or $PRTE_{VH}$ in sent or P_3 is $PINT_H$ or $PINT_{VH}$ or $RREQ_L$ or $RREQ_P$ in sent or P_4 is $PSSR_L$ or $PSSR_P$ or $PRDT_H$ or $PRDT_{VH}$ or P_5 is $CHNO_6$ or P_6 is $SERV_S$ then D = DoS	92.60
3	Message fabrication	If P_3 is $RREP_H$ or $RERR_H$ or $RERR_{VH}$ Sent or $RERR_H$ or $RERR_{VH}$ Received or $RREP_H$ Received then D = Message fabrication	97.25
4	Resource consumption	If P_2 is $PRTE_L$ or $PRTE_P$ or $ECON_H$ or $ECON_{VH}$ or $PRET_H$ or $PRET_{VH}$ or P_5 is $CHNO_7$ or P_6 is $SERV_{FT}$ then D = Resource consumption	95.03
5	Jelly fish	If P_3 is $PRCR_L$ or $PRCR_H$ or $PRCN_L$ or $PDRP_A$ or $PDRP_L$ or P_4 is $TDLY_H$ or $DIMB_P$ then D = Jelly fish	95.73
6	Packet dropping	If P_3 is $PINT_L$ or $RREQ_H$ or $RREP_L$ or $RREQ_H$ then D= Packet dropping	98.40
7	Worm hole	If P_3 is $PRCR_{VH}$ or $PFWR_P$ or $PDRP_{VH}$ then D = Worm hole	92.16
8	Black hole	If P_3 is $PDRP_H$ or $PFWR_L$ or $RREP_P$ or $PINT_P$ then D= Black hole	96.74
9	Repudiation	If P_5 is $CHNO_1$ or $CHNO_2$ or $CHNO_3$ or $CHNO_4$ or $CHNO_5$ then D = Repudiation	96.91

Table 6. Precision, recall and accuracy of various attacks in MANET

Serial Number	Attack Name	T_P	F_P	F_N	T_N	Precision	Recall	Accuracy
1	Jamming	450	071	073	4,100	0.86	0.86	96.93
2	Jelly Fish	145	095	105	4,349	0.60	0.58	95.73
3	Packet Dropping	533	032	043	4,086	0.94	0.93	98.40
4.	Black Hole	220	076	077	4,321	0.74	0.74	96.74
5.	Warm Hole	100	178	190	4,226	0.36	0.34	92.16
6.	Resource Consumption	185	120	113	4,276	0.61	0.62	95.03
7	Repudiation	115	092	053	4,434	0.57	0.68	96.91
8.	Message Fabrication	032	069	060	4,533	0.32	0.35	97.25
9	Denial-of-Service	1,070	180	217	3,277	0.86	0.83	92.60
	Total	2,437	1,167	1,090	37,602	0.68	0.69	94.66

Table 7. Comparative analysis of various models with LARMGRS

Attack Name	SVM	GP	GMM	LM	MLP	KNN	NF	SH	LARMGRS
Jamming	--	--	--	--	--	--	--	96.65	96.93
Jelly Fish	--	--	--	--	--	--	--	87.83	95.73
Packet Dropping	67.11	63.12	62.10	58.02	62.09	98.01	98.10	--	98.40
Black Hole	88.32	82.20	84.14	87.00	87.05	96.02	--	--	96.74
Warm Hole	--	--	--	--	--	--	--	80.76	92.16
Resource Consumption	--	--	--	--	--	90.15	--	--	95.03
Repudiation	--	--	--	--	--	70.34	80.12	86.09	96.91
Message Fabrication	73.88	70.67	71.17	69.80	70.78	--	--	--	97.25
Denial-of-Service	90.01	90.10	89.67	91.10	90.03	--	--	--	92.60

Figure 5. A pictorial representation of the comparative study

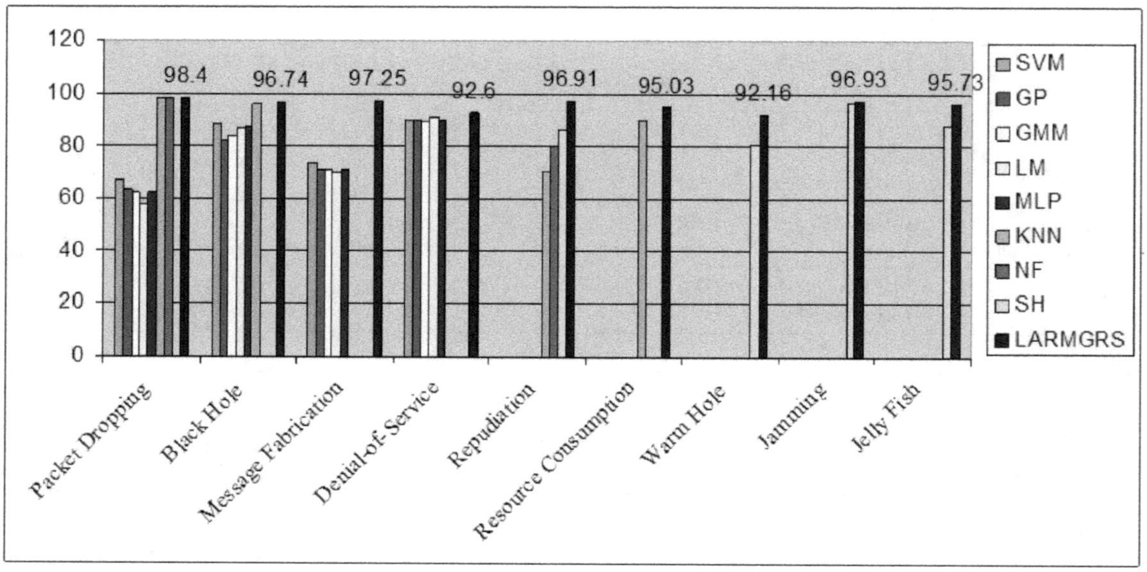

6.2. Comparative Analysis

In this section, we compare results of projected model LARMGRS with eight distinct models such as support vector machine (SVM) (Pastrana, Mitrokotsa, Orfila & Lopez, 2012; Mitrokotsa, Tsagkaris & Douligeris, 2008), genetic programming (GP) (Pastrana, Mitrokotsa, Orfila & Lopez, 2012), gaussian mixture model (GMM) (Pastrana, Mitrokotsa, Orfila & Lopez, 2012; Mitrokotsa, Tsagkaris & Douligeris, 2008), linear model (LM) (Pastrana, Mitrokotsa, Orfila & Lopez, 2012; Mitrokotsa, Tsagkaris & Douligeris, 2008), multi-layer perceptron (MLP) (Pastrana, Mitrokotsa, Orfila & Lopez, 2012; Mitrokotsa, Tsagkaris & Douligeris, 2008), k – nearest neighbour (KNN) (Fattah, Dahalin & Jusoh, 2010), neuro fuzzy (NF) (Chaudhary, Tiwari & Kumar, 2015), and signal hybridization (SH) (Sharma & Singla, 2015). Unlike Table 6, the computation is performed for available case across each model.

The subsequent Table 7 shows the comparative study of all the models mentioned above. The accuracy of the proposed model LARMGRS over the CRAWDAD dataset is higher in all cases as compared to other models mentioned above. It is also observed that in the model SVM, GP, GMM, LM, KNN, and SH is applied to 4 different attacks whereas NF is applied to 2 different attacks. But, LARMGRS is applied to 9 different attacks of MANET simultaneously. In addition, the model LARMGRS works well for numerous attacks of MANET across various layers of TCP/IP. For clear visualization, a graphical representation of the comparative study is depicted in Figure 5.

7. CONCLUSION

Distinct categories of attacks in MANET are key threat protection in modern trends. Protecting from these attacks is a prime concern in MANET like adhoc meeting, adhoc classroom, emergency medical situation, and emergency operations. To deal with these attacks many models are proposed by different authors (Fattah, Dahalin & Jusoh, 2010; Pastrana, Mitrokotsa, Orfila & Lopez, 2012; Mitrokotsa, Tsagkaris & Douligeris, 2008; Sharma & Singla, 2015; Chaudhary, Tiwari & Kumar, 2015). In this paper, we proposed a model LARMGRS for the identification of various categories of attacks in MANET using multigranular rough set. The proposed model is studied over the CRAWDAD dataset and checked for its accuracy. Further, it is compared with other existing models and found better accuracy as compared to other models. The accuracy of the proposed model is highest (98.40%) for identifying packet dropping whereas least (92.16%) for warm hole. But, in all cases of attacks the proposed model LARMGRS is having higher accuracy as compared to other models.

REFERENCES

Acharjya, D. P. (2013). Rough computing based information retrieval in knowledge discovery databases. In *Information and Knowledge Management-Tools, Techniques and Practices* (pp. 123-153).

Acharjya, D. P., & Bhattacharjee, D. (2013). A rough computing based performance evaluation approach for educational institutions. *International Journal of Software Engineering and Its Applications*, *7*(4), 331–348.

Ahmed, N. S. S., & Acharjya, D. P. (2015a). Detection of Denial of Service Attacks in Wireless Network using Dominance based Rough Set. *International Journal of Advanced Computer Science and Applications*, *6*(12), 267–278.

Ahmed, N. S. S., & Acharjya, D. P. (2015b). A dominance based rough set approach for the detection of jamming attack. *International Journal of Philosophies in Computer Science*, *1*(2), 45–66.

Ahmed, N. S. S., Acharjya, D. P., & Sanyal, S. (2017). A framework for phishing attack identification using rough set and formal concept analysis. *International Journal of Communication Networks and Distributed Systems*, *18*(2), 186–212. doi:10.1504/IJCNDS.2017.082105

Chaudhary, A., Tiwari, V. N., & Kumar, A. (2015). A new intrusion detection system based on soft computing techniques using neuro-fuzzy classier for packet dropping attack in MANET. *International Journal of Network Security*, *18*(3), 514–522.

Chaudhary, A., Tiwari, V. N., & Kumar, A. (2015). A co-operative intrusion detection system for sleep deprivation attack using neuro-fuzzy classier in mobile ad hoc networks. *Computational Intelligence in Data*, *2*, 345–353.

Chouhan, N. S., & Yadav, S. (2011). Flooding attacks prevention in MANET. *International Journal of Computer Technology and Electronics Engineering*, *1*(3), 68–72.

Das, T. K., Acharjya, D. P., & Patra, M. R. (2015). Multi criterion decision making using intuitionistic fuzzy rough set on two universal sets. *International Journal of Intelligent Systems and Applications*, *7*(4), 26–33. doi:10.5815/ijisa.2015.04.04

Dubois, D., & Prade, H. (1990). Rough fuzzy sets and fuzzy rough sets. *International Journal of General Systems*, *17*(2-3), 191–209. doi:10.1080/03081079008935107

Fattah, F. A., Dahalin, Z. M., & Jusoh, S. (2010). Dynamic Intrusion Detection Method for Mobile Ad Hoc Network using CPDOD Algorithm. *International Journal of Computer Applications*, 12(5), 22–29.

Garg, R., & Sharma, H. (2014). Proposed Lightweight Sybil Attack Detection Technique in MANET, *International Journal of Advanced Research in Electrical Electronics and Instrumentation Engineering*, *3*(5), 9395–9401.

Gediga, G., & Duntsch, I. (2001). Rough approximation quality revisited. *Artificial Intelligence*, *132*(2), 219–234. doi:10.1016/S0004-3702(01)00147-3

Geetha, M. A., Acharjya, D. P., & Iyengar, N. C. S. (2014). Algebraic properties of rough set on two universal sets based on multigranulation. *International Journal of Rough Sets and Data Analysis*, *1*(2), 49–61. doi:10.4018/ijrsda.2014070104

Kabiri, P., & Aghaei, M. (2011). Feature Analysis for Intrusion Detection in Mobile Ad Hoc Networks. *International Journal of Network Security*, *12*(1), 42–49.

Kataria, J., Dhekne, P. S., & Sanyal, S. (2006). A Scheme to Control Flooding of Fake Route Requests in Ad Hoc Networks. In *Proceedings of the Third International Conference on Computers and Devices for Communications* (pp. 198-201).

Liu, G. (2010). Rough set theory based on two universal sets and its applications. *Knowledge-Based Systems*, *23*(2), 110–115. doi:10.1016/j.knosys.2009.06.011

Mitrokotsa, A., Tsagkaris, M., & Douligeris, C. (2008). Intrusion Detection in Mobile Ad Hoc Networks using Classification Algorithms. *International Journal of Wireless Ad Hoc and Sensor Network*, *5*, 1–12.

Pastrana, S., Mitrokotsa, A., Orfila, A., & Lopez, P. P. (2012). Evaluation of Classification Algorithms for Intrusion Detection in MANETs. *Knowledge Based Systems*, *36*, 217–225. doi:10.1016/j.knosys.2012.06.016

Pawlak, Z. (1982). Rough Sets. *International Journal of Computer and Information Sciences*, *11*(5), 341–356. doi:10.1007/BF01001956

Pawlak, Z. (2002). Rough sets and intelligent data analysis. *Information Sciences*, *147*(1-4), 1–12. doi:10.1016/S0020-0255(02)00197-4

Qian, Y. H., & Liang, J. Y. (2006). Rough Set Method based on Mutli-Granulation. In *Proceedings of the 5th IEEE Conference on Granular Computing* (pp. 297-304). IEEE.

Qian, Y. H., Liang, J. Y., Yao, Y., & Dang, C. (2009). MGRS: A Multi-Granulation Rough Set. *Information Sciences, Elsevier*, *121*, 1–21.

Sharma, M., & Singla, R. (2015). Elimination of Jamming Attack in Ad Hoc Environment, *International Journal of Science. Engineering and Technology Research*, *4*(6), 2206–2210.

Shrivastava, S., & Jain, S. (2013). A Brief Introduction of Different Type of Security Attacks Found in Mobile Ad Hoc Network. *International Journal of Computer Science & Engineering Technology*, *4*(3), 218–222.

Wazid, M., Singh, R. K., & Goudar, R. H. (2011). A Survey of Attacks Happened at Different Layers of Mobile Ad Hoc Network and Some Available Detection Techniques. In *Proceedings of the International Conference on Computer Communication and Networks* (pp. 44-49).

Ziarko, W. (1993). Variable precision rough set model. *Journal of Computer and System Sciences*, *46*(1), 39–59. doi:10.1016/0022-0000(93)90048-2

This research was previously published in the International Journal of Information Security and Privacy (IJISP), 13(4); pages 28-52, copyright year 2019 by IGI Publishing (an imprint of IGI Global).

Chapter 8
A Systematic Study and Analysis of Security Issues in Mobile Ad-hoc Networks

Jhum Swain
Department of Computer Science and Engineering, Siksha 'O' Anusandhan University (Deemed), Bhubaneswar, India

Binod Kumar Pattanayak
Department of Computer Science and Engineering, Siksha 'O' Anusandhan University (Deemed), Bhubaneswar, Odisha, India

Bibudhendu Pati
Department of Computer Science and Engineering, C. V. Raman College of Engineering, Bhubaneswar, India

ABSTRACT

A Mobile Ad Hoc Networks (MANETs) is an assortment of a variety of portable nodes that are linked collectively in a greater number in a wireless medium that has no permanent infrastructure. Here, all the nodes in the network act as both router and host and is in charge for accelerating and forwarding packets to other nodes. Due to the frequent link failure and high speed of mobile nodes, there are many challenges in MANET. As security is the basic challenge in a MANET, this article discusses the various protocols developed for security issues in MANET and IoT and presents a comparative analysis on their performance. After detailed analysis and comparison, it was observed that when security features are imposed at different levels during routing, various protocols perform in different ways. Also, the network performance and network life times vary from one approach to another. So, this article also presents the effect of security features on data drop and delay of prominent protocols after simulation.

DOI: 10.4018/978-1-7998-8545-0.ch008

1. INTRODUCTION

A MANET is a group of autonomous systems which are portable in nature and are linked by wireless links. Each node not only acts as an end system but also has a router to forward packets. As nodes freely move around the system and they organize themselves into a network, so for this reason its structure changes frequently. So, a special kind of algorithm is needed to accommodate the changing topology. As nodes dynamically establish paths among each other its attractive to various types of attackers, so we need secured communication among each other. Dynamically the nodes in MANETs join and leave the network. In MANET nodes openly connect and disappear at any point of time to maintain the connection. Some of the typical applications of MANET include Military applications, emergency and rescue operations, aircrafts, wireless sensor network, medical service, commercial use and personal area network. As MANETs is an infra-structure less network which consists of mobile nodes with wireless network interfaces, so nodes dynamically establish paths among one another. This is one of the reasons that it is attractive to various types of attacks by the intruders. So, a protected communication protocol is needed for this network and hence, security is an important feature in MANET. In this paper a detailed analysis of secured routing protocols in MANET has been carried out with basic consideration being the security mechanism imposed.

2. RELATED WORK

This section presents a detail study and review of advanced approaches proposed by eminent researchers in this field of security in MANET. Transmission of audio data with secured mechanism has been discussed in Athulya and Sheeba's work (2012) where the authors have provided dual safety measures by encrypting and decrypting the audio at each node in the route using stream ciphering method. Security threats and their possible solutions have been discussed in Jan Von Mulert, Ian Weich and Winston K.G. Seah's research (2012), and it also focuses on how modification of data is done using optimization technique for security. Praveen Joshi (2011) discusses about the various security problems and how to prevent them which are adopted in the network layer. Spoofing attack and their prevention techniques has been discussed in Yuseok, Tae-Hyung, and Jong's (2012) work. The author proposed a model i.e. PBM (Policy Based Management) that deals with 4 entities in (Bostjancic & Timecenko) such as QoS, network resources, configuration and security. It has discussed about the vulnerability, challenges and security attacks on ad-hoc routing protocols. A bargained hub may disregard the secrecy rule of security and uncover vital data like private and open keys, status of hub, passwords, ideal course to approved hubs, geographic area of hubs and other control information in parcel headers to unapproved hubs introduce in the system (Dhivya, Karthik & Kumaran). A secured on-demand routing protocol has been proposed in (Wan, Ren, & Gu) that prevents various attacks in MANET and another regression based trust model for MANET has been implemented in (Venkataraman, Pushpalatha, & Rao, 2012) that ensures guaranteed packet delivery in wireless medium. Marchang and Datta (2012) proposed a light weight trust based routing protocol that prevents any external attacks in MANET and a mobile agent based secured platform has been designed in (Rath & Pattanayak, 2017) that ensures improved network life time with better throughput when compared with other similar approaches.

3. STUDY AND ANALYSIS OF SECURED ROUTING PROTOCOLS IN MANET AND IOT

The operation of secure routing plays a very important role in MANETs security due to absence of fixed infrastructure. In this section complete study and analysis of MANET protocols which are basically using some IoT devices and technology are reported. Traditional Internet security methods cannot be applied in MANETs context in terms of not having clear line of defence. In this section a comparative analysis has also been carried out with different secured routing protocols with their security factor as the basic criteria. It is a very challenging task to manage security at the IoT environment as here many heterogeneous devices are networked together. Current research shows that the networks due to IoT are prone to many attacks such as malware, botnets, DoS (Denial of Service) attack, web-based malware, Android malware and spam. Therefore, there is need of developing a standard secured frame work for communication in IoT (Airehrour & Gutierrez, 2016).

Table 1. Security Factors in IoT based Protocols

S.No	Literature	Description	Security Issues
1.	S.L. Keoh et al. (2014)	Here the standard security protocol i.e. CoAP an application protocol is designed	A standardized security protocol is indispensable for the success of IoT.
2.	D. Airehrour et al. (2016)	It manages the security at the IoT environment as there are many heterogeneous devices linked together.	Many attacks such as malware, botnets, DoS (Denial of Service) attack, Web based malware, android malware and spam are analysed.
3.	H.Kumarage et.al. (2016)	Security and privacy is very important in IoT when it is integrated with Cloud Computing.	Security in Cloud based IoT devices has been analysed.
4.	J.Zhou et.al. (2017)	The challenging issue here is how to securely forward the packets.	The query privacy for cloud-based IoT users should be well protected.
5.	K.Zhang et.al. (2017)	There is privacy and security concern while designing smart city applications and architecture.	The ever-growing volume of data and devices in smart city poses open problems for intelligent services and privacy.

4. COMPARATIVE ANALYSIS OF SECURED PROTOCOLS

In this section comparative analysis among different protocols has been done based on their performance. Those factors have been considered for comparison which affect the security issues to a greater extent such as link failure, end-to-end delay, packet drop ratio etc. Four representative protocols have been considered for simulation from each leading category of protocols in MANET such as DSR, AODV, OLSR and a game theory based improved secured protocol. Mechanism of the said protocols has been explained in this section with technical background. The S-DSR (Secure- Dynamic Source Routing) protocol (Echchaachoui, Kobbane & Elkoutbi, 2015) is proposed by the author where the author secures the traffic against potential attacks without decreasing network performance. A game theoretic scheme (Das, Majumdar, & Dasgupta, 2015) that uses LCTF (Least Total Cost Factor) is used to transfer data packets from source to destination have been proposed by the author. A protocol (Ghosh & Datta, 2015) SD-RAC (Secure and Distributed Robust Address Configuration) has been proposed by the author low overhead identity based distributed dynamic address configuration scheme for secure allocation of IP address to authorized nodes of a managed MANET.S-AODV protocol (Cerri & Ghioni, 2008) that analyzes the adaptive strategy and delays the verification of digital signatures has been proposed by the author.

5. SIMULATION RESULTS AND ANALYSIS

The simulation tool Ns 2.35 has been used in this research work to simulate the performance of the above discussed protocols to report their performance during comparison. Table 2 shows the network simulation parameters considered during the simulation process for simulating the protocols considered for evaluation of energy efficiency and security based protocols.

Table 2. Simulation Parameters

Simulation parameters	Data/Value
Channel Type	Wireless Channel
MAC Protocol	802.11
Simulation area	2000 x 2000 m
Protocols	S-DSR, LCTF,SD-RAC,S-AODV
Model for Node Movement	Random way point
Data rate	11 mbps
Simulator	Ns2.35

Figure 1 shows the simulation result after simulating the three representative and prominent protocols in NS 2.35. The simulation results show that S-DSR proposed in (Echchaachoui, Kobbane & Elkoutbi, 2015) performs better in comparison to LCTF (Das, Majumdar, & Dasgupta, 2015) and SD-RAC (Ghosh & Datta, 2015). This is due to the new model used in (Echchaachoui, Kobbane & Elkoutbi, 2015) to prevent the Denial of Service attack where a box plot theory has been used to define a threshold value to get notified for a DoS attack.

Similarly Figure 2 depicts comparison results of three basic protocols considered in this article. We can observe that due to security mechanism there is noticiable delay in all the techniques but out of them the S-AODV proposed in (Cerri & Ghioni, 2008) takes an average delay of 4 seconds during simulation in comparison to S-DSR and SD-RAC security mechanism.

From the comparative analysis, it can be concluded that the S-DSR (Secure- Dynamic Source Routing) protocol secures the traffic against potential attacks without compromising network performance. A game theoretic scheme (Das, Majumdar, & Dasgupta, 2015) that uses LCTF (Least Total Cost Factor) is used to transfer data packets from source to destination and so it shows average data drop in Figure 1. A protocol SD-RAC (Secure and Distributed Robust Address Configuration) in (Ghosh & Datta, 2015) has been proposed by the author with low overhead identity based distributed dynamic address configuration scheme for secure allocation of IP address to authorized nodes of a managed MANET. Similarly, S-AODV protocol (Cerri & Ghioni, 2008) shows minimum delay in Figure 2 that analyzes the adaptive strategy and delays the verification of digital signatures has been proposed by the author.

Figure 1. Data drop comparison due to attack

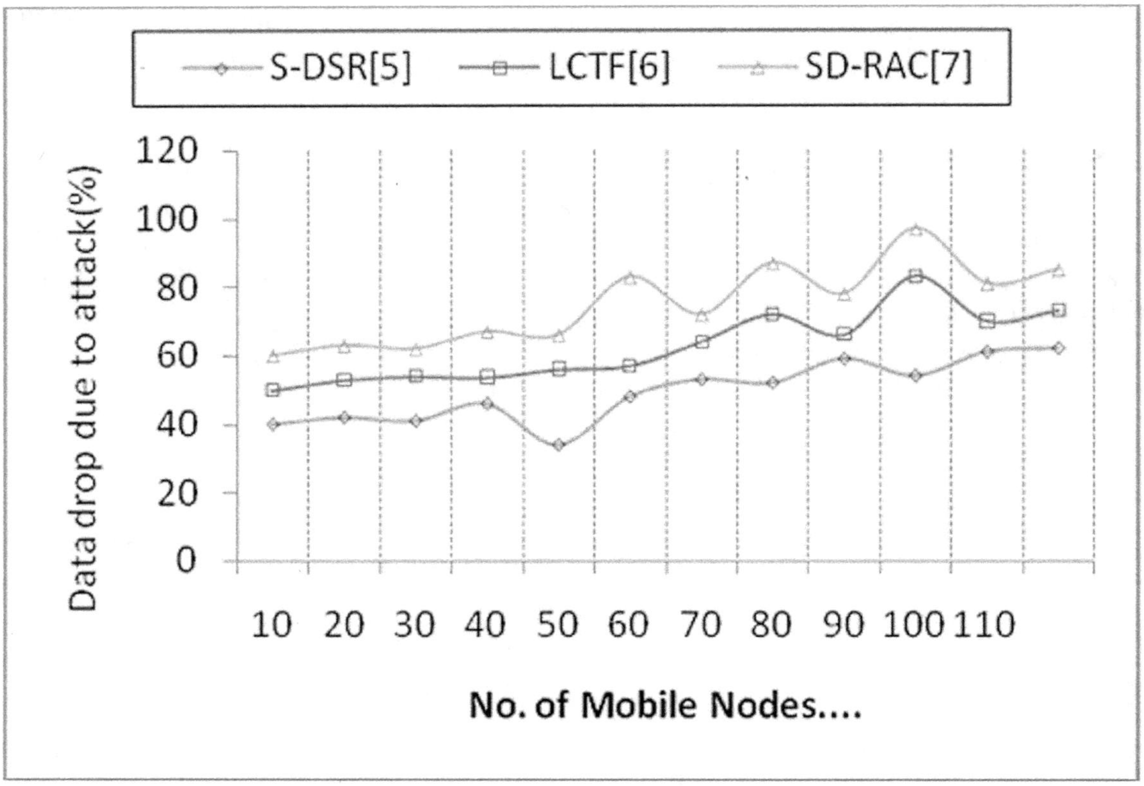

Figure 2. Comparison of Average delay in packet delivery

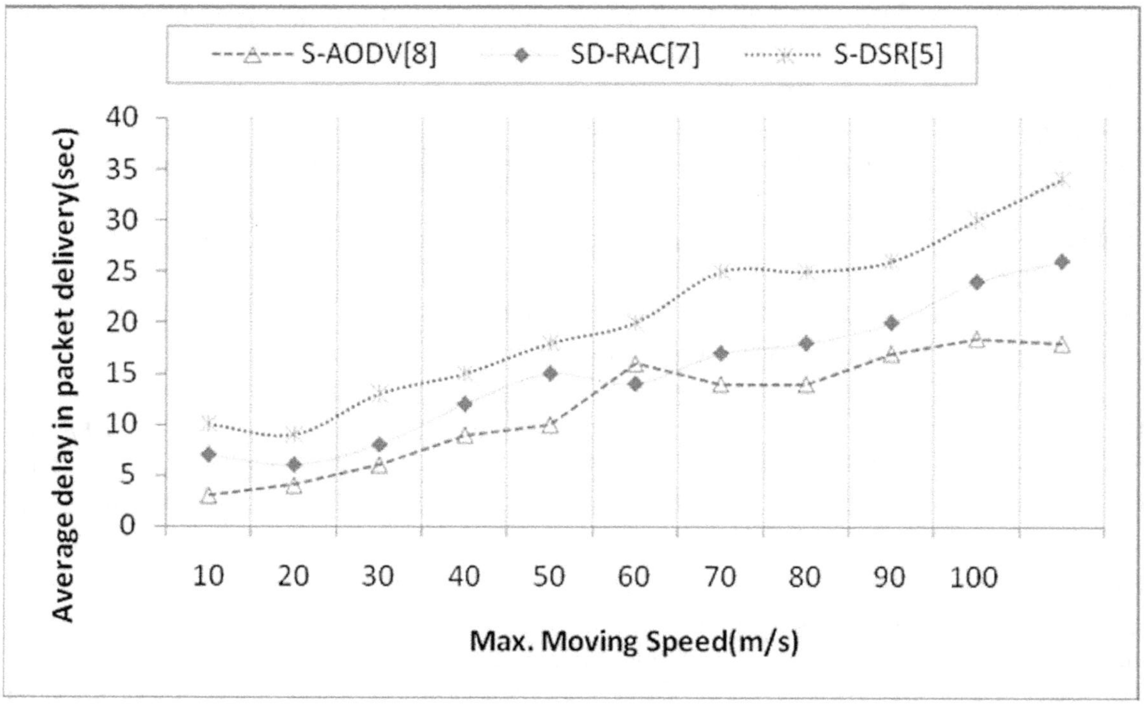

6. CONCLUSION

From the simulation results it can be concluded that as the normal MANET based protocols such as DSR, OLSR,AODV, TORA (Athulya & Sheeba, 2012) and (Praveen, 2011) do not have enough built-in security mechanism to prevent attacks, the considered protocols in this analysis survey, which use modified and improved techniques of the basic MANET protocols, so depending on their functional mechanism, some protocols exhibit higher rate of delay to incorporate the security feature at the level of packet forwarding or sharing the resources. Similarly, about the packet drop rate it can be concluded that definitely there is reduced rate of packet drop than the normal MANET protocols, but when some new platform or model is used with the protocols, definitely there will be minor delay in data transmission. But this packet drops, and delay can be managed with contrast to the amount of data saved due to the vulnerability of MANET. So, the conclusion of this research article is that depending on the security mechanism used in a protocol and as per the complexity level, the delay and data drop will be affected. But minor difference in these factors can be tolerated in a network when there is minor loss in terms of more retransmission attempts in comparison to prevention of the heavy loss due to attack in the network. Future scope of this research article is that it can act as a base platform to impose some advanced security technique during design of a secured routing protocol for MANET and IoT.

REFERENCES

Airehrour, D., & Gutierrez, J. A. (2015). An analysis of secure MANET routing features to maintain confidentiality and integrity in IoT Routing. Retrieved August 2016 from http://www.researchgate.net/publication/277078202

Athulya, M. S., & Sheeba, V. S. (2012). Security in Mobile Ad-Hoc Networks. In *Proceedings of the 2012 Third International Conference on Computing Communication & Networking Technologies (ICCCNT)*, Coimbatore. 10.1109/ICCCNT.2012.6396047

Cerri, D., & Ghioni, A. (2008, February). Securing AODV: The A-SAODV secure routing prototype. *IEEE Communications Magazine*, *46*(2), 120–125. doi:10.1109/MCOM.2008.4473093

Dhivya, P., Karthik, S., & Kalaikumaran, T. (2015). SOA Based Secure Data Transmission over CMEA Protocol in MANET. *In Proceedings of the 2015 Procedia Computer Science* (pp. 434–440).

Echchaachoui, A., Kobbane, A., & Elkoutbi, M. (2015). A new trust model to secure routing protocols against DoS attacks in MANETs. In *Proceedings of the 2015 10th International Conference on Intelligent Systems: Theories and Applications (SITA)*, Rabat. 10.1109/SITA.2015.7358397

Ghosh, U., & Datta, R. (2015). A Secure Addressing Scheme for Large-Scale Managed MANETs. *IEEE eTransactions on Network and Service Management*, *12*(3), 483–495. doi:10.1109/TNSM.2015.2452292

Jeon, Y., Kim, T. H., Kim, Y., & Kim, J. (2012, October). *LT-OLSR: Attack-tolerant OLSR against link spoofing*. In *Proceedings of the 7th Conference on in Local Computer Networks (LCN)* (pp. 216-219).

Joshi, P. (2011). Security Issues in Routing protocols in MANETs at network layer. *Procedia Computer Science, 3*, 954–960. doi:10.1016/j.procs.2010.12.156

Keoh, S. L., Kumar, S. S., & Tschofenig, H. (2014). Securing the Internet of Things: A Standardization Perspective. IEEE Internet of Things Journal, 1(3), 265-275. doi:10.1109/JIOT.2014.2323395

Kumarage, H., Khalil, I., Alabdulatif, A., Tari, Z., & Yi, X. (2016). Secure Data Analytics for Cloud-Integrated Internet of Things Applications. IEEE Cloud Computing, 3(2), 46-56. doi:10.1109/MCC.2016.30

Marchang, N., & Datta, R. (2012, June). Light-weight trust-based routing protocol for mobile ad hoc networks. *IET Information Security*, *6*(2), 77–83. doi:10.1049/iet-ifs.2010.0160

Rakas, S. V. B., & Timcenko, V. V. (2014). Quality of Service and Security Issues in MANET environment. In *Proceedings of the 22nd Telecommunications Forum (TELFOR)*, Serbia, Belgrade.

Rath, M., & Pattanayak, B. K. (2017, January). MAQ: A Mobile Agent Based Quality of Service Platform for MANETs. *International Journal of Business Data Communications and Networking*, *13*(1), 1–8. doi:10.4018/IJBDCN.2017010101

Rath, M., Rout, U. P., Pujari, N., Nanda, S. K., & Panda, S. P. (2017). Congestion Control Mechanism for Real Time Traffic in Mobile Adhoc Networks. In *Computer Communication, Networking and Internet Security* (pp. 149–156). Singapore: Springer.

Das, D., Majumder, K., & Dasgupta, A. (2015). Selfish Node Detection and Low Cost Data Transmission in MANET using Game Theory. In Proceedings of the 11th International Multi-Conference on Information Processing (IMCIP-2015) (pp. 92–101). Science Direct.

Venkataraman, R., Pushpalatha, M., & Rao, T. R. (2012, September). Regression-based trust model for mobile ad hoc networks. *IET Information Security*, *6*(3), 131–140. doi:10.1049/iet-ifs.2011.0234

Von Mulert, J., Weich, I., & Seah, W. K. G. (2012). Security threats and solutions in MANETs: A case study using AODV and SAODV. *Journal of Network and Computer Applications, 35*(4), 1249–1259. doi:10.1016/j.jnca.2012.01.019

Wan, Z., Ren, K., & Gu, M. (2012, May). USOR: An Unobservable Secure On-Demand Routing Protocol for Mobile Ad Hoc Networks. *IEEE Transactions on Wireless Communications*, *11*(5), 1922–1932. doi:10.1109/TWC.2012.030512.111562

Zhang, K., Ni, J., Yang, K., Liang, X., Ren, J., & Shen, X. S. (2017, January). Security and Privacy in Smart City Applications: Challenges and Solutions. *IEEE Communications Magazine*, *55*(1), 122–129. doi:10.1109/MCOM.2017.1600267CM

Zhou, J., Cao, Z., Dong, X., & Vasilakos, A. V. (2017, January). Security and Privacy for Cloud-Based IoT: Challenges. *IEEE Communications Magazine*, *55*(1), 26–33. doi:10.1109/MCOM.2017.1600363CM

This research was previously published in the International Journal of Information Security and Privacy (IJISP), 12(2); pages 38-45, copyright year 2018 by IGI Publishing (an imprint of IGI Global).

Chapter 9
A Survey of Efficient Trust Management Schemes in Mobile Ad-Hoc Network:
Reliable Trust Management Framework of MANET

J. Kaur
J Kaur, Punjabi University, Patiala, India

S. Kaur
PURCITM, Punjab, India

ABSTRACT

Mobile Ad Hoc Networks (MANETs) are comprised of an arrangement of self-sorting mobile hosts furnished with wireless interaction devices gathered in groups without the need of any settled framework as well as centralized organization to maintain a system over radio connections. Every mobile node can react as a host and also, the router freely utilizes the wireless medium inside the correspondence range to deal with the interaction between huge quantities of individual mobile nodes by framing a correspondence system and trading the information among them without using any described group of the base station. A trust-based model in MANET estimates and sets up trust relationship among objectives. Trust-based routing is utilized to keep away data from different attackers like a wormhole, DOS, black-hole, selfish attack and so forth. Trust can be executed in different steps like reputation, subjective rationale and from the supposition of the neighboring node. A trust estimation approach not just watches the behavior of neighbor nodes, additionally it screens the transmission of the information packet in the identification of the route for exact estimation of trust value. A survey is carried out to find some of the limitations behind the existing works which has been done by the researchers to implement various approaches thus to build the trust management framework. Through the survey, it is observed that existing works focused only on the authenticated transmission of the message, how it transmits packets to the destination node securely using a trust-based scheme. And also, it is observed that the routing approach only focused on

DOI: 10.4018/978-1-7998-8545-0.ch009

the key management issues. Certain limitation observed in the implemented approaches of existing work loses the reliability of framework. Thus, to withstand these issues it is necessary to establish a reliable security framework that protects the information exchanged among the users in a network while detecting various misbehaving attacks among the users. Confidentiality, as well as the integrity of information, can be secured by combining context-aware access control with trust management. The performance parameters should be evaluated with the previous works packet delivery ratio, packet drop, detection accuracy, number of false positives, and overhead.

1. INTRODUCTION

PC frameworks were at first made to work by interfacing PCs together with wires and transmitting data over these wires. Framework sizes and occasions extended making a need for system interactions. This headed to the change of the internet and its suite of traditions. The use of the internet and its applications got the opportunity to be inescapable. A prerequisite in terms of network access to entities while not physically joined to the wired framework recently emerged.

Figure 1. System model

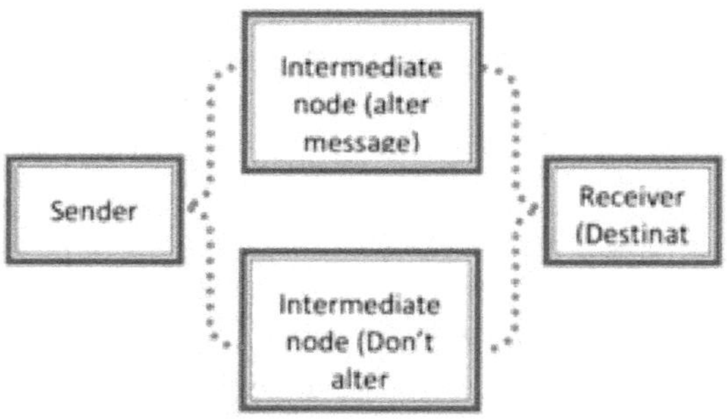

Figure 1 demonstrates the framework model of the MANET. The different modules in the framework model are as per the following.

Module 1 - Sender module (Source node): The objective of this module is to capture the message and afterward partition the message into packets of 48 bytes long, send the packets to the final node with the help of intermediate node and get feedback from the receiver node through the intermediate node.

Module 2 - Intermediate module (Intermediate node): The objective of this module is to get a packet from the sender, modify/don't vary the message and send it to the destination node. Get 2ACK packet from the recipient and send 2ACK packet to the sender.

Module 3 - Receiver module (Destination node): The objective of this module is to get a message from the intermediate node, take out final destination node name and hashcode and decrypted it. Analyze the hash code of the source node and destination node for security reason. Send 2ACK to source node with the help of intermediate node.

To enable this wireless systems administration, traditional devices with techniques are used to co-ordinate with a wired framework, using radio wave progressions through the wireless access point. At the same time, telephone frameworks were encountering a relative change. Cell framework efficiency was delivered to allow cell phones to interface in terms of base stations and confer in a circuit traded environment. Late advancements of remote innovations like Bluetooth, IEEE 802.11 displayed another sort of remote framework known as Mobile Ad-hoc networks. The region of mobile ad-hoc networking manages devices to perform wireless correspondence and systems administration (Marchang & Datta, 2012), which work without a central access point.

2. MANETS

2.1. Security Criteria in MANET

As of now said the mobile ad-hoc systems are defenseless against security issues than the wired systems, in this segment different vulnerabilities are investigated:

- **No predefined Boundary:** In mobile based hoc systems we can't correctly characterize a physical limit of the system. The nodes work in a traveling domain where they are permitted to join and leave the wireless system. When an attack comes in the radio range of a node it will have the capacity to interact with that node. The attacks incorporate eavesdropping; tempering, replay and Denial of Service (DoS) attack (Xia et al., 2013);
- **An adversary inside the Network:** The mobile nodes inside the MANET can uninhibitedly join and leave the system. The nodes inside system may show malicious node behavior. This is difficult to identify that the behavior of the node is malignant. Along these lines, this attack is more threat than the outer attack. These nodes are known as compromised nodes;
- **No Centralized Control approach:** MANETS doesn't include Centralized Control Facility, thus it may lead to numerous security issues. It turns out to be extremely hard to distinguish any attack. Traffic can't be observed from an incorporated point rather the control is shared at every node. The identification turns out to be more troublesome when the advisory varies the attack design and the objective of the attack. To the node, a failure might be brought about by an advisory or because of some system issue. Because of the absence of security links, we can't group the nodes as a trusted node or un-trusted node;
- **Limited Energy Resource:** All the nodes in a mobile ad-hoc system relies upon battery power for their operation. The other force sources are thought to be missing. The advisory can be sent tremendous movement to the objective node. The objective node might be frequently occupied in taking care of these packets; this will bring about the battery energy to be depleted. This will lead to Denial of Service attacks since now the node won't have the capacity to give services inside the network (Pariselvam & Parvathi, 2013);

Sometimes the attackers ask the nodes to perform some useless calculation bringing about its battery energy to be lost. A few nodes may function as selfish nodes. A selfish node does not collaborate when running some normal algorithm. For instance, consider a cluster dependent interruption identification method (Wang, Liu & Jiao, 2011) where a group of nodes agreeably recognize the interruption. A node is chosen as a screen when it wishes to carry this process. A malignant conduct basically neglects from being the detector. At the point when the dominant part of nodes behaves like a selfish node, the entire framework will collapse.

- **Modifying scale:** The scalability approach of the mobile ad-hoc system continues transforming constantly. It is extremely hard to estimate the number of nodes in a mobile ad-hoc system at some future time. The conventions and administrations intended for MANETs must be made perfect for this evolving scalability.

2.1.1 Attacks in MANETS

The attacks in the MANET can be outer or inner. An outer attack may lead to blockage, sends false routing data or causes inaccessibility of services. In an inside attack, the malevolent node from the system increases unapproved access and imitates as honest to the genuine node. It can detect congestion occurrence between different nodes and may take an interest in other network exercises:

- **Denial of Service attack:** This attack intends to destroy the accessibility of a node or on the whole system. In the event that the attack is successful then the administrations won't be accessible. The attacker, for the most part, uses radio signal congestion and the battery depletion technique;
- **Impersonation:** If the authentication approach is not legitimately actualized a malicious node can behave as honest to the genuine node and detect the system congestion. It can likewise send fake forwarding packets and access some secret information (Yang & Sun, 2016);
- **Eavesdropping:** This is one of the passive attacks. The node just infers the secret data. These data can be later utilized by the malicious node. The secret data like area, open (public) key, private key, and password etc. This can be brought by an eavesdropper;
- **Routing Attacks:** The malicious node makes services of routing as an objective. Since it is one of the vital services in MANETs. There are two flavors to this routing attack. One is an attack on routing convention and another is an attack on packet sending or receiving an application. The first is gone for blocking the spread of forwarded data to a node. The last is gone for interrupting the packet delivery against a predefined way (Venkataraman, Pushpalatha & Rao, 2012; Subramaniyan, Johnson & Subramaniyan, 2014; Orwat, Levin & Irvine, 2008; Subbaraj & Savarimuthu, 2014).

2.1.2. Security Solutions in MANET

In this area, we review a few security plans to manage the attacks described in the past areas:

1. **Intrusion Detection:** The Distributed and cooperative system is used to identify the attack. Each node in the MANET takes part at the time of interruption recognition. It identifies the indication of interruption locally and autonomously furthermore propagates this data to different nodes in the network;
2. **Cluster-based Intrusion Detection:** There is a disadvantage of Distributed and co-operated interruption identification system that because of the restricted power limit a portion of the node may act as a selfish node. To dispose of this drawback proposed a cluster dependent interruption

identification plan. The entire MANET is sorted out as an arrangement of groups with the end goal that a node is individual it contains at least one group. One and only node in a cluster will sense the interruption occurrence. The nodes inside a group are inside the same radio extent (Xia et al., 2013);

3. **Defending Wormhole Attack utilizing chain:** In wormhole attack some advisory records of data and replays to another node utilizes tunneling. The receiver node has a trouble of recognizing it from the unique forwarding packet. The attack against wormhole attack is given in (Li, Li & Kato, 2008; Pirzada & McDonald, 2006) utilizing packet chains. A Packet leach is extra data connected to the packet to confine its most extreme permitted transmission distance. Two sorts of chains exist. A geological leash guarantees the most extreme distance amongst sender and recipient. Temporal chain limits the most extreme time of packet excursion. A receiver looks at its time or distance whether it has allowed more than the permitted distance.

2.2. Important Parameters in MANET Security

On account of MANET's exceptional attributes, there are some critical measurements in MANET security that are essential in all security approaches called "Security Parameters" (Parvathy, Sundarakantham & Mercy Shalinie, 2015):

- **Network Overhead:** This parameter alludes to a number of control packets generated by security approaches. Because of shared wireless media, additional control packets may effectively prompt to blockage or impact in MANET. Packet loss is one of the consequences of collision and congestion. Consequently, high packet overhead builds packet lost and the quantity of retransmitted packets. This will effortlessly waste energy of nodes and systems resources;
- **Processing Time:** Every security approach needs time to recognize misbehaviors and to destroy malignant nodes. In terms of MANET's dynamic topology, it is firmly conceivable that paths between different nodes break on account of mobility. In this manner, security approaches must have as low as could be permitted for the processing time thus to build MANET adaptability and avoid rerouting process;
- **Consumption of Energy:** In MANET nodes have limited energy resources. Thus, enhancing energy utilization is significantly challengeable in MANET. High energy utilization reduces nodes as well as the system's lifetime.

2.2.1. Characteristics of MANETS

A MANET comprises of mobile stages (e.g., a switch with different hosts and wireless interaction devices). In this just denoted to as "nodes," which are allowed to move about subjectively. The nodes might be situated in or on planes, ships, trucks, autos, maybe even on individuals or on few devices, and there might be various hosts per switch. A MANET is a self-governing arrangement of the mobile node. The framework may work in the separate environment or may have ways to and interface with an altered system. In the last operational mode, it is normally imagined to work as a "stub" system interfacing with an altered internetwork.

Stub systems deliver traffic activity starting at and/or bound for inside nodes, yet don't allow exogenous movement to "travel" through the stub system. MANET nodes are furnished with remote transmitters and recipients utilizing radio wires which might be omnidirectional and more directional (point-to-point), perhaps routing, etc. (Akbani & Korkmaz, 2011).

At a given point of time, depending upon the nodes' positions and their transmitter and receivers scope designs, transmission power levels, and co-channel impedance levels, a remote connectivity as an irregular, multi-hop graph or Ad-hoc system exists between the nodes. This network arrangement of Ad-hoc may change with time as the nodes move or alter their transmission and gathering parameters.

2.3. MANET Security Challenges

Basically, the vital viewpoints in security are:

1. **Availability:** As shown by this service, each licensed node must have admittance to all information as well as services in the system. Accessibility challenge rises in view of MANET's dynamic topology and open limit. Receiving time, which is the time required for a node to gain the system administrations or information is imperative since time is one of the security parameters. By using loads of security and verification levels, this administration is expelled bypassing security levels requirement period;
2. **Authentication:** The goal of this administration is to give trustable correspondences between two distinct nodes. At a moment that node gains packets from a source, it must ensure about the character of the source node. One way to deal with this administration is using key distributions; certifications and key administration are challengeable. The authors have exhibited another path in view of the trust model and grouping to open the authentication keys;
3. **Data Secrecy:** As demonstrated by this administration, every node or application must have admittance to notified administrations that it has the authorization to receive. The vast majority of administrations that are given by information secretly utilize encryption techniques however in MANET as there is no central administration, key distribution challenges difficulties and it becomes impossible;
4. **Integrity:** According to trustworthiness security benefit, basically licensed nodes can generate editor eliminated packets. So, the attacker can catch all packets and after eliminates or alters them.

2.4. Trust Based Routing In MANET

Trust is characterized in terms of confident dependence of one entity on to the other (Sun et al., 2006), which is a measurement that is estimated on the premise of nodes activity, when it is required and utilized this to take over the decision about the nature of nodes and their administrations, and how to deliver the related access control in terms of its neighbour interaction status then form a reputation for every node by checking the behaviour of nodes. Figure 1 demonstrates the trust-based component. Trust can be actualized in different approaches by reputation, subjective rationale and from the decision of neighboring nodes (Theodorakopoulos & Baras, 2006). The two sorts of trust are accessible, for example, direct trust and in-direct trust.

Direct trust is referred to as main data for neighbors and non-complex to receive. In this proposed approach, each node in the system screens the lead of its neighbors, and if any anomalous activity is detected, it summons an algorithm to distinguish direct trust measure. This module screens neighbor nodes by inactively listening to their correspondence for recognizing dropped packet, more delayed packets, and directed packets. Of course, every one of the nodes while interacting with different nodes, the immediate trust estimation of all the imparting nodes are learned and put away in the trust table of relating nodes with field names like node index node, direct trust esteem and one more aggregate trust estimation of the related node.

The neighbor nodes may go beyond the particular node because of their transmission and also they return back to their transmitted range. On the other hand, the trust measure is computed and the relating entry in the table is upgraded (Li & Li, 2013; Zouridaki, Mark & Hejmo, 2007). Indirect trust is second-hand data about nodes named as a prescribed trust, which is gained from an outsider or neighboring nodes (Wang, Liu & Jiao, 2011; Venkataraman, Pushpalatha & Rao, 2012; Subramaniyan, Johnson & Subramaniyan, 2014). The objective of indirect trust detection is to accumulate or request the trust related data of the target node from the neighboring nodes. While demand for the trust data of the objective node from neighbors, the immediate trust estimation of that neighbor node ought to be considered. This is to maintain a strategic distance from the security attacks like bad-mouthing (Biradar & Manvi, 2012). Figure 2 shows the design of a trust-based detection model in MANET. The monitoring module includes the trust-based mechanism. This mechanism monitors the attackers in the neighboring nodes. Then it finds the most appropriate path. Finally, after resolving the attacks, it transmits the data to the node securely.

Figure 2. Design of trust-based mechanism

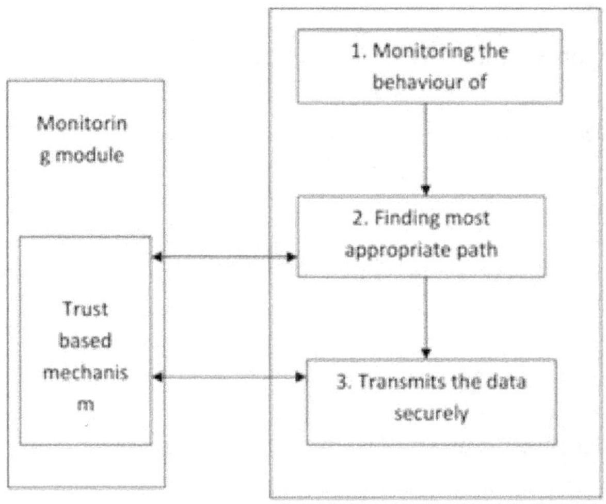

A trust model estimates and sets up trust relationship among objectives. Trust-based routing is utilized to keep away data from different attacks like a wormhole, black hole, Dos, selfish attacks and so forth.

2.5. Need for Trust Based Routing in MANET

Qualities of MANET, for example, dynamic topology changes, non-availability of base station and simplicity of joining or leaving the systems whenever makes the ad-hoc systems into more inclined to physical security dangers and hard to guarantee that a packet has been conveyed to the right destination so there is a requirement to provide secure correspondence and to diminish the risks from malignant nodes by consolidating the trust into MANET.

In trust management applications, "trust" can be characterized as the subjective desire that an objective has about another's future behavior in terms of the historical backdrop of their experiences. Trust can be considered as the dependence of a network node on the capacity to forward packets or offer administrations in a frequent manner, efficiently and reliably. These proposed trust models are basically in terms of the proof created by the past communication among nodes inside a protocol, which basically contains trust estimation and trust application. Inferring nodes behavior is a successful application to Figure out if a node can be trusted or not.

In (Pariselvam & Parvathi, 2013), the framework powerfully upgrades trust levels by utilizing reports from Intrusion Detection Systems (IDS) situated on all nodes in the system. The neighboring nodes to a node displaying suspicious behavior begin trust reports. Utilizing these trust levels as a guide, the source node then chooses a route that full-fills the security prerequisites of the message to be transmitted. The trust is resolved just in terms of interruptions recognized. A few scientists are supporting the utilization of appraisals and incline toward making utilization of rating collection algorithms to estimate the trust from a few perspectives (e.g., CPU usage, remaining energy, data transfer capacity, and so forth.). These perceptions can be joined together to register a trust value.

2.6. Overview

In the proposed method, we present a security framework that ensures the data traded among the clients in a network, while recognizing different misbehavior attacks among the clients. Confidentiality, as well as the integrity of data, is secured by joining Context-Aware Access Control with trust management.

Each node maintains an Access Policy (AP) which comprises of the accompanying access permission:

- Read (R)
- Modify (M)
- Forward (F)
- Process (P)

The client can acquire AP with a particular combination of permission. For instance, a client with strategy F is possible to forward the information (minimum access control), RF can read and forward, RM – Read and Modify. A client with RMFP permissions acquires full access control, which has a place with the top or service level.

An Ontological approach is also presented. Ontologies denote the shape of the domain, i.e. conceptualization which incorporates the model of the area with conceivable limitations. Ontology is the detail of this conceptualization – which is indicated by utilizing specific displaying language and specific terms. Formal indications are required thus to have the capacity to process ontologies and work on ontologies automatically.

For trust administration, a collaborative trust-based routing convention for MANETS (Chen et al., 2014) is utilized.

The direct communications with the neighbors are estimated in terms of a number of legitimate trials made by the node. The level of trustworthiness the client acquires in view of its activities is recorded in a database. Likewise, the suggestions from other associates are gained in terms of the database. Hence the

Ontological trust level is admitted to a node which is a blend of direct communication with its neighbors and the suggestions from its companions (peers).

Along with this Ontological trust, the QoS trust metric (Bo et al., 2011) is viewed as that considers the energy level and degree of nodes co-operation in information transmission.

All these ascribes (attributes) are passed to the Ontology administration mechanism, where the measures are checked against possible ranges and outcomes are returned which are utilized to estimate the general trust of the node.

The nodes are then included to the friend list in view of the trust level. The AP of a friend list is then redesigned in terms of the evaluated trust level. The friends acquiring most elevated trust level can be provided with full access control while the clients having minimum trust level can permit the minimum access control. The nodes damaging the access control authorizations are considered as malignant and can be expelled from the companion (friend) list.

To diminish the false positives and overhauling of the trust level, the weighted voting technique is utilized (Chang & Kuo, 2009). It depends on the reputation of a node to decide how to combine the trust measure getting from each nodes companion list it transmits together with the friend list of the receiver. At the point when entire nodes have a similar reputation of aggregate trust, the algorithm will stop, and the reputation that every one of the nodes holds is viewed as the worldwide reputation.

Apart from the attacks managed of by FACES, the proposed system can secure the specific forwarding attacks, route disturbance attack, packet manufacture attack, node separation attack, battery depletion attacks.

The proposed system will be actualized by utilizing computational model as well feasibility of simulation environment will be chosen continuously. The execution is assessed as far as for packet delivery proportion, packet drop, identification precision, number of false positives and overhead and will advance some more parameters to manufacture semantic profile of node. Different surveys were carried out to overcome the difficulties in a proposed method.

2.6.1 Security Issues in MANET

Mobile networks are basically more inclined to physical security threats than to the constant hardwired systems. Existing link level security methods (e.g. encryption) are regularly connected inside remote systems to diminish these dangers. Missing link-level encryption, behind the network layer, the most problem that needs to be addressed is one of inter-router verification before trading of system control data. A few levels of authentication varying from no security (dependably an alternative) and basic shared-key methodologies, to full open (public) key framework-based validation will be investigated by the group. As an assistant to the working firm trials, a few alternative verification modes might be formatted for use in MANETs.

Different issues identified with security components are organized in Table 1.

2.6.2. Trust Based Routing In MANET

An interrupter node can participate in the system and disturb the routing procedure by dropping the packets it gains for transmitting, i.e. denying to truly coordinated routing technique. Destroying the routing procedure incorporates releasing secret data, packet alteration, interrupting message, hence disrupting the fundamental security requirements.

Table 1. Different issues identified with security components

Author	Topic	Proposal	Advantages	Disadvantages
i) S. Pariselvam ii) R.M.S Parvathi	Trust-Based Security Mechanism For Service Discovery In MANET (Pariselvam & Parvathi, 2013)	A trust-based security mechanism for service discovery in MANET is proposed.	i) Better security against attacks. ii) Reduced packet drops.	i) Affecting service availability. ii) Frequent disconnections of the server.
i) Yan Lindsay Sun ii) Zhu Han iii) K. J. Ray Liu	Information Theoretic Framework of Trust Modelling and Evaluation for Ad Hoc Networks (Sun et al., 2006)	Proposed an information theoretic framework to quantitatively measure trust and model trust propagation in ad hoc networks	i) Improve the network throughput. ii) Effectively detect malicious behaviors in ad hoc networks.	i) Does not deal with security issues occurring in MANET. ii) Difficult to compare or validate these trust metrics. iii) The multipath environment might hinder the performance of the proposed method
i) Jie Li ii) Ruidong Li iii) Tsukuba Jien Kato	Future Trust Management Framework for Mobile Ad Hoc Networks (Li, Li & Kato, 2008)	Trust management framework is proposed to overcome vulnerabilities in the ad-hoc network. Here it proposes theoretical basis and skeleton for this framework.	i) Simplicity and high performance. ii) Achieves the effectiveness and robustness of the OTMF framework	i) Suffers from bad-mouthing, on-off, conflicting Behaviour, Sybil, and newcomer attacks. ii) Link loss
i) Sanjay K. Dhurandher ii) Mohammad S. Obaidat iii) Karan Verma Pushkar Gupta iv) Pravina Dhurandher	FACES: Friend-Based Ad Hoc Routing Using Challenges to Establish Security in MANETs Systems. (Dhurandher et al., 2011)	Proposed an algorithm to determine the trust of nodes in ad-hoc mobile networks called as Friend based Ad hoc routing using Challenges to Establish Security (FACES) by sending challenges and sharing friends list in real life scenarios.	i) Increases the Throughput ii) Reduces the Overhead iii) Detection accuracy iv) Easy mathematical computation	i) Efficiency and security are compromised in a MANET environment
i) Ben-Jye Chang ii) Szu-Liang Kuo	Markov Chain Trust Model for Trust-Value Analysis and Key Management in Distributed Multicast MANETs (Chang & Kuo, 2009)	The multicast MANET is proposed for a sender that sends packets to several receivers through a multicast session.	i) Increases efficiency in mobile ad hoc networks ii) Additional hardware to sense the neighbors iii) Computational complexity in calculating the entropy and trust	i) Less efficient in an unpredicted environment of interest. ii) It does not use either recommendation of the past observations. Hence the trust measurement is totally instantaneous and node dependent.
i) Ing-Ray Chen ii) Jia Guo iii) Fenye Bao iv) Jin-Hee Cho	Trust management in mobile ad hoc networks for bias minimization and application performance maximization (Chen et al., 2014))	Proposed a novel model-based methodology based on SPN techniques for describing the behavior of a mobile group consisting of well-behaved, malicious and uncooperative nodes.	i) Maximize application performance dynamically in response to environmental changes.	i) Less robust in handling heavy traffic and dealing with environmental instability ii) Does not deal with the multipath environment
i) Pallavi ii) Khatri	Using Identity and Trust with Key Management for achieving security in Ad hoc Networks (Khatri, 2014)	Proposed a composite identity and trust-based model (CIDT) which depends on the public key, physical identity, and trust of a node which helps in secure data transfer over wireless channels.	i) Works well for self-certification scheme of a node in the network. ii) The throughput of the network and Packet Delivery Ratio in the network is enhanced by this method.	i) Inefficient in 0situations occurring due to unknown attackers or hijacking of nodes. ii) Inefficient in handling multipath attackers (manipulating data).
i) Ruidong Li ii) Jie Li iii) Peng Liu iv) Hsiao-Hwa Chen	An Objective Trust Management Framework for Mobile Ad Hoc Networks (Li et al., 2007)	Proposed an objective trust management framework (OTMF) for MANETs, by which one node evaluates the trustworthiness of another node objectively based not only on direct observations but on second-hand information.	i) Obtains the more reasonable trust and it can inhibit the selective misbehavior attack. ii) Guarantee message integrity	i) This approach will be ineffective in spare networks.

continues on following page

Table 1. Continued

Author	Topic	Proposal	Advantages	Disadvantages
i) P. Manivannan ii) J. Shakila	Security Establishment in MANET Systems using FACES (Manivannan & Shakila, 2013))	FACES algorithm is proposed. The scheme proposed has been drawn from a network of friends in real life scenarios. The algorithm is divided into four stages, viz. Challenge your neighbor, Rate Friends, Share Friends, and Route through Friends.	i) Enhances network security. ii) Easier detection of malicious nodes in the mobile ad-hoc network. iii) Reduces the overhead on the network.	i) Specific to routing. Nodes should monitor neighbors all the time to construct and update trust relations. ii) Computed trust should be biased.
i) Bo Wang ii) Xunxun Chen iii) Weiling Chang	An Efficient Trust-Based Opportunistic Routing for Ad Hoc Networks (Bo, et al., 2011))	A new trusted opportunistic forwarding model based on the degree of trust and cost of routing is proposed.	i) Improves routing. ii) Enhances network security.	i) Specific to routing. Nodes should monitor neighbors all the time to construct and update trust relations. ii) Computed trust should be biased.
i) S.A. Razak ii) S.M. Furnell iii) N.L. Clarke iv) P.J. Brooke	Friend-assisted intrusion detection and response mechanisms for mobile ad hoc networks (Razak et al., 2008))	Proposed a novel intrusion detection system based upon a friendship concept, which could be used to complement existing prevention mechanisms that have been proposed to secure MANETs	i) Receive audit data gathered by other trusted nodes. ii) Anomaly detection is non-complex	i) Information might be altered by malicious nodes. ii) Difficult to manage/update attack signatures. iii) Unable to detect novel attacks. iv) High false alarms.
i) Rehan Akbani ii) Turgay Korkmaz	Enhancing role-based trust management with a reputation system for MANETs (Akbani & Korkmaz, 2011))	Enhanced RBTM with reputation systems (RSs), and propose a new hybrid trust management system (HTMS).	i) Highly efficient in the transmission of data in a secure manner. ii) Improvement in delivery ratio.	i) Vulnerable to blackmail attacks. ii) Vulnerable to colluding blackmail attackers.
i) R. Saminathan ii) K. Selvakumar	TRUCE –An Adaptive Trust Management Algorithm Over MANET for Service-Based Mobile Computing Environments (Saminathan & Selvakumar, 2011))	Focused on design, development of Trust evaluation, assignment, algorithms termed TRUCE over a dynamic, highly insecure MANET network.	i) Reduced latency. ii) Efficient transmission of data securely.	i) High false alarms. ii) This approach will be ineffective in spare networks.
i)Mukesh Kumar Garg ii) Neeta Singh iii)Poonam Verma	Fuzzy rule-based approach for design and analysis of a Trust-based Secure Routing Protocol for MANETs (Garg, Singh & Verma, 2018))	The fuzzy rule-based approach has been proposed for determining a secure route between sources and destination	i) The throughput of the network has been improved	i) Delay of route selection is should be reduced
i) Shashi Gurung ii) Siddhartha Chauhan	A dynamic threshold based algorithm for improving security and performance of AODV under black-hole attack in MANET (Gurung & Chauhan, 2017)	Dynamic sequence number threshold value based MBDP-AODV protocol has been proposed for mitigating the issue of black-hole attack	i) Delivery of packets has been improved ii) Normalized routing load has been reduced	i) Routing overhead has to be reduced
i) Vallala Sowmya Devi ii) Nagaratna P Hegde	Trust enhanced cluster based multipath routing (TECM) algorithm (Devi & Hegde, 2018)	Lossless data transfer has been achieved by presenting trust enhanced cluster based multipath routing algorithm	i)Routing overhead has been reduced ii) Network lifetime has been improved	i) Information might be altered by malicious nodes.
i) K. Gomathi ii) B. Parvathavarthini iii) C. Saravanakumar	Efficient Secure Group Communication Using Fuzzy Trust Based Clustering and Hierarchical Distributed Group Key Management (Gomathi, Parvathavarthini & Saravanakumar, 2017)	Misbehaving nodes are isolated using the fuzzy trust-based clustering (FTBC)	i)Packets overhead have been reduced	ii) Energy efficiency has to be improved

So malignant attacks on a wireless ad-hoc network can originate from both inside and outside of the system and focus at any node all these imply that each node must be prepared for experience with a malignant node directly or indirectly (Velloso et al., 2010). For that trust dependent routing in terms of trust node valuation answers gives better throughput benefits (Orwat et al., 2008). With the support of trust application mobile nodes can know whether they can trust other mobile nodes directs nodes to eliminate exceedingly unsafe activities, for example, receiving or transmitting packets from (to) the nodes with low reliable levels.

Conventional MANET routing approaches are vulnerable to various security attacks, which combine black-hole as well as grey-hole attacks, flooding and Sybil attackers.

As indicated by latest research, a Secure Routing Protocol for MANETs has been proposed in view of the trust levels of nodes on the network. It uses a trust to segregate black hole routing attacks hence promotes secured routing of information congestion and in addition enhance packet delivery proportion (Li & Li, 2013). The survey behind trust-based directing is organized below in Table 2.

2.6.3. Ontological Framework on MANET

The term ontology, established in theory, depicts the investigation of existence. Software engineering (initially the computerized reasoning group) later embraced the term ontology. The philosophies are 1. "The hypothesis of the described world". 2. "A segment of information frameworks" Thus, other than the philosophical implication of ontology, there are practical purposes behind their use.

Ontology, as a building mechanism, might be further characterized by its utilization. The applications may give the "representational hardware with which to instantiate domain patterns in information bases," permit the questioning of learning based administrations and denotes the outcomes from these inquiries (Li, Li & Kato, 2008). The utilization of ontology has turned out to be significantly more far-reaching since the World Wide Web Consortium (W3C) incorporated the idea as an explicit layer in the model's stacks for the future semantic web (Pirzada & McDonald, 2006).

The semantic web utilizes ontology to determine standard estimated terms. This approach makes information trade in complex and learning databases more available all through the World Wide Web. W3C is utilizing the capacity of ontologies to standardize information into steady terms and to give induction from information because of linkages between normal terms.

The linkages are showed as relationship standards among objects. Inside an ontological system, classes are dynamic groups, sets, or accumulations of objects. Objects are the fundamental individual things in the area. Attributes are defined as properties or attributes that objects can be shared. Relations are ways that items may associate with each other. Ontology is unique in relation to scientific classifications, which don't join the relations concept.

A relationship is a property whose measure is referred to as another entity in the ontology. Ontological approaches can indicate subjectively complex standards about the characteristics of the related items, while a scientific classification only promotes the "is-a" connection. The group of relations taken will completely depict the semantics of the domain.

Much of the surveys as of now aimed around ontologies on the generation of top-level domain ontologies. The main aim of this broad research is to characterize linkages between area particular ontologies in order to provide consistency of the root (widely inclusive) metaphysics. The study behind the ontological approach is arranged below in Table 3.

Table 2. The survey behind trust-based directing

Author	Topic	Proposed	Advantages	Disadvantages
i) Hui Xia ii) Zhiping Jia iii) Xin Li iv) Lei Ju v) Edwin H.-M. Sha	Trust prediction and trust-based source routing in mobile ad hoc networks (Xia et al., 2013)	Proposed a dynamic trust prediction model to evaluate the trustworthiness of nodes.	i) Improves the packet delivery ratio ii) Reduces average end-to-end latency	i) Obstacles are not considered in this scenario ii) Frequent topology changes
i) Surendran Subbaraj ii) Prakash Savarimuthu	Eigen Trust-based non-cooperative game model assisting ACO look-ahead secure routing against selfishness (Subbaraj & Savarimuthu, 2014)	Proposed a QoS-constrained Eigen Trust-based non-cooperative game model for secure fault-tolerant ant look-ahead routing which attempts to identify trusted valid route	i) Improves the accuracy ii) Achieve more packet delivery ratio iii) Reduced delay iv) Reduced overhead v) Reduced packet drop in the network	i) Key generation and management are a tedious task ii) Regression models may create complexity in the nodal computation
i) Fenye Bao ii) Ing-Ray Chen iii) MoonJeong Chang iii) Jin-Hee Cho	Hierarchical Trust Management for Wireless Sensor Networks and its Applications to Trust-Based Routing and Intrusion Detection (Bao et al., 2012)	Proposed a highly scalable cluster-based hierarchical trust management protocol for wireless sensor networks (WSNs) to effectively deal with selfish or malicious nodes.	i) Outperforms traditional anomaly-based IDS techniques in the detection probability ii) Sufficiently low false positives rates	i) Control overhead affects the performance of the routing protocol ii) Security threats are not taken into account
i) Wang Bo ii) Huang Chuanhe iii) Li Layuan iv) Yang Wenzhonga	Trust-based minimum cost opportunistic routing for Ad hoc networks. (Bo et al., 2011))	A new trusted opportunistic forwarding model is proposed by choosing the trusted and highest priority candidate forwarder, then a trusted minimum cost routing algorithm (MCOR) is formally formulated.	i)Increased throughput ii) Reduced delay iii) Expected outcomes of ETX, security-gains and reduces the cost of routing	i) Measurements are totally instantaneous and may not be precise ii) It does not consider the historical behavior of nodes
i) N. Marchang ii) R. Datta	Light-weight trust-based routing protocol for mobile ad hoc-networks. (Marchang, & Datta, 2012)	Proposed a light-weight trust-based routing protocol. The trusted node for a neighbor forms the basic building block of our trust model.	i) Reduces the time consumption ii) More buffer space	i) Limited resource availability ii) In order to monitor traffic that requires a lot of space and time
i) Muthumanickam ii) Gunasekaran iii) Kandhasamy iv) Premalatha	TEAP: a trust-enhanced anonymous on-demand routing protocol for mobile ad hoc networks. (Gunasekaran & Premalatha, 2013)	Introduces the concept of anonymity for an informant who identifies and reports anonymously the misbehavior of the users of the network.	i) Secure communication ii) Effective in providing both anonymity and liability	i) More energy consumption ii) Frequent loss of links
i) Rajashekhar C. Biradar ii) Sunilkumar S. Manvi	Ring Mesh-Based Multicast Routing Scheme in MANET Using Bandwidth Delay Product. (Biradar & Manvi, 2012)	Proposed BDP based multicast routing scheme in MANETs that uses reliable ring mesh backbone.	i) Increase in packet delivery ratio ii) Minimizes control overhead iii) Reduces memory Overhead	i) Suffers from link loss ii) Link breakage at the time of transmission
i) Bo Wang ii) Xunxun iii) Chen Weiling Chang	A Light-weight Trust-based QoS Routing Algorithm for Ad Hoc Networks (Wang, Chen & Chang, 2014)	Introduce the concept of trust and QoS metric estimation into establishing a trust-based QoS model.	i) Increase in packet delivery ratio ii) Minimized end-to-end delay iii) The decrease in routing packet overhead and the ratio of malicious nodes	i) Intrusions occur at the time of transmission ii) Communication overhead

continues on following page

Table 2. Continued

Author	Topic	Proposed	Advantages	Disadvantages
i) Gautam M. Borkar ii) A. R. Mahajan	A secure and trust based on-demand multipath routing scheme for self-organized mobile ad-hoc networks. (Borkar & Mahajan, 2017))	Proposed a novel secure adjacent trust-enhanced routing protocol combined with the trust model, named as AOMDV–SAPTV	i) Enhanced the quality of routing and had found the best path by the optimization algorithm	i) Less robust in a dynamic environment and multipath environment ii) More consumption of energy
i) Menaka Pushpa Arthur ii) Kathiravan Kannan	Cross-layer based multiclass intrusion detection system for secure multicast communication of MANET in military networks (Arthur & Kannan, 2016))	Designed a distributed cross-layer based machine learning anomaly detection system for multicast communication of MANET.	i) Control packet drop in the multicast routing protocol ii) Efficient in data transmission	i) Loss of links at the time of multicasting routing ii) Low-accuracy
i) Jin-Hee Choa ii) Ing-Ray Chenb iii) KevinS Can	Trust threshold based public key management in mobile ad-hoc networks. (Cho, Chen & Chan, 2016))	Proposed a fully distributed trust-based public key management approach for MANETs using a soft security mechanism based on the concept of trust.	i)Resiliency against misbehaving nodes in the network to maintain minimum security vulnerability ii) Availability of service provision in the presence of compromised nodes (iii) Efficiency in minimizing communication overhead incurred by trust and key management operations	i)Leakage of information ii) Congestion occurrence
i) Tameem Eissa ii) Shukor Abdul Razak iii) Rashid Hafeez Khokhar iv) Normalia Samian	Trust-Based Routing Mechanism in MANET: Design and Implementation (Eissa, et al., 2013))	It presents FrAODV, a trust-based scheme for securing AODV routing protocol in MANET using the friendship mechanism.	i)Robust environment ii) Secure community	i) Worm-hole attack ii) Hop-count altering attack iii) Routing messages dropping attack
i) Xiaoqi Li ii) Michael R. Lyu iii) Jiangchuan Liu	A Trust Model Based Routing Protocol for Secure Ad-Hoc Networks (Li, Lyu & Liu, 2004)	Designed a novel secure routing protocol for MANETs. This protocol T AODV extends the widely used AODV routing protocol.	i)Reduces the computation overheads	i) No proper route entry for the destination ii) High Cost
i) Malik N. Ahmed ii) Abdul Hanan Abdullah iii) Hassan Chizari iv) Omprakash Kaiwartya	F3TM: Flooding Factor based Trust Management Framework for secure data transmission in MANETs (Ahmed, Abdullah, Chizari & Kaiwartya, 2017)	Proposed Flooding Factor based Framework for Trust Management (F3TM) in MANETs. True flooding approach is utilized to identify attacker nodes based on the calculation of trust value.	i) Avoids flooding attacks	i) Suffers from interaction overheads ii) Complexity in the estimation of trust measure
i) Radha Krishna Bar ii) Jyotsna Kumar Mandal iii) Moirangthem Marjit Singh	QoS of MANet Through Trust Based AODV Routing Protocol by Exclusion of Black Hole Attack. (Bar, Mandal & Singh, 2013))	Generated a trust value for each node depending upon the packet forwarding ability of the node.	i)Improves the trustworthiness of the nodes ii) Reduces the overhead iii) Highly precision	i) Packet loss in between the links ii) Non-reliable communication

Table 3. The study behind the ontological approach

Author	Topic	Proposed	Advantages	Disadvantages
i) Mark E. Orwat ii) Timothy E. Levin iii) Cynthia E. Irvine	An Ontological Approach to Secure MANET Management (Orwat, Levin & Irvine, 2008)	Introduced the MANET Distributed Functions Ontology, which organizes the commonly used decision parameters and incorporates security parameters	i) Computation complexity is low ii) Reduced delay	i) It is necessary to reduce the cost and communication overhead ii) Do not modify or restrict the route discovery or route maintenance behavior of the underlying protocol
i) Amandeep Verma ii) C. Mcdonald	Ontological engineering approach towards trust-oriented security framework for ad-hoc networks (Pirzada & McDonald, 2006)	Proposed a trust-oriented security framework for an ad-hoc network utilizing ontological engineering approach by modeling ad hoc network, the OLSR (Optimized Link State Routing) protocol and trust model as OWL (Ontology Web language) ontologies, which are integrated using Jena.	i) Combines the trust measurement with the confidence value Hence the trust is represented in a precise way ii) The efficiency of data access was greatly improved iii) Data confidentiality	i) No additional hardware or computations required ii) The convergence time in trust computations and readjustments are high
i) Charikleia Zouridak ii) Brian L. Mark iii) Marek Hejmo	Byzantine robust trust establishment for mobile ad hoc networks (Zouridaki, Mark & Hejmo, 2007)	Proposed new mechanisms to make Hermes robust to Byzantine behavior and introduce a punishment policy that discourages selfish node behavior.	i) Efficient in packet forwarding ii) Reliable routing	i) Limited resources ii) Node fails at the detection of certain types of Byzantine behavior
i) George Theodorakopoulos ii) John S. Baras	On Trust Models and Trust Evaluation Metrics for Ad Hoc Networks (Theodorakopoulos & Baras, 2006)	Focused on the evaluation of trust evidence in ad hoc networks.	i) Enhancement of network performance in terms of throughput and packet delivery ratio ii) Lower interference level and the end-to-end delay were observed	i) Control overhead affects the performance of the routing protocol ii) Security threats are not taken into account
i) Pedro B. Velloso ii) Rafael P. Laufer iii) Daniel de O. Cunha iv) Otto Carlos M. B. Duarte v) Guy Pujolle	Trust Management in Mobile Ad Hoc Networks Using a Scalable Maturity-Based Model (Velloso et al., 2010)	Proposed a human-based model which builds a trust relationship between nodes in an ad-hoc network.	i) Reduces the trust level error ii) Information dissemination, computational complexity, and communication-expensive multi-hop communication can be avoided	i) Un-optimized utilization of network overheads ii) Traffic delay is high
i) Hwan-Seok Yang ii) Jae-Hoon Sun	A study on hybrid trust evaluation model for identifying malicious behavior in mobile P2P (Yang & Sun, 2016)	Proposed hybrid trust evaluation technique based on Trust Zone structure to improve the reliability between nodes.	i) High reliability ii) The selfish behavior of nodes is blocked	i) Does not deal with intrusion or malicious attackers ii) Does not consider environmental factors and their hindrance

continues on following page

Table 3. Continued

Author	Topic	Proposed	Advantages	Disadvantages
i) M. Parvathy ii) K. Sundarakantham iii) S. Mercy Shalinie	Cosine similarity-based clustering and dynamic reputation trust-aware key generation scheme for trusted communication on social networking (Parvathy, Sundarakantham & Mercy Shalinie, 2015)	A novel cosine similarity based clustering and dynamic reputation trust-aware key generation (CSBC-DRT) scheme are proposed.	i) Enhances more speed at the time of transmission ii) More security in terms of key generation	i) Due to the clustering mechanism, it consumes more vitality ii) Suffers from the congestion of nodes
i) Amandeep Verma ii) Manpreet Singh Gujral	Trust oriented security framework for the ad-hoc network (Bar, Mandal & Singh, 2013)	A trust-oriented security framework for an ad-hoc network using ontological engineering approach is proposed by modeling an ad hoc network, the OLSR protocol and trust model as an OWL.	i) Improves the security level of the framework ii) Reduces the communication overhead	i)Channel vulnerability and node vulnerability ii) Power and computational limitations
i) Mohsen Taherian, ii) Rasool Jalili iii) Morteza Amini	PTO: A Trust Ontology for Pervasive Environments (Verma & Gujral, 2012)	Proposed a computational trust model based on the ontology structure, considering the mentioned semantic relations.	i)Improves the lifetime of the network ii) Enhance the processing speed	i)Reduces the memory capacities, size, and performance
i) Bill Tsoumas ii) Stelios Dritsas iii) Dimitris Gritzalis	An Ontology-Based Approach to Information Systems Security Management (Tsoumas, Dritsas, & Gritzalis, 2005)	Proposed the foundations for establishing a knowledge-based, ontology-centric framework with respect to the security management of an arbitrary IS.	i)Support knowledge sharing and reuse in an explicit and mutually agreed manner	i) Degradation of performance
i) Bill Tsoumas ii) Dimitris Gritzalis	Towards an Ontology-based Security Management (Tsoumas, & Gritzalis, 2006)	Proposed a security management framework of an arbitrary information system (IS).	i)Exploitation of a vulnerability leads to the realization of an Unwanted Incident	i) Time complexity
i) Artem V orobiev ii) Nargiza Bekmamedova	An Ontological Approach Applied to Information Security and Trust (Vorobiev, & Bekmamedova, 2007)	Proposed basic security concepts and trust implications respectively.	i)Reduces the security attacks	i) Increases the complexity of the evaluation ii)Communication overhead

3. OBSERVATIONS BASED ON THE SURVEY

In (Pariselvam & Parvathi, 2013; Sun et al., 2006; Li, Li & Kato, 2008; Dhurandher et al., 2011; Chang & Kuo, 2009; Chen et al., 2014; Khatri, 2014; Li et al., 2007; Zouridaki, Mark, & Hejmo, 2007; Theodorakopoulos & Baras, 2006; Velloso et al., 2010; Yang & Sun, 2016; Razak et al., 2008; Akbani & Korkmaz, 2011; Saminathan & Selvakumar, 2011) authors have reported to security issues on MANET. Most of this suffers from the link failure at the time of transmission. In a mission-critical application scenario, when viewed deeply various types of data exchanged between the users may belong to different security levels. Hence those data should be accessed only by selected types of users.

Authors in (Xia et al., 2013; Subbaraj & Savarimuthu, 2014; Manivannan & Shakila, 2013; Bao et al., 2012; Bo et al., 2011; Cho, Chen & Chan, 2016; Eissa et al., 2013; Li, Lyu & Liu, 2004; Ahmed et al, 2017) reported to the trust-based routing approach in MANET. The author in (Subbaraj & Savarimuthu, 2014) observed that the routing approach only focuses on the key management issues. Hence more efficient approaches are needed to build a reliable framework to improve the security.

Authors in (Orwat, Levin & Irvine, 2008; Pirzada & McDonald, 2006; Bar, Mandal & Singh, 2013; Verma & Gujral, 2012; Taherian, Jalili & Amini, 2008; Tsoumas, Dritsas & Gritzalis, 2005; Tsoumas & Gritzalis, 2006; Vorobiev & Bekmamedova, 2007) carried out their research on ontological and recent trends of MANET. Through the survey, it is observed that it focused only on the authenticated transmission of the message, how it transmits packets to the destination node securely using trust-based scheme. The issue behind this was it doesn't view on its reliability of the framework.

In (Dhurandher et al., 2011; Manivannan & Shakila, 2013) authors carried out their research on proposed Friend based Ad hoc routing utilizing Challenges to Establish Security (FACES) in which each observers node include its neighbor node together with its friends list depending on the companionship with different nodes and the measure of information transmission they achieve acquired through Share Your Friends procedure in the system without including the update of the trust measure, the level of energy as well as the co-operative degree. Because of its mobility, level of energy and collaboration may change so need to update the trust estimation of every node. It ought to consider about the creation of false perspectives by pernicious node about the fairness of a node furthermore not able to locate the malicious node on the off chance that malignant node confirms itself with a friendly manner as soon as after it turns as a black node thus to add itself as a friend node in the listener's friend list.

In (Manivannan & Shakila, 2013) every listener node stores distinctive sorts of trust relationship about its neighbor node in its companion list related to its challenges and behavioral themselves. It might give unreliable proof about fair node furthermore not able to discover malignant node that authenticates itself in a friendly manner but immediately it turns out as a black node.

3.1. Inference

Previous works only focused on communication security, such as key management, message authentication etc. From the survey, it is observed that due to these drawbacks it loses the security of the framework. To overcome the shortcomings identified in the survived works, the following points should be considered for the development of trust-based routing in mobile ad hoc networks:

1. Observer node should redesign the trust estimation of the neighbor node at the time of topological changes and when moving to various zone;

2. Should enhance the reliability and robustness of correspondence within the sight of the malignant node;
3. Must offer the impressive security with less overhead;
4. Should estimate the predefined threshold measure to recognize malignant and neighbor node as well as it updates this in terms of dynamic condition;
5. Trust estimation process not only infers the behavior of neighbor node likewise it screens the transmission of information packets in route identification for precise computation of trust measure;
6. Calculated trust pattern must adjust and should match to convoluted situations.

3.2. Challenges

- Trust dependent routing ought to play out the part of secure path detection from source to destination accurately within the sight of malignant nodes;
- The relation between a focal trust authority and different nodes in a system makes confusion in trust establishment of clear ad-hoc networks;
- Computing trust from various levels of inherent learning in different nodes will make trust calculation as heavier and the suitable trust of a node can't be processed (Subbaraj & Savarimuthu, 2014);
- The security arrangement ought to be rapid in terms of the changed trust relationship;
- Nodes in MANET ought to be ready to spend its battery life to keep up its trust measure;
- Trust calculations in MANET are in demand due to a result of various sorts of mobility, for example, low mobility or high mobility;
- Dynamic nature of MANET prompts to instability and inadequacy of trust proof that always shows signs of change after some time (Sun et al., 2006);
- Trust administration mechanism must satisfy the prerequisite of the security, the secrecy, and the life cycle of the entire network.

3.3. Suggested Framework

In above sections, security issues of MANET are survived. Also some existing techniques have been discussed in the previous sections. These techniques might improve the security of the network. By inferring from the survey papers, security of the network is further to be improved. So, new framework has to be suggested that optimal trust based path selection for routing in MANET. Attack nodes in the network can be classified using the optimal classifier technique. Based on the calculated trust parameters, attacked and normal nodes can be classified. After detecting the attack nodes, reliable or optimal path between the source and destination can be selected using the optimization algorithm. Using this suggested framework, security of the network will be improved than that of the previous frameworks.

4. CONCLUSION

It is quite clear that existing trust management approach faces security issues due to the malicious nodes activity. Existing work focuses only on the data transfer mechanism, how it can deliver a message to the destination node using a message authentication technique. Messages communicated between the nodes should be insecure form. Thus, in order to overcome these issues, it is necessary to design a reliable trust management framework for MANET. It is necessary to introduce a security framework that protects the information exchanged among the users in a network while detecting various misbehaving attacks among the users. Confidentiality, as well as the integrity of data, is secured in terms of joining Context-Aware Access Control together with trust management. An Ontological approach is also introduced. The level of honesty the user possesses based on its actions are recorded in a database. Similarly, the recommendations from other peers are obtained based on the database. Thus, the ontological trust level is assigned to a node which is a combination of direct interaction with its neighbors and the recommendations from its peers. It can be implemented using a computational model and feasibility of simulation environment will be decided further. The performance can be evaluated in terms of packet delivery ratio, packet drop, detection accuracy, number of false positives and overhead and will further explore some more parameters to build a semantic profile of node.

REFERENCES

Ahmed, M. N., Abdullah, A. H., Chizari, H., & Kaiwartya, O. (2017). F3TM: Flooding Factor based Trust Management Framework for secure data transmission in MANETs. *Journal of King Saud University-Computer and Information Sciences*, *29*(3), 269–280. doi:10.1016/j.jksuci.2016.03.004

Akbani, R., & Korkmaz, T. (2011). Enhancing role-based trust management with a reputation system for MANETs. *EURASIP Journal on Wireless Communications and Networking*, (1), 90. doi:10.1186/1687-1499-2011-90

Arthur, M. P., & Kannan, K. (2016). Cross-layer based multiclass intrusion detection system for secure multicast communication of MANET in military networks. *Wireless Networks*, *22*(3), 1035–1059. doi:10.100711276-015-1065-2

Bao, F., Chen, R., Chang, M., & Cho, J. H. (2012). Hierarchical trust management for wireless sensor networks and its applications to trust-based routing and intrusion detection. *IEEE eTransactions on Network and Service Management*, *9*(2), 169–183. doi:10.1109/TCOMM.2012.031912.110179

Bar, R. K., Mandal, J. K., & Singh, M. M. (2013). QoS of MANet through trust based AODV routing protocol by exclusion of black hole attack. *Procedia Technology*, *10*, 530–537. doi:10.1016/j.protcy.2013.12.392

Biradar, R. C., & Manvi, S. S. (2012). Ring mesh based multicast routing scheme in MANET using bandwidth delay product. *Wireless Personal Communications*, *66*(1), 117–146. doi:10.100711277-011-0329-0

Bo, W., Chuanhe, H., Layuan, L., & Wenzhong, Y. (2011). Trust-based minimum cost opportunistic routing for Ad hoc networks. *Journal of Systems and Software*, *84*(12), 2107–2122. doi:10.1016/j.jss.2011.05.043

Borkar, G. M., & Mahajan, A. R. (2017). A secure and trust based on-demand multipath routing scheme for self-organized mobile ad-hoc networks. *Wireless Networks*, *23*(8), 2455–2472. doi:10.100711276-016-1287-y

Chang, B. J., & Kuo, S. L. (2009). Markov chain trust model for trust-value analysis and key management in distributed multicast MANETs. *IEEE Transactions on Vehicular Technology*, *58*(4), 1846–1863. doi:10.1109/TVT.2008.2005415

Chen, R., Guo, J., Bao, F., & Cho, J. H. (2014). Trust management in mobile ad hoc networks for bias minimization and application performance maximization. *Ad Hoc Networks*, *19*, 59–74. doi:10.1016/j.adhoc.2014.02.005

Cho, J. H., Chen, R., & Chan, K. S. (2016). Trust threshold based public key management in mobile ad hoc networks. *Ad Hoc Networks*, *44*, 58–75. doi:10.1016/j.adhoc.2016.02.014

Cho, J. H., Swami, A., & Chen, R. (2012). Modeling and analysis of trust management with trust chain optimization in mobile ad hoc networks. *Journal of Network and Computer Applications*, *35*(3), 1001–1012. doi:10.1016/j.jnca.2011.03.016

Devi, V. S., & Hegde, N. P. (2018). Multipath Security Aware Routing Protocol for MANET Based on Trust Enhanced Cluster Mechanism for Lossless Multimedia Data Transfer. *Wireless Personal Communications*, *100*(3), 923–940. doi:10.100711277-018-5358-5

Dhurandher, S. K., Obaidat, M. S., Verma, K., Gupta, P., & Dhurandher, P. (2011). Faces: Friend-based ad hoc routing using challenges to establish security in MANETs systems. *IEEE Systems Journal*, *5*(2), 176–188. doi:10.1109/JSYST.2010.2095910

Eissa, T., Razak, S. A., Khokhar, R. H., & Samian, N. (2013). Trust-based routing mechanism in MANET: Design and implementation. *Mobile Networks and Applications*, *18*(5), 666–677. doi:10.100711036-011-0328-0

Garg, M. K., Singh, N., & Verma, P. (2018). Fuzzy rule-based approach for design and analysis of a Trust-based Secure Routing Protocol for MANETs. *Procedia Computer Science*, *132*, 653–658. doi:10.1016/j.procs.2018.05.064

Gomathi, K., Parvathavarthini, B., & Saravanakumar, C. (2017). An efficient secure group communication in MANET using fuzzy trust based clustering and hierarchical distributed group key management. *Wireless Personal Communications*, *94*(4), 2149–2162. doi:10.100711277-016-3366-x

Gunasekaran, M., & Premalatha, K. (2013). TEAP: Trust-enhanced anonymous on-demand routing protocol for mobile ad hoc networks. *IET Information Security*, *7*(3), 203–211. doi:10.1049/iet-ifs.2012.0141

Gurung, S., & Chauhan, S. (2017). A dynamic threshold based algorithm for improving security and performance of AODV under black-hole attack in MANET. *Wireless Networks*, 1–11.

Khatri, P. (2014, February). Using identity and trust with key management for achieving security in Ad hoc Networks. *In Advance Computing Conference (IACC), 2014 IEEE International* (pp. 271-275). IEEE. 10.1109/IAdCC.2014.6779333

Li, J., Li, R., & Kato, J. (2008). Future trust management framework for mobile ad hoc networks. *IEEE Communications Magazine*, *46*(4).

Li, R., & Li, J. (2013). Requirements and design for neutral trust management framework in unstructured networks. *The Journal of Supercomputing*, *64*(3), 702–716. doi:10.100711227-010-0502-3

Li, R., Li, J., Liu, P., & Chen, H. H. (2007, April). An objective trust management framework for mobile ad hoc networks. In *Vehicular Technology Conference, 2007*. VTC2007-Spring. IEEE 65th (pp. 56-60). IEEE.

Li, X., Lyu, M. R., & Liu, J. (2004, March). A trust model based routing protocol for secure ad hoc networks. *In Proceedings. 2004 IEEE Aerospace Conference* (Vol. 2, pp. 1286-1295). IEEE.

Manivannan, P., & Shakila, J. (2013). Security Establishment in MANET Systems using FACES. *International Journal of Science and Engineering Applications*, *2*(4), 60–66. doi:10.7753/IJSEA0204.1002

Marchang, N., & Datta, R. (2012). Light-weight trust-based routing protocol for mobile ad hoc networks. *IET Information Security*, *6*(2), 77–83. doi:10.1049/iet-ifs.2010.0160

Orwat, M. E., Levin, T. E., & Irvine, C. E. (2008, March). An ontological approach to secure MANET management. In *Third International Conference on Availability, Reliability and Security ARES 08* (pp. 787-794). IEEE.

Pariselvam, S., & Parvathi, R. M. S. (2013). Trust Based Security Mechanism for Service Discovery in MANET. *Journal of Theoretical & Applied Information Technology*, *56*(2), 226–234.

Parvathy, M., Sundarakantham, K., & Mercy Shalinie, S. (2015). Cosine similarity-based clustering and dynamic reputation trust aware key generation scheme for trusted communication on social networking. *Journal of Statistical Computation and Simulation*, *85*(16), 3247–3258. doi:10.1080/00949655.2014.964240

Pirzada, A. A., & McDonald, C. (2006). Trust establishment in pure ad-hoc networks. *Wireless Personal Communications*, *37*(1-2), 139–168. doi:10.100711277-006-1574-5

Razak, S. A., Furnell, S. M., Clarke, N. L., & Brooke, P. J. (2008). Friend-assisted intrusion detection and response mechanisms for mobile ad hoc networks. *Ad Hoc Networks*, *6*(7), 1151–1167. doi:10.1016/j.adhoc.2007.11.004

Saminathan, R., & Selvakumar, K. (2011). TRUCE–An Adaptive Trust Management Algorithm Over MANET for Service-Based Mobile Computing Environments. *Information Security Journal: A Global Perspective*, *20*(4-5), 173-184.

Subbaraj, S., & Savarimuthu, P. (2014). Eigen Trust-based non-cooperative game model assisting ACO look-ahead secure routing against selfishness. *EURASIP Journal on Wireless Communications and Networking*, *2014*(1), 78. doi:10.1186/1687-1499-2014-78

Subramaniyan, S., Johnson, W., & Subramaniyan, K. (2014). A distributed framework for detecting selfish nodes in MANET using Record-and Trust-Based Detection (RTBD) technique. *EURASIP Journal on Wireless Communications and Networking*, *2014*(1), 205. doi:10.1186/1687-1499-2014-205

Sun, Y. L., Yu, W., Han, Z., & Liu, K. R. (2006). Information theoretic framework of trust modeling and evaluation for ad hoc networks. *IEEE Journal on Selected Areas in Communications*, *24*(2), 305–317. doi:10.1109/JSAC.2005.861389

Taherian, M., Jalili, R., & Amini, M. (2008, March). Pto: A trust ontology for pervasive environments. In *22nd International Conference on Advanced Information Networking and Applications-Workshops AINAW 2008* (pp. 301-306). IEEE.

Theodorakopoulos, G., & Baras, J. S. (2006). On trust models and trust evaluation metrics for ad-hoc networks. *IEEE Journal on selected areas in Communications*, *24*, 318-328.

Tsoumas, B., Dritsas, S., & Gritzalis, D. (2005, September). An ontology-based approach to information systems security management. In *International Workshop on Mathematical Methods, Models, and Architectures for Computer Network Security* (pp. 151-164). Springer. 10.1007/11560326_12

Tsoumas, B., & Gritzalis, D. (2006, April). Towards an ontology-based security management. In *20th International Conference on Advanced Information Networking and Applications AINA 2006* (Vol. 1, pp. 985-992). IEEE.

Velloso, P. B., Laufer, R. P., Cunha, D. D. O., Duarte, O. C. M., & Pujolle, G. (2010). Trust management in mobile ad hoc networks using a scalable maturity-based model. *IEEE eTransactions on Network and Service Management*, *7*(3), 172–185. doi:10.1109/TNSM.2010.1009.I9P0339

Venkataraman, R., Pushpalatha, M., & Rao, T. R. (2012). Regression-based trust model for mobile ad hoc networks. *IET Information Security*, *6*(3), 131–140. doi:10.1049/iet-ifs.2011.0234

Verma, A., & Gujral, M. S. (2012). Trust Oriented Security Framework for Ad Hoc Network. *Journal of Computer Science & Information Technology*, *5*, 19–26.

Vorobiev, A., & Bekmamedova, N. (2007). An ontological approach applied to information security and trust. In *ACIS 2007 Proceedings*.

Wang, B., Chen, X., & Chang, W. (2014). A light-weight trust-based QoS routing algorithm for ad hoc networks. *Pervasive and Mobile Computing*, *13*, 164–180. doi:10.1016/j.pmcj.2013.06.004

Wang, J., Liu, Y., & Jiao, Y. (2011). Building a trusted route in a mobile ad hoc network considering communication reliability and path length. *Journal of Network and Computer Applications*, *34*(4), 1138–1149. doi:10.1016/j.jnca.2010.11.007

Wang, W., & Zeng, G. (2010). Bayesian cognitive trust model based self-clustering algorithm for MANETs. *Science China. Information Sciences*, *53*(3), 494–505. doi:10.100711432-010-0050-8

Xia, H., Jia, Z., Li, X., Ju, L., & Sha, E. H. M. (2013). Trust prediction and trust-based source routing in mobile ad hoc networks. *Ad Hoc Networks*, *11*(7), 2096–2114. doi:10.1016/j.adhoc.2012.02.009

Yang, H. S., & Sun, J. H. (2016). A study on hybrid trust evaluation model for identifying malicious behavior in mobile P2P. *Peer-to-Peer Networking and Applications*, *9*(3), 578–587. doi:10.100712083-015-0411-6

Zouridaki, C., Mark, B. L., & Hejmo, M. (2007). Byzantine robust trust establishment for mobile ad hoc networks. *Telecommunication Systems*, *35*(3-4), 189–206. doi:10.100711235-007-9047-z

This research was previously published in the International Journal of Business Data Communications and Networking (IJBDCN), 15(2); pages 35-54, copyright year 2019 by IGI Publishing (an imprint of IGI Global).

Chapter 10
Communiqué Issues in MANET and VANET Protocols With Network Security Disquiet

Mamata Rath
https://orcid.org/0000-0002-2277-1012
Birla Global University, India

Bibudhendu Pati
https://orcid.org/0000-0002-2544-5343
Rama Devi Women's University, India

Jhum Swain
Centurion University of Technology and Management, India

ABSTRACT

With faster expansion of mobile networks and quicker increment of web clients, more individuals approach worldwide data and communication innovation, because of which the issues of utilizing web as a worldwide stage and empowering the savvy protests and machines to coordinate, discuss, register, and ascertain slowly develops. In mobile ad-hoc networks (MANETs) and vehicular ad-hoc networks (VANET), the mobile stations and vehicles are self-reconfigurable as per the difference in network topology. Joint action between convenient centers is more basic because of the way that they confront significant difficulties, for example, frailty to work securely while protecting its advantages and performing secure guiding among center points. In the presence of vindictive hubs, one of the rule challenges in MANET is to plot ground-breaking security course of action that can shield MANET from different routing assaults.

DOI: 10.4018/978-1-7998-8545-0.ch010

INTRODUCTION

When there is some difficulty in communication system in a network environment, the best solution is offered by MANET where it is possible to create communication even in the absence of physical infrastructure. MANET is a specially designed wireless network that supports connectivity in very critical situations such as during hardware break down or where physical connection is a major issue. Because of high effectiveness of MANET and VANET based networks and their self-sufficient nodes, they can provide communication facility smoothly in such cases. Due to multi-jump transfer capability in appropriated way these MANETs(Rath et.al, 2014) are critical in calamity circumstances, for example, combat zone for communication, military condition and cataclysmic events. Headway of technology with situation has been developed in numerous structures in current remote communication((L.K. Shar et.al, 2015). Because of unstable network topology, and developments of nodes dependably change powerfully in numerous ways, it is absolutely hard to control and oversee such an incautious circumstance and communicating with neighbor nodes with propose co-activity is another dreary issue. Administration of Quality of Service and its reasonable control is additionally empowering.

Figure 1. A general MANET scenario with device connectivity

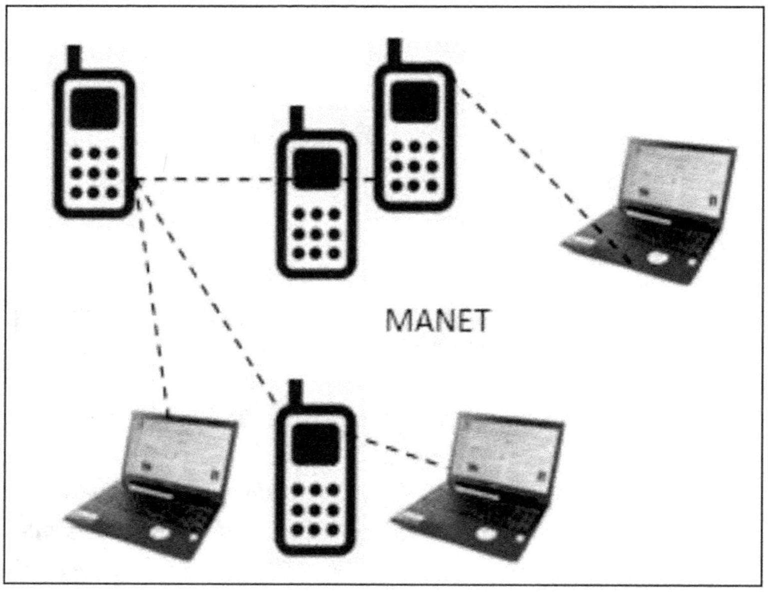

Figure 1 A general MANET scenario with device connectivity where a sender device sends information to its neighbor device and then the messages is forwarded through intermediate nodes. During data communication from sender to beneficiary, there are numerous requirements in the way containing way discovering, node choosing, identification of connection disappointments, route upkeep, route repair, holding steering tables and to take amend choice of parcel quickening towards the heading of correct goal. Motivation behind the current survey article has been drawn from various security challenges (Rath et.al, 2018) and vulnerability in Ad-hoc Networks both in MANET and VANET.

Organisation of the Chapter

Section 1 presents the introduction part that describes the MANET and VANET concept with challenges in their routing strategies. Section 2 presents security protocols with different techniques used for routing in MANET and VANET. Section 3 offers a description of Key Factors Used in MANET and VANET for Security Evaluation. Next section 4 narrates Safekeeping Approaches Used in MANET and VANET Protocols. Section 5 offers explanation of Protection and Confidentiality preserving scheme in VANET and the last segment offers the conclusion part.

Security Protocols with Different Approaches in MANET and VANET

Unicast routing protocols in MANET provide right methods of creating routes from source node to destination node and they also support route maintenance during topology change with reliability of the messages (Rath et.al, 2016). Classification of various unicast routing protocols are presented in Figure 2. Though many routing protocols based on these classification has been developed by experts in networking, still security during communication (Rath et.al, 2019) remains a major challenge specifically during real time transmission of data.

Figure 2. Classification of MANET Routing Protocols

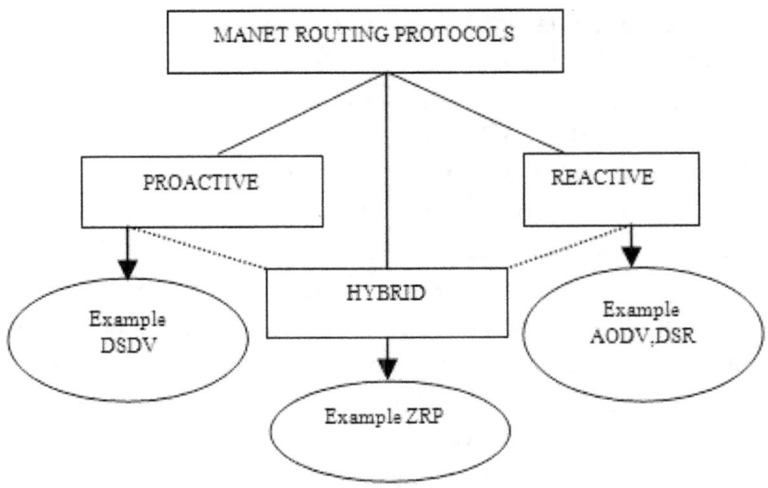

In VANET the design of an effective routing protocol largely depends on communication between vehicles and from mobile vehicle to road side units(Rath et.al, 2018). In VANET, the routing technique is different than MANET because of the high level of mobility of running vehicles and sudden change of topology. Reducing end to end delay during data transmission always remains the basic challenge in both MANET and VANET.

Based on many factors such as handling the network overhead, speed and accuracy, the correct protocol design should be carried out to satisfy the communication need with better quality of service(Rath et.al, 2019). According to the dynamic features of VANET, and considering the effective response in diverse environment, the VANET protocols have been classified into various categories such as ad-hoc or topology driven, location based, cluster based, broadcast and geo-cast protocols. They are diagrammatically represented in Figure 3.

Figure 3. VANET Routing Protocols

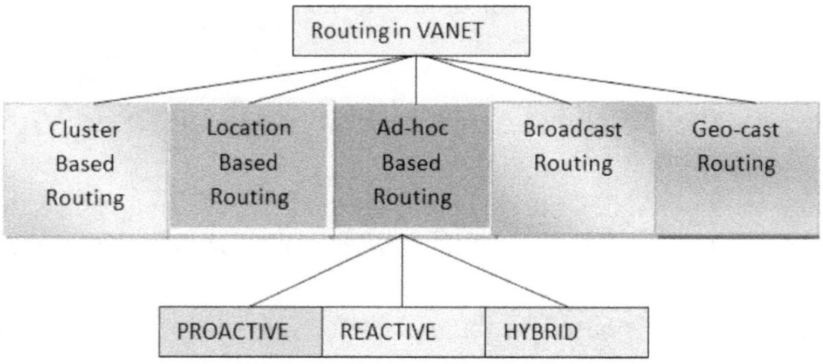

PREVENTION OF ATTACK USING 2ACK PLAN

Malicious attack during routing in MANETs has been contemplated strongly (K.Liu et.al, 2007) with agreeable work stations. When the network platform is all set, routing conventions for MANETs are planned securely dependent on the suspicion that every single partaking node are completely helpful. 2ACK plan has been suggested that fills in as an extra method for routing plans to distinguish routing bad conduct and to moderate their unfriendly impact. The principle thought of the 2ACK plan is to send two-jump affirmation parcels the other way of the routing way. With the end goal to decrease extra routing overhead, just a small amount of the got information parcels are recognized in the 2ACK plan. Investigative and recreation results are exhibited to assess the execution of the proposed plan.K closest neighbor (kNN) questions, which recover the k closest sensor information things related with an (area subordinate sensor information) from the area of the inquiry guarantor, are helpful for area based administrations in mobile conditions. However, Y.Komai et.al, (2015) centers around the kNN question handling in mobile impromptu networks (MANETs). Enter difficulties in structuring framework conventions for the MANETs incorporate low-overhead versatility to network topology changes because of node mobility, and inquiry preparing that accomplishes high precision of the question result without a concentrated server. The filling region (FA) has been proposed utilizing a technique.

Network Layer Security Protocol E2-SCAN

A plan that can effectively recognize malevolent nodes has been arranged (Khan et.al, 2017) utilizing network parameters to decide if parcel misfortunes are because of line floods or node mobility in MANETs. The commitments incorporate the FGA conspire for parcel misfortune and the improvement of an extensive trust demonstrate for malevolent node distinguishing proof and detachment. FGA conspire is assessed as far as viability and execution measurements under various network parameters and setups. The test results demonstrate that the proposed trust display accomplishes a critical decrease in false positives rate and an expansion in the rate of discovery of genuinely pernicious nodes contrasted and customary non-FGA plans.

AASR Security Protocol

Inexplicable connection are critical for a large number of the utilizations of mobile adhoc networks conveyed in enemy conditions. A noteworthy necessity on the network is the capacity to give unidentifiability and unlinkability to mobile nodes and their activity. In spite of the fact that various unknown secure routing conventions have been proposed, the prerequisite isn't completely fulfilled. The current conventions are helpless against the assaults of phony routing parcels or refusal of-benefit broadcasting, even the node characters are ensured by pen names. Another routing convention, i.e., Authenticated Anonymous Secure Routing (AASR) has been offered (W. Liu et.al, 2014), to fulfill the prerequisite and safeguard against the assaults. Further, in this scheme, the packets which are transmitted are verified by an authentication process indicating a stamp to ensure protection against attack.

A suitable routing protocol is needed in challenging security threatening situation which not only improves life time of network but also makes the data delivery faster among mobile nodes keeping in constraint of limited energy of nodes (Oreku et.al, 2018) . There are many routing protocols proposed so far by taking into consideration power and delay factors. Battery energy of the mobile nodes are consumed and slowly reduces as and when the node in its active state dynamically sends packets to its neighboring node or receives packets from other nodes or even when it stays idle and slowly doing the job of listening the channel in absence of wired medium for any probable message requirements from different nodes. Therefore power aware routing protocols lessen the communication power required to send and receive packets. An efficient power control strategy determines the optimal routing path which also minimizes the total transmission energy which is needed to send packets to the destination. As per another approach a node can go to inactive state by switching off or can go to into sleep mode or power-down mode the node has no data packet to send or receive. Load distribution technique is the most important method for enhancing active communication power . Objective of these methods is to balance the use of power among the nodes in the network and to improve e the network lifetime by not considering over-drained nodes when choosing a suitable route. Therefore the routing protocol has to be well-defined in order to manage in extreme cases when most nodes have gone sleep mode and do not forward packets for other stations.

KEY FACTORS USED IN MANET AND VANET FOR SECURITY EVALUATION

There are many challenges faced by VANET due to high mobility of computing devices and sudden change of dynamic network structure. As the wireless network is open to all the recipients so there is more chance of hacking and intruders attacking the network. Similarly developing a really secured routing protocol has been always a challenge for network engineers due to which many a times there are cyber attacks in the network. Poor deign of intrusion Detection System enforces hacking by the intruders due to vulnerability in encryption algorithm. Fig 4. depicts such factors in a diagrammatic view. There are some fault detection algorithms used in VANET that can study normal behaviour of network through a neural network and apply some machine learning strategy to catch the intruders.

Figure 4. Security Challenges in VANET

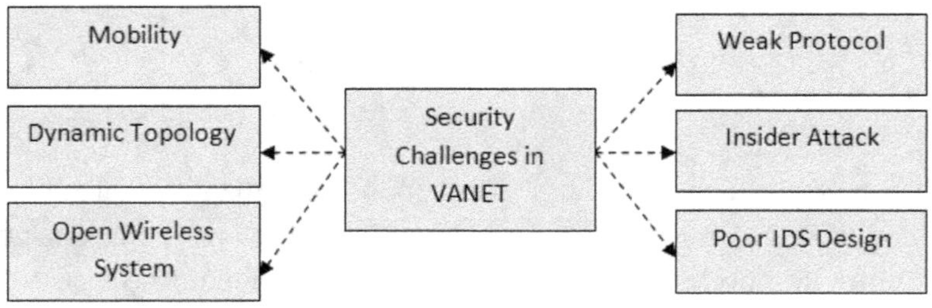

Signal-to-Noise Ratio and Delay

Some routing protocols in MANET and VANET are developed to improve the throughput performance in comparison to chance of error and SNR (Signal to Noise Ratio). Some protocols suffer a lot due to poor connectivity and unpredictable link connection as sometimes the topology suddenly change. Many propagation vector model for mobile smart phone OS are also proposed by researchers .The problem related to functionality of existing models are discussed and the optimal approach to save a infected system with some healer agents with NP-COMPLETE problems are also defined. It is shown through simulation in NS3 that despite less knowledge of the exact future, these healers can get recovery time within the future arrival time of all the infected nodes.

Real Time Information Processing in MANET

Information handling applications are growing rapidly now a days as there is large demand of technology-centric applications in the Internet (Rath M. et.al, 2019), therefore information-oriented networking has given attention to the research community. Power-efficient systems, domains and distributed platforms for pervasive computing, efficient fusion, and real-time secured delivery of multimedia applications are very importance in developing new applications for green wireless media. These areas are most challenging in MWSN solicitations to be compatible with the green wireless media prototype ((L.K.Shar et.al, 2015). A review has been done about the current state of the art in green QoE MWSNs, in relation to mobile users and inaccessible uses which are present in remote area.Power systems which are to be used in future have a high rate of complexity with more number of smart devices, dealing with real time information. So reliable and real-time capable communication networks are necessary to allow control, monitoring, management and to guarantee constant power system handling. Here the authors present and propose a smooth and dynamic approach to control the network which is based on Software-Defined Networking (SDN) to satisfy specific communication needs of both transmission power grid and distribution power grid. So, here a testbed is proposed, which will enable the evaluation of multiple failure situations such as link failure and congestion (N.Pujari et.al, 2017) by studying the situation with fast recovery and optimized solutions (Rath et.al, 2019).A detailed study has been carried out about the problem of using minimum power level to offer satisfactory performance for real time applications over fading wireless channels which are not reliable (Rath et.al, 2016).

Medium Access Control Using 802.11p Standard

VANETs are ending up increasingly prevalent as an approach to build the movement safety and solace. The IEEE 802.11p standard, particularly the 802.11p MAC convention (Rath et.al, 2017), has pulled in much consideration as a major aspect of the WAVE convention in VANETs. Safety applications, as one of the fundamental classifications of applications in VANETs, is extremely trying for the plan of a MAC convention because of their low dormancy and high unwavering quality prerequisites. The CCH interim is likewise a key parameter for the 802.11p MAC convention since it can influence the execution of safety message conveyance fundamentally. In [9] a re-enactment based assessment is proposed to assess the execution of the 802.11p MAC convention with different vehicle densities and CCH interim settings. The assessment comes about demonstrate that the 802.11p MAC convention can be enhanced by means of broadening the CCH interim. Nonetheless, the unwavering quality is still extremely difficult because of high impact rates.

Vehicular Cooperation in VANET Using 802.11p Convention

In a vehicular ad hoc network (VANET), the execution of the communication convention is vigorously impacted by the vehicular thickness flow. Be that as it may, the vast majority of the past deals with VANET execution demonstrating gave careful consideration to vehicle dispersion or essentially accepted homogeneous auto circulation. Clearly vehicles are disseminated no homogeneously along a road section because of activity flags and speed limits at various bits of the road, and in addition vehicle collaborations that are critical on occupied avenues. In light of the inadequacy, an unique technique has been presented to think about the broadcasting execution of 802.11p VANETs with down to earth vehicle dispersion in urban conditions.

Figure 5. A VANET Scenario with Attack by Intruder

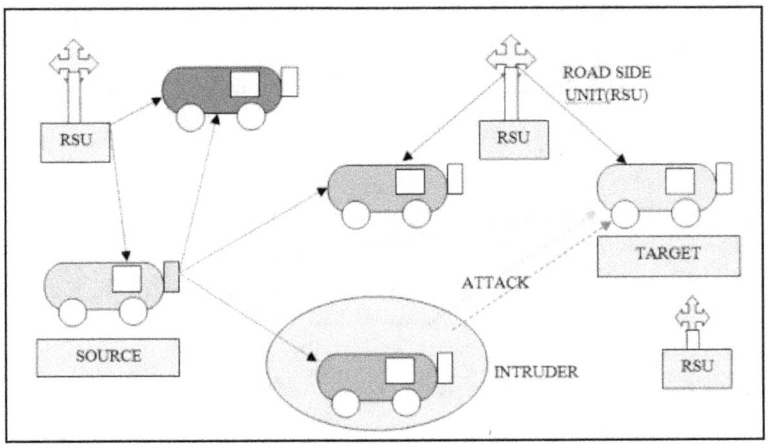

Figure 5 demonstrates a traffic scene from anticipated approach by Rath et.al, (2018). In this approach an Intrusion Detection VANET routing protocol is used that functions according to the notion of efficient routing in vehicular ad-hoc networks. The planned framework deals with Smart Intelligent protocol as the core module and it has different parts to control the communication between vehicles. Road Side Units (RSUs) communicate with vehicles depending on the needs of applications. RSUs are powered on for longer durations, they have undermines lifetime and reliability. There are various methods of managing power availability to RSU such as Power-down RSUs, Use low-power proxies on behalf of RSUs. When an intruder vehicle silently communicates with another vehicle in an authoritative method, the protocol checks the authenticity back to know the origin of Request and hence using a technique, it identifies the Intruder vehicle. Then registration of that vehicle is cancelled and it can not operate in VANET any more. To start with, it adopts the observationally checked stochastic movement models, which consolidate the impact of urban settings, (for example, activity lights and vehicle associations) on auto appropriation and produce down to earth vehicular thickness profiles. Relating 802.11p convention and execution models are then created. At the point when combined with the activity models, they can anticipate broadcasting productivity, postponement, and throughput exhibitions of 802.11p VANETs in light of the information of auto thickness at every area on the road. Broad recreation is led to check the exactness of the created scientific models with the thought of vehicle cooperation. When all is said in done, our outcomes show the relevance of the proposed technique on displaying convention execution in down to earth signalized road networks and shed bits of knowledge into the outline and advancement of future communication conventions and networking capacities for VANETs.

SAFEKEEPING APPROACHES USED IN MANET AND VANET PROTOCOLS

Hidden Channel Attack in MANET

As Mobile ad-hoc networks are badly suffered by higher mobility rate of nodes in a network, so communication hand-off between stations is a complex challenge. Taking advantage of this there are hidden attacks takes place in many channels. However, to address this issue, a security display custom model to the particular necessities of MANETs was presented by Acs, Buttyan, and Vajda (M. Burmester et.al, 2009) . Figure 6 shows self-organising and de-centralising nodes in MANET architecture

Figure 6. Self-organising and de-centralising nodes in MANET architecture

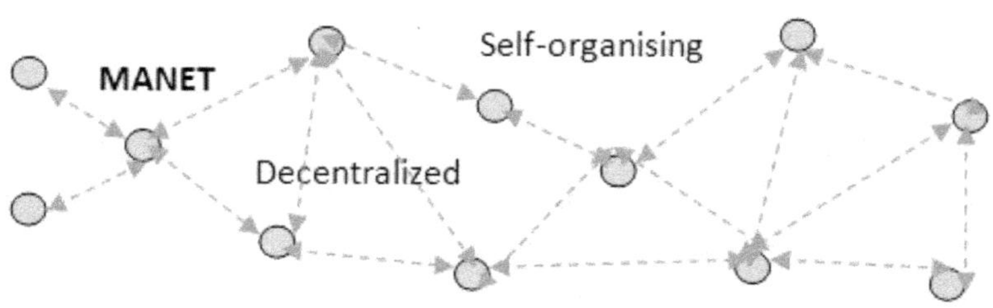

Among the novel attributes of this security show is that it guarantees security ensure under simultaneous executions, a component of essential down to earth suggestion for this sort of conveyed calculation. A novel course disclosure calculation called endairA was additionally proposed, together with an asserted security confirmation inside a similar model. The security evidence for the course disclosure calculation endairA is defective, and in addition, this calculation is helpless against a concealed channel assault. The security structure that was utilized for course revelation and its calculability is a basic element for omnipresent applications. So some exchange was done with respect to a portion of the significant security challenges for course revelation in MANETs.

Security Threats in Intelligent Transport System

A standout amongst the most basic security tasks network scientists have is staying aware of new advancements and considering the security suggestions that accompany them-basically, guaranteeing that security has enhanced to execute new thoughts from the soonest conceivable minute (M. Carvalho et.al, 2008) . Along these lines, specialists have had critical enthusiasm for the field of mobile impromptu networks (Manets). Such networks are habitually seen as a key communications innovation empowering agent for network-driven fighting and catastrophe alleviation tasks, and as the innovation develops, Manets are progressively achieving numerous different applications in regions, for example, dedicated transportation frameworks and blame tolerant mobile sensor lattices.

Fully Non-Interactive Admission Control Technique

Transmission through decentralized computing stations is a basic and major security benefit in mobile networks. It is expected to safely adapt to dynamic participation and topology and to bootstrap other vital security natives, (for example, key administration) and administrations, (for example, secure routing) without the help of any brought together confided in power. A perfect confirmation strategy must include negligible communication among MANET nodes, since network can Dbe temperamental. Likewise, since MANETs are regularly made out of feeble or asset constrained devices, affirmation must be productive regarding calculation and communication(N.Saxena et.al, 2009) . Most already proposed affirmation conventions are restrictively costly and require substantial communication among MANET nodes. Spotlight on normal kind of MANET has been executed on this region and a protected, productive, and a completely noninteractive confirmation procedure has been anticipated and intended for such sort of network. The Admission convention depends on mystery sharing procedures utilizing bivariate polynomials. Once more, another plan has been arranged that enables any match of MANET nodes to productively set up an on-the-fly secure communication channel.

Game Theory Based Software Tool for State Information of MANET Nodes

Game theory can be applied in critical situations where all participants in a game pursue their best possible techniques where all other participant's best techniques are also used. There is a dominant strategy which is considered to be the most prominent one where a single strategy is best for a player irrespective of what technique the other players decide to implement. Tacit Collusion in Game Theory means where firms undertake actions that are likely to minimize a competitive response. When one or more agents in a collusive agreement report it to authorities, it is known as whistle blowing. Similarly a zero sum game refers to an intermediate transaction where whatever is gained by one node / party must be lost by the other. As per this principle B. Paramasivan et.al, (2015) proposed a security aspect based on game theory where two nodes participate in game model. Figure 7 illustrates the key concepts used in the game theory as discussed above.

Figure 7. Key concepts of game theory

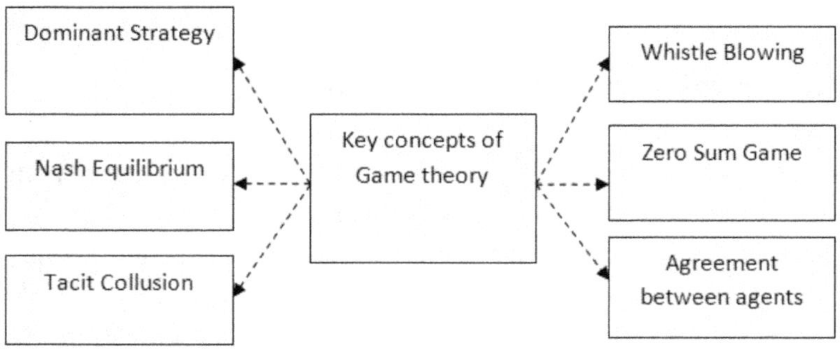

The mean field game theory gives a ground-breaking numerical instrument to issues with countless ((Y.Wang et.al, 2014). This plan can empower an individual node in MANETs to settle on vital security safeguard choices without incorporated organization. Likewise, since security resistance components expend valuable framework assets (e.g., vitality), the proposed plan considers the security prerequisite of MANETs as well as the framework assets(H.Jain et.al, 2014). Besides, every node in the proposed plan just has to know its very own state data and the total impact of alternate nodes in the MANET. Subsequently, this plan is a completely appropriated plan. Reproduction results additionally affirm to show the viability of the proposed plan.

In Mobile ad-hoc networks, there are also many concerns similar to VANETs. They are Wireless network link problem, Dynamic topology change, lack of understanding between nodes, Lack of a clear rule of defense, very Limited power back up, Ad hoc Networks system and Security problems. Figure 8 displays Security challenges in Mobile Ad-hoc Networks with various security confronts.

Figure 8. Security challenges in mobile Ad-hoc networks

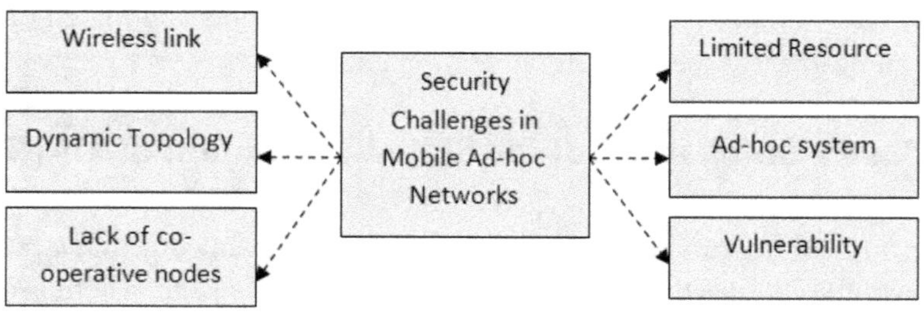

QoS Constrained Fault Tolerant ant Look Ahead Routing Algorithm

In the very challenging network of MANET, it is highly essential that routing pattern must be selected much before the sender sends the packet towards the destination (Rath et.al, 2017) .Quick choices dependably repay network execution (S.Surendran et.al, 2015) . What's more, most MANET routing conventions expect a well disposed and agreeable condition, and thus are helpless against different assaults. Trust and Reputation would fill in as a noteworthy answer for these issues. Taking in the network attributes and picking right routing choices at right occasions would be a noteworthy arrangement. A broad study of blame tolerant conventions and subterranean insect state calculations has been done which are connected to routing in MANETs. A Quality of Service (QoS) compelled fault tolerant profound forward routing calculation which endeavors to distinguish legitimate route and look-ahead route sets which may help in picking the substitute way if there should be an occurrence of substantial route disappointment.

PROTECTION AND CONFIDENTIALITY PRESERVING SCHEME IN VANET

In different routing approaches it is found that there should be many number of fields to correctly forward the routing packets in a mobile-ad-hoc network.Before a mobile station advances a packect, if the beneficiary's trust level does not meet the bundle's prerequisite level, at that point the beneficiary must play out certain security affiliation methodology, for example, re-verification (D. Q. Nguyen et.al, 2016) .An investigation of the pestilence broadcast delay in such setting has been readied where the network, versatility and trust models are introduced which are very conventional and permit to acquire the defer part actuated just by the security relationship along a way.

Figure 9. Applications of VANET

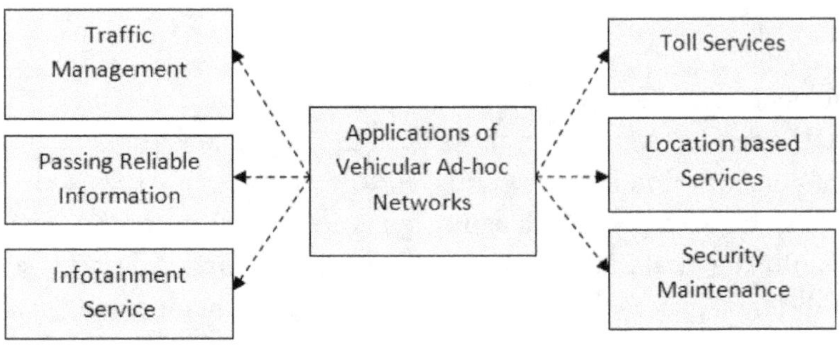

Figure 9 depicts various applications of VANET. All these are basically related to safety, security and reliability of vehicular network systems (Rath et.al, 2017). It includes traffic management, communication of reliable information over the wireless networks, infotainment services, toll services, location based services and security maintenance through out the communication system with the support of various network protocols and their conventions (Rath et al, 2018).

Numerical outcomes gotten by simulations additionally affirm the precision of the examination (D. Q. Nguyen et.al, 2016). Specifically, it tends to be seen from both simulation's and examination results that, for extensive and meagerly associated networks, the postponement caused by security affiliations is little contrasted with the aggregate deferral of a bundle. This additionally implies parameters like network thickness and nodes' speed, instead of any trust demonstrate parameter, have more effect on the general deferral. Table 1 describes various security issues as handled in VANET using improved techniques.

Table 1. Various security issues in VANET using improved techniques.

Literature	Year	Related Topic	Related Domain
S.Morteza et al.	2018	Privacy Preserving Authentication scheme for VANET	Security Privacy iVANET
S.Sattar et al.	2018	Energy efficiency Analysis and reliability of safety messages broadcast in VANET	Reliability and Safety VANET
H.Hasrouny et al.	2017	VANET Security Challenges & Solutions	Security VANET
R. Baiad et al.	2016	Cross layer detection scheme to detect black hole attack in VANET	Network Security Safety VANET
S.Sharma et al.	2018	VANET Based Intrusion detection System using Cluster Head	Security Cluster Computing VANET
R.oliveira et al.	2017	Reliable Protocol in VANET with Safety Applications	Safety Security VANET
M.N. Mejri et al,	2017	GDVAN – Improved algorithm for VANET	Greedy Behaviour attack detection algotithm
M.Mehdi et al.	2017	Game Theory Based Trust Model for VANET	Optimization Theory VANET

VANET is an eminent innovation with promising future and also incredible difficulties particularly in its security. VANET security systems are introduced in three noteworthy practical parts (Rath et.al, 2018) . At that point consideration has been attracted to various open issues and specialized difficulties identified with VANET security, which can help numerous different scientists for sometime later. Miniature drones (R.oliveira et al., 2017) are being used in checking, transport, security and disaster administration, and different spaces. Imagining that drones frame autonomous networks consolidated into the air activity, An abnormal state engineering for the outline of a collective flying system comprising of drones with on-load up sensors and inserted preparing, coordination, and networking capacities (R.oliveira et al., 2017) . A multi-drone system has been proposed comprising of quad copters and exhibit its potential in disaster help, pursuit and protect, and airborne checking. Moreover, illustration of configuration difficulties and present potential solutions have been actually looked into.

VANETS are standout and brilliant networks amongst the most concrete and testing research zones in vehicle organizations and Information and Telecommunication System (ITS) planners. The applicability of such networks opens the route for an extensive variety of uses, such as health care applications, portability and availability for both driver and travelers to employ the vehicle frameworks in an easily, proficiently and more secured way.

CONCLUSION

A detailed review and analysis has been completed in the above research work about the magnificent MANET and VAVET routing strategies with proficient convention. Suitable determination of nodes for sending the data packets in a way which deals with the power proficiency as well as timely delivery is indeed a promising task.

Out of many research works, protocol of S. Umamaheswari et.al, (2015) called Enhanced ANTSEC framework with cluster based cooperative caching has been selected as an efficient work that is proposed to protect cluster based MANET from attack. This exploration exhibits exertion plans to reduce the issue of reasonable load adjusting among every one of the nodes with the goal that any halfway found nodes does not get overburdened. The above segment of research article presents a complete analysis on current state of the art of the spectacular network of VANET and MANET communication and the vibrant routing schemes used by different networks and in research work as well as provides a learning platform to the new researchers to extend their study further in various security concern in the network.

REFERENCES

Al-Nahari. (2013). Receiver-based AODV routing protocol for MANETs. *13th International Conference on Intelligent Systems Design and Applications (ISDA)*. doi:10.1109/ICCCI.2013.6466148

Ali & Huiqiang. (2012). Node Centric Load Balancing Routing Protocol for Mobile Ad Hoc Networks. *Proceedings of the International Conference of Engineers and Computer Scientists,* 1.

Baiad, R., Alhussein, O., Otrok, H., & Muhaidat, S. (2016). Novel cross layer detection schemes to detect blackhole attack against QoS-OLSR protocol in VANET. *Vehicular Communications*, *5*, 9–17. doi:10.1016/j.vehcom.2016.09.001

Bhatsangave & Chirchi. (2012). OAODV Routing Algorithm for Improving Energy Efficiency in MANET. *International Journal of Computers and Applications*, *51*, 15–22.

Burmester, M., & de Medeiros, B. (2009). On the Security of Route Discovery in MANETs. *IEEE Transactions on Mobile Computing*, *8*(9), 1180–1188. doi:10.1109/TMC.2009.13

Carvalho, M. (2008). Security in Mobile Ad Hoc Networks. *IEEE Security and Privacy*, *6*(2), 72–75. doi:10.1109/MSP.2008.44

Chang, J., Tsou, P., Woungang, I., Chao, H., & Lai, C. (2015). Defending Against Collaborative Attacks by Malicious Nodes in MANETs: A Cooperative Bait Detection Approach. *IEEE Systems Journal*, *9*(1), 65–75. doi:10.1109/JSYST.2013.2296197

Choudhury, Ragha, & Marathe. (2015). Implementing and improving the performance of AODV by receive reply method and securing it from Black hole attack. *Procedia Computer Science, 45*, 564 – 570.

Defrawy, K. E., & Tsudik, G. (2011). Privacy-Preserving Location-Based On-Demand Routing in MANETs. *IEEE Journal on Selected Areas in Communications*, *29*(10), 1926–1934. doi:10.1109/JSAC.2011.111203

Dhurandher, S. K., Misra, S., Ahlawat, S., Gupta, N., & Gupta, N. (2009). E2-SCAN: An extended credit strategy-based energy-efficient security scheme for wireless ad hoc networks. *IET Communications*, *3*(5), 808–819. doi:10.1049/iet-com.2008.0315

Dhurandher, S. K., Obaidat, M. S., Verma, K., Gupta, P., & Dhurandher, P. (2011). FACES: Friend-Based Ad Hoc Routing Using Challenges to Establish Security in MANETs Systems. *IEEE Systems Journal*, *5*(2), 176–188. doi:10.1109/JSYST.2010.2095910

Dolk, V. S., Tesi, P., De Persis, C., & Heemels, W. P. M. H. (2017). Event-Triggered Control Systems Under Denial-of-Service Attacks. IEEE Transactions on Control of Network Systems, 4(1), 93-105. doi:10.1109/TCNS.2016.2613445

Gaeta, R., Grangetto, M., & Loti, R. (2014). Exploiting Rateless Codes and Belief Propagation to Infer Identity of Polluters in MANET. *IEEE Transactions on Mobile Computing*, *13*(7), 1482–1494. doi:10.1109/TMC.2013.137

Ghosh, U., & Datta, R. (2015). A Secure Addressing Scheme for Large-Scale Managed MANETs. *IEEE eTransactions on Network and Service Management*, *12*(3), 483–495. doi:10.1109/TNSM.2015.2452292

Gupta, B. B., Joshi, R. C., Misra, M., Meena, D. L., Shrivastava, G., & Sharma, K. (2010). Detecting a wide range of flooding DDoS attacks using linear prediction model. In *IEEE 2nd International Conference on Information and Multimedia Technology (ICIMT 2010)* (Vol. 2, pp. 535-539). IEEE.

Hasrouny, H., Samhat, A. E., Bassil, C., & Laouiti, A. (2017). VANet security challenges and solutions: A survey. *Vehicular Communications*, *7*, 7–20. doi:10.1016/ j.vehcom.2017.01.002

Hurley-Smith, D., Wetherall, J., & Adekunle, A. (2017). SUPERMAN: Security Using Pre-Existing Routing for Mobile Ad hoc Networks. *IEEE Transactions on Mobile Computing*, *16*(10), 2927–2940. doi:10.1109/TMC.2017.2649527

Jain, H. R., & Sharma, S. K. (2014). Improved energy efficient secure multipath AODV routing protocol for MANET. *Advances in Engineering and Technology Research (ICAETR), 2014 International Conference on,* 1-9. 10.1109/ICAETR.2014.7012847

Joshi, P. (2011). Security Issues in Routing protocols in MANETs at network layer. *Procedia Computer Science, Elsevier*, *3*, 954–960. doi:10.1016/j.procs.2010.12.156

Khan, M. S., Midi, D., Khan, M. I., & Bertino, E. (2017). Fine-Grained Analysis of Packet Loss in MANETs. *IEEE Access: Practical Innovations, Open Solutions*, *5*, 7798–7807. doi:10.1109/ACCESS.2017.2694467

Komai, Y., Sasaki, Y., Hara, T., & Nishio, S. (2015). k Nearest Neighbor Search for Location-Dependent Sensor Data in MANETs. *IEEE Access: Practical Innovations, Open Solutions*, *3*, 942–954. doi:10.1109/ACCESS.2015.2445323

Kumari, R., & Sharma, K. (2018). Cross-layer based intrusion detection and prevention for network. In *Handbook of Research on Network Forensics and Analysis Techniques* (pp. 38–56). IGI Global. doi:10.4018/978-1-5225-4100-4.ch003

Liu, K., Deng, J., Varshney, P. K., & Balakrishnan, K. (2007). An Acknowledgment-Based Approach for the Detection of Routing Misbehavior in MANETs. *IEEE Transactions on Mobile Computing*, *6*(5), 536–550. doi:10.1109/TMC.2007.1036

Liu, W., & Yu, M. (2014). AASR: Authenticated Anonymous Secure Routing for MANETs in Adversarial Environments. *IEEE Transactions on Vehicular Technology*, *63*(9), 4585–4593. doi:10.1109/TVT.2014.2313180

Mamata, R. (2018). An Analytical Study of Security and Challenging Issues in Social Networking as an Emerging Connected Technolog. In *Proceedings of 3rd International Conference on Internet of Things and Connected Technologies*. Malaviya National Institute of Technology. Retrieved from https://ssrn.com/abstract=3166509

Mamata, R. B. P. (2018). Communication Improvement and Traffic Control Based on V2I in Smart City Framework. *International Journal of Vehicular Telematics and Infotainment Systems*, *2*(1).

Mehdi, M. M., Raza, I., & Hussain, S. A. (2017). A game theory based trust model for Vehicular Ad hoc Networks (VANETs). *Computer Networks*, *121*, 152–172. doi:10.1016/j.comnet.2017.04.024

Mejri, M. N., & Ben-Othman, J. (2017). GDVAN: A New Greedy Behavior Attack Detection Algorithm for VANETs. *IEEE Transactions on Mobile Computing*, *16*(3), 759–771. doi:10.1109/TMC.2016.2577035

Miglani, A., Bhatia, T., Sharma, G., & Shrivastava, G. (2017). An Energy Efficient and Trust Aware Framework for Secure Routing in LEACH for Wireless Sensor Networks. Scalable Computing. *Practice and Experience*, *18*(3), 207–218.

Mohan & Selvakumar. (2013). Power controlled routing in wireless ad hoc networks using crosslayer approach. Egyptian Informatics Journal, 13, 95–101.

Nguyen, D. Q., Toulgoat, M., & Lamont, L. (2016). Impact of trust-based security association and mobility on the delay metric in MANET. *Journal of Communications and Networks (Seoul)*, *18*(1), 105–111. doi:10.1109/JCN.2016.000013

Nishanthini, C., Rajkumar, G., & Jayabhavani, G. N. (2013). Enhanced performance of AODV with power boosted alternate path. *Computer Communication and Informatics (ICCCI),2013 International Conference on*, 1-4.

Oliveira, R., Montez, C., Boukerche, A., & Wangham, M. S. (2017). Reliable data dissemination protocol for VANET traffic safety applications. *Ad Hoc Networks*, *63*, 30–44. doi:10.1016/j.adhoc.2017.05.002

Paramasivan, B., Prakash, M. J. V., & Kaliappan, M. (2015). Development of a secure routing protocol using game theory model in mobile ad hoc networks. *Journal of Communications and Networks (Seoul)*, *17*(1), 75–83. doi:10.1109/JCN.2015.000012

Patil, P., Marathe, N., & Jethani, V. (2016). Improved ALERT protocol in MANET with strategies to prevent DOS & MITM attacks. *2016 International Conference on Automatic Control and Dynamic Optimization Techniques (ICACDOT)*, 372-377. 10.1109/ICACDOT.2016.7877611

Patil, V. P. (2012). Efficient AODV Routing Protocol for MANET with enhanced packet delivery ratio and minimized end to end delay. International Journal of Scientific and Research Publication, 2(8).

Pattanayak, B., & Rath, M. (2014). A Mobile Agent Based Intrusion Detection System Architecture For Mobile Ad Hoc Networks. *Journal of Computational Science*, *10*(6), 970–975. doi:10.3844/jcssp.2014.970.975

Pournaghi, S. M., Zahednejad, B., Bayat, M., & Farjami, Y. (2018). NECPPA: A novel and efficient conditional privacy-preserving authentication scheme for VANET. *Computer Networks*, *134*, 78–92. doi:10.1016/ j.comnet.2018.01.015

Rath & Oreku. (2018), Security Issues in Mobile Devices and Mobile Adhoc Networks. In Mobile Technologies and Socio-Economic Development in Emerging Nations. IGI Global. doi:10.4018/978-1-5225-4029-8.ch009

Rath. (2018c). Effective Routing in Mobile Ad-hoc Networks With Power and End-to-End Delay Optimization: Well Matched With Modern Digital IoT Technology Attacks and Control in MANET. In Advances in Data Communications and Networking for Digital Business Transformation. IGI Global. Doi:10.4018/978-1-5225-5323-6.ch007

Rath. (2018f). Real Time Analysis Based on Intelligent Applications of Big Data and IoT in Smart Health Care Systems. International Journal of Big Data and Analytics in Healthcare, 3(2), 45-61. Doi:10.4018/IJBDAH.2018070104

Rath. (2018g). Technical and Operational Utility of Ubiquitous Devices with Challenging Issues in Emerging Ubiquitous Computing. International Journal of Mobile Devices, Wearable Technology, and Flexible Electronics, 9(1), 16-35. Doi:10.4018/IJMDWTFE.2018010102

Rath. (2019a). Spectrum Access Issues and Security in Cognitive Radio Network. International Journal of Organizational and Collective Intelligence, 9(2), 31-44. DOI: . doi:10.4018/IJOCI.2019040103

Rath, M. (2017). Resource provision and QoS support with added security for client side applications in cloud computing. *International Journal of Information Technology*, *9*(3), 1–8.

Rath, M. (2018a). An Exhaustive Study and Analysis of Assorted Application and Challenges in Fog Computing and Emerging Ubiquitous Computing Technology. *International Journal of Applied Evolutionary Computation, 9*(2), 17-32. Retrieved from www.igi-global.com/ijaec

Rath, M. (2018b). A Methodical Analysis of Application of Emerging Ubiquitous Computing Technology With Fog Computing and IoT in Diversified Fields and Challenges of Cloud Computing. *International Journal of Information Communication Technologies and Human Development*, *10*(2). doi:10.4018/978-1-5225-4100-4.ch002

Rath, M. (2018d). Smart Traffic Management System for Traffic Control using Automated Mechanical and Electronic Devices. *I Mater. Sci. Eng., 377.* /377/1/01220110.1088/1757-899X/377/1/012201

Rath, M. (2018e). An Analytical Study of Security and Challenging Issues in Social Networking as an Emerging Connected Technology. *Proceedings of 3rd International Conference on Internet of Things and Connected Technologies (ICIoTCT)*. Retrieved from https://ssrn.com/abstract=3166509

Rath, M. (2019b). *Intelligent Information System for Academic Institutions: Using Big Data Analytic Approach. In Interdisciplinary Approaches to Information Systems and Software Engineering* (pp. 207–232). IGI Global. doi:10.4018/978-1-5225-7784-3.ch009

Rath, M. (2019c). Information Security Problem and Solution Approaches in Social Applications. In Cyber-Physical Systems for Social Applications. IGI Global.

Rath, M. (2019d). *Security in IoT and Big Data With Technical Challenges. In Handbook of Research on Big Data and the IoT* (pp. 197–210). IGI Global; doi:10.4018/978-1-5225-7432-3.ch011

Rath, M., & Mishra, S. (2019e). *Advanced-Level Security in Network and Real-Time Applications Using Machine Learning Approaches*. Machine Learning and Cognitive Science Applications in Cyber Security. doi:10.4018/978-1-5225-8100-0.ch003

Rath, M., & Panda, M. R. (2017). MAQ system development in mobile ad-hoc networks using mobile agents. *IEEE 2nd International Conference on Contemporary Computing and Informatics (IC3I)*, 794-798.

Rath, M., & Panigrahi, C. (2016). Prioritization of Security Measures at the Junction of MANET and IoT. In *Second International Conference on Information and Communication Technology for Competitive Strategies*. ACM Publication. 10.1145/2905055.2905187

Rath, M., & Pati, B. (2017). *Load balanced routing scheme for MANETs with power and delay optimization. International Journal of Communication Network and Distributed Systems* , 19.

Rath, M., & Pati, B. (2019). Security Assertion of IoT Devices Using Cloud of Things Perception. *International Journal of Interdisciplinary Telecommunications and Networking*, *11*(2), 2.

Rath, M., Pati, B., Panigrahi, C. R., & Sarkar, J. L. (2019). QTM: A QoS Task Monitoring System for Mobile Ad hoc Networks. In *Recent Findings in Intelligent Computing Techniques*. Springer. Retrieved from https://link.springer.com/chapter/10.1007/978-981-10-8639-7_57

Rath, M., Pati, B., & Pattanayak, B. (2015a). Energy Competent Routing Protocol Design in MANET with Real time Application Provision. *International Journal of Business Data Communications and Networking, IGI Global*, *11*(1), 50–60. doi:10.4018/IJBDCN.2015010105

Rath, M., Pati, B., & Pattanayak, B. (2015b). Delay and power based network assessment of network layer protocols in MANET. *2015 International Conference on Control, Instrumentation, Communication and Computational Technologies (IEEE ICCICCT)*, 682-686. 10.1109/ICCICCT.2015.7475365

Rath, M., Pati, B., & Pattanayak, B. (2016a). Energy Efficient MANET Protocol Using Cross Layer Design for Military Applications. *Defence Science Journal*, *66*(2), 146. doi:10.14429/dsj.66.9705

Rath, M., Pati, B., & Pattanayak, B. (2016b), Comparative analysis of AODV routing protocols based on network performance parameters in Mobile Adhoc Networks. In Foundations and Frontiers in Computer, Communication and Electrical Engineering (pp. 461-466). CRC Press, Taylor & Francis.

Rath, M., Pati, B., & Pattanayak, B. (2016c). Resource Reservation and Improved QoS for Real Time Applications in MANET. *Indian Journal of Science and Technology*, *9*(36). doi:10.17485/ijst/2016/v9i36/100910

Rath, M., Pati, B., & Pattanayak, B. (2016d). QoS Satisfaction in MANET Based Real Time Applications. *International Journal of Control Theory and Applications*, *9*(7), 3069–3083.

Rath, M., Pati, B., & Pattanayak, B. (2019). Manifold Surveillance Issues in Wireless Network and the Secured Protocol. *International Journal of Information Security and Privacy*, *13*(3).

Rath, M., Pati, B., & Pattanayak, B. K. (2016). Inter-Layer Communication Based QoS Platform for Real Time Multimedia Applications in MANET. Wireless Communications, Signal Processing and Networking (IEEE WiSPNET), 613-617. doi:10.1109/WiSPNET.2016.7566203

Rath, M., Pati, B., & Pattanayak, B. K. (2017). Cross layer based QoS platform for multimedia transmission in MANET. *11th International Conference on Intelligent Systems and Control (ISCO)*, 402-407. 10.1109/ISCO.2017.7856026

Rath, M., Pati, B., & Pattanayak, B. K. (2018a). Relevance of Soft Computing Techniques in the Significant Management of Wireless Sensor Networks. In Soft Computing in Wireless Sensor Networks (pp. 86-106). Chapman and Hall/CRC, Taylor & Francis Group. doi:10.1201/9780429438639-4

Rath, M., Pati, B., & Pattanayak, B. K. (2018b). An Overview on Social Networking: Design, Issues, Emerging Trends,and Security. In *Social Network Analytics: Computational Research Methods and Techniques*. Academic Press. Retrieved from https:// www. elsevier.com/ books/ social network-analytics/ dey/978-0-12-815458-8

Rath, M., Pati, B., & Pattanayak, B. K. (2019a). Mobile Agent-Based Improved Traffic Control System in VANET. In A. Krishna, K. Srikantaiah, & C. Naveena (Eds.), *Integrated Intelligent Computing, Communication and Security. Studies in Computational Intelligence* (Vol. 771). Singapore: Springer. doi:10.1007/978-981-10-8797-4_28

Rath, M., Pati, B., & Pattanayak, B. K. (2019b). *Design and Development of Secured Framework for Efficient Routing in Vehicular Ad-hoc Network. Int'l Journal of Business and Data Communication Network.*

Rath, M., & Pattanayak, B. (2016). A Contemporary Survey and Analysis of Delay and Power Based Routing Protocols in MANET. *Journal of Engineering and Applied Sciences (Asian Research Publishing Network)*, *11*(1), 536–540.

Rath, M., & Pattanayak, B. (2017). MAQ:A Mobile Agent Based QoS Platform for MANETs. *International Journal of Business Data Communications and Networking, IGI Global*, *13*(1), 1–8. doi:10.4018/IJBDCN.2017010101

Rath, M., & Pattanayak, B. (2018a). Technological improvement in modern health care applications using Internet of Things (IoT) and proposal of novel health care approach. *International Journal of Human Rights in Healthcare*. Retrieved from http://emeraldgrouppublishing.com/products/journals/journals.htm?id=ijhrh

Rath, M., & Pattanayak, B. K. (2014). A methodical survey on real time applications in MANETS: Focussing On Key Issues. *International Conference on, High Performance Computing and Applications (IEEE ICHPCA)*, 22-24. 10.1109/ICHPCA.2014.7045301

Rath, M., & Pattanayak, B. K. (2018b). Monitoring of QoS in MANET Based Real Time Applications. Smart Innovation, Systems and Technologies, 84, 579-586. doi:10.1007/978-3-319-63645-0_64

Rath, M., & Pattanayak, B. K. (2018c). SCICS: A Soft Computing Based Intelligent Communication System in VANET. Smart Secure Systems – IoT and Analytics Perspective. *Communications in Computer and Information Science*, *808*, 255–261. doi:10.1007/978-981-10-7635-0_19

Rath, M., & Pattanayak, B. K. (2019a). Performance Evaluation of Optimized Protocol in MANET. *International Journal of Information and Computer Security*. Retrieved from http://www.inderscience.com /info/ingeneral/ forthcoming.php?jcode=ijics

Rath, M., & Pattanayak, B. K. (2019b). Prevention of Replay Attack Using Intrusion Detection System Framework. In *Progress in Advanced Computing and Intelligent Engineering*. Springer. Retrieved from https://link.springer.com/ chapter/10.1007/978-981-13-1708-8_32

Rath, M., & Pattanayak, B. K. (2019c). Security Protocol with IDS Framework Using Mobile Agent in Robotic MANET. *International Journal of Information Security and Privacy*, *13*(1), 46–58. doi:10.4018/IJISP.2019010104

Rath, M., Pattanayak, B. K., & Pati, B. (2017). *Energetic Routing Protocol Design for Real-time Transmission in Mobile Ad hoc Network. In Computing and Network Sustainability, Lecture Notes in Networks and Systems* (Vol. 12). Singapore: Springer.

Rath, M., Rout, U. P., & Pujari, N. (2017). *Congestion Control Mechanism for Real Time Traffic in Mobile Adhoc Networks. In Computer Communication, Networking and Internet Security. Lecture Notes in Networks and Systems* (Vol. 5, pp. 149–156). Singapore: Springer.

Rath, M., & Solanki, V. K. (2019a). Handbook of IoT and Big Data. CRC Press.

Rath, M., & Solanki, V. K. (2019b). Contribution of IoT and Big Data in Modern Health Care Applications in Smart City. In Handbook of IoT and Big Data. CRC Press.

Rath, M., Swain, J., Pati, B., & Pattanayak, B. K. (2018). *Attacks and Control in MANET. In Handbook of Research on Network Forensics and Analysis Techniques* (pp. 19–37). IGI Global.

Reshmi, T. R., & Murugan, K. (2017). Light weight cryptographic address generation (LW-CGA) using system state entropy gathering for IPv6 based MANETs. *China Communications*, *14*(9), 114–126. doi:10.1109/CC.2017.8068770

Rong, B., Chen, H., Qian, Y., Lu, K., Hu, R. Q., & Guizani, S. (2009). A Pyramidal Security Model for Large-Scale Group-Oriented Computing in Mobile Ad Hoc Networks: The Key Management Study. *IEEE Transactions on Vehicular Technology*, *58*(1), 398–408. doi:10.1109/TVT.2008.923666

Sahoo, J., & Rath, M. (2017). Study and Analysis of Smart Applications in Smart City Context. *2017 International Conference on Information Technology (ICIT)*, 225-228. 10.1109/ICIT.2017.38

Sattar, S., Qureshi, H. K., Saleem, M., Mumtaz, S., & Rodriguez, J. (2018). Reliability and energy-efficiency analysis of safety message broadcast in VANETs. *Computer Communications*, *119*, 118–126. doi:10.1016/j.comcom.2018.01.006

Saxena, N., Tsudik, G., & Yi, J. H. (2009). Efficient Node Admission and Certificateless Secure Communication in Short-Lived MANETs. *IEEE Transactions on Parallel and Distributed Systems*, *20*(2), 158–170. doi:10.1109/TPDS.2008.77

Shakshuki, E. M., Kang, N., & Sheltami, T. R. (2013). EAACK—A Secure Intrusion-Detection System for MANETs. *IEEE Transactions on Industrial Electronics*, *60*(3), 1089–1098. doi:10.1109/TIE.2012.2196010

Shar, L. K., Briand, L. C., & Tan, H. B. K. (2015). Web Application Vulnerability Prediction Using Hybrid Program Analysis and Machine Learning. *IEEE Transactions on Dependable and Secure Computing*, *12*(6), 688–707. doi:10.1109/TDSC.2014.2373377

Sharma, K., & Gupta, B. B. (2016). Multi-layer defense against malware attacks on smartphone wi-fi access channel. *Procedia Computer Science*, *78*, 19–25.

Sharma, K., & Gupta, B. B. (2018). Taxonomy of distributed denial of service (ddos) attacks and defense mechanisms in present era of smartphone devices. *International Journal of E-Services and Mobile Applications*, *10*(2), 58–74. doi:10.4018/IJESMA.2018040104

Sharma, S. (2018). Hybrid fuzzy multi-criteria decision making based multi cluster head dolphin swarm optimized IDS for VANET. *Vehicular Communications*, *12*, 23–38. doi:10.1016/j.vehcom.2017.12.003

Sharma & Patheja. (2011). Improving AODV Routing Protocol with Priority and Power Efficiency in Mobile Ad hoc WiMAX Network. *International Journal of Computer Technology and Electronics Engineering*, *2*(1), 87–93.

Shrivastava, G. (2016, March). Network forensics: Methodical literature review. In *2016 3rd International Conference on Computing for Sustainable Global Development (INDIACom)* (pp. 2203-2208). IEEE.

Shrivastava, G., & Bhatnagar, V. (2011). Analyses of algorithms and complexity for secure association rule mining of distributed level hierarchy in web. *International Journal of Advanced Research in Computer Science*, *2*(4).

Shrivastava, G., Kumar, P., Gupta, B. B., Bala, S., & Dey, N. (Eds.). (2018). *Handbook of Research on Network Forensics and Analysis Techniques*. IGI Global. doi:10.4018/978-1-5225-4100-4

Shrivastava, G., Sharma, K., Khari, M., & Zohora, S. E. (2018). Role of Cyber Security and Cyber Forensics in India. In *Handbook of Research on Network Forensics and Analysis Techniques* (pp. 143–161). IGI Global. doi:10.4018/978-1-5225-4100-4.ch009

Sridhar, S., Baskaran, R., & Chandrasekar, P. (2013). Energy supported AODV (EN-AODV) for QoS routing in MANET. *The 2nd International Conference on Integrated Information, Social and Behavioral Sciences, 73*, 294 – 301.

Surendran, S., & Prakash, S. (2015). An ACO look-ahead approach to QOS enabled fault- tolerant routing in MANETs. *China Communications*, *12*(8), 93–110. doi:10.1109/CC.2015.7224693

Tsuda, T., Komai, Y., Hara, T., & Nishio, S. (2016). Top-k Query Processing and Malicious Node Identification Based on Node Grouping in MANETs. *IEEE Access: Practical Innovations, Open Solutions*, *4*, 993–1007. doi:10.1109/ACCESS.2016.2541864

Umamaheswari, S., & Radhamani, G. (2015). Enhanced ANTSEC framework with cluster based cooperative caching in mobile ad hoc networks. *Journal of Communications and Networks (Seoul)*, *17*(1), 40–46. doi:10.1109/JCN.2015.000008

Wang, Y., Yu, F. R., Tang, H., & Huang, M. (2014). A Mean Field Game Theoretic Approach for Security Enhancements in Mobile Ad hoc Networks. *IEEE Transactions on Wireless Communications*, *13*(3), 1616–1627. doi:10.1109/TWC.2013.122313.131118

Wei, Z., Tang, H., Yu, F. R., Wang, M., & Mason, P. (2014). Security Enhancements for Mobile Ad Hoc Networks With Trust Management Using Uncertain Reasoning. *IEEE Transactions on Vehicular Technology*, *63*(9), 4647–4658. doi:10.1109/TVT.2014.2313865

Yu, M., Zhou, M., & Su, W. (2009). A Secure Routing Protocol Against Byzantine Attacks for MANETs in Adversarial Environments. *IEEE Transactions on Vehicular Technology*, *58*(1), 449–460. doi:10.1109/TVT.2008.923683

This research was previously published in Forensic Investigations and Risk Management in Mobile and Wireless Communications; pages 195-220, copyright year 2020 by Information Science Reference (an imprint of IGI Global).

Chapter 11
Optimized Three-Dimensional Security Framework to Mitigate Risks Arising From BYOD-Enabled Business Environment

Lizzy Oluwatoyin Ofusori
University of KwaZulu-Natal, South Africa

Ncamiso Nkululeko Jahalenkhosi Dlamini
University of KwaZulu-Natal, South Africa

Prabhakar Rontala Subramaniam
University of KwaZulu-Natal, South Africa

ABSTRACT

Bring your own device (BYOD) has infiltrated the work environment and businesses are enjoying the benefits coupled with the adoption of the trend. At the same time, the adoption of BYOD has introduced a number of security threats that businesses are failing to match. In the pursuit of addressing security threats that are introduced by the adoption of this technology trend, this chapter recommends a three-dimensional (3-D) security framework that can be used to mitigate the risks emanating from a BYOD enabled environment. Data was collected from the employees of two banks in Africa that supported BYOD trend to measure individual and organizational practices. This chapter investigates further on these practices in addressing implications of BYOD. Three categories of security threats are delineated in this chapter. A review of existing security frameworks is presented in this chapter. The chapter concludes by outlining a 3-D security framework as a potential solution to protect BYOD enabled business environment.

DOI: 10.4018/978-1-7998-8545-0.ch011

INTRODUCTION AND BACKGROUND

Mobile technologies are increasingly becoming a preferred way of communication as a result of the evolution of the fourth Industrial revolution (Niesen, Houy, Fettke, & Loos, 2016). Most manufacturing organizations continuously evolve into smart and interconnected production systems. As mobile devices are becoming more dependable, they are also becoming prevalent in workplaces (Astani, Ready, & Tessema, 2013). This pervasiveness has enabled Bring Your Own Device (BYOD) in most organizations. BYOD is a growing trend that allows individuals of an organization to bring their own devices such as smartphones, laptops and tablets to work. This interconnectivity of smart devices has resulted in massive amounts of data from individuals, organizations and society (Niesen et al., 2016). Hence, the likelihoods of millions of people connected by mobile devices, with unprecedented storage capacity, processing power and access to knowledge, are unlimited. This data from individuals results in lack of privacy, data from organizations leads to lack of integrity while data from society leads to security issues. Presently, the number of mobile devices connected to the internet is more than the population of the world. This margin will keep expanding and thus create an avenue for cyber-attack in large scale.

The effect of allowing individuals to access any network with their mobile devices has impacts on network access, network control, helpdesk resources and even information ownership (Astani et al., 2013). Similarly, protection of organizational data becomes a challenge because these personal gadgets are mobile, hence employee are able to carry company data wherever they go. Some other security challenges such as keeping the network malware-free, understanding who and what is on the network, lack of control over the amount of information that should be stored at the endpoint on a BYOD device and giving the appropriate access policies to enforce compliance and audit requirements were identified by Astani et al. (2013). As technological development continues to grow, the threat to individual, organizations and society also increases (Lee, 2015). The major concern is how the connectivity can be harnessed for productive use without affecting privacy, security and integrity.

This chapter provides an overview of the BYOD trend and some key attributes that led to BYOD becoming a business standard. This chapter further presents the security threats and legal issues that confront organizations together with current practices that have been adopted to cushion against these threats. A framework is proposed to illustrate the distinction amongst threats that continuously affect businesses and the society at large through embracing the BYOD trend. A discussion explaining the difference between threats emanating from BYOD adoption and the Cyberspace is provided. Thus, the chapter proposes a 3-D BYOD security framework that can be used to prioritize awareness in order to ensure data integrity is maintained by all means.

In order to measure the BYOD security threats, some data were collected from the banking sectors in the African continent: Nigeria and Swaziland. These two countries are considered from two different extremes: Nigeria is one of the countries where the use of technology is accepted and is being used by the population at large. There are policies and a regulatory framework that have been incorporated to support the use of technology (Umar, 2015). On the other hand, Swaziland is one of the economically under developed countries in the continent. At present, there are no policies that have been implemented that specifically regulate the use of personal gadgets in the work environment. Swaziland has inadequate policy and regulatory frameworks; consequently, the integration of information and communication technologies (ICT) is currently uncoordinated and uncontrollable regardless of the current initiatives (Madzima, Dube, & Mashwama, 2013). As a result, there is lack of adequate planning, technical support and inadequate infrastructure which affects the introduction and adoption of ICT in the country (Madzima et al., 2013).

LITERATURE REVIEW

Evolution of BYOD

The infiltration of personally owned mobile devices like smartphones gave birth to the Bring Your Own Device (BYOD) phenomenon, where personal gadgets started entering the workspace. According to Copeland and Crespi (2012), the technological world witnessed a new paradigm shift in consumerization, where individuals were not just bringing personal devices but started using web applications for work tasks. Zahadat, Blessner, Blackburn, and Olson (2015) define BYOD as a movement that has been around for some time, characterized by individuals bringing their own personal gadgets to the workplace and installing preferred programs to accomplish tasks assigned to them. Broomhead (2013) redefined this movement using statement like "the rise of mobility and marginalization of the PC" and "move-and–do culture".

Efforts were made to stop the infiltration. For example, IT departments were disabling USB ports on company PCs and they used central software based on policy controls to monitor desired workstations to extents that were deemed necessary (Zahadat et al., 2015). Organizations' efforts to stop these privately-owned devices from entering their environments were not successful but they had no option but to acknowledge that employees would always want to use their mobile devices in their working area. Employees would always make a case and argue that consumer apps were cheaper (Harris, Ives, & Junglas, 2012). This resulted in mobile devices owned by employees, and the organizations provided financial and technical support for these devices (Vignesh & Asha, 2015). IT departments had to adjust their budgets and increase security measures to accommodate personal devices entering the work environment. Gartner (2012) considers the use of mobile devices in the workplace to be among the ten most important strategic trends. Broomhead (2013) echoes Gartner by mentioning that BYOD has become disruptive in the sense that employees always want to bring personal devices to the organization and want to connect to everything. Consequently, these connections are without proper accountability or oversight.

Moavenzadeh (2016) a member of the Management Committee of the World Economic Forum and head of mobility industries, raised the following concerns about the fourth industrial revolution. Firstly, the agility and fast pace in technology innovation has led to there not being a clear cut meaning of privacy and security as a result of real time availability of information. Secondly, how will the technology world collaborate to build regulatory frameworks and standards that promote growth and adoption of new technologies? According to Schwab (2016), the fourth industrial revolution profoundly affects the nature of security in businesses. Thus, the recommended framework in this chapter can be the first step in dealing with regulatory mandates that have strict security measures in order to promote compliance in businesses. Basically, the framework can help in data management since the fourth industrial revolution has shifted the focus to develop cyber-physical systems to monitor and control physical processes (Stancioiu, 2017). Understanding the security domains covered in this chapter can also serve as a decision framework or tool that can help businesses comprehend and prioritize the needs of employees and goals of the organization at large.

Benefits of BYOD in the Fourth Industrial Revolution

As mobile technologies are increasingly becoming a preferred way of communication in the fourth industrial revolution (Niesen et al., 2016), the possibilities of billions of people connected by mobile devices, with unprecedented storage capacity, processing power, and access to knowledge, are unlimited. These possibilities are increased by the emerging trend of BYOD in most organizations. Copeland and Crespi (2012), outline different steps that most of the organizations took in adopting BYOD. Firstly, companies started encouraging personal devices and connecting them to corporate internet servers. Secondly, companies started connecting personal devices to corporate applications. Finally, the companies stopped providing phones and laptops to employees thereby adopting the option of BYOD strategy. According to Bello, Armarego, and Murray (2015), BYOD became a compelling trend because organizations embraced the following benefits: lower costs, increased productivity and less technical training to employees. Lebek, Degirmenci, and Breitner (2013) mentioned that by adopting BYOD, organizations realized that there was greater freedom and flexibility, increased motivation and easier adoption of the technology. Another benefit identified by Baldassari and Roux (2017) is smart connectivity. Through smart connectivity, mobile devices and company systems can now interact to deliver greater efficiencies and capabilities, hence, there is optimized software, hardware and people (Baldassari & Roux, 2017). Understanding and realizing the benefits of the BYOD trend and the recommended framework delineated in this chapter can contribute to coming up with initiatives that can improve all business dimensions.

Individual Practices

The sudden rise in the usage of mobile devices has significantly increased the figure of interconnected devices to 12.5 billion while the population of the world is at 6.8 billion as at 2010 (Cisco, 2013). As a result of the high rate of growth of personal mobile devices connecting to corporate networks, it is unsurprising that the number of security incidents is also expected to grow. According to Company85 Briefing (2012), most of the personally owned devices connect to many unregulated networks such as restaurants, cafes and bars. Another study conducted by Cisco (2013) also revealed that 69 per cent of mobile device users had unapproved applications installed on their mobile device. However prior to the installation of these apps, mobile device users sometimes grant access requests to device resources as part of terms and conditions to install (Disterer & Kleiner, 2013). This leads to privacy invasion as users' private information can now be used without their knowledge. Also some of these downloaded applications are malicious and thus pave the way for hackers to penetrate and hack into the system(Pratt Jr & Jones, 2013). A recent study by Uz, (2014) points out that the adoption of cloud storage services as a concern for security of corporate data. Most employees are now using cloud services (e.g. iCloud, Drop box and Google drive) because of the insufficient memory on their mobile devices. These cloud storage services enable employees to copy files into the cloud for later retrieval but corporate information residing in such services may pose a security risk since they no longer reside in the protected corporate boundaries (Uz, 2014). The various mobile devices brought to work by employees make it challenging for the IT department to track the applications running on these devices. Similarly, most employees devices are not configured by the organization and thus pose a security threat (Disterer & Kleiner, 2013). The individual practices would have been stopped but most organizations have adopted this practice because of the profound benefits they derived from it and this has gradually led to the practice in organizations.

Organization's Practices

The organization's principles, policies and values define the practices of employees within the organizations. Some organizations carry out training to ensure employees' compliance to laid down principles and security policies governing the sector. This helps to avoid any form of security threats that may arise in the course of employees discharging their duties. However, Osterman Research (2012) points out that allowing personal devices to be used within the BYOD environment implies catering for the risks that it can bring into the organization. Tu and Yuan (2015) add that by allowing the use of personal devices, information security becomes exposed to new risks. One of the alarming concerns is the protection of valuable assets of organizations. Valuable assets refer to the information that is recorded, processed or stored and which the organization holds the ownership of (Bello et al., 2015). Protection of organizational data becomes a challenge because these personal gadgets are mobile; hence employees are able to carry company data wherever they go. According to report by Dimensional Research (2014), among organizations that allow personal devices to connect to corporate networks 93 per cent face challenges adopting BYOD policies, 94 per cent raise a concern over lost or stolen device, 66 per cent admit higher security risk for careless employees than cybercriminals. Another survey by Aruba Network (2013) pointed out that 75 per cent of IT managers have expressed concern about BYOD devices. Since BYOD enables employees to access corporate data any place, any time, it is important to ensure the security of corporate information and assets.

It is essential to protect confidentiality of corporate data on devices used within a BYOD environment. Furthermore, it is not only essential to protect confidentiality of corporate data, but also important to monitor and reject unauthorized and illegal access of corporate data. Unauthorized access comes from insiders (employees) when they are not supposed to access corporate data. Illegal access comes from outsiders when they want to recover corporate data stored on a device, e.g. malicious users try to steal data from a lost device. Once data is downloaded to a device, it is easy to make copies and transfer files from the device to other media. Thus, unauthorized and illegal data access on personal devices should be monitored and rejected.

Legal Issues in BYOD

Allowing employees to use their mobile devices for work purposes has raised a number of legal concerns. In a BYOD context, legal concerns emphasize properly addressing the existing policies, regulations and legislations between employers and employees (Lebek et al., 2013). The BYOD philosophy causes violations of working hour regulations (Lebek et al., 2013) because employees are forever connected to jobs even after working hours. Employees are able to access work material on weekends even on vacations. Consequently, this can lead to employees demanding compensation for the expanded working hours (Lebek et al., 2013). In addition to that, there is an assumption that employees are concerned about being liable when corporate information gets lost and when employees lose or damage their devices.

Implications of BYOD

In recent years, companies have enjoyed the compelling advantages of BYOD (Bello et al., 2015). Firstly, there is the low corporate cost whereby the acquisition and maintenance of these devices is solely the responsibility of the employee. Secondly, simplicity of use and the availability of content is another factor that has prompted the adoption of the BYOD (Dunnett, 2012). There is less technical training required for employees due to employees using mobile devices they are familiar with (Bello et al., 2015). Thirdly, BYOD has proved to increase employee productivity and efficiency as employees are no longer confined to do work only when they are on the organizational premises. They can now carry corporate data wherever they are (Garba, Armarego, Murray, & Kenworthy, 2015). Bello et al. (2015) concur that BYOD has become common in the corporate place and it is re-organizing ownership and control of organizational information systems and resources. BYOD has infiltrated a number of companies, there has been one major drawback: making end users or employees aware of the different threats introduced by using personally owned mobile devices for work purposes.

The issue of privacy has been one of the major concerns since BYOD has made the line between work and private become less distinct. Mora et.al. (2015), assert that different studies reveal that it becomes a challenge to distinguish between personal and organizational data once employees use their personal devices in an organizational context. From a business point of view, it raises the question of data integrity as company data is moved from corporate network to a personal device which the company has no control over. In a BYOD aspect, the privacy aspect focuses on employees' concern that their personal information (e.g. photos, emails, GPS data) becomes at the employer's disposal (Lebek et al., 2013). There is privacy invasion once an employer wants access to an employee's personal device hence such actions can lead to litigation if the subject of privacy is not properly addressed. Securing personally owned devices has become a nightmare for ICT departments because each new device brought into the company leaves the company with less control over what is transmitted and most importantly how it is transmitted (Twinomurinzi & Mawela, 2014). This difficulty emanates from the fact that personal devices are not locked down or configured to corporate standards to determine the amount of access that can be given to each personal device. BYODs are extensions of enterprise networks (Wang, Wei, & Vangury, 2014). However, it is difficult to ensure compliance with enterprise security policies for BYODs (Wang et al., 2014). Although there are some BYOD security measures, they all have limitations when applied on a BYOD environment. Furthermore, BYODs are personal devices which are mostly adopted for business use. It is practically not possible to manually review an employee's personal device due to the large number of devices and also the ownership. Alternative automatic options should be explored to enforce enterprise security policies on a BYOD (Xu, Saïdi, & Anderson, 2012)

BYOD SECURITY THREATS AND CYBER THREATS

A cyber threat is any malicious act that attempts to gain access to a control system and/or computer network without permission or authorization from the owners (Olasanmi, 2010). Today, cyberspace is witnessing the advent of a complete range of mobile devices and applications that have made organizations susceptible to security threats from all types of miscreants. Cybercriminals take advantage of the fact that almost everyone uses a mobile device and as such make it easy to spread threats through the pervasive technology (Wada & Odulaja, 2012). A typical example is the recent security threat called "wannacry ransomware" that has affected about 74 countries (Ehrenfeld, 2017). This security threat must have started from an individual, then organizations, before affecting the society which involves about 74 countries. As BYOD evolves, cyber criminals have now targeted mobile devices. The main difference between BYOD security threats and cyber threats is found in what the literature describes as concerns (Brodin, 2016). For BYOD there is an extra factor which creates even more concern; risk of mixing the organization's and the employee's personal data, software licensing issues, shared use of devices with non-employees and when the employee leaves the organization. There is a high risk that the organization's data is mixed with private data, and in the end, it will be difficult to distinguish who owns what. Similarly, when an employee leaves the organization, the person keeps the device where the data is stored. The responsibility for protecting BYODs has always been the users' responsibility. However, the move of organizational control of the device to private individuals raises security concerns and increases the demand on the user for security awareness (Brodin, 2016).

Security Threats Resulting From BYOD

While organizations in an effort to boost productivity and increase service delivery use ICT, they are confronted with inherent security threats frustrating the advancement of its progress (Ehimen & Bola, 2010). However, these threats can be categorized under technical, mobility and social threats. Figure 1, shows the categories security threats.

Figure 1. Categories of security threats (Author's own)

Visual Perspective of the Risk Levels

Figure 2 provides a visual perspective of how the three categories of the security threats are interwoven and constitute a security risk to any organization. Low risk (LR) represents those threats that are specific to the each domain. The threats contributing to the intersection of two domains represents medium risk (MR). While the threats contributing to the intersections of three domains in the 'Venn diagram' represent high risk (HR); which is invisible to the current security measures (refer Figure 2)

HR constitutes those threats that cut across the technical, social and mobility domains. They can be very harmful to the banking sector as they bring down the three domains at the same time. MR are security threats that are harmful but not to the same extent as the HR and are more harmful than the LR. These threats are only related to two domains and can bring down the two domains if the right security measures are not put in place. Meanwhile, LR are security threats that are harmful but not to the same scale as the other two risks (HR and MR). These security threats are only related to a particular domain; it does not affect more than one domain. The following sections identify the threats under each domain and classify them within the appropriate category. Although the MR is the intersection of two domains, they are categorized under a particular domain based on their characteristics.

Figure 2. Classification of the risk levels (Author's own)

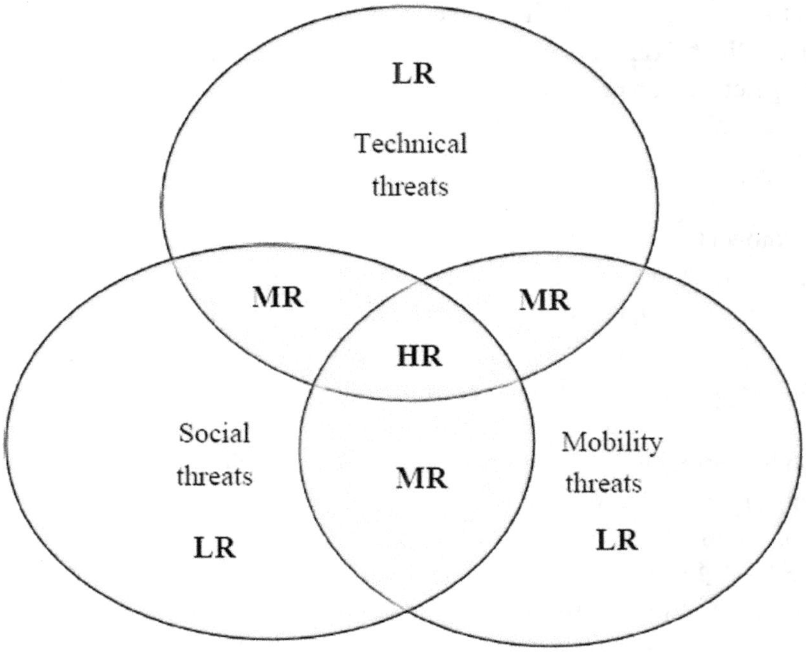

Technical Threats

In the context of this chapter, the technical threats represent threats emanating from the technical knowledge in the use of mobile device. It is important to identify these threats because by their very nature, they are complex to detect yet harmful to mobile device users and organizations to the extent they expose them to other threats that require a separate security management. The following section presents some of the key technical threats related to mobile device users in any organizations.

Phishing

One of the major concern is the issue of phishing, where hackers are using any means to gain access, and intercept personal and company sensitive data. This is not just any attack, but a selective attack known as spear phishing where specific personnel and their company are being targeted once the attacker has specific knowledge about the company and the individual (Ngoqo & Flowerday, 2015). A hacker may send an email to management asking them to respond to an attached survey on how the company executes performance management. The managers may fall for this phish and download the survey, which automatically installs malware to the company network. Phishing can take many forms; attackers have shifted from using emails to instant messaging, social networking sites and short message service (SMS). According to Hong (2012), spear phishing attacks have victimized big entities like the Australian Prime Minister's Office and the Canadian government between the year 2010 and 2011. Thus, the study highlights the importance of awareness mainly because phishing is not just a technical issue but also a social issue. Phishing is capable of infiltrating any organization regardless of the powerful firewalls, proxy servers and encryptions the company may have. The main points of entry into an organization are individuals not the systems put in place. What it takes is to deceive a person and that person innocently gives information to the wrong person. It is imperative for employees and employers to be educated on such issues to prevent them from falling for a phish. Thus, phishing can be regarded as LR as it requires the consent of a targeted person or organization to respond or download an attached email.

Malware

Malware is a well-known threat to mobile devices and is speedily increasing as a result of BYOD implementation (Felt, Finifter, Chin, Hanna, & Wagner, 2011). Malware is often disguised as a patch, game, utility, or third-party software application that can be installed, which may include viruses, spyware and Trojans (Gharibi, 2012). As soon as it is installed on mobile devices, it can cause massive attacks (Pratt Jr & Jones, 2013). The malicious application can function in numerous ways such as accessing sensitive information, accessing location information, activating the device's camera or microphone in order to record information, initiating phone calls or gaining access to user's browsing history (Felt et al., 2011). For example, Wannacry is a type of malware that gets into the computer or mobile device through email attachments or WhatsApp messages, and automatically encrypts every file (Ehrenfeld, 2017). Malware can also be regarded as LR because it requires the cooperation of the targeted individual or organization to comply by installing disguised software or opening an attachment. This can be prevented if the user is aware of what constitutes a potential threat.

Keystroke Logging

This is when an attacker monitors and archives keystrokes in order to access sensitive information (Boyles, Smith, & Madden, 2012). Key loggers' common practice is that they transmit information gathered to a cybercriminal's website or email address (Boyles et al., 2012). Thus the study emphasizes employee awareness and education on security threats in order for employees to understand the consequences of their mobile device activities. Keystroke logging, also known as key logging, is used to record typed characters on mobile devices in order to capture valuable or sensitive information such as user's ID, password and credit card numbers (Ladakis, Koromilas, Vasiliadis, Polychronakis, & Ioannidis, 2013). The captured information is usually transferred to a cyber-criminal email address or website (Pratt Jr & Jones, 2013). Keystroke logging can be considered LR because it only records typed characters on mobile devices, however, if employees are well informed, it can be avoided.

Rogue Device

A rogue device can be defined as an unauthorized connection of mobile devices to the network which pose a security threat to the organization (Golde, Redon, & Borgaonkar, 2012). It is used to breach the key areas of security for mobile subscribers such as intercepting communication, impersonating traffic, and tracking phones (Chen, Chen, Lin, & Sun, 2014; Golde et al., 2012). Rogue device can be considered LR because it can easily be detected if the organization puts the right security measures in place.

Data Interception

Data interception refers to the obstruction of data transmission to and from the device, and remotely altering the messages (Evripidis, 2008). This can be a major threat to wireless networks used by BYOD devices (Bello et al., 2015). If information is sent over the internet by a mobile user and it is intercepted, it becomes a serious concern, because of the risk that the data can be accessed, edited, or destroyed. Data such as personal information, chats, emails and passwords are being intercepted using a tool called packet sniffer or packet analyzer (Wu, 2009). Mobile devices such as smartphones, tablets and laptops are easily commjacked. This is when an attacker spies on data exchanges being transmitted to a mobile device (Evripidis, 2008). Public Wi-Fi has been good medium exploited by attackers. Attackers will capture and alter data packets between devices when mobile devices connect to unsecure Wi-Fi networks; this is referred to as a man-in-the-middle attack (Bello et al., 2015). Thus data interception can be regarded as MR because it remotely intercepts and alters messages sent to and from devices. In some instances, the user may not be aware that the message has been altered.

Network Exploit

Mobile systems that operate on cellular or local (e.g. Wi-Fi or Bluetooth) networks often experience software flaws (Pratt Jr & Jones, 2013). Network exploits take advantage of such flaws to propagate malware and in most cases they succeed without user interference. Attackers make use of special tools to find users on a Wi-Fi network and hijack the users' information which is then used to impersonate a user online (NEEDHAM & Lampson, 2008). Furthermore, there is a possible attack known as bluesnarfing which gives attackers access to sensitive information (e.g. credit card details) by exploiting a software flaw in a Bluetooth-enabled device (Pratt Jr & Jones, 2013). Thus network exploit can be considered MR because it takes the advantage of software flaws to propagate malware and in most cases they succeed without user interference.

Mobility Threats

Mobility threats refer to those threats associated with employees' traveling with their mobile devices that contain work documents. These devices are connected to secure and unsecure networks where the security policies differ.

Lost/Stolen Device

Studies have shown that in every year there are million cases of data breaches as a result of lost or stolen mobile devices (Juniper Network, 2011; Karen, 2015). However, many mobile device users do not realize these security loopholes. While the portability of these mobile devices allows continuous access to work related functions and personal information from any location, valuable information can easily fall into the wrong hands, and may be used for fraudulent and other illegal purposes when stolen or lost (Juniper Network, 2011). Thus, the implication of such lost or stolen mobile devices is that sensitive information can be compromised by a malicious hacker.

Organizations are most worried about the loss of data when employees are allowed to access corporate networks using personal devices. This is due to devices getting lost or stolen. According to Alfreds (2013) the escalating number of mobile devices users lost has led to criminals targeting the mass market of mobile devices with the intention of stealing personal information like banking details. According to Leavitt (2013), passwords can be intercepted by hackers with the intention of stealing credentials, collecting sensitive corporate data and installing malware that can take over devices. Information leaks to unauthorized users because organizations have difficulty to detect the leakage before it happens. As a result IT departments hardly have an opportunity of taking preventive measures to minimize the impact which is usually possible at the early stage of diagnosis (Pillay et al., 2013). This problem is propelled by the challenge of not being able to wipe out company sensitive information when the devices get lost hence such devices are not configured or locked down by the IT department. Thus lost/stolen devices can be regarded as LR because the individual or organization has the ability to wipe out personal or confidential information once the right security measures are put in place.

Wi-Fi Eavesdropping

Accessing the Internet on Wi-Fi networks at any locations such as airports, shops, street is not secure (Du & Zhang, 2006). Hackers take advantage of such wireless networks to eavesdrop on conversations and remotely modify messages from the mobile device. The hackers (cyber attacker) often create a hotspot with a device and such device is used to compromise a legitimate WiFi network in order to steal the user's information and in turn commit online fraud or hack into the bank's database (Balachandran, Voelker, & Bahl, 2005). Wi-Fi eavesdropping can be considered LR because if mobile device users do not access free Wi-Fi at any location it can easily be prevented.

Unauthorized Location Tracking

Location tracking enables the position of registered mobile devices to be recognized and tracked. While it is used openly for legitimate purposes in the banking sector, it can also be used under wraps (Enck et al., 2014). A legitimate software application or malware installed on the user's mobile device can be used to obtain the location data (Pratt Jr & Jones, 2013). Current smartphones are endowed with various sensors that can be used to deduce the user's whereabouts (Nguyen et al., 2013). Unauthorized location tracking can be considered MR because it is used under wraps to deduce a user's location without the knowledge of the user.

Social Threats

In the context of this chapter, social threats represent users' attitudes and awareness levels in using mobile devices. Whilst these threats are normally not well addressed because of their invisibility compared to other forms of threats, it is important that organizations recognize their influence on their security system. Otherwise if they are unchecked they have the potential to expose organizations to other security threats due to their association with people. The following section briefly discusses some of the main social threats that pose a risk to any organization.

Malicious Insider Threats

Malicious insider threats occur when individuals in a position of trust abuse the trust for personal benefits (Bowen, Salem, Hershkop, Keromytis, & Stolfo, 2009). Malicious insiders can be a current employee, a former employee, contractor, consultant or a trusted partner taking advantage of the special access privileges and knowledge to commit security breaches (Mathew, Upadhyaya, Ha, & Ngo, 2008). BYOD employees who are malicious insiders have the likelihood of performing phishing, malware attacks, spoofing and data interception. Similarly theft of mobile devices can easily occur without notice in the organizations. Most employees ignorantly share sensitive information which can be hijacked by a malicious insider before getting to the intended user (Mathew et al., 2008). Malicious insiders can be considered LR because the threat can be avoided if the working conditions from employers accommodate employees' demands which are associated with BYOD.

User Policy Violations

BYOD devices are easily exposed to vulnerabilities due to user policy violation. Carelessness and ignorance such as disabling firewall and antivirus applications to increase performance and speed, downloading and accessing unsecured websites that might contain malware, can expose BYOD devices to vulnerabilities and threats. Most organizations are constantly facing challenges of ensuring their employees adhere with user policies (Vance, Siponen, & Pahnila, 2012). Bello et al. (2015) noted that "*no matter how well developed and structured organizational policies are, they are rendered useless if not used adequately by employees*". A recent study has pointed out a breach that exposed over 5000 customers' data as a result of employees using a peer-to-peer (P2P) file sharing application from a BYOD laptop device on the network (Bello et al., 2015). According to Dimensional Research (2013) only 38 per cent of organizations actually enforce their policies by blocking these file-sharing applications on the corporate network, while 28 per cent admit that some employees don't follow the policy. Thus user policy violations can be regarded as LR because they can only occur as a result of users' ignorance or carelessness which inadvertently exposes the device to security threats.

Data Ownership Violation

A recent study has pointed out the adoption of file-sharing sites such as iCloud, Drop box and Google drive, as a concern for security of corporate data (Uz, 2014). These file sharing services enable employees to copy files into the cloud for later retrieval but corporate information residing in such services may pose a security risk since they no longer reside in the protected corporate boundaries. According to Dimensional_Research (2013) only 38 per cent actually enforce their policies by blocking these file-sharing sites on the corporate network, while 28 per cent admit that some employees don't follow the policy. Data ownership violation can be considered as MR because any data stored outside the protected boundary of an organization is susceptible to attack.

Data Privacy Violation

Most employees of banks use social media such as Facebook, Twitter, Instagram, WhatsApp and LinkedIn as a platform to interact with other colleagues or other users (Aula, 2010; Chanda & Zaorski,

2013). Unfortunately some of the personal and company's information that is made available on the social media is being hijacked by knowledgeable hackers who buy and sell the personal and company's information in order to commit security breaches (Aula, 2010). Thus data privacy violation can also be regarded as MR because once confidential or personal information is made available on social media; it is no longer within the users' control.

Mitigating all these risks is impossible due to lack of awareness. Lack of awareness is capable of infiltrating all the risk domains regardless of the powerful firewalls, proxy servers and encryption the organization may have (Kathleen, 2015). Astani et al. (2013) have also argued that the security awareness of BYOD devices is so poor leaving businesses vulnerable to security threats. Thus, lack of awareness is major security threat associated with HR domain. Organization needs to be aware of this and acknowledge that employees can be "weakest link" in the security environment because they fail to perform specified security behavior due to insufficient awareness.

PROMINENT EXISTING SECURITY MEASURES FOR BYOD

There are various types of security measures available to address BYOD security threats and many more are being developed. According to Shazmeen and Prasad (2012) the on-going development is as a result of new measures to attack existing technologies which are also being developed at a similar rate. Thus, the existing security measures are not sufficient as mobile devices often create different sets of threats that require additional or special control. However, it is necessary to review some of the existing technologies available and their effectiveness.

Password Authentication

To increase mobile device security through authentication, the concept of user name and password has been a fundamental way of protecting sensitive and confidential information (González, Tapiador, & Garnacho, 2008; Sree, 2008). Although, there are different tools used to verify the identities electronically, the most common authentication method is the use of user ID and passwords. According to González et al. (2008) a properly designed and implemented multifactor authentication method (e.g. password) is more reliable and stronger against any fraud attempt. A study by Bilal and Sankar (2011) reveals a more effective alternative method which is to use digital signatures together with supplementary devices such as smart cards. However, such methods by themselves have significant weaknesses in practice and are dependent upon strict enforcement of password security (Gui-Hong, Hua, & Gui-Zhi, 2010).

Anti-Malware

Anti-malware software is used as a signature-based detection on mobile devices to identify, prevent and take action to remove malicious software programs, such as malware, viruses and worms (Friedman & Hoffman, 2008). There is an auto-update feature in most antivirus programs that enables the program to download profiles of new viruses. The anti-virus program is known to scan several files on a user's system to identify matches between each file's code and those in the signature database. Such identified matches are flagged as malware. According to Srinivasan (2007) documents must always be authenticated by antivirus software before downloading. More importantly, they must be checked to be certain they originated from a trusted and reliable source and cannot be altered and thus good antivirus software is essential to protect the devices from viruses. However, Friedman and Hoffman (2008) have identified some problems associated with signature-based malware detection: first it can only detect known malware. This implies that unknown malware cannot be detected. Second, the authors of malware create self-modifying malware that alters its own signature every time. Third, encryption can disguise the signature of a malware program. Thus, signature based detection is a must but not a completely reliable form of protection.

Anti-Phishing

According to James and Philip (2012), anti-phishing can be defined as an application integrated into the browser as an extra plug-in. Presently, many anti-phishing solutions have been proposed and some approaches have attempted to solve the phishing problem at the e-mail level (Gharibi, 2012). Such techniques are closely related to anti-spam research, although anti-spam techniques have proved to be very effective in also intercepting phishing e-mails (James & Philip, 2012). However the effectiveness of anti-spam techniques mostly depends on many critical factors such as regular filter training and the availability of anti-spam tools that are currently not used by the majority of Internet users (Gharibi, 2012).

Firewall

A firewall is a software program or piece of hardware that is used to protect corporate resources from outside intruders (e.g. hackers, viruses, and worms) that try to reach the mobile devices over the internet. According to Clark (2013) a firewall can be referred to as a security system that controls access to a protected network, such as a corporate network. It examines all messages entering or leaving the internet and blocks those that do not meet the specified security requirement. Friedman and Hoffman (2008) have also identified an important role of firewall on mobile devices which is to block the use of Wi-Fi, Bluetooth and phone communication. This explains the reason why firewall is not mostly use in mobile devices but is used in computer systems. However, while firewalls can play an important role in detecting malware, they can be compromised by an unauthorized intruder (Kahate, 2013).

Encryption

Data encryption at rest and in motion helps to prevent data loss in case of stolen/lost devices (Gharibi, 2012). Sensitive data such as passwords, login information and accounts must by no means travel unencrypted over a wireless system because wireless systems can be easily sniffed and thus compromised (Gui-Hong et al., 2010). This is to protect the device from hackers breaking into the personal device.

Intrusion Detection System (IDS)

Intrusion Detection System (IDS) technology is used to detect both pre-mortem and or post-mortem security threat (Amer & Hamilton, 2010). In IDS a monitoring component is included to specifically arrest network packets moving to and from the IDS. This also helps to determine any malicious usage or activity by automatically sending a malicious detection signal (Scheidell, 2009). This prompt signal can include barricading the network transmission coming from the attacker's internet protocol and also the transmission of an electronic mail message to a system administrator

Mobile Device Management

A Mobile Device Management (MDM) is basically used for the management of a mobile device (e.g. bring your own technology (BYOD) and also to enforce some specific security policies on those devices (Wang et al., 2014). A few examples of MDM are: Zenprise MobileManager, FiberLink Maas360, Air Watch MDM, and Mobile Iron. However, there are several limitations and drawbacks to this approach. First, MDM does not separate corporate and personal space on a mobile device. Additionally, due to lack of space isolation, the security policies administered by MDM are on the entire device (Wang et al., 2014). Thus if MDM is used, employees might lose the flexibilities associated with the personal space

Training on Acceptable Use of ICT Policy

Employees are being trained on acceptable use of ICT policy in the banking sector and access to the bank's information must be carried out within the specified acceptable boundaries (Broughton, Higgins, Hicks, & Cox, 2009). Similarly employees are given a copy of the ICT policy and it is the responsibility of the employee to get familiar with the terms of policy. Unfortunately not all employees get acquainted with this policy (Mulligan & Gordon, 2002)

Enforcement of Security Policies

Most banks enforce security policy on the employee to avoid any form of security breaches. Employees are forced to comply with the terms of the policy and where it is confirmed that an employee has violated the policy, such employee is disciplined or reprimanded (Herath & Rao, 2009). However despite the policy enforcement there are still recurring social security threats such as insider attack, social media attack and file sharing attack (Bulgurcu, Cavusoglu, & Benbasat, 2010). Thus the social security measures are incapable of completely mitigating these security threats.

EXISTING SECURITY FRAMEWORKS

Security frameworks refer to series of documented processes that are used to define procedures and policies around the ongoing and implementation of information security controls in an organization (Granneman, 2013). These frameworks come in various degrees of complexity and are used to build an information security program to reduce vulnerabilities and manage risks.

ISO/IEC 27000 Series

The International Standards Organization established the ISO 27000 series of standards. The main focus of ISO 27000 series is on information security issues. It provides an extensive information security framework that can be adopted by any organization (Granneman, 2013). However for the purpose of this research, the strengths and weaknesses of ISO 27001 and ISO 27005 will be explored in line with the prevailing demands for successful and robust security frameworks.

1. **ISO/IEC 27001:** The specification for an Information Security Management System (ISMS) is the ISO 27001 standard. The standard states the requirements for establishing, implementing, operating, monitoring, reviewing, maintaining, and improving an Information Security Management System within an organization for overall business risks (ISO, 2005). This will ensure the selection of ample and commensurate security controls to safeguard information assets. The standard introduces a model called the "Plan-Do-Check-Act" (PDCA) which helps in establishing, implementing, monitoring and improving the effectiveness of an organization's ISMS. However, some organizations basically adopt ISO 27001 in order to pursue an effective means for operational information security risk management but ignore the fact that the standard was not designed to be used at operational level but at the high level as an ISMS framework. The ISO 27000 series also consists of ISO 27002 which is considered as an extension to ISO 27001 (Barlette & Fomin, 2010). It provides recommended controls that an organization can adopt to address information security risks.
2. **ISO/IEC 27005:** The existing gaps in terms of information security risk management in ISO 27001 and ISO 27002 were closed by the introduction of ISO 27005 standard. The standard was used to elaborate and build up on the core that was introduced in ISO27001 by means of identifying inputs, actions, guidelines for implementation and outputs for every single statement. However, studies have shown that the adoption of this standard as a means for information security risk management is minimal (Dobson & Hietala, 2011), because it is not intended to be used as an information security risk assessment methodology (ISO, 2005).

While ISO/IEC 27000 series framework is helpful in establishing, implementing, monitoring and improving the effectiveness of an organization's ISMS, its adoption for security management is minimal (Dobson & Hietala, 2011). Al-Ahmad and Mohammad (2012) have also argued that the implementation is minimal because it was not designed to be used as methodology for information security risk assessment. Sequel to the arguments by Al-Ahmad and Mohammad (2013), this framework is completely inadequate to protect the banking sector from security threats arising from the use of mobile devices.

PCI DSS

Payment Card Industry Data Security Standard (PCI DSS) assists to protect cardholder data, maintain a vulnerability management program, build and maintain a secure network and assist to implement solid access control measures (Council Payment Card Industry, 2010). Every user that transmits or stores debit or credit card data is mandated to comply with PCI requirements (Al-Ahmad & Mohammad, 2012). This helps organizations to reduce losses resulting from fraud, manage information security risks and safeguard consumer data (Council Payment Card Industry, 2010). However, studies have shown that PCI DSS is not designed to be used as an information security framework or assessment framework (Al-Ahmad & Mohammad, 2013; Council Payment Card Industry, 2010) and therefore does not fit into the objectives of this study.

COBIT

One of the most increasingly adopted information technology frameworks for IT Governance is COBIT (Control Objectives for Information and related Technology) developed by the Information Systems Audit &Control Association (ISACA, 2011). The focus of COBIT is to define IT control objectives and develop the controls to meet the objectives. It comprises thirty four (34) processes that manage and control information and the technology (Brand & Boonen, 2007). A study has shown that COBIT can be used to create the IT governance framework, create the needed alignment between business and IT, establish the IT risk management organization and improve IT processes (Tambotoh & Latuperissa, 2014). Some studies have also shown that COBIT can be used to meet the organization's requirements, compliance needs and to create IT policies and procedures (Al-Ahmad & Mohammad, 2013; Barlette & Fomin, 2010). In a similar view, Parvizi, Oghbaei, and Khayami (2013) reveal that COBIT is used as a means to conduct audit, standardize IT processes and increase organization effectiveness and maturity levels. Security threats can be minimized with proper IT governance, however, some studies have argued that COBIT only establishes the foundation for having a solid IT organization in an organization but does not provide a methodology to conduct information security risk assessments (Al-Ahmad & Mohammad, 2012; ISACA, 2011). Similarly, the framework work did not emphasize user awareness as a means to avoid or reduce these security threats. This makes the COBIT framework inadequate in mitigating these security threats.

NIST SP 800 Series

The National Institute of Standards and Technology (NIST) Special Publication 800 series is one of the frameworks used to properly address information security (NIST, 2012). It was published in 1990 and has grown to provide an extensive variety of risk factors needed to determine information security risk. Its major activities include the following; assess the security controls using appropriate methods and procedures; categorize information systems and the information within those systems; authorize information systems operation based on a determination of the risk to organizational operations and assets; implement security controls in the systems; monitor and assess selected security controls in information systems on a continuous basis (Al-Ahmad & Mohammad, 2012). For ease of use, ideal assessment scales, tables and templates for common risk factors were included. This enables maximum flexibility in the design of risk assessments based on the organization's established scope, purpose, assumptions and constraints (Ross, 2011). However, to ensure that ultimate value is provided by security assessments, analysis must be conducted which implies that organizations must address the technical weaknesses as well as weaknesses in organizational process and procedure (Stouffer, Falco, & Scarfone, 2008). This technical weakness as well as organizational process and procedure weakness is one of the major issues why this framework is inadequate to address security. Similarly, user awareness on these security threats was not fully addressed.

CISCO SCF (Security Control Framework)

Cisco security control framework is made up of a model, control structure/set and methodology basically designed to assist in assessing the technical risk in infrastructure architecture. The 'model' explains the security objectives and security actions as a means to address the underlying principles and to organize individual controls. The control structure elucidates the individual controls and control sets in an organization and its relationship with the Cisco SCF model and methodology. The methodology outlines the requirements needed to perform an assessment on the security architecture using the Cisco SCF. However, Cisco SCF is designed based on the definition and use of appropriate control sets for specific business environments and assessment objectives. Thus it only consists of a set of rules used to assess architecture in an orderly manner.

IBM Security Framework

IBM has created a bridge to help with security challenges that addresses communication gaps between the technical and business perspectives. The IBM security framework addresses security challenges in the following domains: data and information, people and identity, application and process, physical infrastructure, network, server and endpoint, security governance, risk management and compliance. However, according to Buecker, Borrett, Lorenz, and Powers (2010), IBM security framework, only focuses on the 'what' not the 'how' and can only help to interpret users requirements into coarse-grained business solutions, not into specific IT solutions or IT components

Following the review of these frameworks, the limitations identified are a deterrent to the fourth industrial revolution. With the fourth industrial revolution, the possibilities of billions of people connected by mobile devices, with unprecedented storage capacity, processing power, and access to knowledge, are unlimited. These possibilities will be multiplied by emerging trends such as BYOIoT. However, all of these existing frameworks have their shortcomings and do not adequately address the significant impact of the fourth industrial revolution to accommodate the security threat to domains (Figure 1). Hence, a security framework is required to address these domains in the fourth industrial revolution.

DISCUSSION

Two similar studies were conducted in the banking sector of Nigeria and Swaziland to find out the implication of a BYOD enabled environment. Due to the type of transactions and sensitive information processed within the banking system, risk management is a critical aspect that must be addressed when developing a security framework (Wang et al., 2014).This study intended to contribute towards developing a security framework for BYOD environment. The findings from a study conducted in Nigerian banks shows that despite the security measures put in place, security threats are still on the increase. Furthermore, BYOD security awareness is not provided for the employees. Likewise those that provide security awareness training do not ensure employees' compliance to these security policies. Supporting these findings, Elwess (2015) states that as much as organizations may have knowledge on the subject of BYOD, research findings show that individuals are either still not aware of the potential threats or decide to ignore them. The data on individual practices (employees' practises) reveals that 77 per cent of mobile device users configured their mobile device to automatically connect to wireless network and 78.8 per cent click on links, images, advertisement, videos and games respectively on social media. This implies that organization data residing on employees' personal devices becomes exposed to all these unprotected environments. Mora et.al. (2015), assert that it has become a challenge to distinguish between private and organizational data once employees use their privately owned devices in an organizational context. The study further reveals that 83 per cent of Nigerian bank employees backup work documents on internal memory and cloud services of their mobile device as against the stipulated policy to back up on the bank's server.

Similarly, data collected from Swazi Bank in Swaziland revealed that a significant number of employees (93%) still connect to public Wi-Fi which makes mobile devices vulnerable to cyber-attacks. In addition to that, a majority of employees (92%) save company data on cloud based sharing services. Such findings show lack of awareness and comprehension amongst employees on the severity and vulnerability of using mobile devices in a work context. Thus, it is important for any organization to have a measure of awareness. This means management need to acknowledge that employees can be the weakest link in the security environment because they fail to perform specified security behaviours due to insufficient awareness (Johnston, Warkentin, McBride, & Carter, 2016).

Similarly as outlined in Figure 2, HR domains are security threats that can be very harmful to the organization as they affect the three threats domains (technical, mobility and social) at the same time. While MR domains are security threats that are pertaining to two threats domains and can bring down the two domains, and LR domains are security threats that are pertaining to a particular domain: it does not affect more than one domain. It is very important for an organization to be aware of these risk domains particularly the HR domain which is invisible to the current security measures. However, there is no suitable security measure to overcome lack of awareness. Thus this chapter proposes a 3-D security framework that will assess individual, organization and third party knowledge on BYOD threat. This chapter recommends that awareness is viewed holistically in three different levels: individual; organization; and third party.

Individual Perspective

The studies conducted in Nigeria and Swaziland revealed user perception as a common factor that influences technology. User perception can be explained by perceived severity of security threats, perceived vulnerability of mobile devices to security threats, security measure efficacy and self- efficacy in executing security policies. Perceived severity of security threats focuses on employees' perception of the magnitude or seriousness of possible threats that they are exposed to when using their mobile devices in the workplace. Perceived Vulnerability on the other hand looks at assessing how employees understand the chances of falling victim to security threats. Security Measure Efficacy focuses on how an individual feels or perceives that security measures put in place can effectively protect the company and the user. Self-efficacy looks at if individuals feel comfortable executing security policies required for mobile devices in the workplace.

The results show that a significant number of employees agreed that the organization would be prone to security threats if they did not adhere to security controls. Therefore, the greater the exposure the greater they perceived the vulnerability of mobile devices to security threats. The results indicate that a significant number of employees believed that existing policies could protect them. The results show differences in the way employees think of the effectiveness of security measures. This suggests that it is imperative for organizations to be cautious when aligning training programs and policies to promote compliance in the organization. Moreover, results highlighted that employees felt comfortable executing security policies required for mobile devices in the workplace. Thus, the framework emphasizes that organizations must have policies that employees can easily comprehend and execute. This is to ensure that employees can be able to follow security policies even with minimum assistance. The findings support Alexandrou's (2016) argument that an individual will be successful in adopting security controls on a mobile device only if there is high confidence in applying the security controls.

Studies also highlighted that attitude had a significant influence on employees 'awareness. The determining factors in attitude toward security response cost, regulatory concern and reaction to security policies. When looking at the security response cost of using mobile devices results revealed that employees are mostly concerned about time, convenience, and comprehension, which can be barriers when adopting security measures for their mobile devices. Employees feel such barriers may result in frustration and reduce motivation to comply with security policies. Results suggest that there are higher chances for employees not to follow policies that are too restrictive which may result in low employee job satisfaction. This further shows that employees are looking for simplicity; they need something they can easily comprehend and implement.

Organizational Perspective

The recommended framework postulates that organizations need to invest in educating and training the workforce. Creating awareness on the subject of organizational assets is key. Organizational assets include Corporate Intellectually property, Classified Information, Financial Assets, Device and service availability and functionality and Political reputation. Failing to do so means employees cannot comprehend the importance of data or information to the organization. The organization cannot achieve employee commitment in securing company information. As a result, employees can hardly understand what exactly will be at stake if information is compromised. Consequently, this can lead to regulatory fines, damage to company reputation and loss of intellectual property. For example, both studies revealed that there were no sufficient efforts done in addressing the subject of organizational assets; hence, some employees could not fully comprehend the consequences of non-compliance.

A point to bear in mind is that learning is a continuum; it begins with awareness, builds to training and leads to education (Wilson & Hash, 2003).Organizations may have all the necessary resources to protect and improve the security of corporate information assets but employees still remain one of the threats that need to be dealt with. According to Freedman (2015)), information systems managers are more interested on technical issues and solutions such as intrusion, routers and firewalls and pay less attention to soft issues caused by users' lack of security awareness. Rose (2013) states that it has been demonstrated that employees who do not adhere to company security policies are the main hazard for company security. This implies that organizations need to engage employees through training and awareness programs to enlighten them on how to protect company data. As employees are increasingly accessing privileged corporate information and applications, it is vital for organizations to understand and address the implications caused by unknown devices entering the organizational environment (Twinomurinzi & Mawela, 2014). Awareness programs are designed to allow employees the flexibility to use their personal devices and at the same time mitigate the security and concerns of the ICT department (Twinomurinzi & Mawela, 2014).

Third Party

The framework further recommends that the third party (e.g. Internet service provider, device manufacturer etc) needs to ensure that organizations comply with stipulated control system. It may be common knowledge that requirements for secure integration of mobile devices can be derived from standards such as COBiT or ISO 2700 (Disterer & Kleiner, 2013). The fact remains that users first need to buy into the idea of information security. This control system is meant to validate data integrity and keep the organizations in check. Cross data validation helps to assess the organization for non-compliance.

CONCLUSION AND RECOMMENDATIONS

The benefits of BYOD to most organizations include increased productivity, flexibility and efficiency. In spite of these benefits, privacy invasion, security and data integrity have been a major concern for individuals, society and organizations that adopt BYOD. Securing personally owned devices has become a nightmare for ICT departments because each new device brought into the company leaves the company with less control over what is transmitted and most importantly how it is transmitted (Twinomurinzi & Mawela, 2014). Miller, Voas, and Hurlburt (2012), distinguish two main threats to corporate information security that can be introduced by the BYOD philosophy. The first one is malware intrusion (worms, viruses, trojans) as privately owned devices are integrated into corporate network facilities (Miller et al., 2012). Exchanging personal and professional information in personal devices will enable the user to navigate inside social networks which can result in important organizational data being leaked in the event of a security incident (Mora et.al, 2015). The second threat is the increase in the possibility of data loss and theft (Miller et al., 2012). Corporate information may fall into the wrong hands when an employee misplaces or loses his or her personal device.

The validity and integrity of data is lost through data contamination. This is due to personal information and confidential business information sharing the same storage space on personally owned mobile devices. According to Romer (2014), contamination happens when personal files accidentally end up on corporate file servers, even worse, personal files containing malware spread to business files from the mobile device to corporate internal file servers. Data is no longer under control once information migrates to a device that is not controlled by the company (Miller et al., 2012).

The adoption of BYOD resulted in IT departments having the challenge of how to secure data and prevent information systems from being misused or compromised. The departments confronted the question of how to secure sensitive information and how to address privacy concerns when personal devices enter work space (Alexandrou, 2016, p. 1). Furthermore, with the advent of the fourth industrial revolution, there is an acceleration in innovation and an influx of mobile device entering the work space. Thus productivity has been improved (Schwab, 2016). However, the rapid pace of change in innovation has led to adverse effects in legislation and regulations (Schwab, 2016). This implies that regulators have to adapt to the fast changing environment, thus no single framework is sufficient to address the drawbacks of the fourth industrial revolution. Matinde (2015) points out that the magnitude of BYOD challenges will always intensify as long as business models continuously evolve to increase mobility especially when there is evident growth in the mix of users and geographically diverse business offices. Elwess (2015) states that as much as organizations may have knowledge on the subject of BYOD, research findings show that individuals are either still not aware of the potential threats or decide to ignore them. Similarly, some of the organization's management who are involved in decision making are also employees who are lacking in awareness. Thus there is an urgent need for a 'unified single point of context aware' security framework outlined in Figure 3 to help curb these security threats.

Figure 3. 3-D BYOD security framework (Author's own)

Dimensions	Motivating factors	Role	Response
Individual	End user perception	Learn	Attitude
Organization	Security assurance	Train	Control disruptive behaviour.
Third party	Compliance to device and service	Device usage control	Data integrity support

As employees are increasingly accessing privileged corporate information and applications, it is vital for organizations to understand and address the implications caused by unknown devices entering the organizational environment (Twinomurinzi & Mawela, 2014). Awareness programs are designed to allow employees the flexibility to use their personal devices and at the same time mitigate the security and concerns of the ICT department (Twinomurinzi & Mawela, 2014).

The end process in the model is to conduct an evaluation, where employees are expected to show full commitment (prescriptive commitment) in adhering to company security policies when using mobile devices in the work place. Sipponen (2000) laments that companies are still struggling to reach a point where the workforce would internalize and follow given guidelines. As a result you find that employees are still unaware of policies or they fail to apply them. Prescriptive commitment can take different forms like having employees who will avoid installing unnecessary applications, avoid sharing company sensitive data especially over unprotected networks, avoid jail-breaking, have good physical control of mobile devices. On the part of the organization, the objective should be enforcing security policy and awareness compliance, while the third party is expected to carryout data cross validation to ensure data integrity.

Looking to the future Mark (2014) has predicted that by 2020, BYOD will evolve into BYOIoT (Bring Your Own Internet of Things), where we have so many inter-connected devices. According to Mark (2014), BYOIoT elements are now working their way into the office environment because they do have the potential to make businesses even more efficient in day-to-day operations. It is however easy to see the potential benefits along with the possible pitfalls. While BYOIoT has the potential to make businesses even more efficient in day-to-day operations, they pose more security threats to the organizations by the huge amount of data coming from the smart devices that is not controlled or checked by the organizations. However, the fourth industrial revolution has proved to have effects on national and international security whereby there have been conflicts involving different states (cyberwar) which have become increasingly "hybrid" in nature. It is worth mentioning that the developed framework does not address fourth industrial revolution issues relating to conflicts between two states, which requires further research.

REFERENCES

Al-Ahmad, W., & Mohammad, B. (2012). Can A Single Security Framework Address Information Security Risks Adequately? *International Journal of Digital Information and Wireless Communications*, *2*(3), 222–230.

Al-Ahmad, W., & Mohammad, B. (2013). Addressing information security risks by adopting standards. *International Journal of Information Security Science*, *2*(2), 28–43.

Alexandrou, A. (2016). *A security risk perception model for the adoption of mobile devices in the healthcare industry*. Pace University.

Alfreds, D. (2013). *News 24. BYOD security threats to SA firms*. Retrieved April 2, 2016, from http://www.news24.com

Amer & Hamilton. (2010). Intrusion detection systems (IDS) taxonomy-a short review. *Defense Cyber Security, 13*(2).

Aruba Network. (2013). *Conquering today's bring your own device challenges: A framework for successful BYOD initiatives*. Aruba White Paper.

Astani, M., Ready, K., & Tessema, M. (2013). BYOD Issues and strategies in organizations. *Issues in Information Systems*, *14*(2).

Aula, P. (2010). Social media, reputation risk and ambient publicity management. *Strategy and Leadership*, *38*(6), 43–49. doi:10.1108/10878571011088069

Balachandran, A., Voelker, G. M., & Bahl, P. (2005). Wireless hotspots: Current challenges and future directions. *Mobile Networks and Applications*, *10*(3), 265–274. doi:10.100711036-005-6421-5

Baldassari, P., & Roux, J. (2017). Industry 4.0: Preparing for the Future of Work. *People & Strategy*, *40*(3), 20–23.

Barlette & Fomin. (2010). The Adoption of Information Security Management Standards. *Information Resources Management: Concepts, Methodologies, Tools and Applications: Concepts, Methodologies, Tools and Applications*, 69.

Bello, A., Armarego, J., & Murray, D. (2015). Bring your own device organizational information security and privacy. *Journal of Engineering and Applied Sciences (Asian Research Publishing Network)*, *10*(3), 1279–1287.

Bilal, M., & Sankar, G. (2011). Trust & Security issues in Mobile banking and its effect on Customers. *School of Computing. Blekinge Institute of Technology, SE-371*, 79.

Bowen, B. M., Salem, M. B., Hershkop, S., Keromytis, A. D., & Stolfo, S. (2009). Designing host and network sensors to mitigate the insider threat. *IEEE Security and Privacy*, *7*(6), 22–29. doi:10.1109/MSP.2009.109

Boyles, Smith, & Madden. (2012). Privacy and data management on mobile devices. *Pew Internet & American Life Project, 4*.

Brand, K., & Boonen, H. (2007). *IT Governance based on CobiT® 4.1-A Management Guide*. Van Haren.

Brodin, M. (2016). *BYOD vs. CYOD: What is the difference?* Paper presented at the 9th IADIS International Conference Information Systems 2016.

Broomhead, S. (2013). *Gartner says BYOD is disruptive*. Retrieved February 23, 2016, from http://www.ehna.acfee.org/read/art-54e605c41el4a

Broughton, A., Higgins, T., Hicks, B., & Cox, A. (2009). Workplaces and social networking-The implications for employment relations. Institute for Employment Studies, Brighton.

Buecker, A., Borrett, M., Lorenz, C., & Powers, C. (2010). *Introducing the ibm security framework and ibm security blueprint to realize business-driven security*. International Technical Support Organization.

Bulgurcu, B., Cavusoglu, H., & Benbasat, I. (2010). Information security policy compliance: An empirical study of rationality-based beliefs and information security awareness. *Management Information Systems Quarterly*, *34*(3), 523–548. doi:10.2307/25750690

Chanda, R. A. J. I. B., & Zaorski, S. T. E. V. E. (2013). Social media usage in the financial services industry: Toward a business-driven compliance approach. *Journal of Taxation and Regulation of Financial Institutions*, *26*(5), 5–20.

Chen, C.-M., Chen, Y.-H., Lin, Y.-H., & Sun, H.-M. (2014). Eliminating rouge femtocells based on distance bounding protocol and geographic information. *Expert Systems with Applications*, *41*(2), 426–433. doi:10.1016/j.eswa.2013.07.068

Cisco. (2013). *BYOD Smart Solution*. Retrieved 7 January, 2015, from http://www.cisco.com/web/solutions/trends/byod_smart_solutions/index.html

Clark, P. G. (2013). *Firewall policy diagram: Novel data structures and algorithms for modeling, analysis, and comprehension of network firewalls*. Academic Press.

Company85 Briefing. (2012). *BYOD and the security implications of consumerization*. Retrieved May 5, 2016, from http://safebridge.pt/Whitepapers/BYOD/Company85%20BYOD%20and%20the%20security%20implications%20of%20consumerisation.pdf

Copeland, R., & Crespi, N. (2012). *Analyzing consumerization-Should enterprise business context determine session policy?* Paper presented at the Intelligence in Next Generation Networks (ICIN), 2012 16th International Conference on. 10.1109/ICIN.2012.6376024

Council Payment Card Industry. (2010). *PCI DSS 2.0*. PCI Council Publication/United States.

Device, Y. O. (n.d.). Retrieved March 10, 2015, from http://www.bringyourownit.com/2012/06/25/byod-bring-your-own-device/

Dimensional Research. (2013). *The impact of mobile devices on information security: A survey of professionals*. Retrieved 30 December, 2014, from http://www.dimensionalresearch.com

Dimensional_Research. (2013). *The impact of mobile devices on information security: A survey of professionals*. Retrieved 30 December, 2014, from http://www.dimensionalresearch.com

Disterer, G., & Kleiner, C. (2013). *BYOD Bring Your Own Device*. Retrieved 19 December, 2014, from http://www.dx.doi.org/10.1016/j.protcy

Dobson, I., & Hietala, J. (2011). *Risk Management: The Open Group Guide*. Van Haren Pub.

Du & Zhang. (2006). *Risks and Risk Control of Wi-Fi Network Systems*. Academic Press.

Dunnett, R. (2012). *Information Security, Mobile Security and Internet of Things*. BYOD- Bring.

Ehimen & Bola. (2010, January). Cybercrime in Nigeria. *Business Intelligence Journal*, 93-98.

Ehrenfeld, J. M. (2017). WannaCry, Cybersecurity and Health Information Technology: A Time to Act. *Journal of Medical Systems*, *41*(7), 104. doi:10.100710916-017-0752-1 PMID:28540616

Elwess, T. (2015). *Bring your own device* (Doctoral dissertation). Utica College.

Enck, W., Gilbert, P., Han, S., Tendulkar, V., Chun, B.-G., Cox, L. P., ... Sheth, A. N. (2014). TaintDroid: An information-flow tracking system for realtime privacy monitoring on smartphones. *ACM Transactions on Computer Systems*, *32*(2), 5. doi:10.1145/2619091

Evripidis, R. (2008). *Lawful Interception and Countermeasures: In the era of Internet Telephony*. Academic Press.

Felt, A. P., Finifter, M., Chin, E., Hanna, S., & Wagner, D. (2011). A survey of mobile malware in the wild. *Proceedings of the 1st ACM workshop on Security and privacy in smartphones and mobile devices*. 10.1145/2046614.2046618

Freedman, A. (2015). Managing personal device use in the workplace: How to avoid data security issues and to dig yourself out of your failed BYOD policy. *Suffolk J. Trial & App. Adv.*, *20*, 284–361.

Friedman, J., & Hoffman, D. V. (2008). Protecting data on mobile devices: A taxonomy of security threats to mobile computing and review of applicable defenses. *Information, Knowledge, Systems Management*, *7*(1-2), 159–180.

Garba, Armarego, Murray, & Kenworthy. (2015). Review of the information security and privacy challenges in Bring Your Own Device (BYOD) environments. *Journal of Information Privacy and Security*, *11*(1), 38-54.

Gartner. (2012). *Creating a Bring Your Own Device (BYOD) Policy*. Retrieved April 4, 2016, from https://www.gartner.com/doc/1983515/creating-bring-device-byod-policy

Gharibi, W. (2012). *Some Recommended Protection Technologies for Cyber Crime Based on Social Engineering Techniques--Phishing*. arXiv preprint arXiv:1201.0949

Golde, N., Redon, K., & Borgaonkar, R. (2012). *Weaponizing Femtocells: The Effect of Rogue Devices on Mobile Telecommunications*. Paper presented at the NDSS.

González, Tapiador, Estévez, & Garnacho. (2008). *Content authentication and access control in pure peer-to-peer networks*. Universidad Carlos Iii De Madrid.

Granneman, J. (2013). IT security frameworks and standards: Choosing the right one. *TechTarget Network*. Available: http://searchsecuri ty. tech target. comlti p/IT-securi ty frameworks-and-standards-Choosing-the-right-one

Gui-Hong, L., Hua, Z., & Gui-Zhi, L. (2010). *Building a Secure Web Server Based on OpenSSL and Apache*. Paper presented at the 2010 International Conference on E-Business and E-Government. 10.1109/ICEE.2010.334

Harris, J., Ives, B., & Junglas, I. (2012). IT Consumerization: When Gadgets Turn Into Enterprise IT Tools. *MIS Quarterly Executive, 11*(3).

Herath, T., & Rao, H. R. (2009). Protection motivation and deterrence: A framework for security policy compliance in organisations. *European Journal of Information Systems, 18*(2), 106–125. doi:10.1057/ejis.2009.6

Hong, J. (2012). The state of phishing attacks. *Communications of the ACM, 55*(1), 74–81. doi:10.1145/2063176.2063197

ISACA. (2011). *COBIT 4.1*. Retrieved 24 May, 2016, from www.isaca.org

ISO. (2005). *Information technology–Security techniques–Information security management systems–Requirements*. ISO.

James, D., & Philip, M. (2012). *A novel anti phishing framework based on visual cryptography*. Paper presented at the Power, Signals, Controls and Computation (EPSCICON), 2012 International Conference on. 10.1109/EPSCICON.2012.6175228

Johnston, A. C., Warkentin, M., McBride, M., & Carter, L. (2016). Dispositional and situational factors: Influences on information security policy violations. *European Journal of Information Systems, 25*(3), 231–251. doi:10.1057/ejis.2015.15

Pratt Jr. & Jones. (2013). Mobile device management in the DoD enterprise network. *Practice, 44*(3), 179-196.

Juniper Network. (2011). *Mobile Device Security- Emerging threats, Essentials Strategies*. Retrieved 29 December, 2015, from www.google.co.uk/url?sa=t&rct=j&q=&esrc=s&source=web&cd=1&ved=0ahUKEwj60-6g5oDKAhWMWBoKHWjtASwQFggyMAA&url=http%3A%2F%2Fwww.bytes.co.uk%2Fdownload_file%2Fview%2F943%2F537%2F&usg=AFQjCNGVmC_t7GmVQqdw36AgYt-v9orttwg

Kahate, A. (2013). *Cryptography and network security*. Tata McGraw-Hill Education.

Karen, T. (2015). *Device Debacles – Lost, Stolen, and Neglected Data Risks*. Retrieved from https://www.allclearid.com/blog/device-debacles-lost-stolen-and-neglected-data-risks

Kathleen, R. (2015). *Lack of cybersecurity awareness linked to CIOs*. Retrieved from www.searchsecurity.techtarget.com/opinion/Lack-of-cybersecurity-awareness-linked-to-CIOs

Ladakis, E., Koromilas, L., Vasiliadis, G., Polychronakis, M., & Ioannidis, S. (2013). You can type, but you can't hide: A stealthy GPU-based keylogger. *Proceedings of the 6th European Workshop on System Security (EuroSec)*.

Leavitt, N. (2013). Today's mobile security requires a new approach. *Computer, 46*(11), 16–19. doi:10.1109/MC.2013.400

Lebek, Degirmenci, & Breitner. (2013). *Investigating the Influence of Security, Privacy, and Legal Concerns on Employees' Intention to Use BYOD Mobile Devices*. Academic Press.

Lee, N. (2015). *Cyber attacks, prevention, and countermeasures. In Counterterrorism and Cybersecurity* (pp. 249–286). Springer.

Madzima, K., Dube, E.L., & Mashwama, P.M. (2013). *ICT Education in Swaziland Secondary Schools: Opportunities and Challenges*. Academic Press.

Mark, O. (2014). *Bring your own internet of things*. Retrieved 01 January, 2015, from http://www.bringyourowninternetofthingscomingtobusinessin2015/

Mathew, S., Upadhyaya, S., Ha, D., & Ngo, H. Q. (2008). *Insider abuse comprehension through capability acquisition graphs*. Paper presented at the Information Fusion, 2008 11th International Conference on.

Matinde, V. (2015). *The rise of BYOD and corporate data threats*. Academic Press.

Miller, Voas, & Hurlburt. (2012). BYOD: Security and privacy considerations. *IT Professional, 14*(5), 0053-0055.

Moavenzadeh, J. (2016). *The fourth industrial revolution: Reshaping the future of production*. Retrieved 20 November, 2017, from www.eiseverywhere.com/file_uploads/fe238270f05e2dbf187e2a60cbcdd68e_2_Keynote_John_Moavenzadeh_World

Mulligan, P., & Gordon, S. R. (2002). The impact of information technology on customer and supplier relationships in the financial services. *International Journal of Service Industry Management, 13*(1), 29–46. doi:10.1108/09564230210421146

Needham & Lampson. (2008). *Network Attack and Defense*. White Paper.

Ngoqo, B., & Flowerday, S. V. (2015). Information Security Behaviour Profiling Framework (ISBPF) for student mobile phone users. *Computers & Security, 53*, 132–142. doi:10.1016/j.cose.2015.05.011

Nguyen, Tian, Cho, Kwak, Parab, Kim, Yuseung, . . . Zhang. (2013). *UnLocIn: Unauthorized location inference on smartphones without being caught*. Paper presented at the Privacy and Security in Mobile Systems (PRISMS), 2013 International Conference on.

Niesen, T., Houy, C., Fettke, P., & Loos, P. (2016). *Towards an integrative big data analysis framework for data-driven risk management in industry 4.0*. Paper presented at the System Sciences (HICSS), 2016 49th Hawaii International Conference on.

NIST. (2012). *NIST Special Publication 800-30, Revision 1, "Guide for Conducting Risk Assessments"*. Retrieved 8 June, 2016, from www.nist.org

Olasanmi, O. O. (2010). Computer crimes and counter measures in the Nigerian banking sector. *Journal of Internet Banking and Commerce, 15*(1), 1.

Osterman Research. (2012). *The Byod (Bring Your Own Device) Trend- Putting IT in control of BYOD*. Retrieved May 13, 2017, from http://www.hyperoffice.com/byod-whitepaper/

Parvizi, R., Oghbaei, F., & Khayami, S. R. (2013). *Using COBIT and ITIL frameworks to establish the alignment of business and IT organizations as one of the critical success factors in ERP implementation*. Paper presented at the Information and Knowledge Technology (IKT), 2013 5th Conference on. 10.1109/IKT.2013.6620078

Pillay, Nham, Tan, Diaki, Senanayake, & Deshpande. (2013). *Does BYOD increase risks or drive benefits?* Academic Press.

Romer, H. (2014). Best practices for BYOD security. *Computer Fraud & Security, 2014*(1), 13–15. doi:10.1016/S1361-3723(14)70007-7

Rose, C. (2013). BYOD: An examination of bring your own device in business. *The Review of Business Information Systems (Online), 17*(2), 65. doi:10.19030/rbis.v17i2.7846

Ross, R. S. (2011). Guide for conducting risk assessments. *NIST Special Publication*, 800-830.

Scheidell, M. (2009). *Intrusion detection system*. Google Patents.

Schwab, K. (2016). Navigating the fourth industrial revolution. *BIZNEWS*. Retrieved from http//www.biznews.com/wef/davos-2016/01/20/Klaus-schwab-navigating-the-fourth-industrial-revolution/

Shazmeen & Prasad. (2012). A Practical Approach for Secure Internet Banking based on Cryptography. *International Journal of Scientific and Research Publications, 2*(12), 1-6.

Sree. (2008). Exploring a novel approach for providing software security using soft computing systems. *International Journal of Security and Its Applications, 2*, 51-58.

Srinivasan, R. (2007). *Protecting anti-virus software under viral attacks*. Citeseer.

Stancioiu, A. (2017). The fourth industrial revolution "industry 4.0". *Fiability & Durability/Fiabilitate si Durabilitate,* (1).

Stouffer, K., Falco, J., & Scarfone, K. (2008). NIST SP 800-115: Technical Guide to Information Security Testing and Assessment. National Institute of Standards and Technology.

Tambotoh & Latuperissa. (2014). *The Application for Measuring the Maturity Level of Information Technology Governance on Indonesian Government Agencies Using COBIT 4.1 Framework*. Intelligent Information Management.

Tu & Yuan. (2015). *Coping with BYOD Security Threat: From Management Perspective*. Academic Press.

Twinomurinzi, H., & Mawela, T. (2014). Employee perceptions of BYOD in South Africa: Employers are turning a blind eye? *Proceedings of the Southern African Institute for Computer Scientist and Information Technologists Annual Conference 2014 on SAICSIT 2014 Empowered by Technology*. 10.1145/2664591.2664607

Umar. (2015). Financial Regulations and the Nigeria's Banking Sector. *Journal of Research in Business and Management, 3*(11),05-13.

Uz, Ali. (2014). *The Effectiveness of Remote Wipe as a Valid Defense for Enterprises Implementing a BYOD Policy*. Academic Press.

Vance, A., Siponen, M., & Pahnila, S. (2012). Motivating IS security compliance: Insights from habit and protection motivation theory. *Information & Management, 49*(3), 190–198. doi:10.1016/j.im.2012.04.002

Vignesh, U., & Asha, S. (2015). Modifying security policies towards BYOD. *Procedia Computer Science, 50*, 511–516. doi:10.1016/j.procs.2015.04.023

Wada, F., & Odulaja, G.O. (2012). *Electronic Banking and Cyber Crime In Nigeria-A Theoretical Policy Perspective on Causation*. Academic Press.

Wang, Y., Wei, J., & Vangury, K. (2014). *Bring your own device security issues and challenges.* Paper presented at the Consumer Communications and Networking Conference (CCNC), 2014 IEEE 11th. 10.1109/CCNC.2014.6866552

Wilson & Hash. (2003). Building an information technology security awareness and training program. *NIST Special Publication, 800*(50), 1-39.

Wu, X. (2009). *SIP on an Overlay Network*. Academic Press.

Xu, R., Saïdi, H., & Anderson, R. J. (2012). *Aurasium: practical policy enforcement for android applications.* Paper presented at the USENIX Security Symposium.

Zahadat, N., Blessner, P., Blackburn, T., & Olson, B. A. (2015). BYOD security engineering: A framework and its analysis. *Computers & Security, 55*, 81–99. doi:10.1016/j.cose.2015.06.011

KEY TERMS AND DEFINITIONS

BYOD: A trend that allows employees to bring their personal mobile devices to the workplace. They have the freedom to use mobile devices (such as laptops, tablets, or smartphones) for work-related purpose.

BYOIoT: A trend where we have so many inter-connected devices. These devices are sometimes connected to corporate network and they communicate with each other.

Rogue Device: An unauthorized connection of mobile devices to the network which pose a security threat to the organization.

Data Interception: Refers to the obstruction of data transmission to and from the device, and remotely altering the messages.

Eavesdropping: An unauthorized real-time interception of a private communication, such as instant message, a phone call, or videoconference.

Intrusion Detection System (IDS): A technology used to detect both pre-mortem and or post-mortem security threats.

Keystroke Logger: This is also known as key logging is used to record typed characters on mobile devices in order to capture valuable or sensitive information such as user's ID, password, and credit card numbers.

Mobile Device Management (MDM): Used for the management of mobile device and also to enforce some specific security policies on those devices.

WiFi: Facility that allows computers, smartphones, or other devices to connect to the internet or communicate with one another wirelessly within a particular area.

This research was previously published in the Handbook of Research on Information and Cyber Security in the Fourth Industrial Revolution; pages 205-233, copyright year 2018 by Information Science Reference (an imprint of IGI Global).

Chapter 12
Simple Screen Locking Method Using Randomly Generated Number Grid on Image

Hazleen Aris
Universiti Tenaga Nasional, Kajang, Malaysia

Zailani Ibrahim
Universiti Tenaga Nasional, Kajang, Malaysia

Ariff Azman
Universiti Tenaga Nasional, Kajang, Malaysia

ABSTRACT

Smartphones have become part and parcel of our daily life. Due to the more and more advanced features incorporated, its role has extended beyond calls and text messages, and it has become a place where important personal information is being stored. Thus, it needs to be protected from unauthorised users, which is usually achieved by using screen locking. However, improved screen locking security often compromises other aspects, such as usability and cost. In this article, a new screen locking method (SyS) that is both simple, secure from shoulder surf and smudge attacks, and not expensive to implement is presented. It uses an image and a chosen number to create the screen locking code. Evaluation results show that the SyS screen lock method scores best in terms of usability and security compared to the other three common screen locking methods, namely personal identification number or PIN, password, and pattern. It also performs well theoretically when compared against existing screen locking methods that aim at defending against similar attacks.

DOI: 10.4018/978-1-7998-8545-0.ch012

INTRODUCTION

The use of smartphones and similar mobile devices nowadays has extended beyond phone calls and text messages. As smartphones become smarter and able to run applications and perform many functions of the personal computers, people are using them widely as their personal organisers, resulting in many important and confidential information being stored in the smartphones. This has thus increased the need to protect the smartphones from unauthorised access. Realising that no device as sophisticated as a smartphone or a tablet is completely secure in situ, device manufacturers and developers are doing their best to ensure the safety of users' information stored in their smartphones. One of the means to do this is through the provision of screen locking, which comes in a number of types including slide lock, keypad lock and pattern lock (Shin, Park, Lee, & Park, 2012). In slide lock, screen unlocking is achieved by tapping and sliding the screen in a pre-defined direction, often horizontal. Another implementation of slide lock allows sliding to be made in any direction, also known as glass lock (Shin et al., 2012). Keypad lock, on the other hand, uses a set of pre-defined and ordered personal identification numbers (PINs) or password to unlock the screen. The former is also known as digit lock. It often requires users to enter their PINs by tapping on the numbers displayed on the screen. As opposed to PINs, passwords can be alphanumeric. Finally, in pattern lock, a pre-defined pattern drawn over a grid of dots is used to unlock the screen. Pattern lock provides bigger password space size than say, a four-digit PIN.

In terms of security, slide lock is obviously not secure as it only requires the user to slide the screen to unlock the phone. The purpose of its invention is to keep the phone from accidentally turning itself on and activating its applications while it is in your pocket or purse. Keypad lock and pattern lock, while more secure, are somehow vulnerable to attacks, such as smudge attack, shoulder surf attack and camera attack. Smudge attack is caused by oily residues, that is, smudges, left on the touch screen surface. The presence of residual oils on touchscreen devices increases the likelihood of an attacker gaining unauthorised access to the device due to the smudges' persistence that makes them difficult to be obscured or deleted, especially with the availability of equipment to collect and analyse them (Aviv, Gibson, Mossop, Blaze, & Smith, 2010). Shoulder surf attack on the other hand uses direct observation techniques, such as looking over someone's shoulder, to obtain the unlock combination. Shoulder surfing is highly likely in crowded places where it is relatively easy to stand next to people and watch unnoticed as they unlock their devices. With the aid of binoculars and other vision-enhancing devices, longer distance shoulder surfing can also be done (Rouse, 2005) and with the help of a camera, it is also known as camera attack (Winkler et al., 2015). While shoulder surf is a risk for both touch screen and keypad-based screen locking, smudge attack is only a threat for touch screen devices.

In the attempts to make screen locking more personalised and hence, less vulnerable, biometric features were incorporated in the screen locking methods. In biometric screen locking methods, biometrics information of the users is used to unlock the screen. One example is face unlock, which was introduced back in the Ice Cream Sandwich Android operating system. To use this feature, a user has to ensure that his face is captured inside a face-shaped ring of dots using the front facing camera until the device has gathered enough information to be able to unlock with it. However, the face unlock method is vulnerable to photo attack. It can be spoofed by using the owner's photo (Racoma, 2012). Another widely used biometric screen locking method is finger scan. More often used in the iOS and increasingly gaining popularity in the more advanced Android smartphones, this method scans and records the user's fingerprint using a dedicated scanner embedded into the smartphone. While the accuracy of latent fingerprint recognition is said to be able to reach 99.8% (Ulery, Hicklin, Buscaglia, & Roberts, 2011), incorporation of sensors such as fingerprint scanner and its analyser comes with additional cost, which does not justify if its use is limited to only screen locking function. The same concern also applies to the other screen locking methods that use biometrics information. Furthermore, some of the even more secure biometric screen locking methods are very complicated that they compromise usability. More importantly, as pointed out in in Harbach, von Zezschwitz, Fichtner, Luca, & Smith (2014), shoulder surfing was perceived to be a relevant risk in only 11 out of 3410 situations and the users favour protection more than authentication. The finding also showed that only 26% of the perceived worst-case risks could have been avoided by authentication. This means that additional cost and effort that the users are willing to spend need to be carefully assessed as it is unlikely that the manufacturers, developers or users would be willing to invest more when the return is not deemed significant.

As such, there is a need to design and develop a new screen locking method for mobile devices that is comparatively secure and easy to use, and at the same time is also affordable. To address this need, a *simple* yet *secure* (SyS) screen (un)locking method is proposed in this research. The objective is to design and develop a new screen locking method that is able to provide an optimum trade-off between security and usability without having to increase the production cost that may result in expensive mobile device's price. A grid of numbers (integers) of varying size and an image are the main components of the SyS screen locking method. The integers are randomly generated to populate the grid that overlays the image. To unlock the screen, a user is required to drag a chosen integer and land it on a specified location on the image. The number and the location are both pre-determined during setup. The screen will unlock the moment it detects that the integer is placed at the supposed location on the image. The method is deemed simple in the sense that its use does not require high cognitive load, as the user only needs to remember one number and one location where the number should be placed. It is considered secure as it is still able to protect the users reasonably well from the two types of attack included in the research scope, which are smudge and shoulder surf. Furthermore, the SyS method is also not expensive to implement for two reasons. First, it does not involve complicated algorithm or programming effort and secondly, it does not require additional hardware such as sensors to be embedded in the mobile devices in order to be used. Therefore, its application is not expected to cause significant increase to the production cost of the devices. For the purpose of evaluation, prototype implementation of the SyS screen locking method was implemented in an Android smartphone. The evaluation compared the SyS method against three common screen locking methods, which are PIN, pattern and password. Analysis of the results, which was performed using a multi-criteria decision-making method involving a number of parameters showed that the SyS method ranked the best in terms of both security and usability. Findings from this research contribute to the existing body of knowledge by introducing a secure and usable new screen locking method that is empirically evaluated. The remaining of this paper is organised as follows. The next section outlines the current research in improving the security of screen locking methods, and their impacts towards usability. The development of the SyS screen locking method is then described, followed by the evaluation, results and discussion.

RELATED WORK

Various screen locking methods have been proposed to authenticate smartphone users, hence, securing the information stored in the smartphone from unauthorised users. One of the most widely used method is digit lock, which offers 10,000 unique four-digit lock (Arif, Pahud, Hinckley, & Buxton, 2013). Another method that gains most users' attention because of its usability is the pattern lock. It offers 389,112 distinct patterns (Aviv et al., 2010), which is a lot more than the four-digit lock. Despite the increased password space, these popular methods have been criticised for their vulnerability due to the limited number of possible combinations that makes them easily guessed or smudged (Skillen & Mannan, 2013). In the efforts to improve smartphone security, a number of other screen locking methods have been proposed, which are either extensions of existing digit lock and pattern lock methods, or entirely new methods. The new methods can be categorised into the following.

- Visual or graphical screen locking
- Biometric screen locking

Extension of Digit Lock

In an attempt to improve the security of digit lock, a hybrid method for authenticating the smartphone users was introduced (Arif et al., 2013). It increases the number of unlock combinations by adding gestures to the conventional digit lock. Four simple gestures were used; up (↑), down (↓), left (←) and right (→). An example of unlock combination using this method is 9←6•5↓7• where a • simply means a tap. To unlock the screen using this code, user has to make a left swipe from the 9 key, followed by a tap on the 6 key, followed by a downward swipe on the 5 key and then a tap on the 7 key. With this method, a total of 6,250,000 unique combinations can be used, and since it has the look-and-feel of the conventional digit lock, it somehow encourages the users to give it a try. Pilot testing performed on the method showed that it had lower entry speed and was more error prone than the traditional digit lock method, which was expected to improve with practice. User testing of the method also discovered that the users were comfortable and felt more secure using the method.

A similar approach to the above is the SwiPIN method, which is based on a random assignment of simple touch gestures to specific digits (von Zezschwitz, De Luca, Brunkow, & Hussmann, 2015). In this method, the digits are divided into two regions on the screen with five digits in each region. All digits in each group except one are marked with an arrow that indicate four directions; up, down, left and right. To unlock, the users will need to swipe the screen in the direction shown by the arrow on the number, for example, if the first digit is 3 and the arrow on number 3 is pointing upwards, user will need to swipe the screen upwards without having to touch the number. They just need to ensure that the swipe is done in the respective region where the digit belongs. Feasibility evaluation was performed by comparing SwiPIN and traditional PIN in terms of performance, security and likeability. Evaluation results showed that traditional PIN was more preferred by the users and they can perform it faster. Thus, SwiPIN is an option only in situations when the risk of exposing the user's PIN is present.

Extension of Pattern Lock

At least two methods were proposed to improve the security of pattern lock, a screen locking method introduced in Android smartphones. In TinyLock (Kwon & Na, 2014), which aimed to resist smudge attacks by tweaking the user interface design of the pattern lock system, three new design elements were added to the conventional pattern lock. Firstly, the pattern grid size was reduced according to an optimal size as a result of a preliminary study. Secondly, another tiny grid of pattern was added on top of the actual grid, hence, there are two tiny pattern grids on the screen. The starting point selected by the user will flicker in the top grid. This can help the user to confirm his/her selection of the initial dot without being visually blocked by their fingers. Finally, a virtual wheel of the size of the tiny grid was introduced that appears after a correct pattern was drawn. The users have to complete drawing the cycle following the wheel either clockwise or anti-clockwise. The purpose is to unconsciously enforce them to distort the smudges left on the touchscreen during unlocking. Evaluation of the TinyLock method was performed against the conventional Android pattern lock system. Results showed that security against smudge attacked was improved, but at the expense of usability.

Another screen locking method that can be regarded as an extension or a variation of pattern lock method is the XSide, an authentication mechanism that uses the front and the back of the smartphones to enter stroke-based passwords (De Luca et al., 2014). Strokes at the back of the smartphone were captured using the touch sensitive area on the back of the device, a feature that was claimed to hit the market soon. Evaluation was performed by only comparing amongst the different combinations of the XSide screen locking method, which were front strokes only, back strokes only, one side switch and two side switches. Evaluation results showed that the front strokes not only was the easiest but the least secure and the two side switches was the best in terms of security but worst in terms of usability, that is, taking the longest time to unlock.

Graphical Screen Locking

Moving away from the conventional digit and pattern locks, screen locking methods that are not based on either of these two were also found. They are based on graphics and known as graphical screen locking methods. In reviewing the first twelve years of graphical password implementation, three types of graphical password were identified; (unassisted) recall, recognition and cued recall (Biddle, Chiasson, & Van Oorschot, 2012). Recall requires a person to remember the password without any cues such as in the form of free drawing (Jermyn, Mayer, Monrose, Reiter, & Rubin, 1999). Recognition presents the user with the information that requires him to decide if it matches the information previously memorised, for example, memorising a portfolio of images during password creation and recognising the images from among decoys to unlock. In cued recall, an external cue is provided to help in remembering the information. It typically requires the users to remember and target specific locations on an image. Various graphical authentication methods belonging to either one of these three types were proposed ever since. In PassPoints, a user's password consists of any arbitrarily chosen sequence of points on the image (Wiedenbeck, Waters, Birget, Brodskiy, & Memon, 2005). To unlock, a user has to click close to the chosen click points or within some (adjustable) tolerance distance. The evaluation results showed that the graphical password users created valid passwords with fewer difficulties than the alphanumeric password users. However, the graphical password users took longer and made more invalid password inputs than the alphanumeric password users while practicing their passwords. Although the two groups performed similarly with regard to the memorability of their passwords, the graphical group took more time to input a password. However, PassPoints method was implemented and evaluated on personal computers rather than smartphones, which makes it somewhat less relevant to our scope of study. In Shin et al. (2012), the unlock combination of its graphical password is made up of a hexagon of six circles. A circle can be touched more than once, up to a maximum of seven times. When a circle is touched more than once, its colour changes to allow users to identify the correct inputs. This method was claimed to be good in terms of convenience and security. However, the evaluation details were not presented.

An authentication system based on colour-code was also introduced (Bandre, 2015). The method depends on the user-defined colour-code pairing and sequence. Two sets of colour pallets appear on each side of the phone screen upon its activation. To unlock, a user selects a preferred colour to generate colour-code sequence passkey from any set and slides it towards the opposite side on the preferred colour of another set. The user can increase the colour sequence up to a single pair or traverse complete bipartite graph in each phase. The colour of circles arrangement in both sets are randomly changed each time the user tries to unlock the phone to reduce the possibility of brute force attacks. Based on a multi-phase security scheme, the method provides the users with multi-level security so that a user could either utilise single level to generate passkey or utilise all levels to generate complete passkey. This method is said to be able to approximately generate ten million passkeys, which means having the theoretical password space of 10^7. Evaluation of the method was only theoretically performed by comparing this method with slide lock, digit lock and pattern lock based on the size of the theoretical password space.

The next method in this category, SmudgeSafe, is an authentication system that uses random geometric image transformations to increase the security of the cued-recall graphical passwords against smudge attack (Schneegass, Steimle, Bulling, Alt, & Schmidt, 2014). The transformations include translation, rotation, scaling, shearing and flipping of the chosen graphic. A user first sets up his password by choosing an image to be used. Next, he sets up a graphical password consisting of a series of password points within the image. Each time the screen is activated, the chosen image will be randomly transformed so that the traces will be different during each unlock. Evaluation of this method was performed against pattern lock and digit lock methods. Similar to a number of previously presented methods above, SmudgeSafe was found to be more secure against smudge attack at the expense of usability. A few other graphical screen locking methods are also available including those proposed in H.-M. Sun, Chen, Fang, & Chang (2012) and Weiss & De Luca (2008). However, it is not the intention of this paper to exhaustively cover all methods. Interested readers can also refer to the list in Suo, Zhu, & Owen (2005).

Biometric Screen Locking

As presented above, in order to improve security, the number of possible unlock combinations was increased by adding more steps to the screen locking methods, hence increasing the theoretical password space. This subsequently increases the complexity of the unlock codes, which degrades usability, as shown by the evaluation results of a number of methods described earlier. Extra cognitive load is required to remember the extra steps. Given that 64% of the users do not use passwords on their smartphones, which might be caused by the usability factor ("Keep your phone safe," 2013), the additional steps required to unlock the screen might not appeal to the users. To address this issue, biometric screen locking methods that use sensors completely were introduced. Two types of biometrics information can be used; behavioural and physiological. The sensors are used to read these biometrics information and users therefore do not have to remember any pattern or password. In Lee & Lee (2015), an implicit and continuous authentication method was proposed. It is a multi-sensor-based method that learns owner's behavioural patterns and environment characteristics. Thus, authentication is performed without direct involvement of the users. It builds the owner's profile based on the measurements taken from the built-in sensors in smartphones such as accelerometer, orientation sensor and magnetometer. Evaluation performed showed that the method is efficient and requires less than 10 seconds to train the model and about 20 seconds to detect an abnormal user. Although the system seems easy to be used by the authorised user, change in user's behaviour because of certain circumstances, for instance injury from an accident that might affect the user's movement, can cause the system to be unable to recognise the behaviour patterns. On another circumstance, when the phone is stolen and used by a different user, the method might be learning the new 'owner's' profile and wrongfully grants access to him.

Screen locking methods that are based on physiological biometrics include a method that uses periocular biometrics to authenticate users (Raja, Raghavendra, Stokkenes, & Busch, 2014). The periocular information is captured by phone camera. It employs three well known feature extraction methods namely, scale invariant feature transforms (SIFT), speeded up robust features (SURF) and binarised statistical image features (BSIF) to recognise the periocular information. The method has been identified to be robust, and study shows that BSIF method consistently performed well in all phone conditions. However, this method requires additional hardware or image recognition software to be installed that may increase the smartphone cost.

In an attempt to secure the access of Apple's iPhones from the observation attack such as shoulder surf, an observation-resistant authentication system that uses either audio or haptic cues that obfuscate a PIN entry process was proposed (Bianchi, Oakley, Kostakos, & Kwon, 2011). In this method, PIN items are drawn from a limited set of tactile or audio cues, and selection of a sequence of these cues composes a password. It relies on the touch screen for the input, uses the phone's built-in capabilities for audio output and connects via Bluetooth to a SHAKE SK6 device to generate the tactile stimuli. The evaluation of the method showed its feasibility in terms of the unlocking time, error rate and user acceptance. However, it suffers from both usability and cost due to the complexity in creating the password of cues and the need to mount additional device on the back of the phone to generate the haptic stimuli.

From the summary of the existing screen locking methods presented above, it can be seen that the creation of new screen locking methods were largely motivated by the security issues with the conventional methods (PIN, pattern and password). Evaluation of the methods, whenever performed, was made against either shoulder surf or smudge attack, and was accompanied by the usability evaluation. With regard to usability evaluation, three common attributes of usability were used; unlocking time, error rate and user rating in terms of ease-of-use. In overall, it can be seen that while security is improving based on the evaluation performed on the existing approaches, this was done at the expense of usability, which is implied in terms of increased average time to unlock, increased error rate and decrease in user rating compared to the conventional methods. While biometric screen locking methods aim at improving usability and memorability by keeping the unlocking process as natural as possible, they suffer in terms of complexity and cost of implementation. Therefore, in the proposed method that will be described in the next section, a new screen locking method is designed to improve security with minimum compromise to usability and cost.

SYS METHOD

Our proposed simple yet secure (SyS) screen locking method aims to be both secure and simple, that is, usable. Secure against at least smudge and shoulder surf attacks and usable in the sense that it does not require heavy cognitive load such as remembering a set of numbers or alphanumeric password. A password in the SyS method is made up of two main components; a randomly generated number grid and an image. A number grid is evenly divided and equally spaced squares containing a number in each square. It is usually ordered either ascending or descending. However, in our method, the number grid is randomly generated. Hence, it does not follow any particular order. The second component of the SyS screen locking method is a digital image. A digital image is made up of picture elements called pixels, which enable specific location area on an image to be detected and recorded.

Setup Algorithm

During the setup phase, the system requires the user to perform an initial setup of the components needed for the screen locking to work. At this phase, the user needs to determine two items; a number and the location to place the number on the chosen image. Similar to Wiedenbeck et al. (2005), the SyS screen locking method allows any image to be used. The images could be provided by the system or chosen by the user. Figure 1 shows the algorithm to perform the initial setup of the SyS screen locking method.

The choose_image process requires the users to identify an image to be used for their screen locking password. They can choose any image that is stored in their smartphone. The choose_number process requires the user to choose any one number within the range of 0 to 9 as the number that is going to be used to unlock the screen. The set_grid_size process allows the users to choose a specific size for the number grid to be used during the unlocking process or to use a random grid size. If a random grid size is chosen, the grid size will change during each unlock. Additionally, if a specific grid size is chosen, the get_grid_size process will be invoked where the user is required to select a preferred grid size to be used. The options given include 4×4, 5×5, 6×6, 7×7 and 9×9 grids. For this option, the size of the number grid will be the same during each unlock.

Next, the set_number_position process requires the user to position the chosen number at a specific location on the chosen image. The verify_number_position process that follows forces the user to repeat placing the chosen number at the identified location on the image for confirmation. Finally, the set_backup_lock process enables the user to specify an existing type of unlocking method to be used in case he fails to unlock the device using the SyS screen lock method later. For example, user will be able to select either pattern lock method or digit lock method as backup. The chosen backup method will be prompted to the user after failing to unlock the device using the SyS method for a predetermined maximum number of consecutive tries.

Unlock Algorithm

To unlock a screen using the SyS method, the users are required to match the two items set up earlier. The chosen number will appear in the randomly generated number grid displayed when the phone is activated. The user needs to drag the grid and land the number at the right position on top of the chosen image. Figure 2 shows the algorithm to unlock a smartphone screen using the SyS method. The display_image process displays the image chosen as the wallpaper for the locked screen when the phone is activated. The display_random_grid process displays a separate layer of a number grid on top of the image. The whole of the number grid layer can be dragged around by the user in the process of placing the selected number at the predefined location on the image. After the correct placement, the user needs to release the number grid in order to unlock the screen. The get_number_position process then fetches the position of the number at the point where the user releases the grid and verifies whether it matches the position defined during setup. If it does, the screen will be unlocked. Otherwise, the user will be given up to a certain number of times (*max* - 1) to retry. When the number of tries (*try*) has exceeded the maximum allowable, the user will be prompted with the backup unlock method. Additionally, to increase the security further, it can also be set that certain number of matches should be achieved before

Figure 1. Setup algorithm of the SyS method

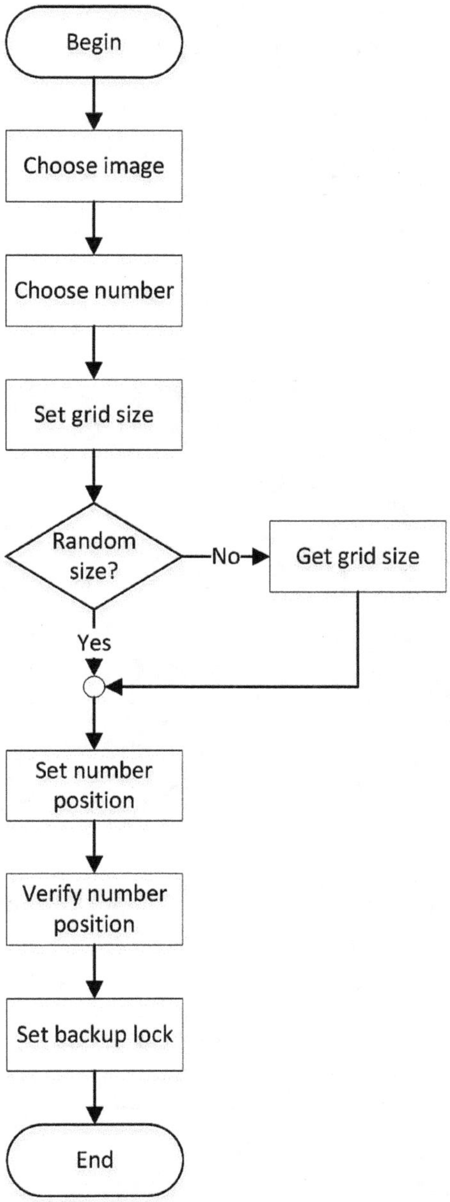

the phone unlocks (*cfm*). For this reason, the number of matches (*match*) has to be tracked as shown in the algorithm.

Theoretically, the SyS screen locking method is kept simple by having the user to remember only two things; the single number chosen, and the location where the number has to be placed on the chosen image. This significantly reduces the cognitive load compared to having to remember a series of PIN or alphanumeric password. In terms or security, the number grid that appears during each unlock is randomly generated, which means that the number chosen by the user will be swiped or dragged from different direction during each unlocking. This will improve the security of the method against smudge and shoulder surf attacks as it will be difficult for attackers to guess the swipes that come from different directions. Furthermore, the grid can be dragged by touching at any point on the grid, not necessarily on the chosen number. This can further disguise the chosen number from the attackers.

Figure 2. Unlock algorithm of the SyS method

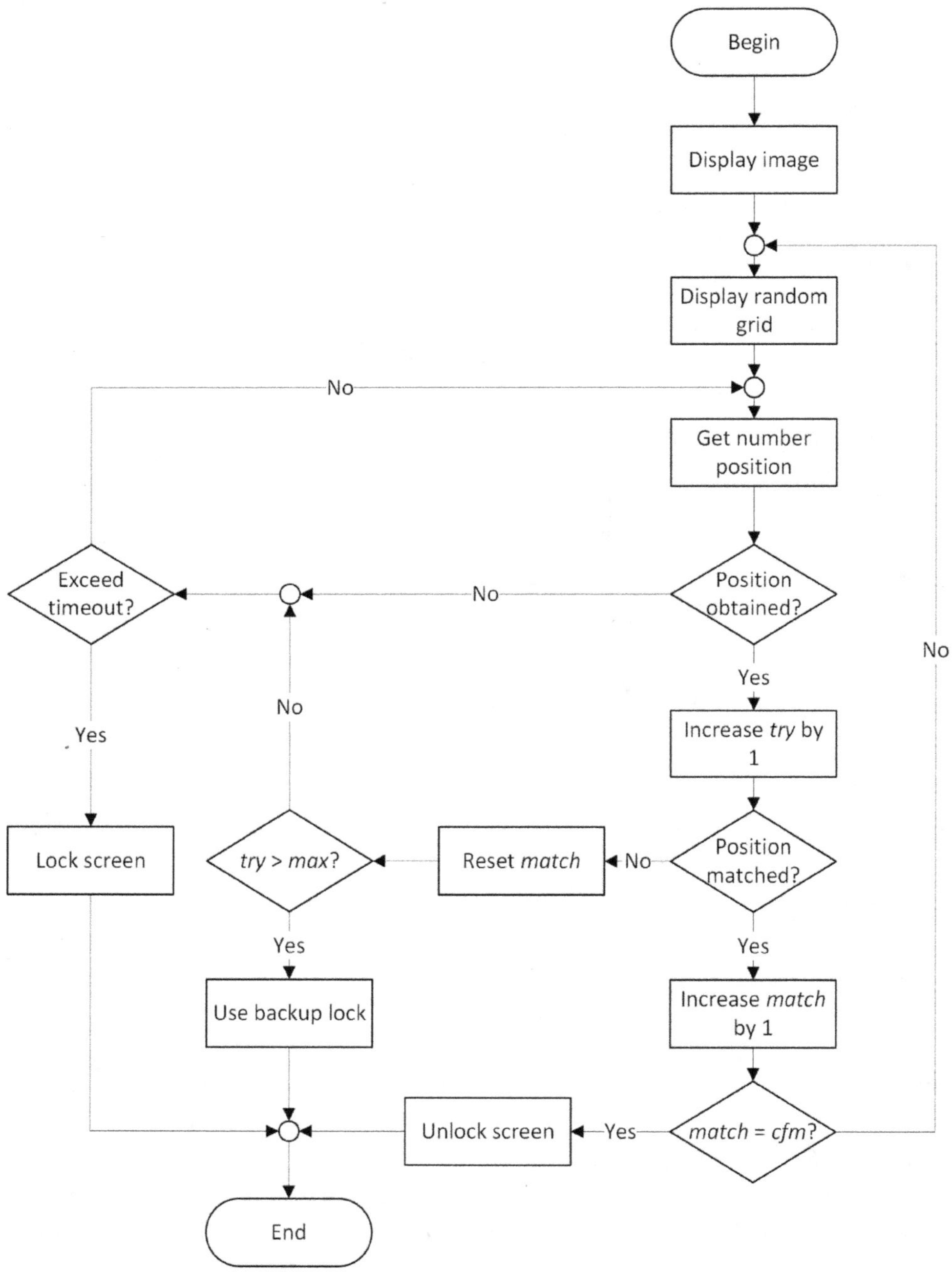

EVALUATION

For the purpose of assessing the security and usability of the SyS screen locking method, we performed comparative evaluation of the SyS method against three widely used screen locking methods; PIN (digit), pattern and password. Two aspects of evaluation were performed; security and usability. Mobile security is the protection of smartphones, tablets, laptops and other portable computing devices, and the networks they connect to, from threats and vulnerabilities associated with wireless computing (Jeon, Kim, Lee, & Won, 2011; Rouse, Wigmore, & Westervelt, 2012). Security in the context of our evaluation is the protection against successful breach or unauthorised access using the two types of attack; smudge and shoulder surf. On the other hand, usability is the ease of use and learnability of a smartphone. In the context of our research, we evaluated usability in terms of the number of unlocking attempts, unlocking speed and rating by authorised users.

Prototype Development

For the purpose of the evaluation, a prototype of the SyS screen locking method was first developed. To do this, an unofficial replacement of one of the existing lock methods that was not used in the study, which is the Android face unlock, was made by rooting the device. Rooting basically means adding a standard Linux function called super user (SU), which grants permissions to accomplish anything to any file at any place in the system. Once rooted, the device is completely controlled by the user, enabling addition or removal of files and access to functions that are not permissible before. The face unlock application package (APK) file was retrieved from the system folder and decoded with a Java decompiler tool to get the source code. After successfully decompiling the APK, the source code was used as the base file for adding the code to implement the SyS screen locking method. When all of the required functions were properly integrated, the code was recompiled into a modified version of the face unlock APK file and installed back onto the root system of the Android device. The resulting interfaces of the SyS screen locking method are shown in Figure 3, Figure 4, Figure 5, Figure 6 and Figure 7.

Participants

An experiment was performed to evaluate the security and usability of the SyS screen locking method using role playing method involving 30 participants, which was about the same number or more compared to other similar studies (H.-M. Sun et al., 2012; von Zezschwitz et al., 2015; Wiedenbeck et al., 2005). The participants, who were involved in the evaluation on voluntary basis, comprised 19 undergraduate university students and 11 working adults. Their age ranged between 20 and 40 years old, consisting of about the same number of males (16) and females (14). All of them were experienced smartphone users and had experience using the three existing screen locking methods before.

Figure 3. Interface to choose preferred image

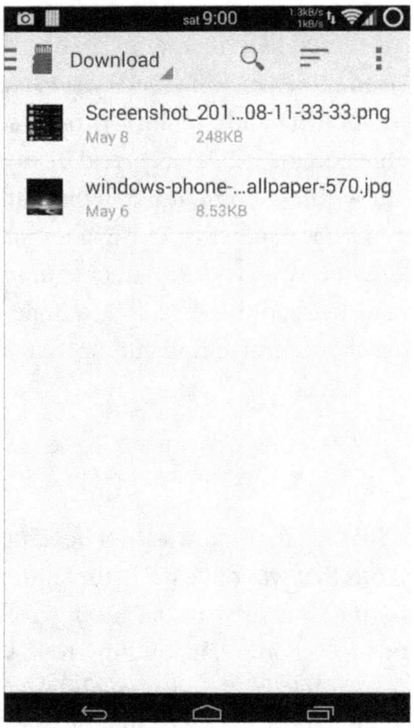

Figure 4. Interface to choose preferred number

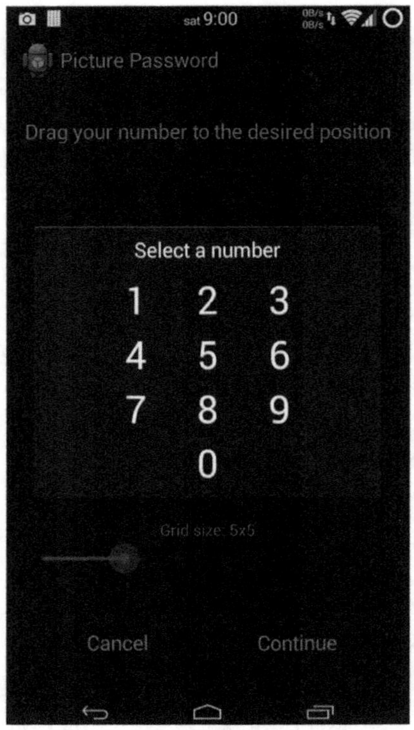

Figure 5. Interface to set the chosen number at a location on the image

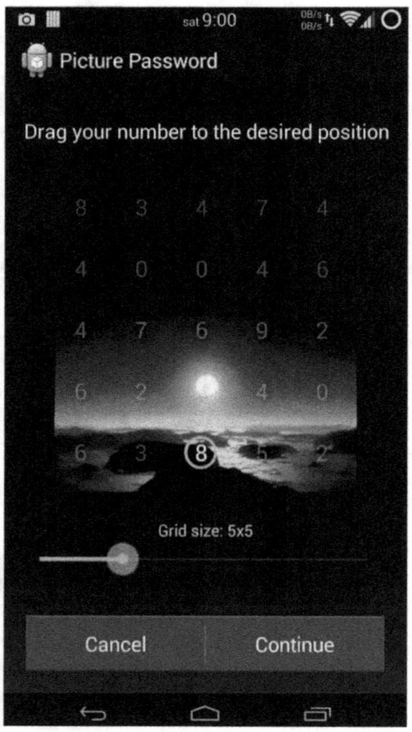

Figure 6. The locked screen

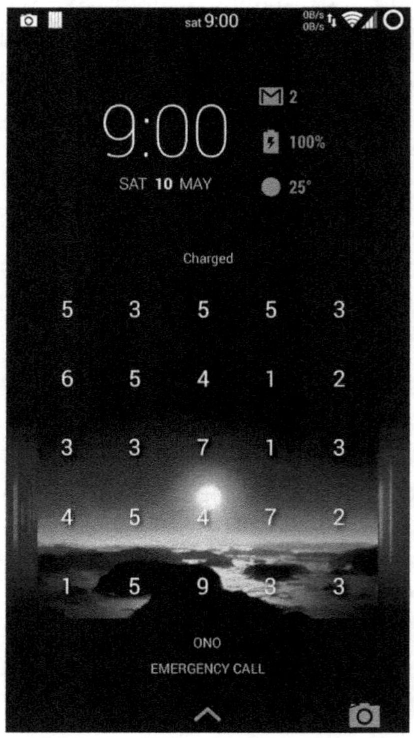

Figure 7. Interface to set the preferred grid size

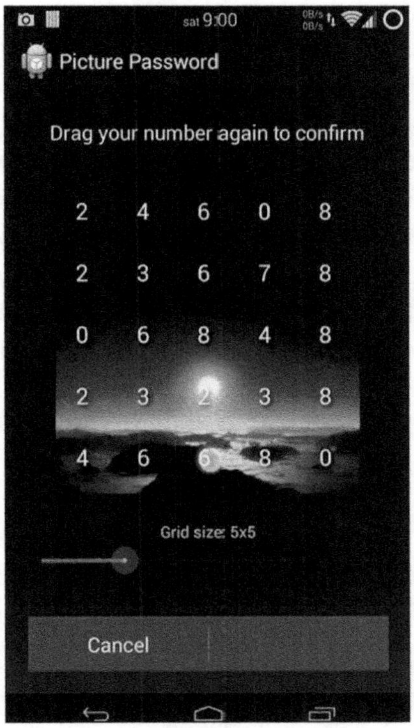

Procedures

Prior to the evaluation session, the participants were briefed about the SyS method and were given sufficient time to train using the method. The training period ended when the participants admitted that they had understood the method sufficiently well that they were comfortable and confident to use it for the experiment. The evaluation began after the training period was over. For security evaluation, each of the four screen locking methods was evaluated by dividing the 30 participants equally into two groups. The first group played the role as the attackers and the second group played the role as the (authorised) users. Each attacker was paired with a user and was asked to attempt the unlock breach using shoulder surf attack by observing from a close proximity. Their roles were then swapped to perform breach attempts using smudge attack by scrutinising the traces left after a user successfully unlocked the phone. These resulted in 15 shoulder surf attack results and 15 smudge attack results for each of the four screen locking methods under evaluation. The order of the four screen locking methods to be attempted was randomised for each pair of attacker and user to avoid factors such as fatigue from affecting the results. For the security evaluation, the number of attempts made to successfully breach the locked screen for each attacker and each method was recorded. Each attacker was given a maximum of three attempts to crack each type of password before the breach attempt was considered fail.

For the usability evaluation of the SyS screen locking method, each participant was asked to perform the four types of the screen locking method. For each method, the user was first asked to set up the screen locking code and then unlock the screen using the newly set up code. The codes are all self-selected. The number of attempts and the time taken to unlock the screen (for successful unlock) using each method was recorded for each participant. Finally, each participant was also asked to rate the four types of screen locking method that they had tried in terms of their ease-of-use using a Likert scale with 5 being the easiest and 1 the hardest. In both aspects of evaluation, the same image downloaded earlier from the Internet was used to reduce potential bias in the results due to the image complexity. The image was chosen in such a way that there is no prominent part of the image that can have a high potential of being obviously chosen as the landing point for the chosen number. The grid number size used for the evaluation was also fixed at 9 x 9, a size that in our opinion, is an optimum compromise between usability and security. However, each (authorised) user was allowed to pick his own number and determine the point where the number should be placed on the image. Users were also allowed to freely choose their passwords, patterns and PINs to enable the experiment to run naturally. We also did not control the hand posture (e.g. one handed) and users could use whatever style that they felt natural. Furthermore, no supporting device such as camera was used during the experiment.

RESULTS AND ANALYSIS

The final results consisted of 30 responses for each of the four screen locking methods included in the experiment, resulting in a total of 120 responses. Result of the security evaluation is shown in Table 1. As can be seen from the table, PIN and pattern screen locking method recorded between 80% and 100% breaching success using shoulder surf and smudge attacks in our experiment, which confirmed the lack of security of the methods. Password screen locking method was found to be more vulnerable to shoulder surf attack, which recorded 70% breaching success. Using smudge attack, 50% of the attackers managed to breach the passwords. For the SyS method, three attackers (20%) managed to unlock the screen using shoulder surf attack and only one (7%) managed to do so using smudge attack. For each successful attack, we also recorded the number of attempts made. Table 1 shows that the SyS method obtained the lowest number of successful breaches with only 3 and 1 respectively for shoulder surf and smudge attacks. For the number of successful breaches, password method scored the highest average of 2.30 (SD = 0.48) for shoulder surf attack while the SyS method scored 1.67, but from only three successful breaches compared to 10 successful breaches for the password method. For smudge attack, the SyS method recorded the highest average of 3 from only one successful breach.

Table 1. Result of the security test

Unlocking Method	Number of Successful Breach		Average Number of Breach Attempt*	
	Shoulder Surf	Smudge	Shoulder Surf	Smudge
PIN	15	12	1.73	2.67
Pattern	15	15	1.27	1.60
Password	10	7	2.30	2.57
SyS	3	1	1.67	3.00

* From successful breaches

As mentioned earlier, three types of information were collected for the purpose of measuring the SyS method usability; number of unlock attempts, time taken to unlock, and a rating given by the users. Despite choosing the subjects using convenience sampling for the evaluation, the data obtained passed the normality tests performed. Subsequently, one-way ANOVA test was used to determine the statistical significance of the results. It can be seen from Table 2 that the SyS method scored the lowest average of 1.23 ($\rho = 0.0438$, $F_{3, 116} = 2.7883$) for the average number of unlock attempts by authorised users, which is slightly below the average for the PIN method. With regard to the average time taken to unlock the screen, the SyS method also placed first at 1.62 seconds ($\rho = 0.0000$, $F_{3, 116} = 18.0475$). It also scored the best mean of 5 in terms of the usability rating given by the users. Kruskal-Wallis test performed on the usability rating showed a statistically significant relationship between the user rating and the unlocking method ($H(3) = 72.34$, $\rho < 0.0001$), with the median values as shown in Table 2.

Table 2. Result of the usability test

Method	Average Number of Unlock Attempt	Average Time Taken to Unlock (s)	Median Rating by User
PIN	1.30	1.64	4
Pattern	1.47	1.90	4
Password	1.63	2.60	4
SyS	1.23	1.62	5

Due to a number of attributes being used in measuring the security and the usability of the screen locking methods under evaluation, and the difference in terms of the data direction, technique for order of preference by similarity to ideal solution (TOPSIS), a multi-criteria decision making (MCDM) method was used to 'fuse' the results together in order to rank the compared screen locking methods. TOPSIS selects a solution that is the closest to the ideal solution (Lai, Liu, & Hwang, 1994) and in the context of our research, this means the most secure and the most usable method. Computation results using the TOPSIS method for security attributes of the smudge and shoulder surf attacks are shown in Table 3 and Table 4 respectively. In the tables, $S_i{*}$ is the distance from the ideal case, S_i' is the distance from the worst case and $C_i{*}$ is the closeness to the ideal case. As can be seen from both tables, the SyS method ranks first in terms of security against the smudge and shoulder surf attacks, compared to other methods. The TOPSIS method was also applied on the usability attributes and the result showed that SyS method scored the best in terms of usability too as shown in Table 5.

Table 3. TOPSIS result for security attributes (Smudge)

Method	$S_i{*}$	S_i'	$C_i{*}$	Rank
PIN	0.33698	0.14005	0.29359	3
Pattern	0.44893	0	0	4
Password	0.18781	0.26232	0.58276	2
SyS	0	0.44893	1	1

Table 4. TOPSIS result for security attributes (Shoulder surf)

Method	S_i*	S_i'	C_i*	Rank
PIN	0.46835	0.06554	0.12275	3
Pattern	0.48381	0	0	4
Password	0.26923	0.24092	0.47225	2
SyS	0.08894	0.46494	0.83942	1

Table 5. TOPSIS result for usability attributes

Method	S_i*	S_i'	C_i*	Rank
PIN	0.11481	0.33251	0.74333	2
Pattern	0.21774	0.21548	0.49738	3
Password	0.41993	0	0	4
SyS	0	0.41993	1	1

DISCUSSION

The Secure yet Simple (SyS) screen (un)locking method was designed and developed with the aim to attain an optimum trade-off between security and usability, and at the same time is not expensive to implement. The method uses a number grid over an image, which makes it different from the other conventional and graphical methods. It was evaluated from two perspectives; security and usability. Evaluation results that focused on the two types of attack, which are shoulder surf and smudge, found that the SyS method ranked highest, that is, the most secure from both types of attack. This means that the smartphone screens locked using the SyS method is the most difficult to be breached, compared to those locked using the three commonly available screen locking methods, which are PIN, pattern and password. These three methods were chosen for comparison due to their widespread availability and hence, familiarity amongst the users. Almost all smartphones have either one or all of these methods readily available for use. Therefore, their use in our experimental evaluation did not require additional effort to programme and install the methods on tested smartphones. These methods are also seen used by other similar studies. For example, PIN and pattern were also used for comparison in Schneegass et al. (2014) and Guerar, Merlo, & Migliardi (2017) and pattern was used in von Zezschwitz, Koslow, De Luca, & Hussmann (2013).

Smudge attack happens when the attacker is able to see the traces left by the owner when unlocking the phone. The smudge can appear to be very visible if, for example, special lighting condition is applied. In our experiment, the SyS screen locking method was proven to be reliably resistant to smudge attack with evaluation results showing only one successful breach that only happened after the third attempt. This has made the ranking of the SyS method against the other methods compared to become quite straightforward and as expected, the SyS method scored the best for the distance from the ideal solution (S_i*=0), distance from the worst solution (S_i'=0.44893) and closeness to the ideal solution (C_i*=1). Perfect 0 for S_i* and perfect 1 for C_i* are strong indication of the SyS method's resistance against the smudge attack compared to the PIN, pattern and password methods. Other approaches that attempted at overcoming or reducing the risk of smudge attack mostly involves adding more steps to the unlocking process. For example, an additional step was explicitly added in the TinyLock method (Kwon & Na, 2014) after the users have correctly drawn their authentication patterns. The additional step that requires a user to rotate the virtual wheel that appears either clockwise or anti-clockwise forces the user to distort the smudges left when drawing the authentication pattern earlier. This additional step adds more time to the unlocking process that can give negative impact on its usability. Furthermore, the grid size in the TinyLock method was also reduced, which may cause difficulty to the users with large size fingers. The Pattern 90 method in von Zezschwitz, Koslow, De Luca, & Hussmann (2013) does not add additional step to the unlocking process. However the pattern grid is randomly rotated clockwise by 90° or 360° prior to the unlocking and the location of the pattern grid on the screen is also randomly assigned to either top, middle or bottom of the screen. These random moves have severely degraded the usability of the method as indicated by its evaluation results. Random move is also applied in SmudgeSafe (Schneegass et al., 2014), where the background image underneath the pattern grid is randomly translated (shown at a different location), rotated, scaled, sheared, or flipped horizontally or vertically during each unlock. This had impacted usability, especially the flipped image that received the worst evaluation in terms of performance. Rotation, which has a similar concept as Pattern 90, received the worst evaluation in terms of login time. The other two anti-smudge-attack methods, which are Marble and its variation, Marble Gap (von Zezschwitz, Koslow, et al., 2013), though are moving away from the pattern grid, are still based on a number of password points in the form of coloured circles called marbles, that appear in random positions during each unlock. It can be deduced that the basis for all screen locking methods that aim to protect users against smudge attack is *randomisation*. Randomly changing the position of a pattern grid or the location of the password points during each unlock causes the traces left by the users after unlocking their phones to be useless to the attackers because the latter will be presented with a different unlock setting that requires different movement of fingertips in the next attempt. This was shown true even under the most extreme assumptions (von Zezschwitz, Koslow, et al., 2013). However, randomisation cannot be too excessive as it may cause confusion to the users, increase complexity and degrade usability. Along the same line, randomisation is also used in the SyS method where integers in the grid are randomly arranged during each unlock. However, the randomisation is kept to the minimum and the grid does not randomly transform into another translation or rotate during each unlock, which avoids the complexity of Pattern 90 and SmudgeSafe methods. However, as mentioned earlier, for added security, the user is given the option to randomise the grid size, which is another advantage of the SyS method that put such control at user's hand. Furthermore, unlike the above methods, which still require a number of password points to be used, the SyS method users only need to drag their finger from one point on the screen to another once. This is an advantage in terms of memorability as the user only needs to remember one integer and the location on the image where the integer should be dropped. Above all, since the above methods were only designed to protect the users against the smudge attack, some of them are very susceptible to the other type of attack, which is shoulder surf. For example, the TinyLock method will completely reveal the unlock pattern to the shoulder surf attacker with the presence of a mirror grid on top of the screen. On the contrary, the SyS method is also designed to protect the users against the shoulder surf attack, which is discussed next.

Unlike smudge attack, shoulder surf attack happens when the attackers secretly watch as the users are unlocking their phones, which can happen in a crowded area. In our experiment, the SyS screen locking method placed third behind the PIN and password methods with the average number of attempts equals to 1.67 from three (20%) successful breaches. However, this is less worrying compared to the number of successful breaches of 15 (100%) for PIN and pattern, and 10 (67%) for password. This has made the ranking of the SyS method against the other methods under comparison with respect to the resistance against shoulder surf attack to become less straightforward as it did not score the best for all of the attributes. Nevertheless, with the score $S_i^* = 0.08894$, $S_i' = 0.46494$ and $C_i^* = 0.83942$, the SyS method emerged to be the most resistant against shoulder surf attack compared to the PIN, pattern and password methods. Other methods designed with the specific aim of combatting the shoulder surf attack includes the SwiPIN (von Zezschwitz et al., 2015), PassMatrix (H. M. Sun, Chen, Yeh, & Cheng, 2018) and XSide (De Luca et al., 2014). SwiPIN tried to make screen locking more robust by combining the swiping motion in the pattern method and tapping motion in the PIN and password methods. In principle, it is still based on the PIN method. However some numbers need to be tapped and some numbers need to be swiped upward, downward, left or right depending on the directions of arrows on the respective numbers. The way numbers are equally and exclusively separated into two regions on the screen made it longer for the user to unlock as they need to first identify the region of the number that they want to use next. Once the number is located, the user needs to subsequently identify if it needs to be tapped or swiped in certain direction, which further increases the unlock time. Thus, it is not surprising when the mean time to unlock for the SwiPIN method is relatively high at 3.66 seconds. The method has therefore significantly impacted its usability. In terms of security, an independent study also showed that the SwiPIN method can be reliably breached with between six and 11 attempts (Wiese & Roth, 2016). However, this is acceptable in our opinion as a mechanism can be implemented that allows a maximum three attempts before the phone is permanently locked or locked for a longer period of time, which is enough to demotivate the attacker. However, an even longer average unlock time of 31 seconds was recorded by the PassMatrix method (H. M. Sun et al., 2018), which is similar to the SyS method in terms of the use of a background image. Though the average unlock time is considered acceptable by its designers, it is far too long compared to the other methods as shown in Table 7. Furthermore, the method was not compared against the more common methods such as the PIN or pattern in the study. The error rate is also the third highest at 13.3%. Thus, in our opinion, this kind of method is mostly suitable for systems with explicitly high security requirements. Finally, a very different approach was used in XSide where both sides (front and back) of a smartphone are used to reduce the chances of successful breach in a shoulder surf attack (De Luca et al., 2014). The basis of the method is the pattern lock and the users can switch from the front to the back to complete the required strokes. The strokes made at the back of the phone are therefore not visible to the attackers. The switching from the front to the back and vice versa causes additional delay as the users need to reorient themselves each time. However, the error rate was found to be surprisingly low at 1.2%. The users rated the method as moderate with one side switch and poor with two side switches. It can be expected that the rating can become poorer with more number of side switches. Furthermore, XSide needs additional installation and configuration to the device to enable the other side of the phone to be used. The success is based on the assumption that mobile devices with touch sensitive back sides will hit the market soon, which until today, about four years later, we have not seen much of such devices. More recent smartphones hardly have anything at the back other than a camera lens and a flash light. Their evaluation results also showed that the front only screen locking method was still preferred by the users and seen as the most usable and easiest. In summary, it can be seen that randomisation is still the basis for screen locking methods that aim to overcome the shoulder surf attack, except for XSide. However, they have usability problems with reports showing longer average time to unlock and lower rating by the users. Although the SyS method recorded a higher rate of 20% for successful shoulder surf attacks, it had the lowest average unlocking time. Successful breach rate is dependent on the sample size and with 15 cases used in our security evaluation, each case contributed to a larger percentage. The three successful shoulder surf breach attempts can be attributed to a number of reasons such as too obvious location at where the chosen number is placed. For example, the chosen picture may have a ball in the middle and the user chose to place the chosen number in the middle of the ball. Thus, it is recommended that user avoids this kind of obvious landing place for improved security, a similar recommendation made by H. M. Sun, Chen, Yeh, & Cheng (2018). Furthermore, in the SyS method, the users do not necessarily have to place their finger tip on the chosen number. They can place it anywhere on the grid to drag it as long as the number is dropped on the supposed location. This can further obscure the unlock code from the shoulder surf attacker. Though it can be argued that the phone might be unlocked if an attacker accidentally releases the number on the right location, our evaluation results showed that chances for this to happen are very low with zero successful breach that happened this way in our experiment.

Finally, the usability evaluation results of the SyS method also found it to be the most usable compared to the PIN, pattern and password methods as far as the three attributes used in the usability evaluation are concerned. The attributes, which are the average time to unlock, the average number of attempts to unlock and the rating given by the users, are used in measuring the usability by similar studies. The SyS method scored the lowest for the average time to unlock, with a marginal difference of 0.02 second from the average time to unlock for the PIN method. In terms of the ease-of-use, most of the users rated the SyS method as the most usable and gave it the best rating of 5, which is interesting and considered encouraging despite it being new and less familiar compared to the other three methods. This is, in our opinion, a strong indication of the SyS potential for widespread application. By scoring the best in the three attributes used to measure usability, the TOPSIS analysis for the SyS method yielded a perfect 0 S_i^*, a perfect 1 for C_i^* and 0.41993 for $S_i^{'}$, which put the SyS method at the top of the rank for usability. To avoid the potential bias and to help strengthen the validity of our experiment, we used the usability attribute values from an independent study that compared the usability of the PIN and pattern methods (von Zezschwitz, Dunphy, & De Luca, 2013) as benchmark. As shown in Table 6, we obtained about the same value for mean time to unlock for the PIN method. Other values are better from our experiment. Other studies were also found that included error rate as an attribute for usability. Even though error rate was not explicitly included as one of the attributes in our usability analysis, the record is available for comparison with similar methods. Based on the categorisation of errors into critical and non-critical in von Zezschwitz, Dunphy, et al. (2013), the SyS method attained 0% critical error rate and all subjects were able to unlock their phones after at most two attempts. No failed attempts were recorded. With 0% error rate, the SyS method is at par with PassMatrix, but with much faster average unlock time and better usability rating by the users.

Table 6. Comparison of usability evaluation data of PIN and pattern methods

Method	(von Zezschwitz, Dunphy, et al., 2013)		Our Work	
	PIN	Pattern	PIN	Pattern
Mean time to unlock (*s*)	1.50	3.14	1.64	1.9
Error rate (%)	15.0	4.2	0	0
User rating	4.0	5.0	4.0	4.0

Nevertheless, comparison based on error rate results across similar studies has to be performed with caution because they were reported in the form of percentage. Percentage value is affected by a number of factors such as the number of participants involved in the evaluation, hence the number of cases. The percentage value represented by one participant may differ significantly due to this reason. Other values such as the average time to unlock and the average rating given by users may also be affected by the sampling size and the experiment settings. Therefore, related work on the screen locking methods to overcome smudge and/or shoulder surf attacks can only be theoretically compared with the SyS method. Table 7 summarises the common evaluation results obtained from these work, including the sample size as reference. In some cases, user ratings were linearly transformed using the same Likert scale with five points where 5 being the best to enable uniform comparison. Apart from these descriptive analyses, TOPSIS or other MCDM or statistical analyses to rank the methods or evaluate their statistical significance were not performed.

Table 7. Summary of security and usability evaluation of existing methods

Unlocking Method	Security		Usability			Sample Size (N)
	Shoulder Surf (Successful Breach, %)	Smudge (Successful Breach, %)	Mean Unlocking Time (Mean, *s*)	Error Rate (%)	User Rating (Median)	
TinyLock	*	0	2.11	0	3.0	18
SmudgeSafe	*	30	3.64	26.0	4.0	20
Pattern 90	*	46	1.66	14.0	4.0	24
Marbles	*	0	5.23	3.1	5.0	
Marble Gap	*	0	5.98	2.1	4.5	
XSide	9.0	*	4.10	1.2	3.0 (for one switch) 2.0 (for two switches)	32
SwiPIN	14.0	*	3.66	3.1	4.0	16
PassMatrix	9.6	*	31.00	0	4.0	30
SyS	20.0	6	1.62	0	5.0	30

* Test evidence not found/aspect not evaluated

In a nutshell, security and usability are the most relevant aspects to be measured in order to determine the performance of the SyS method. Other work in the area were also seen measuring the performance of their methods from these two aspects. Usability and security were also the foci in the review on the first 12 years of graphical passwords development made by Biddle et al., (2012). However, in addressing the security and usability, it is believed that some flexibility should also be given. Reasonable amount of customisation should be given to the users as they are the ones who know most about their security requirements and usability constraints. In the SyS method, users are allowed to predetermine the size of the number grid to be used or to make it random. In principle, the larger the grid size, the better the security. However, larger grid size means smaller box size for each number, which may impact usability. With this kind of flexibility, the users can choose accordingly depending on their needs. As the SyS method does not require specialised hardware or complicated programming, additional cost is not an issue and implementation of the method is not expected to significantly increase the smartphone price.

Limitations

Even with the carefully designed experiment, there are still a number of limitations we would like to mention. Since the study was performed for a short term within a limited period of time, memorability of the SyS method was not assessed. All of the unlock codes used by the participants were created by them during the experiment and it was assumed that the training period given was sufficient for them to become accustomed to the new method and achieve the same familiarity with the other three methods compared. It was also assumed that the users had similar familiarity with the PIN, pattern and password methods. In the actual application, users are unlikely to forget the code if it is being used during each unlock. Furthermore, less effort is presumably needed to remember one number and one location on an image where it is supposed to be placed, compared to remembering a few numbers or password points on a screen. In terms of generalisability, the results are presented based on the evaluation performed involving 30 participants only and evaluation in the wild was not performed. However, we were careful in making sure that we have a mixture of students and non-students with good spread of age values. Furthermore, to keep experiment settings as close to natural as possible, we let the users to use self-selected unlock codes of varying complexity. With regard to the types of security attack evaluated, we did not test against Brute-force attack. As defined in the scope earlier, the SyS method was designed with the aim to protect against smudge and shoulder surf attacks, also known as soft side channel attack (Andriotis, Tryfonas, Oikonomou, & Yildiz, 2013). In addition, Brute-force attack in our opinion, was already handled by most smartphones, for example, by limiting the number of attempts before the phone is permanently locked or requiring more difficult process to unlock. Another thing that we could and did not evaluate was on the correlation between the types of users and the unlock performance. This was done on purpose because the focus of the evaluation was on the security and usability in general. Therefore, all types of users were treated the same. This was also the case in all similar studies.

CONCLUSION

With the more evident roles played by the smartphones and other mobile devices in our daily life, the need to ensure that the information stored does not fall into the wrong hands is of paramount importance. Screen locking can be considered as the first line of defence for the smartphones. However, to have screen locking method that is secure, not complicated and cheap is challenging. Obtaining optimum trade-off between the two is a real challenge. More often than not, either one has to be compromised to make way for the other and vice versa. For example, when password space is increased for better security, usability in terms of cognitive load will suffer. In this paper, a new screen locking method, the SyS, that is both easy to use and safe from smudge and shoulder surf attacks is presented. In the method, the two conflicting criteria were successfully fulfilled, as shown by the results presented. Furthermore, it does not require any additional devices or sensors to be used, which will not cause significant additional cost to its implementation. Evaluation results showed that the SyS screen locking method scored the best in terms of the usability and security compared to the other three common screen locking methods namely the PIN, password and pattern. It also performed well in comparison with the other work in this area. This method offers an alternative approach to improve the screen locking security without increasing the theoretical password space size as our evaluation results showed that despite having comparatively low theoretical password space size, the SyS method is still the most difficult to be breached. However, the theoretical password space size cannot be too small either for fear that the method may succumb to the Brute force attack. Brute force attack was not included in the study as it primarily focuses on the smudge and shoulder surf attacks. Future work includes considering improving the SyS method by using *n* levels authentication and to study the impact of this change on the security and usability. A longer term evaluation involving more participants is also planned to enable us to quantitatively assess the memorability of the method.

REFERENCES

Andriotis, P., Tryfonas, T., Oikonomou, G., & Yildiz, C. (2013). A Pilot Study on the Security of Pattern Screen-lock Methods and Soft Side Channel Attacks. In *Proceedings of the Sixth ACM Conference on Security and Privacy in Wireless and Mobile Networks* (pp. 1–6). New York, NY: ACM. 10.1145/2462096.2462098

Arif, A., Pahud, M., Hinckley, K., & Buxton, Wi. (2013). A Tap and Gesture Hybrid Method for Authenticating Smartphone Users. In *Proceedings of the 15th International Conference on Human-computer Interaction with Mobile Devices and Services* (pp. 486–491). New York, NY: ACM. 10.1145/2493190.2494435

Aviv, A. J., Gibson, K., Mossop, E., Blaze, M., & Smith, J. M. (2010). Smudge Attacks on Smartphone Touch Screens. In *Proceedings of the 4th USENIX Conference on Offensive Technologies* (pp. 1–7). Berkeley, CA: USENIX Association. Retrieved from http://dl.acm.org/citation.cfm?id=1925004.1925009

Bandre, S. R. (2015). Design and implementation of smartphone authentication system based on Color-Code. In *2015 International Conference on Pervasive Computing (ICPC)* (pp. 1–5). 10.1109/PERVASIVE.2015.7087038

Bianchi, A., Oakley, I., Kostakos, V., & Kwon, D. S. (2011). The Phone Lock: Audio and Haptic Shoulder-surfing Resistant PIN Entry Methods for Mobile Devices. In *Proceedings of the Fifth International Conference on Tangible, Embedded, and Embodied Interaction* (pp. 197–200). New York, NY: ACM. 10.1145/1935701.1935740

Biddle, R., Chiasson, S., & Van Oorschot, P. C. (2012). Graphical passwords: Learning from the first twelve years. *ACM Comput. Surv., 44*(4), 19:1-19:41. Doi:10.1145/2333112.2333114

De Luca, A., Harbach, M., von Zezschwitz, E., Maurer, M.-E., Slawik, B. E., Hussmann, H., & Smith, M. (2014). Now You See Me, Now You Don't: Protecting Smartphone Authentication from Shoulder Surfers. In *Proceedings of the 32Nd Annual ACM Conference on Human Factors in Computing Systems* (pp. 2937–2946). New York, NY: ACM. doi:10.1145/2556288.2557097

Guerar, M., Merlo, A., & Migliardi, M. (2017). ClickPattern: A Pattern Lock System Resilient to Smudge and Side-channel Attacks. *Journal of Wireless Mobile Networks, Ubiquitous Computing and Dependable Applications, 8*(2), 64–78.

Harbach, M., von Zezschwitz, E., Fichtner, A., De Luca, A., & Smith, M. (2014). It's a Hard Lock Life: A Field Study of Smartphone (Un)Locking Behavior and Risk Perception. In *Symposium On Usable Privacy and Security (SOUPS 2014)* (pp. 213–230). Menlo Park, CA: USENIX Association. Retrieved from https://www.usenix.org/conference/soups2014/proceedings/presentation/harbach

Jeon, W., Kim, J., Lee, Y., & Won, D. (2011). A Practical Analysis of Smartphone Security. In M. J. Smith & G. Salvendy (Eds.), Human Interface and the Management of Information. In *Interacting with Information: Symposium on Human Interface 2011, Held as Part of HCI International 2011, Orlando, FL, July 9-14, 2011* (pp. 311–320). Springer Berlin Heidelberg. doi:10.1007/978-3-642-21793-7_35

Jermyn, I., Mayer, A., Monrose, F., Reiter, M. K., & Rubin, A. D. (1999). The design and analysis of Graphical Passwords. In *Eighth USENIX Security Symposium* (pp. 1–14).

Keep your phone safe. (2013). *Consumer Reports*. Retrieved from http://www.consumerreports.org/cro/magazine/2013/06/keep-your-phone-safe/index.htm

Kwon, T., & Na, S. (2014). TinyLock: Affordable defense against smudge attacks on smartphone pattern lock systems. *Computers & Security, 42*, 137–150. doi:10.1016/j.cose.2013.12.001

Lai, Y.-J., Liu, T.-Y., & Hwang, C.-L. (1994). TOPSIS for MODM. *European Journal of Operational Research, 76*(3), 486–500. doi:10.1016/0377-2217(94)90282-8

Lee, W. H., & Lee, R. B. (2015). Multi-sensor authentication to improve smartphone security. In *2015 International Conference on Information Systems Security and Privacy (ICISSP)* (pp. 1–11).

Racoma, J. A. (2012). Android Jelly Bean Face Unlock "liveness" check easily hacked with photo editing. Retrieved from http://www.androidauthority.com/android-jelly-bean-face-unlock-blink-hacking-105556/

Raja, K. B., Raghavendra, R., Stokkenes, M., & Busch, C. (2014). Smartphone authentication system using periocular biometrics. In *2014 International Conference of the Biometrics Special Interest Group (BIOSIG)* (pp. 1–8).

RouseM. (2005, September). Shoulder Surfing. Retrieved from http://searchsecurity.techtarget.com/definition/shoulder-surfing

Rouse, M., Wigmore, I., & Westervelt, R. (2012). Mobile Security (Wireless Security). In *TechTarget Whatis.com*. Retrieved from http://whatis.techtarget.com/definition/mobile-security

Schneegass, S., Steimle, F., Bulling, A., Alt, F., & Schmidt, A. (2014). SmudgeSafe: Geometric Image Transformations for Smudge-resistant User Authentication. In *Proceedings of the 2014 ACM International Joint Conference on Pervasive and Ubiquitous Computing* (pp. 775–786). New York, NY: ACM. 10.1145/2632048.2636090

Shin, K. I., Park, J. S., Lee, J. Y., & Park, J. H. (2012). Design and Implementation of Improved Authentication System for Android Smartphone Users. In *2012 26th International Conference on Advanced Information Networking and Applications Workshops* (pp. 704–707). 10.1109/WAINA.2012.31

Skillen, A., & Mannan, M. (2013). *Myphrase: Passwords from your Own Words*. Unpublished. Retrieved from http://spectrum.library.concordia.ca/976791/

Sun, H. M., Chen, S. T., Yeh, J. H., & Cheng, C. Y. (2018). A Shoulder Surfing Resistant Graphical Authentication System. *IEEE Transactions on Dependable and Secure Computing, 15*(2), 180–193. doi:10.1109/TDSC.2016.2539942

Sun, H.-M., Chen, Y.-H., Fang, C.-C., & Chang, S.-Y. (2012). PassMap: A Map Based Graphical-password Authentication System. In *Proceedings of the 7th ACM Symposium on Information, Computer and Communications Security* (pp. 99–100). New York, NY: ACM. 10.1145/2414456.2414513

Suo, X., Zhu, Y., & Owen, G. S. (2005). Graphical passwords: a survey. In *21st Annual Computer Security Applications Conference (ACSAC'05)* (p. 10 pp.-pp.472). doi:10.1109/CSAC.2005.27

Ulery, B. T., Hicklin, R. A., Buscaglia, J., & Roberts, M. A. (2011). Accuracy and reliability of forensic latent fingerprint decisions. *Proceedings of the National Academy of Sciences of the United States of America, 108*(19), 7733–7738. doi:10.1073/pnas.1018707108 PMID:21518906

von Zezschwitz, E., De Luca, A., Brunkow, B., & Hussmann, H. (2015). SwiPIN: Fast and Secure PIN-Entry on Smartphones. In M. Brenner, N. Christin, B. Johnson, & K. Rohloff (Eds.), *Proceedings of the 33rd Annual ACM Conference on Human Factors in Computing Systems* (pp. 1–8). New York, NY: ACM. 10.1145/2702123.2702212

von Zezschwitz, E., Dunphy, P., & De Luca, A. (2013). Patterns in the Wild: A Field Study of the Usability of Pattern and Pin-based Authentication on Mobile Devices. In *Proceedings of the 15th International Conference on Human-computer Interaction with Mobile Devices and Services* (pp. 261–270). New York, NY: ACM. 10.1145/2493190.2493231

von Zezschwitz, E., Koslow, A., De Luca, A., & Hussmann, H. (2013). Making Graphic-based Authentication Secure Against Smudge Attacks. In *Proceedings of the 2013 International Conference on Intelligent User Interfaces* (pp. 277–286). New York, NY: ACM. 10.1145/2449396.2449432

Weiss, R., & De Luca, A. (2008). PassShapes: Utilizing Stroke Based Authentication to Increase Password Memorability. In *Proceedings of the 5th Nordic Conference on Human-computer Interaction: Building Bridges* (pp. 383–392). New York, NY: ACM. 10.1145/1463160.1463202

Wiedenbeck, S., Waters, J., Birget, J.-C., Brodskiy, A., & Memon, N. (2005). PassPoints: Design and longitudinal evaluation of a graphical password system. *International Journal of Human-Computer Studies, 63*(1-2), 102–127. doi:10.1016/j.ijhcs.2005.04.010

Wiese, O., & Roth, V. (2016). See You Next Time: A Model for Modern Shoulder Surfers. In *Proceedings of the 18th International Conference on Human-Computer Interaction with Mobile Devices and Services* (pp. 453–464). New York, NY: ACM. 10.1145/2935334.2935388

Winkler, C., Gugenheimer, J., De Luca, A., Haas, G., Speidel, P., Dobbelstein, D., & Rukzio, E. (2015). Glass Unlock: Enhancing Security of Smartphone Unlocking Through Leveraging a Private Near-eye Display. In *Proceedings of the 33rd Annual ACM Conference on Human Factors in Computing Systems* (pp. 1407–1410). New York, NY: ACM. 10.1145/2702123.2702316

This research was previously published in the International Journal of Mobile Human Computer Interaction (IJMHCI), 10(4); pages 42-71, copyright year 2018 by IGI Publishing (an imprint of IGI Global).

Chapter 13
Phishing Attacks in Mobile Platforms

Thangavel M.
Thiagarajar College of Engineering, India

Yaamine A. M.
Thiagarajar College of Engineering, India

Nandhini J. T.
Thiagarajar College of Engineering, India

ABSTRACT

On facing this fast-moving world every bit of the world is leading by technology. Mobile phones are the major part of our day to day life. Mobile platforms are said to be the hardware and software environment for laptops, tablets, smartphones and also for other portable devices. Mobile operating systems and browsers lack secure application identity indicators, so the user cannot always identify whether a link takes to the expected application. Phishing is a most popular attack and it gets targeted mostly at financial organizations. It is a type of online theft where the sensitive information is obtained by redirecting the victim to malicious websites. This chapter discusses about Phishing attacks in Mobile Platforms.

INTRODUCTION

Mobile phones are mainly targeted by malware and they can also be used in botnets. Mobile devices are equipped with high – end, power-intensive resources with more memory, and variety of sensors. These are mainly used to provide a majestic, rich and for a sophisticated purpose. As far as Windows and Mac dominate the laptop world, where Apple and Android rule the smartphone and the tablet universe. Mobile apps were offered for the general productivity and for information retrieval which includes email, calendar, contacts, messages, and weather information. Now a day's mobile phones are also targeted by phishing attacks. So we can't say that mobile phones are not only used to make calls. We can say that mobile phones are a small computer. Smartphones are mainly used for social media which usually means to affect the bottom line security.

DOI: 10.4018/978-1-7998-8545-0.ch013

In mobile devices, user interfaces are constrained in a very small screen. In mobile operating systems, there is a lack of secure application identity indicators. The end user definitely can't tell what the website he/she is interacting with most of the end users i.e. 60% of mobile users will enter a password at least twice a day. By launching an application stored through app store phishing application can be done. Most of the phishing applications are launched with the help of the Android app store. By reliving the sensitive information, the attacker can easily hack the details of the users. Trojan Activity should in a suitable way. This will direct the end user to make a mistake in the malicious application of a trusted one.

The goal behind phishing is data, money and personal information glomming through the fictitiously unauthentic website. The best strategy for eschewing the contact with the phishing web site is to detect authentic time malignant URL. Phishing websites can be tenacious on the substructure of their domains. They customarily are cognate to URL which needs to be registered (low-level domain and upper-level domain, path, query). Recently acquired status of intra-URL relationship is utilized to evaluate it utilizing distinctive properties extracted from words that compose a URL predicated on query data from sundry search engines such as Google and Yahoo.

Now a day the threat of phishing attacks on mobile computing platforms is getting incremented as many of the users use mobile phones to access the net banking, Gmail, Face book and some other application. MobiFish is automated bulwark scheme, where users no desideratum to make the final decision, but it is the users who conclusively abstract the phishing app. Authentically, they do not require to explicitly make the decision at all, since the only explication for the authenticate failure is a phishing attack. Here there is no desideratum of developer to design the browser UIs, MobiFish is compatible with all subsisting websites and apps. In this paper, we propose specialized form of phishing attacks which target at the sedulous account registry function of mobile OSs. We employ the optical character apperception (OCR) technique to extract text from the screenshot of a authenticate interface, which achieves better performance on mobile phones than on PCs.

Due to the lack of secure identity indicators which means that an inter-application link could be subverted. Next, the user will be directed to the wrong target. In a direct phishing attack, the sender will have some malicious application which links the user to the spoofed screen instead of the real target application. We are in need to address the problem of phishing by implementing the Trojan which commits phishing via preinstalled mobile applications. In a man-in-the-middle attack, the sender will be benign, but some other malicious party intercepts the link and he/she will load a spoofed target application in the place of the target application.

Trojan activity is a kind of malicious type which can be implemented in the built-in application. The trojan activity asks the user to enter some data that will appear in the form of a popup dialog. This popup dialog will collect some sensitive information. After finishing these steps the details collected will send to the hacker. The most vulnerable activity is sending the smishing messages this is also one of the phishing techniques. For example, a person gets a message that he needs to enter the bank details to claim some prizes. Through this activity, the attacker will get the sensitive information from the user so he can easily hack the mobile.

The other ways to hack the mobile by using Wi-Fi phishing. Wi-Fi Phishing will occur when a person gets connected to the internet through hotspots. If it is wireless, then an attacker will use evil twin for eavesdropping. If it is wired, then the attacker will use Starbucks for hacking the information. To overcome this, we can move to content-based filtering. It will have a set of rules which can easily predict the suspect the phishing contents.

Figure 1. Vulnerable areas in mobile business environment

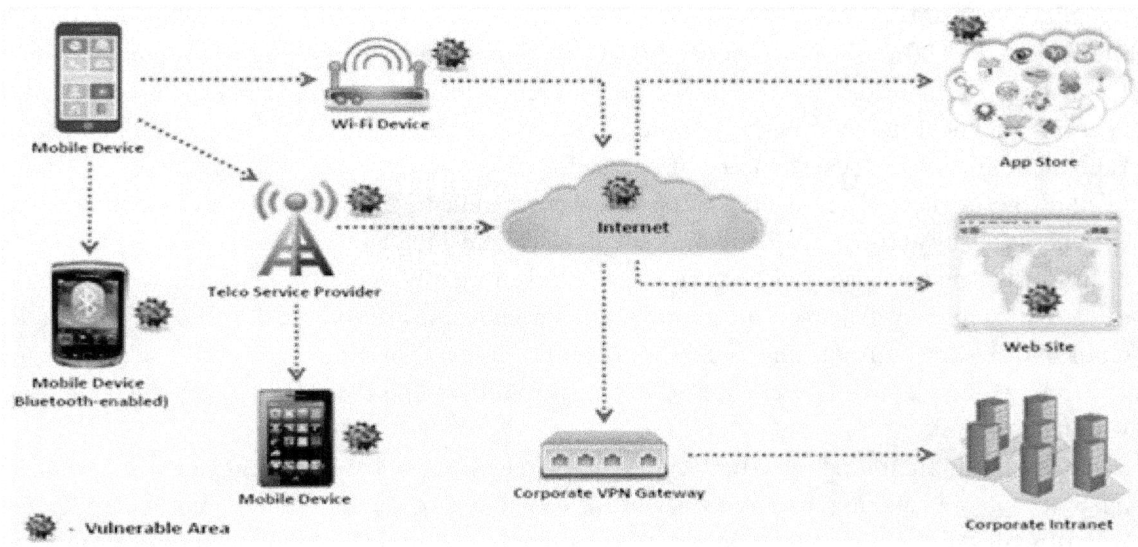

Websites and mobile applications which commonly link the user to the social network which has password protected facility and obviously, payment applications. After it will ask the user to reflexively enter their basic information for gaining the credentials of the user to perform a phishing attack. By this analysis of common behavior, we could identify a number of new phishing attacks are performed against mobile platforms.

Mobile phishing is also dangerous due to some hardware limitations of the mobile devices and also, we can include mobile user habits. In this, we did a deep study on the security vulnerabilities (Niu 2008) which is caused by mobile phishing attacks, including the web page.

Figure 2. Types of phishing attack

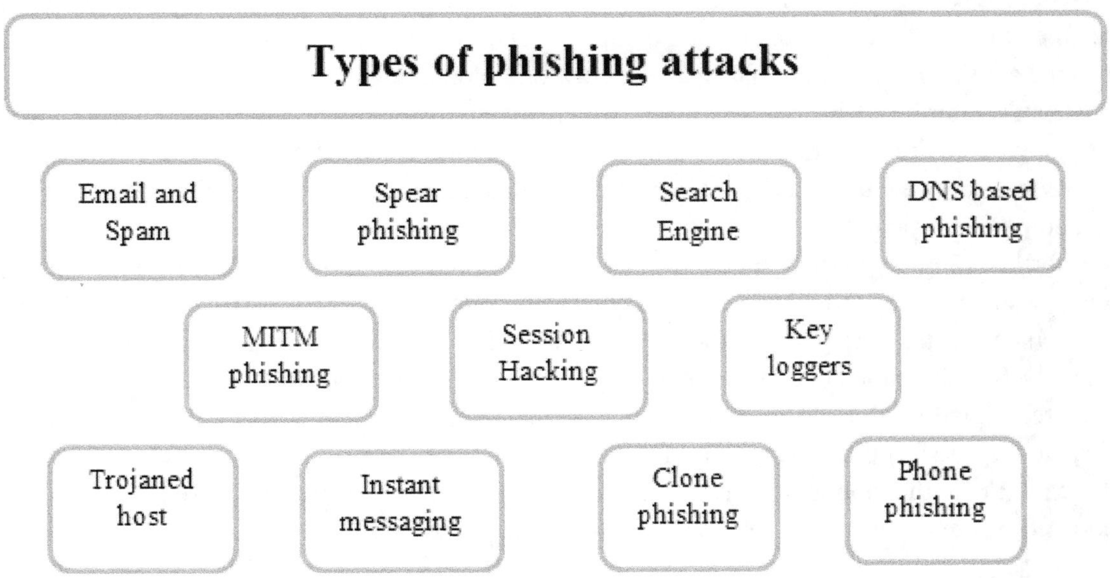

Phishing attacks, the application of phishing attacks, and the account registry phishing attacks. Existing schemes which are designed for PCs which can't effectively address the various phishing attacks on mobile devices (Felt 2011). We propose MobiFish is a novel automated lightweight anti-phishing scheme (Anti-Phishing 2006) for the mobile platforms. MobiFish which verifies the validity of the web pages, applications, and persistent accounts by comparing it with the actual identity to the spoofed identity.

The aim of Phishing attacks is to steal private information such as usernames, passwords, and credit card details by the method of impersonating a legitimate entity. Most of the security researchers have been proposed many anti-phishing schemes, the threat of phishing attacks is not reduced. On the other hand, lots of the phishing sites were expired and revived rapidly. By the other side, Phishing attackers are doing an extraordinary job for improving their techniques so that their new attacks are able to overcome the existing anti-phishing tools (Carnor 2007). Mobile phishing (Micro 2012) is an upcoming threat that targeting the mobile users of financial institutions, online shoppers, and mainly on the social networking companies.

Due to the hardware limitations, such as the small screen size and the inconvenience of user input and application switching and mainly on the lack of identity indicators, mobile user habits and preferences, etc... Mostly all phishing attacks on mobiles are in the form of bogus websites (i.e. Malicious sites). Nowadays browsers are the powerful tool which gives them enough to support for all kinds of Internet services, and the people who are accustomed to online banking, online shopping, online socializing, etc. Users are being requested to provide the personal information, and of course providing private information and credentials to websites. Through this information, an attacker can easily hack the information and perform some unwanted matters.

Phishing detection schemes can be hardly divided into two categories: heuristics-based schemes and blacklist based schemes. Blacklist-based schemes will be able to detect only those phishing sites which are listed in the blacklist and manly it can't detect the zero-day phishing attacks such as those that have only appeared for days or for an hour. It is true that new phishing sites will already steal the user credentials or have expired before being added into the blacklist. Heuristics-based schemes largely depend on features extracted from URL and HTML source code it is also based on the other techniques such as machine learning which is used to determine the validity.

To reduce Phishing attacks there are some security concerns for the fake mobile applications. It will reduce the phishing attacks on the mobile platforms. The first security is that the users must download the official apps from app stores. The user should aware of clicking the unknown links. Next, the users must use only safer browsers. Safer browsers which contain the security features and it also safeguard the users by eliminating the malicious and phishing sites for the security purpose. It is difficult to see the URLs in a complete manner. To recover from this bookmarking is a perfect solution to preserve the users landing on unwanted sites.

The forthcoming approach is by controlling the app stores. Vendors must keep eyes on developers before they upload their app to the public. Many security concerns have provided the anti-virus programs for the personal computers to avoid the malicious activities using some apps (e.g. COMODO). Next, by using the identity indicator, we can remove the header. The header is removable because the mobile screen size is small. Mobile devices have small screens, so users are unable to see the entire URLs, this will lead to the possible phishing attacks. This paper gives the clear overview of various types of mobile phishing attacks in the mobile platforms and it also discusses some mitigation approaches and security concerns. Users must be more careful while downloading apps and when they run on their devices.

DISCUSSIONS

AntiPhish is utilized to evade users from utilizing fraudulent web sites which in turn may lead to phishing attack. Here, AntiPhish traces the sensitive information to be filled by the utilizer and alerts the utilizer whenever he/she is endeavouring to apportion his/her information to a untrusted web site. The much efficacious elucidation for this is cultivating the users to approach only for trusted websites. However, this approach is fictitious. Anyhow, the utilizer may get chicaned. Hence, it becomes obligatory for the associates to present such explications to surmount the quandary of phishing. Widely accepted alternatives are predicated on the creepy websites for the identification of "clones" and maintenance of records of phishing websites which are in hit list.

Web users have been suffering from phishing attacks since their first appearance in 2003. Researchers have proposed many solutions such as vigilant aegis and phishing detection to bulwark against phishing attacks. Mobile phishing assailment could additionally be in the form of Emails or Short Messaging Accommodations (SMS). Phishing emails (Bergholz 2010) customarily request users to click a link to a fictitiously unauthentic website where the utilizer is prompted to enter authenticate credentials. In additament, many phishing detection implements have been designed for phishing on PC web pages. Predicated on the methods utilized, they can be generally categorized into two groups: heuristics schemes and blacklist schemes. Heuristics schemes outperform blacklist schemes since they can deal with incipient phishing sites without having to wait for an update. Customarily, heuristics schemes for phishing detection utilize other techniques such as machine learning techniques and search engine.

Phishing websites, mainly utilized for e-banking accommodations, are very intricate and dynamic to be identified and relegated. Due to the involution of sundry ambiguities in the detection, certain crucial data mining techniques may prove an efficacious designates in keeping the e-commerce websites safe since it deals with considering sundry quality factors rather than exact values. An efficacious approach to surmount the fuzziness in the e-banking phishing website assessment is utilized a keenly intellective resilient and efficacious model for detecting e-banking phishing websites is put forth.

BACKGROUND

There is a great need to investigate issues related to credit cards misuses, bank accounts hacking, security threats, particularly loosing of an important credentials by Phishing attack. The major critical issue is stealing the account numbers and passwords stealing without the knowledge of the users, as trading over internet increasing day by day. Hence user must be aware of various phishing techniques that hackers use to gain the illegal access of important credentials while user accessing e-commerce-based services using their personal computers or mobile phones. The following are brief explanation about various phishing attack models.

SMS Phishing Attack

SMS phishing is an act of trying to acquire personal and financial information by sending SMS (instant Message or IM) containing a deceptive link. Most of the consumers access the internet through a mobile. The mobile user is not conditioned to received spam text message on their mobile. This SMS phishing attack is more effective since it is easy to set up and difficult to detect and stop before they cause harm. And also, most of the mobile anti-virus does not check the SMS.

Figure 3. SMS Phishing attack working process

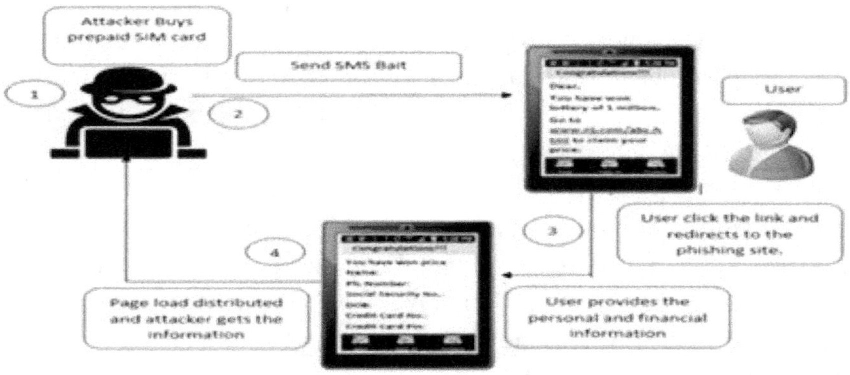

Mobile Spam

Unsolicited text/email message is sent to mobile device from known/unknown phone number or email IDs. Spam message contains advertisements or malicious links that can trick users to reveal confidential information. A significant amount of bandwidth is wasted by this spam message. This kind of the attack is done for financial gain.

Figure 4. Sample mobile spam message

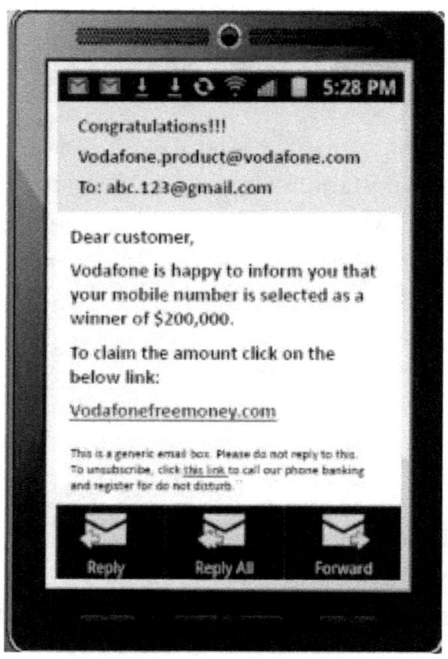

Bluetooth and Wi-Fi Connection Attack

Mobile device pairing on an open connection using public Wi-Fi or any encrypted Wi-Fi routers which allows the attacker to eavesdrop and intercept data transmission using a technique such as;

- BlueSnarfing attack is the process of stealing the information via Bluetooth.
- Blue Bugging attack is the process of gaining control of the device via Bluetooth.

The sharing of data from malicious devices can infect and breach data on the recipient device.

Generally, browsers have many practical features and the convenient functions which are truncated during their adaptation to the hardware-constrained mobile platform. This will definitely result in an unpleasant experience for users. To improve this services most well-known enterprises have published the mobile applications (apps) for the major mobile platforms. This blows the new light on phishing scams, some of the phishing attackers develop fake apps and repackaging the legitimate apps. Next, they are uploading these phishing applications(apps) to unofficial app markets. By after using these apps the user's mobiles are gettingattacked. The next process is when the attack succeeds, the user's credentials and information's will be sent to the phishing server.

Phishing web pages are easier to detect than the phishing apps because,for many web pages, he/she can be able to judge the destination of form-data from HTML source code i.e. Action attribute in the form tag. But if we think of mobile apps, there is no way to check whether the user credentials and also the information's are sent to the legitimate authentication server or to the attacker's server. Hence, phishing attacks on mobile phones are more complicated than which is performed on the PCs. Thus, we need to design effective mobile phishing Défense schemes for both the web pages and the applications.

Figure 5. Pairing mobile devices on open Bluetooth and Wi-Fi connection

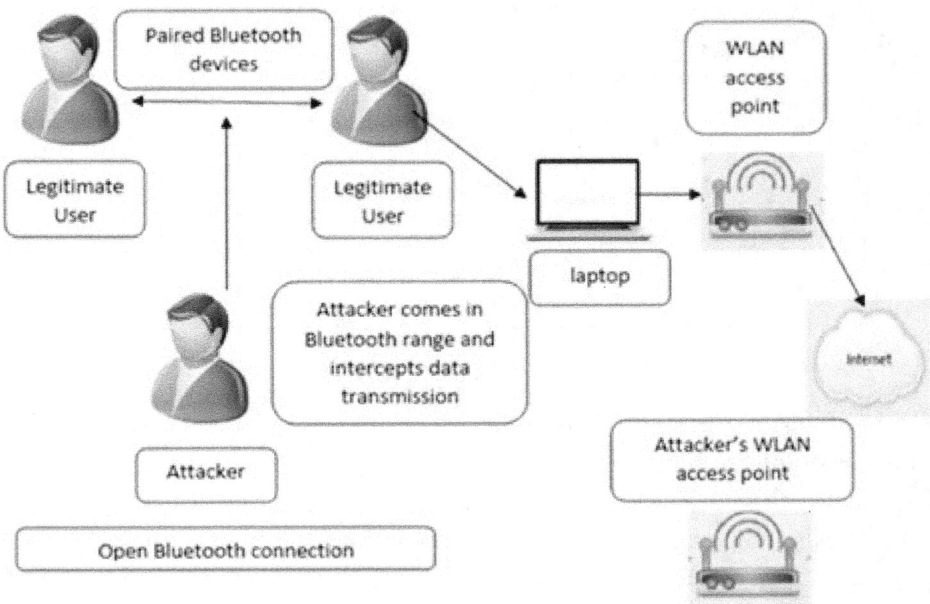

Mobile Web Page Phishing Attack

Here we need to describe the three types of phishing attack models. First and Foremost, we are going to discuss the Mobile Web Page Phishing Attack. In PC browsers, the most widely applied method is Web page phishing detection. For the Web page phishing Detection, there are two existing methods used that is Blacklist-based matching and Heuristics-based.

The blacklist method is mainly used to search a suspicious site in a list of reported phishing sites. The blacklist can achieve higher accuracy for the cost of human verification but if it delays in updating the phishing sites it will definitely degrade its effectiveness. Blacklist-based methods cannot defend the new phishing sites which are not listed for example zero-day attacks. Next detection based method is Heuristic detection methods which are based upon the features which are extracted from URL and HTML source code this will often work with the help of search engine and machine learning techniques. These two features are summarized from previously reported phishing sites.

Figure 6. comparison of OSR performance in mobile phone and PC

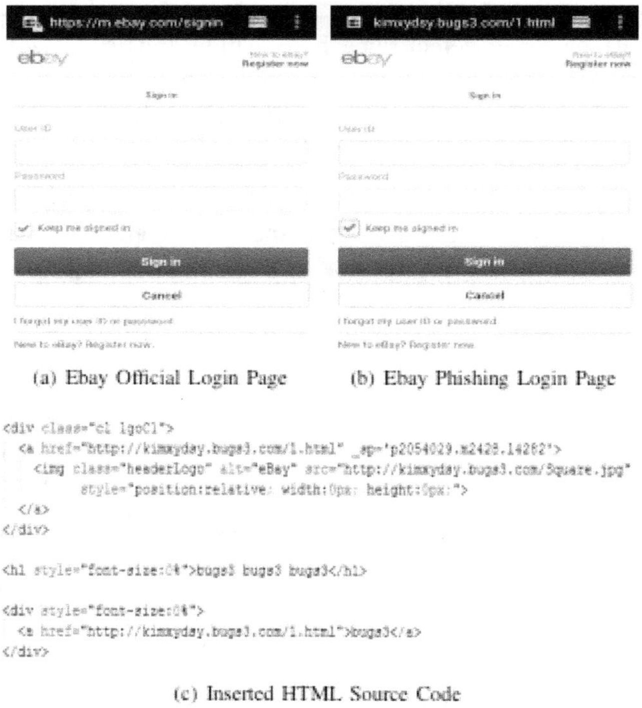

(a) Ebay Official Login Page (b) Ebay Phishing Login Page

(c) Inserted HTML Source Code

Moreover, a new phishing site may not contain these features at all because most of the features will only appear in some of the phishing samples. This defines that the most carefully constructed phishing sites which remove all of the suspicious features which could able to bypass the heuristics-based detections which means that the heuristics based approaches definitely can't achieve a 100% detection rate. Information which is extracted from HTML source code which may not be able to reflect the web page displayed to the user. This is the best way attackers can add some text, images which link into HTML source code with this they can also do some undesirable content which disappears from a webpage it by covering with other images and simply changing its size.

From this we can attackers can easily manipulate the word frequency, brand name, and company logo. This can be explained easily by real snapdeal. The attacker will copy the code of the original site and moved it to some websites. The attacker will also upload the image components to that website and he will change some other links to the corresponding places on our website. He will mainly focus on the form data submission URL. Moreover, the modified web page looks exactly the same as the official Snapdeal site. As a result, no user will suspect his validity without looking at its URL. Phishers or attackers can insert as many errors as needed into the HTML source code for obfuscating the conventional identity extractors. Mostly a large number of errors are extracted to convince the identity extractor the webpage which claims error instead of snapdeal. The webpage title will not be able to the user.

Mobile Application Phishing Attack

Applications which are based on Phishing attacks are quite uncommon in computers. Application-oriented phishing attacks are further classified into two types: First wayis the most of the phishing applications will attempt to hijack the existing legitimate apps. Phishing apps will continuously perform the task polling and this will launch themselves so that they can detect the launch of the target apps. As a whole, the fake login interface layers will be overcome at the top of the real one. Next, the phishing app will appear as the target app. The users could not know what is happening when the window is in the switching process. We can solve this by checking the identity of the foreground app which is available in the task list.

The second way is that phishing app will directly show the original app. This will definitely occur when the mobile users download the fake apps from the unofficial app market sits. There are various methods to steal the user credentials but the essential attack pattern will be ended with the transmission of credentials a personal information to the attacker. By this, we can come to the conclusion both the runtime monitoring and blocking the communication of the suspicious apps which can effectively defeat the app phishing attacks.

Mobile Account Registry Phishing Attack

The Android system will provide a facility for accessing the centralized account registry and management function. This will allow the mobile users to log on to their online accounts (e.g. Instagram, Facebook, Twitter, etc.). So generally, this function will be called as Persistent Account Registry. Account Registry will be very convenient for the user who will not be bothered to enter their credentials and personal information each time for launching a social networking application. Mostly the mobiles are used for the personal purpose which can be accessed only by the mobile holding person. This can be guaranteed by simply setting by overall screen locking password. In the Android mobile, the persistent account registry can be done through Add account w option which is present in the Settings directory, this shows a list of applications which support persistent account registry.

Basically, any app can register itself to show up in the account list as a new account type. The application must contain an authentication service which is in need to be present in the AccountManagerService of the Android system. The persistent account registry is vulnerable to a phishing attack. The icon and label which are used in the main menu will not be the same which are displayed in the account list. So, the third-party app will register itself as another entity in the account list. This can be

Figure 7. persistent account registry function

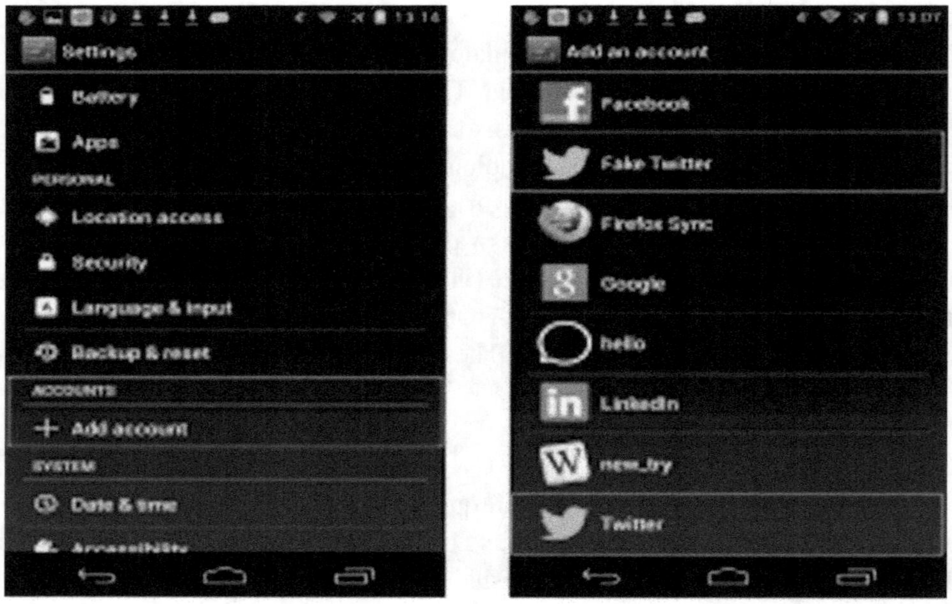

(a) Settings Directory (b) Account List

explained by an example where malicious game app which is pretending to be an application of a social networking company or financial institution. So, we can consider the app which is a fakeFacebook app in the account list. The GUI of that app will look as such like the original Facebookapp. When the user enters into the login page and he will enter all the credentials. Next, this credentials will be sent to the attacker or phisher. Here they have dealt with a fake login interface i.e.Appfish. Moreover, the account phishing attack is more stealthy.

MOTIVATION FOR ACCESSING THE PHISHINGSITES

Phishers or attackers are using some fancy tactics to direct users to their phishing sites or applications, which masquerade as trustworthy entities. To solve this problem, we need to find the discrepancy between the identity it needs to be claimed and where the actual identity is placed. By having the HTML source code, we cannot find the claimed identity of a phishing site, so this is not only a reliable clue for identifying the phishing site. As an alternative, we should also focus on the screen which is presented to the user because users are directly spoofed by what they see. BY using the existing anti-phishing schemes, we can't able to detect the app phishing attacks and account registry phishing attacks. Thus we are in a strong need to find an effective Défense scheme against the phishing attacks which are playing on the mobile platforms (Wu, 2016).

Extracting the Identity

The screen which is being presented to the mobile users should be the exact place of the claimed identity which is extracted from. The best way is to take a screenshot i.e. capturing the screen content. We can observe that most of the people will believe that screenshots can work well in identity extraction and verification.

The first way is that most of the login interfaces of the legitimate mobile sites and apps are very simple. The entire login page and the majority of the page can be captured in a screenshot. Another way is that the brand names and the company logo (identity)can be located at the apparent places in the login page this can be easily captured and extracted from the screenshot. Screenshots are useful the phishing detection in both web pages and applications. The source code of apps is not available. So, there is no possible way to acquire the content of an app login interface other than taking a screenshot. Next process is to obtain the claimed identity so for this, the OCR technique is utilized to convert the captured screenshot into text.

SLD is known as second-level domain where we can see the actual identity of the mobile page. Mostly famous enterprises will make their brand names as the second-level domain name (SLD) of their official websites. This can be explained with some example. Now lets us take some banking sites. We can take Bank of America. Bank of America will be using the entire brand name as the second level domain name. Hardly in most of the cases, the brand names of the bank cannot be same as the second level domain name. The brand names will contain a symbol 'AT&T' where most of the content based schemes will be failing in this case due to the mismatch of the brand name 'AT&T'. In the second level domain, it will be showing this symbols as 'att' in legitimate URL most of the special symbols will not include in the domain name. This inconsistency can be solved by mapping it with the whitelist. Here the brand name 'AT&T' which will be mapped to the second level domain 'att' before the identity verification, the actual identity of mobile apps cannot be decided because of the unknown source code. This will enter all the information and credentials of the user. Suspicious apps will be cleared before the information of the users is allowed to transmit.

OCR Techniques

OCR is defined as the Optical character recognition (OCR). Optical character recognition which is the process where the mechanical and electronic conversion of an image will be converted to the machine-encoded text. Optical character recognition is the process which has been utilized to extract the text from the text only logo. Goldfish is the process where it uses the optical character recognition which is to read the text from the web page especially from the company logo. We can definitely trust that Optical Character Recognition technique will be achieving the better performance on the mobile platforms because mobile is limited with a small screen size and with a relatively higher pixel density. Now we introduced the Optical Character recognition technique on the mobile platform. This technique will achieve the better performance and it will have a greater effect on the mobile phones by using some real experiments.

THE MOBIFISH SCHEME

MobiFish consists of three major components

1. WebFish
2. AppFish
3. AccountFish

These are designed to protect the mobile web pages, applications, and persistent accounts.

WebFish

In Web-scheme the defense scheme will be intimate with the URL loading. The first process is bowser tyrs to load a webpage, firstly web fish will scan its URL When a browser attempts to load a web page, WebFish first scans its URL to check whether the domain name is an IP address. Most of the Legitimate websites will always use their domain names as a verification of their identities. Attackers and phishers will be likely to use only an IP address for disguising their fake identities. WebFish will be containing the HTML source code of that loading page. Next, it will check if there is any form like the loading page. Phishing web pages must also need login pages with an input tag for entering the credential s and the personal details of the users and the user will submit the details to that phishing websites.

As the whole WebFishwill be checking the existence of forms so that not every page need not go through the checking. Moreover, the identity extraction module will be not relying on any part of the HTML source code. If any of the forms is found WebFish will Start to identify the extraction and verification. Firstly, it will extract the SecondLevelDomainName from the URL, which will be representing the actual identity of the site. Next SecondLevelDomainName will be indexed in the in the Mapping White-List (MWL). This will match the SecondLevelDomainName of the Brand name which gets recorded in the Mapping-White list. Next, the actual SecondLevelDomainNamewill be replaced by the corresponding brand name. On anotherside, the screencap native function will be called to take a screenshot of the login page which will extract the text with the OCR tool. Here we need to note something if we take a screenshot the URL which is shown in the URL bar which will also get captured in the screenshot.

Before doing the identity verification the captured URL must be removed from the text because it will be containing the actual identity of the site. By seeing the existence of a URL bar which is captured in the screenshot will be determined by checking with the first line contains one of the top-level domain names such as gov, edu, com, and org. For speeding up the DetectionProcess There is a need to search for sensitive terms such as username, password and debit card, credit card number in the text. If nothing was found in that page it will generally use for the data input purposes and the page will be marked as a safe page. At last, the step is we need to search the SecondLevelDomainName in the text. If this is not found in it then it will be marked as a phishing site.

The AppFish Scheme

AppFishwill maintain its database which is called as Suspicious App Set (SAS). Suspicious App set which will be containing the profiles of untrusted apps including the user ID (Uid), the launching time, and the screenshot text. If the user is getting suspected with some of the apps then they can name that app into SAS. The apps which are listed in the Suspicious App Set will be monitored by AppFish.

These apps can be classified into the two forms:

- Specified for a particular company

Figure 8. Workflow of Webfish

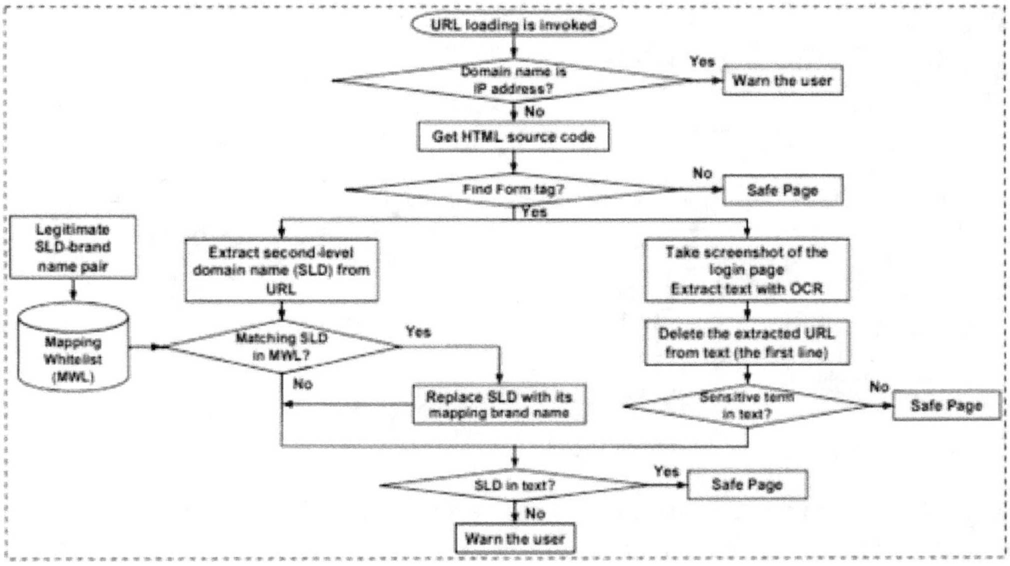

- Required user sign in.

First, they will be collecting the domain names of the collaborators. The domain names will get added into the SAS profile in advance. Owning multiple domains which will often happen to websites which needs an extra storage. This can be explained by taking the example of Facebook. Facebook will request the data from the domains like fbcdn.net. These sites will be used as a content delivery network(CDN). Most of the users will be uploading the photos on the facebook. Facebook can't store a largenumber of photos in facebook.com. So it will be using akamaihd.net. If the user needs to view a photo the request will be sent to the akamailhd server. Secondly, there are most apps where the users need not do login in browses and apps for news,music,maps,etc. Mostly in these cases, phishing attacks will not happen at all. Moreover, browsers i.e. webpage login will be protected by WebFish.

The AppFishdefense scheme works will be categorized in two phases:

- Launching phase
- Authentication phase

In launching phaseAppFish will be obtaining the name of each launching application then it will search for it in SAS.Once the logging process begins, the AppFish will be taking a screenshot of the logging process and it will be extracting the text by using the OCR tool. The text which contains the application userid and also it note the launching time when the user is logged into the profile. When the user has entered the credentials and the personal information. Next, he will click the sign in button so that the authentication phase begins. A trusted or legitimate applications(like Facebook and Twitter) will usually send the users credentials to a remote server for the authentication purpose through HttpGetor HttpPost. After the credentials get verified the applications will be loading the data belonging to that account. On the other side phishing apps can't load the original usersdata .Only one thing they can do

Figure 9. Workflow ofappfish

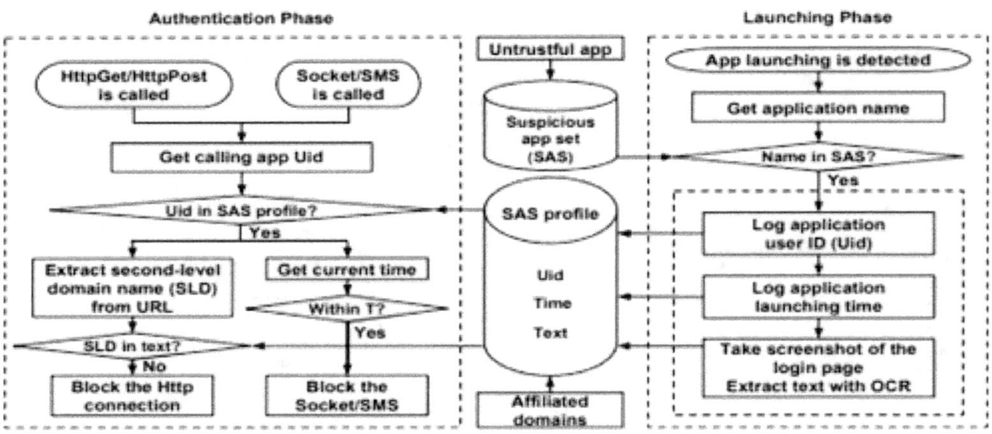

is they can track the users that they have entered the wrong password. The user will suspect about the validity of the application, he/she will uninstall it. Hence, a phishing app is able to send out the user credentials only during the period from submission to uninstallation. Appfish will be monitoring the possible paths for a phishing app to transmit data to the outside world, which basically include HttpGet/ HttpPost, socket, SMS and email (email which is also based upon the socket). If an application uses any of these things to send out an information, AppFish will be checking whether it is get suspected in SAS. If it is confirmed, then the HTTP connections (HttpGet/HttpPost) will get filtered while other communications (socket/SMS) are blocked.

The suspicious apps will be requested to connect with the URLs. AppFish will be ensuring that the SLD name which appears in the screenshot text and the affiliated domain names will be stored in the SAS profile. Moreover, socket and SMS functions will be blocked for a certain amount of time. This process could be enough for the user to notify the abnormality and the user will uninstall the malicious apps. Thus in phishing applications, they could not be able to send out credentials and the personal details before being removed by the user. Next, we need to search the second level domain name with the text which is being extracted from the login page. the reason, why we extracted the text instead of the application name, is, in phishing attacks which are based on the task interception i.e. hijacking, phishing app will be having the same app name as that of the phishing server, which will pop up a fake login page in some other entity. For an example, if the users download the phishing app named ‘abc’. The user will download it because of its fancy operations and functions. Moreover, the malicious app will be pop up a fake Facebook login interface as soon as the original Facebook launches. Once the user gets spoofed, the app “cdf” immediately sends the credentials to the phishing server “cdf.com”. In this example, the foreground i.e. Fake application name “cdf” is the same as the SecondLevelDomainName of phishing server, but the user cannot be about the Facebook phishing login interface.

The AccountFish Scheme

Due to the unlimited registry among the 3rd party apps will be possessing a huge threat to the privacy of the user and the account security. According to the survey of 2014, Yahoo stated that the average number of apps which is installed by an Android user is 95%. Users could not remember the app that has been installed so it is vulnerable to account registry phishing attacks. So they have proposed AccountFish to defend against the phishing attack which will mainly target at persistent accounts. The account registry phishing attacks can be categorized into three types. With the help of the malicious app will be appeared in the main menu and the account list. The first type is Attack; the malicious app will be appearing to be a different app of the targeted account. For example, a game app registers a into account. In the second type, B attack the malicious app will not appear in the main menu. Third type is C attack; the malicious app will directly showup the target app i.e. repackaged app that means they will be having the same application name as that which is shown in the main menu as the account name which appears in the account list.

Figure 10. Workflow of Accountfish

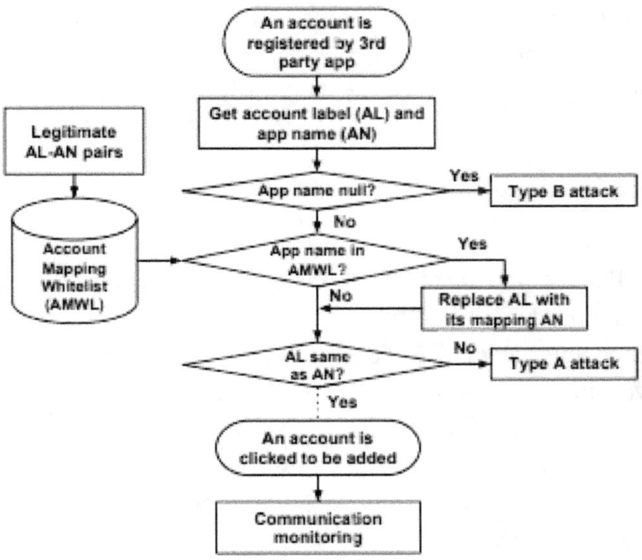

Detection mechanism for the first type and second type phishing attacks which is used to compare the app name in the main menu and the account label in the account list. Malicious apps can be dynamically registered and it can change the account information. The accountfishscheme must be able to inspect the registration of accounts which are held during the runtime. This can only have accomplished by changing the Android source code. Here we can check with the app name and account label. whether these two things i.e. app name and the account label are inconsistent the app will be highly a malicious app. Still, there are some legitimate apps where their account labels are not exactly the same as the corresponding app names. This problem can be solved by using an account mapping whitelist(AMWL) which contains all the inconsistent legitimate apps. Suspicious Apps for suspicious apps which are not available in the whitelist, a warning will be contained for that app and the account names will be provided instantly. The apps which are being presented in the AMWL and those apps which are having consistent names will go for further checks. Type C Attack which is similar to AppFish, the decision cannot be made until the transmission of credentials happens over the apps for the AccountFish, the screenshot is unnecessary for the 3rd party apps so that account login interface cannot be hacked.

Another Hacking Scheme

Since 2003 web users are suffering from phishing attacks. Till now many types of research have been proposed many solutions.

i)Alert protection
ii)Phishing detection

First, we can see about the alert protection. The alert notification which is a simple notification which is used when the end user needs to enter his/her sensitive information. Antiphishing technique which was proposed by kid. Antiphishing is a method which will detect the sensitive information of the end user. This method will warn the end user when he tries to give the information to the website which seems to be untrusted. Moreover, this scheme will not automatically check and detect the phishing attacks. The end user needs to judge selfly after they warned. More detection tools which have been designed for phishing attacks for the PC web pages. As we seen earlier it has been classified into two groups:

- Heuristics schemes
- Blacklist schemes

Heuristics schemes will overcome the blacklist schemes because they will deal with the new phishing sites. Mostly heuristics schemes for the phishing detection will utilize the machine learning techniques and search engine. The next approach is CANTINA. CANTINA which is a content-based approach for detecting the phishing websites, and also it adopts TF-IDF information retrieval algorithms.

Next Garera who proposed a heuristics based scheme which is used to identify the several generic features of phishing URLs and these features will be used in a logistic regression classifier. CANTINA method which is used for a comprehensive feature-based solution for the web page phishing which used to combine the machine learning and search engine techniques. HTML source code could not be trusted because it may not reflect the actual content to the end user.

Most of the phishing sites look very identical which are very similar to authentic sites.

Spoof Guard is a method which uses URLs, images, links and domain names. These are used to check the similarity between the present page and the page which is stored previously. Another method is PhishZoo. Phish Zoo which uses the profiles of the trusted websites appearances which are built with fuzzy hashing techniques. These are used to detect the phishing attack. PhishZoo will make the profiles of sites. This profile will consist of fuzzy hashes of common elements for example URL, frequently used text, HTML codes, images, etc. These are the common things are which are commonly related to the appearances of the sites. These will enhance the phishing detection scheme by adding the images into the profiles by utilizing SIFT image matching algorithm. If it depends on the different approaches the phishing sites cannot detect the sites. Goldphish is a method where the OCR i.e. optical character recognition for the phishing detection in PC browsers. OCR is technique is used to extract the text from images which are found in web pages (e.g., the logo of the company). By using this it will be compared with the pranked domains from a google search.

MOBILE PHISHING DETECTION

The major threat for the mobile users is a mobile phishing attack. Here we mainly discussed the hardware constraints of mobile devices. Browsers are advised to be redesign to make the authenticity and the origin of the user to be apparent to the users. It is a tough job to add the more features to the mobile because of its small screen size. They proposed an identity bar will display an identity bar will show the domain name or foreground application of the web page. Another phishing defense technique which employs a trusted ad unified login for UI apps. ScreenPass which provides a trusted software keyboard. This will allows the end users to store their passwords domains together with their credentials. The OCR which is used to ensure about the passwords that are entered by the users will be done through the trusted software keyboard. There is another detection method where they will be using an attested login for the user credentials by the use of the certificate. This certificate will be holding both the software and the hardware credentials and This also needs the modifications for the client apps. However, a proxy service which is designed for performing anti-phishing filtering against the URLs, page content, or user context. This is a need to be downloaded and configured manually in the browser. Users must be able to authenticate the identity of the proxies because the attacker can also set up some fake proxies. This idea which is similar to Antiphishing. Antiphishing is a scheme where it will give the warning of credential rendering instead of phishing vulnerability. Another method is Trusted activity chains this is to protect the activities from spoofing preventions. It is the developer duty to annotate the chain of activities that should not be interrupted. This show that existing apps are not protected and the developers must take high responsibility in this area.

SOLUTION AND RECOMMENDATIONS

MobiFish can be defined as which is capable of defending against phishing attacks on mobile web pages, apps, and persistent accounts. The aim of the MobiFish is to solve the essential problem of identity masquerade without reliance on HTML source code, on the search engine and for the machine learning techniques. So, for this inconvenience, we can use the optical character recognition (OCR) technique to extract the text from the screenshot of a login interface. This will definitely achieve better performance on the mobile phones than on PCs. We could able to find the spoofed identity from the original text and the actual identity which from the URL of a web page or the remote server mainly for mobile apps. The two identities are different then it must send a warning message to the user. For the account registry phishing attacks, MobiFish will effectively detect all three variants from checking the consistency of account label to app name and the destination of credential transmission.

Hardware Limitation

Due to the small size of the mobile screens and limited computational power, browsers which are working in the mobile systems need to be removed or we can degrade some of the features to make more space for web contents and to maintain a smooth user experience (e.g. loading speed). Phishing web pages which could be detected and blocked on computer browsers which can still be accessible from the mobile browsers. In addition to this, the user interface which is designed for mobile browsers is also simplified, this simplified design could instead help the phishing sites to bypass user inspection. For accommodating the small phone screen size most of the mobile browsers need to remove the status bar and need to hide the URL bar when the web pages' finish loading.

When the loading process is being done, long URLs are discarded to fit the browser frame. There is a risk associated here for the user because the ability for reading and verifying the URLs is very critical for detecting the phishing attack. Since the ability to read and verify URLs is crucial in detecting phishing attacks and the partial URL which means especially a URL with only partial domain name displayed this will lead the user to the spoofed page this will increase the phishing attack. We can explain this by example, think now we are having the official website of "bank of India" which shows the URL bar with a partial domain name. This will definitely lead to a successful phishing attack. The legitimate users are convinced by the partial URL and they are ready to submit their credentials and personal information. Here the phishing attack will happen the user will think that URL will move to the http://securebankofindia.com will move to the "https://secure.bankofinda.com.phishing.com". This phishing attack would fail if the entire URL and the full domain name is displayed.

There are two possible ways, the first way is once the user can view the URL completely and another possible way to which the user could view the URL completely just by clicking the address bar and also by scrolling all the way to the end. The second way is to view the actual destination of a link, this can be done by holding the link for about two seconds. The destination URL which will display the domain name with many as 31 characters instead of showing the 19 characters in the URL bar.

Most of the legitimate sites, the domain names of the websites which can be easily mimicked by replacing the letters. This can be explained by distinguishing 'I' from the capital 'i' for example, it is hard to distinguish 'l' from the capital 'i' because mobile browsers will display both the characters in the vertical slash shape. For both the capital 'I' and capital 'i' together and also to find the small difference in height which is very difficult to discern by human eyes. So, for this kind of letter replacement phishing attacks both the attentive users and observant users who will always check the entire URL i.e. domain name which is likely to be fooled. Mostly the lack of identity indicator was another issue in the mobile phishing.

User Habits and Preferences

Mostly the user habits and the preferences will further increase the vulnerability to mobile phishing attacks. For the past few years' touchscreen smartphones are becoming more dominant in the mobile phone market. Moreover, by typing in a virtual keyboard is not that much easy as on a physical keyboard because of the inability of the lower input accuracy, particularly in the actions walking or sitting in a moving vehicle. Because of this reason, it is tempting to follow the links in web pages and e-mails rather than typing the links manually. The other factor is that on smartphones, switching over the applications and shifting to some other pages within a browser which is more complicated and dangerous than being performed on a PC. Phishing attacks can be succeeded because users will get trust and accustomed to entering their personal information and credentials in the most familiar which has a repeated login interface. If users frequently check their legitimate links whose targets prompt them for private data of the user and they will be supplied the requested data.

FUTURE RESEARCH DIRECTIONS

Although traditional phishing attacks mainly rely on hacking the mobile environment, they focus mainly on taking advantage of the limited (or lack of) security in mobile devices. Several detection schemes are available for desktop and wired computers but they are not readily available across wireless and mobile devices. This is due to the limitations of mobile devices, namely power, processing, and storage. Implementing traditional ant phishing solutions, such as machine learning approaches, in a mobile environment is inapplicable since some of these solutions are heavy in nature. Ant phishing solutions in a mobile environment should take advantage of the high predictive accuracy of machine learning approaches and reduce high overhead associated with such approaches by building a distributed client-server framework.

CONCLUSION

In this, we went through the issue which is presented in the mobile phishing detection. Here We proposed MobiFish. MobiFish is a novel automated phishing defense scheme for the mobile platform. Hence we identified the inefficiency of the heuristics-based anti-phishing schemes which highly rely on the HTML source code of web pages. Recent years have witnessed the incrementing threat of phishing attacks on mobile computing platforms. In this project we proposed MobiFish, a novel automated phishing bulwark scheme for mobile phones with android OS. In this work first we identified the impuissances of the heuristics-predicated anti-phishing schemes that highly rely on the HTML source code of web pages. MobiFish solve this issue by utilizing OCR that accurately extracts text from the screenshot of the authenticate page so that the claimed identity can be verified. Compared to subsisting anti-phishing schemes, MobiFish is lightweight as it works without utilizing external search engines or machine learning techniques. MobiFish can additionally detect account phishing attacks. We are implementing MobiFish on Smartphone running the Android OS. MobiFish resolves these issue by using OCR.OCR (optical character recognition) where we can accurately extract the text from the screenshot of the login interface. From this, we can be verified with the claimed identity. When we compare the MobiFish with the Existing OCR based anti-phishing schemes, MobiFish is considered to be the lightweight because it works with by using external search engines or the machine learning algorithms. MobiFish can also detect the app phishing attacks and account phishing attacks. In research perspective, the plan is to implement this proposed approach on Smartphones running with different Android OS. Further complexity of this proposed scheme needs to be analyzed with respect to performance and efficiency.

REFERENCES

Anti-Phishing Working Group. (2006). *Phishing activity trends report*. http://www.antiphishing.org/reports/apwg

Bergholz, A., De Beer, J., Glahn, S., Moens, M., Paaß, G., & Strobel, S. (2010). New filtering approaches for a phishing email. *Journal of Computer Security*, *18*(1), 7–35. doi:10.3233/JCS-2010-0371

Cranor, L. F., Egelman, S., Hong, J. I., & Zhang, Y. (2007). PhindingPhish: Evaluating anti-phishing tools. *Proceedings of The 14th Annual Network and Distributed System Security Symposium (NDSS).*

Felt, A.P., & Wagner, D. (2011) Phishing on mobile devices. *Proceedings of W2SP'11: WEB 2.0 Security and Privacy.*

Karlof, C., Tygar, J. D., & Wagner, D. (2009). Conditioned-safe ceremonies and a user study of an application to web authentication. *Proceedings of the 5th Symposium on Usable Privacy and Security (SOUPS).* 10.1145/1572532.1572578

Niu, Y., Hsu, F., & Chen, H. (2008). iphish: phishing vulnerabilities on consumer electronics. *Proceedings of the 1st Conference on Usability, Psychology, and Security.*

Trend Micro. (2012). *Mobile phishing: A problem on the horizon.* Author.

Wu, L., Du, X., & Wu, J. (2016). Effective Defense Schemes for Phishing Attacks on Mobile Computing Platforms. *IEEE Transactions on Vehicular Technology, 65*(8), 6678–6691. doi:10.1109/TVT.2015.2472993

KEY TERMS AND DEFINITIONS

MWL: A whitelist is a list of e-mail addresses or domain names from which an electronic-mail blocking program will sanction messages to be received.

OCR: The fundamental process of OCR involves examining the text of a document and translating the characters into code that can be utilized for data processing.

SLD: A second-level domain is a categorical part of a website, page domain name or URL address that complements a top-level domain.

SMS: It is the most rudimental communications technology for mobile data transfer and is characterized by the exchange of short alphanumeric text messages between digital line and mobile contrivances.

URL: A uniform resource locator (URL) is the address of a resource on the Internet. A URL denotes the location of a resource as well as the protocol used to access it.

This research was previously published in the Encyclopedia of Organizational Knowledge, Administration, and Technology; pages 1228-1246, copyright year 2021 by Business Science Reference (an imprint of IGI Global).

Chapter 14
A Novel Approach to Detect Spam and Smishing SMS using Machine Learning Techniques

Ankit Kumar Jain
https://orcid.org/0000-0002-9482-6991
National Institute of Technology, Kurukshetra, India

Sumit Kumar Yadav
Income Tax Department, Government of India, India

Neelam Choudhary
National Institute of Technology, Kurukshetra, India

ABSTRACT

Smishing attack is generally performed by sending a fake short message service (SMS) that contains a link of the malicious webpage or application. Smishing messages are the subclass of spam SMS and these are more harmful compared to spam messages. There are various solutions available to detect the spam messages. However, no existing solution, filters the smishing message from the spam message. Therefore, this article presents a novel method to filter smishing message from spam message. The proposed approach is divided into two phases. The first phase filters the spam messages and ham messages. The second phase filters smishing messages from spam messages. The performance of the proposed method is evaluated on various machine learning classifiers using the dataset of ham and spam messages. The simulation results indicate that the proposed approach can detect spam messages with the accuracy of 94.9% and it can filter smishing messages with the accuracy of 96% on neural network classifier.

DOI: 10.4018/978-1-7998-8545-0.ch014

1. INTRODUCTION

1.1. Contextualization

Smishing is a cyber-security attack in which the mobile user is deceived into installing malicious software into their mobile phone. Smishing word is constructed by the combination of two words i.e. SMS + Phishing = Smishing. In other words, smishing message is a harmful spam message which aims at stealing mobile users' sensitive data (Goel & Jain, 2018). These messages may contain a link. On following the link, the user is asked to enter their details for verification purpose (Choudhary & Jain, 2017). Attackers also lure victims by sending messages that look like they originate from an authentic bank, stating that their account has been locked and to unlock the account, the victim is asked to follow the link in the message. The purpose of spam messages is generally to promote some product or to disturb the user with useless messages. On the other hand, phishing message always have some criminal intent. The effect of smishing attack is financial loss and identity theft.

1.2. Importance/Relevance of the Theme

Smishing attacks are increasing day by day as attackers find SMS, a cheaper and more convenient way to communicate with victims ("Why do phishing attacks work better on mobile phones," 2011). Smartphone users are considered to be three times more likely to fall victims of phishing attacks than desktop users (MOBILE THREAT REPORT, 2012). In 2014, cloud mark report (Hacked Hotel Phones Fueled Bank Phishing Scams, 2015) identified a phishing SMS, which attempted to steal user's secret information like credit card number, bank account details, etc. A smishing message was sent to thousands of people with different bank affiliations (SMiShing & Vishing News, 2017).

1.3. Research Question

There are various security measures available to control SMS Spam problem, but these are not so mature. Many Android apps (Spam Blocker App; Mr. Number - Caller ID & Spam Protection) are also available on play store. However, their detection accuracy is not up to the mark. Moreover, existing text filtering techniques mainly focus on email spam (Diale, Celik, & Walt, 2019). However, with the popularity of mobile devices, SMS spam and smishing is the one of the major issues these days. Existing approaches mainly focus only on SMS Spam detection. However, there is no efficient technique developed which can efficiently filter out smishing text messages from SMS Spam.

1.4. Objectives

This paper presents a novel machine learning based approach to detect the spam and smishing messages. The paper used effective feature set for detection of spam and smishing messages. The proposed approach is divided into two phases. The first phase distinguishes the spam messages from ham messages using eleven basic features. The second phase filters smishing message from the spam message using four phishing features. The performance of our proposed method is evaluated on various machine learning classifiers using the dataset of ham and spam messages. Followings are the major contributions of the proposed approach:

- Implementation of a classifier for spam and smishing messages.
- Identification of four outstanding phishing features suitable for mobile smishing detection.
- Detection of zero day mobile phishing attacks using the proposed features.
- Utilization of information gain values to get the best features for SMS spam and smishing detection.

1.5. Structure of the Paper

The rest of the paper is organized as follows. The related work is presented in Section 2. The system architecture of our proposed scheme is described in Section 3. The experimental details including various experimental results are presented in Section 4. Finally, the conclusion and future work is presented in Section 5.

2. RELATED WORK

These days various tools and techniques are developed to ensure safety and security in the mobile devices. However, problem of security still arises. Smishing attacks are rapidly growing on mobile devices (Goel & Jain, 2017).

A mechanism to normalize and expand short and noisy messages is given by Almeida et al. (2011). It is a novel approach that can highly improve the quality of the text messages before classification, due to which better attributes are obtained in turn improving the classification accuracy. Two dictionaries - lexicographic and semantic dictionaries are used. It has been seen that text processing removes the basic text representation problems, improving the classification accuracy.

To filter spam SMS text messages, Ma et al. (2016) proposed a message topic model (MTM). MTM is based on probability topic model. To appropriately represent SMS message, pre-processing is used along with some background and symbol term. K-means algorithm is used to avoid sparse problem. They clustered spam text messages into rough classes and then accumulated spam messages in single file to identify word co-occurrence patterns.

Joo et al. (2017) used Naïve Bayes classification algorithm to improve the efficiency of the smishing detection scheme. They have statistical learning method to filter simple text messages and smishing text messages. They have used morphological analyzer to extract noun words and Bayesian classifier to represent the uncertainties related to the models and parameter values.

Chan et al. (2015) presented two methods for SMS Spam filtering, namely, feature reweighting method and good word attack. Both approaches focus on the length of the message along with considering the weight of the message.

Etaiwi and Awajan (2017) investigated the comparison of two features implementing them separately on the same database. The two features used in this paper include a bag of words and word count, which are compared using various algorithms like SVM, random forests, etc. However, this approach achieved only 80-90% detection accuracy in the case of a bag of words and 60-70% for word count.

Yadav et al. (2011) developed an application "SMSAssassin" which can be used over mobile phones for filtering spam messages in real time. Naïve Bayesian Classifier and SVM are two classifiers used to classify the messages as ham or spam. Crowd-sourcing is used to keep track of new patterns and keywords, and to update the list of spam patterns and keywords. Authors have used two lists - GlobalSpamKeywordsFreq and SpamKeywordsFreq list at the server and the mobile phones, respectively to keep track of spam keywords. There is a UserPreferencesList in the application where a user can specify ham and spam keywords of their choice.

Jain and Gupta (2018) proposed a rule-based framework for detection and prevention of smishing messages. The authors have identified nine rules that effectively differentiate smishing messages from ham messages. These rules are extracted from the content of the message. The scheme was analyzed on three different classification algorithms – decision tree, PRISM and RIPPER. RIPPER gave best results as compared to other algorithms. Additionally, the scheme is able to detect zero hour smishing attacks.

El-Alfy and AlHasan (2016) proposed a text based message filtering mechanism for SMS as well emails. They have used 11 features for the detection of malicious messages and evaluated their approach on five standard SMS and email dataset.

Silva et al. (2017) proposed a technique for detecting spam text messages on the basis of Minimum Description Length (MDL) principle. MDL based technique provides effective and efficient message classification approach. This technique gives low computational cost even with large amount of data.

Ali and Maqsood (2018) developed a NET library written in C# which provides a cross-platform solution for spam detection. They have discussed some algorithms and have implemented Random Forests. The .NET library developed can run on any platform as per requirements. They have trained the C# library efficiently resulting in a good accuracy.

Idris et al. (2015) proposed an email spam detection algorithm that uses particle swarm optimization algorithm to detect spam messages. The designed system is an improvement of random detector generation in negative selection algorithm. The accuracy achieved by the system is 83.20%, while, for negative selection algorithm, the accuracy achieved is 68.86%.

3. PROPOSED APPROACH

Smishing message is a harmful spam message that steals personal credentials. As per our observation, our research work finds the followings characteristics of a smishing message:

1. Contains the malicious fake links or mobile number
2. Advertises something like providing free minutes, etc.
3. Asks the user to subscribe to some service.
4. Announces, the user as a winner of some fake contest and luring him using the prize money.
5. Intends to spread some fake news.

Therefore, this article considers these characteristics while making the proposed feature set. It is difficult to detect smishing message, therefore, we have analyzed fake messages in depth to filter out smishing messages from spam messages. Our scheme is able to effectively detect spam as well as smishing messages with high accuracy. The proposed approach first filters out the spam messages and then filters the smishing messages from these spam messages with the help of identified features. The architecture and working of the proposed framework is discussed below:

Figure 1. Architecture of proposed scheme

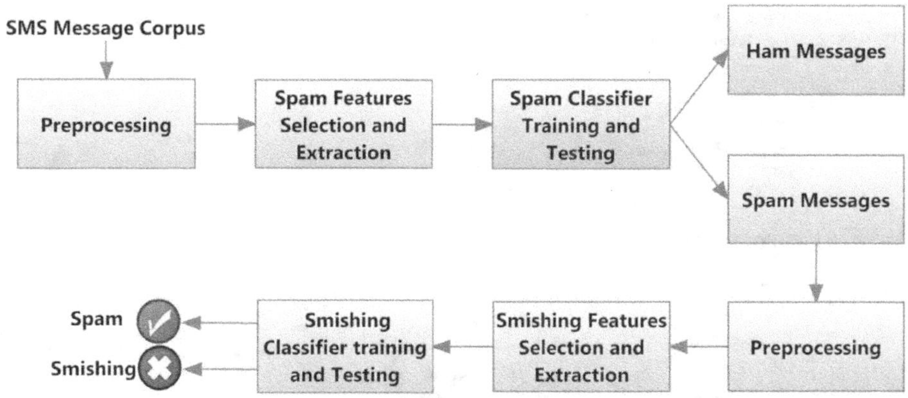

3.1. System Architecture

Figure 1 shows the architecture of proposed scheme. The first phase of our method extracts spam features from the SMS. After filtering spam SMS, the second phase classifies smishing messages. Detailed flow chart of proposed approach is presented in figure 2.

Phase 1: Filtering Spam messages - In the first phase, each incoming message is classified into spam and ham categories. The mechanism of filtering spam messages is discussed as follows:-

1. Preprocessing - Preprocessing the entire SMS Spam Corpus is done to remove the redundancy. The SMS Spam Corpus is stored and handled using Microsoft Excel to clean the data from noise.
2. Spam Features Selection and Extraction – The proposed approach uses 11 basic features to filter out spam messages. A feature vector value 1 is set for spam messages and 0 for ham messages.
3. Classification Training- The labeled dataset is used to train the machine learning classifier. Further, the cross validation set improves the parameters, and evaluates the performance of the method.

Phase 2: Filtering Smishing messages – Once the approach classifies ham and spam messages, the next phase is used for filtering the smishing SMS from spam ones. The mechanism of filtering smishing messages from spam messages is discussed as follows:-

1. Preprocessing - This phase labels the message as spam or smishing. We select smishing messages manually by reading all spam messages as no benchmark dataset is available for smishing messages. Afterwards, a feature vector value 1 is set for smishing messages and 0 is set for spam messages.
2. Smishing Features Selection and Extraction- In this, our approach selects four outstanding features based on Smishing characteristics in order to filter out Smishing messages from spam messages.
3. Smishing Classifier Training and Testing- The trained classifier separates the message into two classes namely, spam and smishing. The rest of the process is similar for all the stages. In this, a message is checked for smishing. First, the URL is checked for the message. If the URL is present, it is a phishing message otherwise, the second feature is checked to determine whether award token is present or not and so on. After filtering all our features, if a message is smishing, then a warning message is shown to the user otherwise the message is declared as spam.

Figure 2. Flowchart of proposed scheme

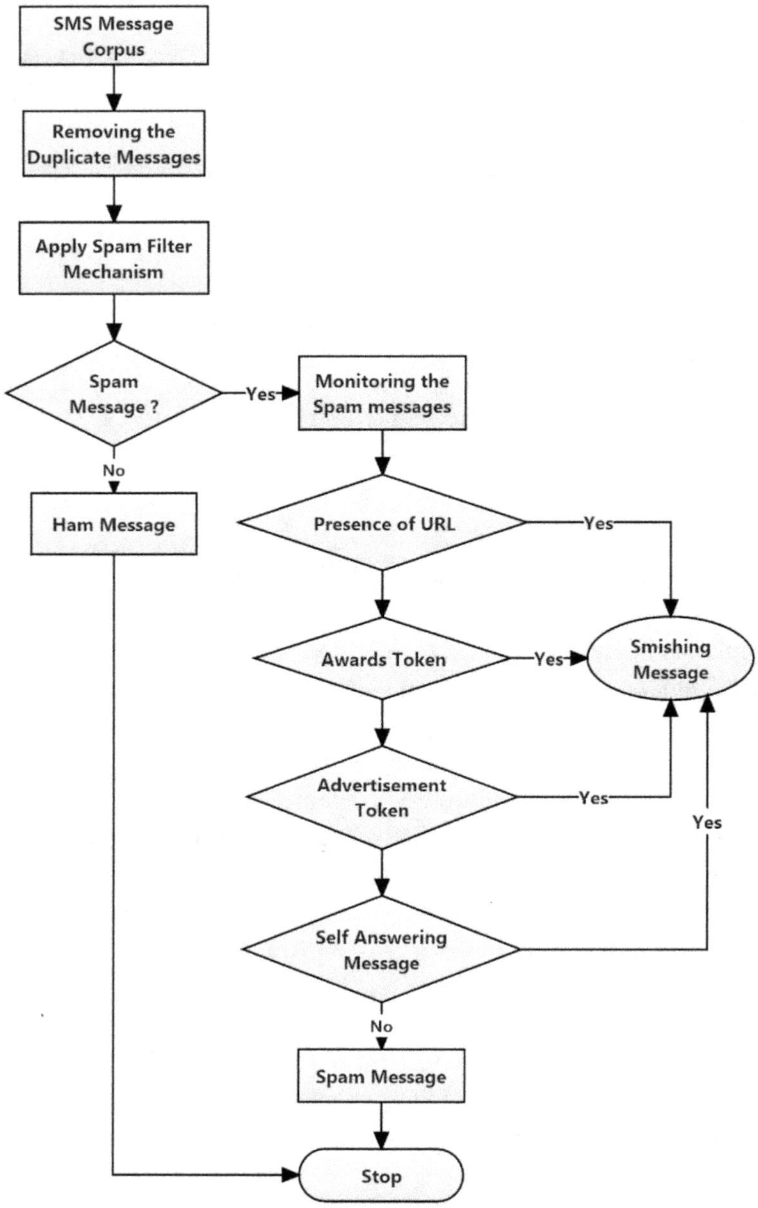

3.2. Feature Selection

Feature selection is a critical task for our approach. It does not require sophisticated analysis to avoid the delay in message service. Moreover, to increase the spam and smishing detection, accuracy features extraction should be highly correlated with the message type. In our proposed approach, the feature set is divided into two categories i.e. Basic Features and Phishing Features. The diagrammatic view of proposed features is presented in figure 3. These features are discussed in detail below:-

Figure 3. Proposed feature set

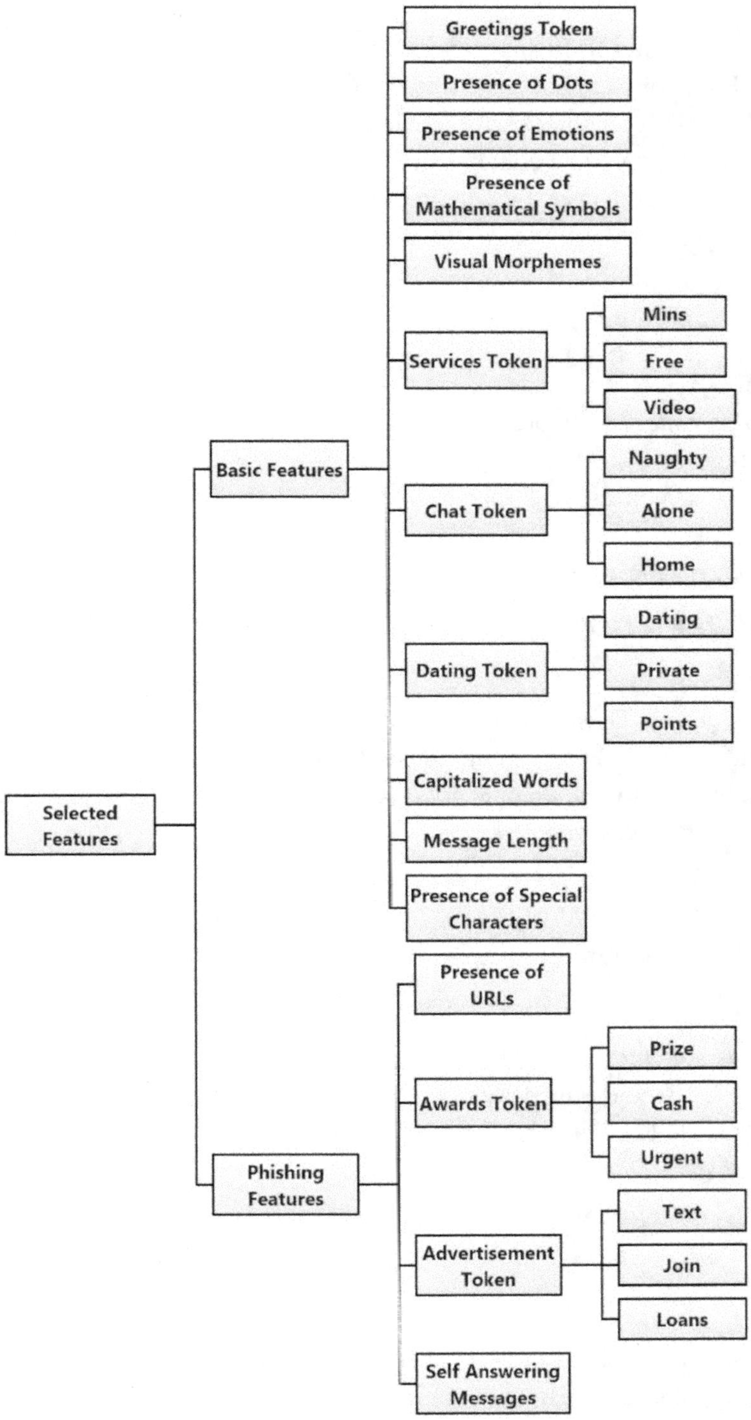

3.2.1. Basic Features

These features are used to filter SMS spam from message corpus. Our scheme used a set of 11 features $\{F_1, F_2, F_3, \ldots, F_{11}\}$. Following are the basic feature set:

F_1: *Greetings Token*- This feature checks for the presence of greeting keywords like Good Morning, Good Night, Dear, etc. If a message contains greetings token, then it falls under the category of ham message. 3% messages in our dataset contain greeting tokens.

$$F_1 = \text{If} \begin{cases} message\, contains\, any\, greeting\, token \rightarrow Ham \\ otherwise \rightarrow Spam \end{cases} \tag{1}$$

F_2: *Presence of dot symbol*- The presence of dot symbol is a good sign for ham messages as users use dots while chatting over social networking sites or on various chatting apps. Moreover, users often use dots to separate the sentences.

$$F_2 = \text{If} \begin{cases} message\, contains\, dot\, symbol \rightarrow Ham \\ otherwise \rightarrow spam \end{cases} \tag{2}$$

F_3: *Presence of emotions*- These symbols are generally used while chatting. Emotions can be of two types i.e. positive and negative. Positive emotions like love, nice, sweet, happy, etc. and negative emotions like sad, angry, hurt, nasty, etc. People uses these emotions in the form of symbols to express their feelings or views about the message.

$$F_3 = \text{If} \begin{cases} message\, contains\, positive\, or\, negative\, emotions \rightarrow Ham \\ otherwise \rightarrow spam \end{cases} \tag{3}$$

F_4: *Presence of special character* - The special characters which our approach checks are !, $, &, # and ~. The character "$" denotes currency in the fake award SMS, and character "!" shows special attention of user like "CONGRATULATIONS!", "WINNER!", etc.

$$F_4 = \text{If} \begin{cases} message\, contains\, special\, characters \rightarrow spam \\ otherwise \rightarrow Ham \end{cases} \tag{4}$$

F_5: *Presence of Mathematical Symbols*- Mathematical symbols are not used in genuine text messages because people use simple language to communicate rather than mathematical symbols. In this feature, the approach checks whether the SMS contains +,-, /, ^ and % mathematical symbols.

$$F_5 = \text{If} \begin{cases} message\, contains\, mathematical\, symbol \rightarrow spam \\ otherwise \rightarrow Ham \end{cases} \tag{5}$$

F_6: *Services Token*- The presence of suspicious keywords like text, join, loans are considered to be spam keywords. These keywords are very common in spam messages. Spammers lure users by using these keywords, asking them to join their services or obtain loans through them. After reviewing our dataset, the approach was able to find that 5% of the messages contains services token.

$$F_6 = \text{If} \begin{cases} message\, contains\, services\, token \rightarrow spam \\ otherwise \rightarrow Ham \end{cases} \tag{6}$$

F_7: *Visual Morphemes*- The visual morphemes like xx, xxx, xxxx are considered to be spam keywords. In our dataset, only 2% of messages contain the visual morphemes tokens.

$$F_7 = \text{If} \begin{cases} message\, contains\, visual\, morphemes \rightarrow spam \\ otherwise \rightarrow Ham \end{cases} \tag{7}$$

F_8: *Message length*- In most of the countries, text messages contain a maximum of 160 characters only. Chatting SMS are generally small as compared to spam message. This feature results in spam if the message length is greater than 160 characters.

$$F_8 = \text{If} \begin{cases} message\, length \geq 160 \rightarrow spam \\ otherwise \rightarrow Ham \end{cases} \tag{8}$$

F_9: *Chat Token*- The presence of suspicious keywords like naughty, alone, and home are considered to be spam keywords. These kinds of words are not present in normal genuine messages.

$$F_9 = \text{If} \begin{cases} message\, contains\, chat\, token \rightarrow spam \\ otherwise \rightarrow Ham \end{cases} \tag{9}$$

F_{10}: *Dating Token*- The presence of suspicious keywords like dating, points and private are considered to be spam keywords. After reviewing our dataset, only 1% of spam messages contains dating token.

$$F_{10} = \text{If} \begin{cases} message\, contains\, dating\, token \rightarrow spam \\ otherwise \rightarrow Ham \end{cases} \tag{10}$$

F_{11}: *Capitalized Words*- The spammer generally uses uppercase keywords to pursue mobile user's attention.

$$F_{11} = \text{If} \begin{cases} message\, uses\, uppercase\, word \rightarrow spam \\ otherwise \rightarrow Ham \end{cases} \tag{11}$$

3.2.2. Smishing Features

These features are used to filter smishing messages from spam messages. Proposed scheme used four smishing features (F_{12} - F_{15}). These features are as follows:

F_{12}: Presence of URL- Approach considers the presence of URL in the message as a smishing feature, since phishers may ask the user to click on a link to visit a website for winning a prize, fake update or to provide some personal information on the fake webpage. After reviewing our dataset, 14% smishing messages contain URL.

$$F_{12} = \text{If} \begin{cases} URL\ is\ present\ in\ the\ S \rightarrow Smishing \\ otherwise \rightarrow Spam \end{cases} \tag{12}$$

F_{13}: Awards token- The token like prize, cash and urgent are used to lure users in terms of cash prize or award. In our dataset, 21% of smishing messages contains awards keywords.

$$F_{13} = \text{If} \begin{cases} message\ contains\ awards\ token \rightarrow Smishing \\ otherwise \rightarrow Spam \end{cases} \tag{13}$$

F_{14}: Advertisement Token- The presence of suspicious keywords like mins, free and video are considered to be smishing keywords. Phishers use these keywords to attract a user by advertising some product or service for free. In our dataset 21% of smishing messages contains advertisement token.

$$F_{14} = \text{If} \begin{cases} message\ contains\ advertisement\ token \rightarrow Smishing \\ otherwise \rightarrow Spam \end{cases} \tag{14}$$

F_{15}: Self Answering SMS- The existence of self-answering message like asking user to subscribe to any facility falls under the category of smishing message.

$$F_{15} = \text{If} \begin{cases} message\ is\ self\ answered \rightarrow Smishing \\ otherwise \rightarrow Spam \end{cases} \tag{15}$$

4. IMPLEMENTATION DETAILS AND EXPERIMENTAL RESULTS

4.1. Performance Metrics

The evaluation measures set for our experiment are true positive rate (TPR), true negative rate (TNR), false positive rate (FPR), false negative rate (FNR), precision, recall, accuracy, and receiver operating characteristics (ROC) area (Jain & Gupta, 2019). The way to calculate these values can be explained through a confusion matrix as displayed in Table 1.

Table 1. Confusion Matrix

	True Results	
	Positive Classification	**Negative Classification**
Prediction		
Positive	True Positive Rate	False Positive Rate
Negative	False Negative Rate	True Negative Rate

4.2. Dataset Collection

The proposed approach uses SMS dataset v.1 from SMS Spam research work (Almeida, Hidalgo, & Yamakami, 2011). This dataset contains 5574 text messages in the English language. This corpus has been collected from following sources:-

1. A set of 425 spam SMS extracted from the Grumbletext Website.
2. A list of 450 ham SMS collected from Caroline Tag's PhD Thesis.
3. A subset of 3,375 ham SMS collected from NUS SMS Corpus.
4. A set of 1,002 ham and 322 spam messages are included from the SMS Spam Corpus v.0.1 Big.

Moreover, we have also collected 71 unique smishing message images from pinterest.com and converted these images into text and added them into the dataset.

4.3. Classification Algorithms

Various classification algorithms are used in our approach to find out best prediction accuracy. These classification algorithms are discussed as follows:

1. Naïve Bayes (NB): It is a simple probabilistic classifier based on the Bayes theorem that uses independent feature model along with a decision rule.
2. Neural Network (NN): It is a feed forward neural network which is used to map the input dataset to an output dataset. It uses back propagation which is a supervised learning technique. There are multiple layers of hidden nodes with a nonlinear activation function.
3. Logistic Regression (LR): It assumes a logistic and a normal distribution function, it finds the relationship of the dependent variable with respect to one or more independent variables.

4.4. Experiment 1: Filter Spam messages from Ham messages

There are six classification algorithms used for spam classification, namely NB, LR, NN, DT, J48 and RF. Proposed work used 10-fold cross-validation in the experiment. Table 2 shows the outcomes of our scheme for filtering spam messages on various classification algorithms. From Table 2, we find that Neural Network gives best results in filtering of spam messages.

Table 2. Classification Results for spam classification

Algorithm	TPR	FPR	Precision	Recall	ROC	Accuracy
Naïve Bayes (NB)	0.979	0.292	0.959	0.979	0.961	0.945
Logistic Regression (LR)	0.982	0.300	0.958	0.982	0.957	0.946
Neural Network (NN)	0.985	0.294	0.959	0.985	0.953	0.949
Decision Table (DT)	0.988	0.351	0.951	0.988	0.945	0.946
J48	0.989	0.335	0.953	0.989	0.916	0.948
Random Forest (RF)	0.977	0.288	0.959	0.977	0.929	0.947

As our approach gets the best results for Neural Network, so we have analyzed this classification algorithm, for various cross folds. Table 3 presents the numeric values of various folds for Neural Network and Figure 4 shows the performance results.

Table 3. Performance Results for Neural Network

Neural Network	TPR	FPR	Precision	Recall	ROC	Accuracy
5:5	0.988	0.315	0.956	0.988	0.955	0.950
4:6	0.988	0.315	0.956	0.988	0.958	0.949
3:7	0.990	0.309	0.957	0.990	0.955	0.952
2:8	0.989	0.315	0.956	0.989	0.955	0.951
1:9	0.985	0.294	0.959	0.985	0.953	0.949

Figure 4. Neural Network result for various folds

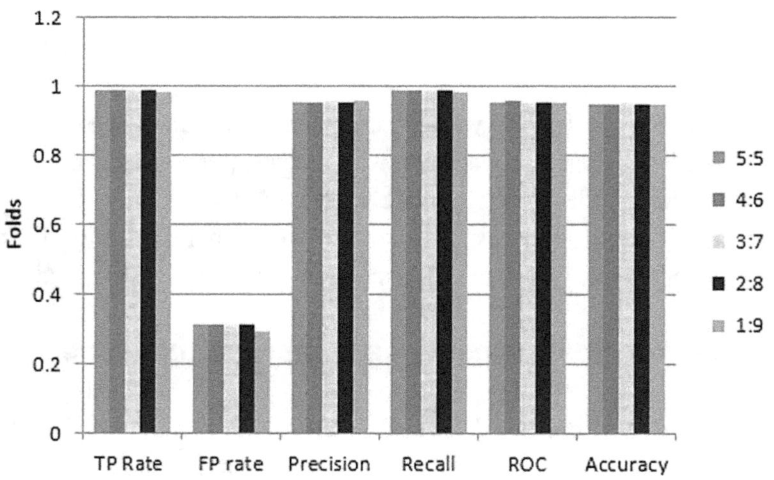

The histograms of the basic feature set are shown in Figure 5. Each feature of histogram shows every possible value of each feature instance. The feature value is divided into two types - spam messages are colored as red and ham messages are colored as blue. The overlapping between feature values can be easily noticed. This indicates that it is not possible to differentiate between spam and ham messages using a single feature. Thus, each feature has its own importance in detecting spam and ham messages.

Figure 5. Histograms of spam features

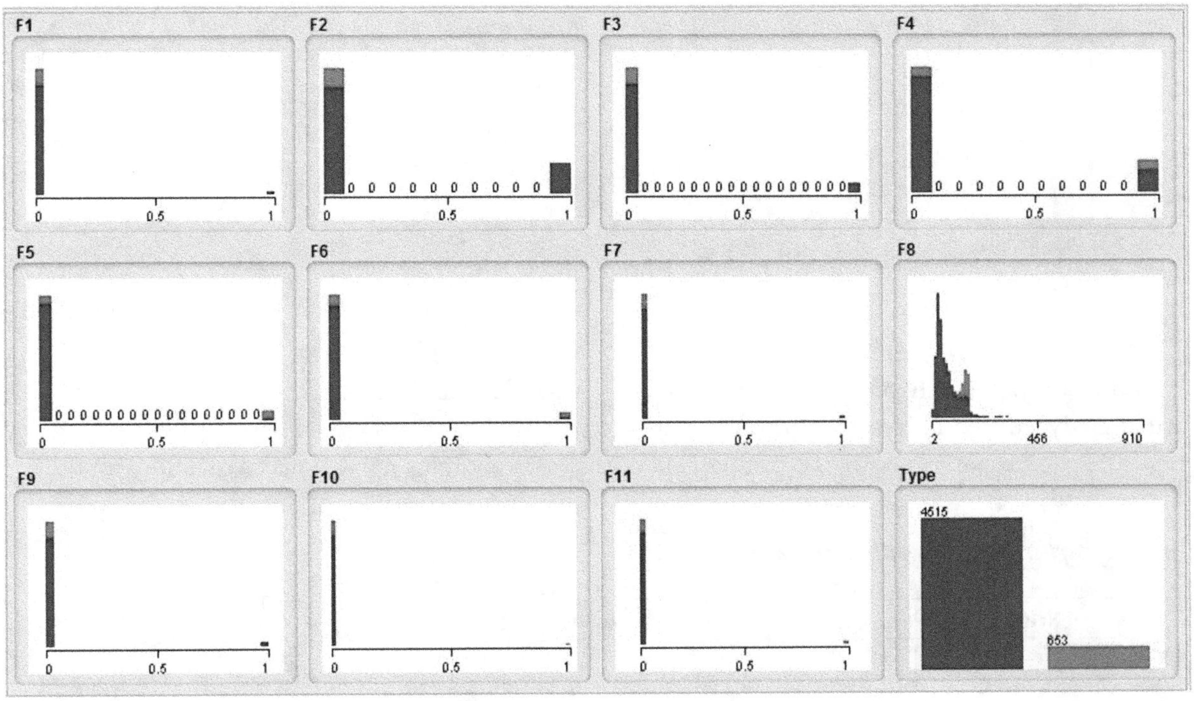

4.5. Experiment 2: Filtering Smishing Messages

In this experiment, proposed approach filters smishing messages from spam messages. Table 4 presents classification results of various machine learning classifiers for filtering smishing messages.

Table 4. Classification results for smishing messages classification

Algorithm	TPR	FPR	Precision	Recall	ROC	Accuracy
Naïve Bayes	0.923	0.010	0.994	0.923	0.960	0.950
Logistic	0.910	0.035	0.989	0.910	0.945	0.946
Neural Network	0.932	0.007	0.994	0.932	0.958	0.960
Decision Table	0.922	0.013	0.990	0.922	0.953	0.952
J48	0.922	0.013	0.990	0.922	0.953	0.952
Random Forest	0.925	0.015	0.991	0.925	0.954	0.958

Figure 6 shows the ROC curve for Neural Network where x-axis shows the FPR and y-axis presents the TPR for the class type. Here, class type differentiates smishing from spam messages.

Figure 6. ROC curve for Neural Network classification algorithm

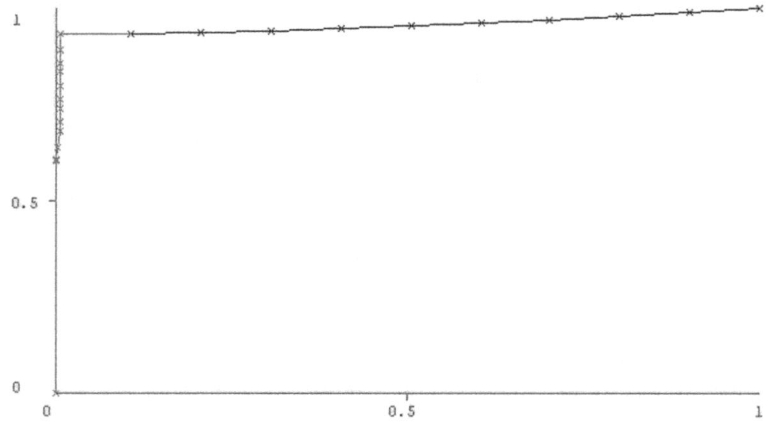

In this, our approach gets the best results for Neural Network with an accuracy of 96%. We have further analyzed Neural Network for various cross folds. Table 5 shows the numeric results for different cross folds and Figure 7 shows the comparison of the different rates for various cross folds.

Table 5. Performance Results for Neural Network

Neural Network	TPR	FPR	Precision	Recall	ROC	Accuracy
5:5	0.988	0.315	0.956	0.988	0.955	0.950
4:6	0.988	0.315	0.956	0.988	0.958	0.949
3:7	0.990	0.309	0.957	0.990	0.955	0.952
2:8	0.932	0.007	0.994	0.932	0.958	0.960
1:9	0.932	0.007	0.994	0.932	0.958	0.960

The importance of three main smishing features for smishing message classification is shown in Figure 8. After extracting features we get the list of spam, ham and phishing messages.

4.6. Importance of Features

Information gain helps in analyzing the quality of each feature and removing redundant features, thus, helping in selecting the best feature. We run information gain feature selection algorithm to choose the ten important features for classification of ham and spam messages and three important features for smishing and spam classification. In Table 6, we present top ten features for filtering spam messages with their respective information gain values and Table 7 shows the top three features for filtering smishing messages with their IG values.

Figure 7. Smishing detection result on various folds for neural network

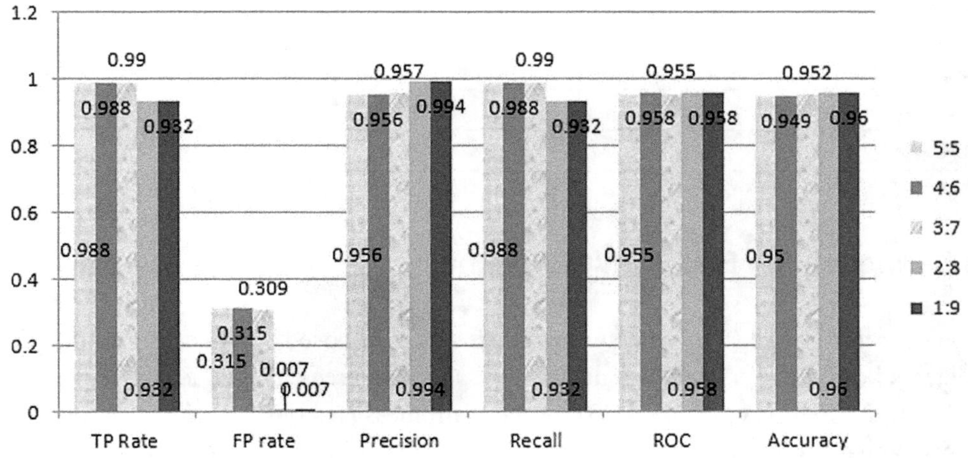

Figure 8. Importance of smishing feature set for smishing message classification

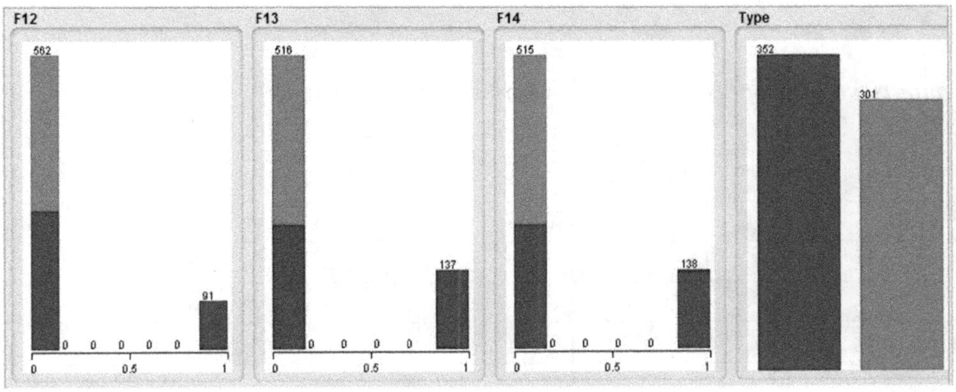

Table 6. Feature set for filtering spam messages

Feature Name	IG Value
Message length	0.20866
Presence of Mathematical Symbols	0.16135
Services Token	0.08021
Capitalized Words	0.05418
Presence of special characters	0.04318
Presence of Dots	0.2786
Dating Token	0.01531
Presence of Emotions	0.01505
Visual Morphemes	0.00578
Chat Token	0.00322

Table 7. Feature set for filtering phishing messages

Feature	IG Value
Awards Token	0.2213
Advertisements Token	0.1987
Presence of URL	0.1381

4.7. Comparisons with Existing Approaches

Table 8 shows a comparative analysis between existing and proposed approach for mobile devices. Our approach detects both spam as well as smishing messages whereas techniques by Yadav et al. (2011), Delany et al. (2012), Eshmawi et al. (2013) detect spam messages only. We have employed feature re-weighting mechanism which is not used by Bottazzi et al. (2015), Delany et al. (2012). Although we have used various algorithms for spam and smishing detection but the maximum performance is achieved by neural network algorithm. Accuracy achieved by our scheme is 96% which is comparatively more than the accuracy achieved by other techniques.

Table 8. Comparison of proposed approach with existing approaches

Author	Accuracy	Spam Detection	Smishing Detection	Feature Reweighting	Classification Algorithm
Bottazzi et al. (2015)	89.2%	√	√	×	J48, SMO, BayesNet, IBK, SGD
Yadav et al. (2011)	86%	√	×	√	Bayesian, SVM
Delany et al. (2012)	93.31%	√	×	×	SVM
Eshmawi and Nai (2013)	88%	√	×	√	SVM, RF NB, CART
Idris et al. (2015)	83.20%	√	×	×	PSO
Etaiwi and Awajan (2017)	87.30	√	×	√	SVM, NB, RF, DT
Ma et al. (2016)	96%	√	×	×	SVM
Proposed Approach	96%	√	√	√	Neural Network

4.8. Limitations of Approach

Following are the few limitations of our proposed approach:

- The feature set of our phishing detection approach depends entirely on the text content of the message, and it can detect the smishing message only if it is written in English language.
- Nowadays, various malicious applications can steal the user's data, and send it to the phishers. Consequently, our approach is not capable enough to detect such malicious applications.

5. CONCLUSION AND FUTURE WORK

5.1. Conclusion

This paper presented a scheme to classify spam and smishing messages efficiently. The main contribution of this article is the filtering of smishing messages from spam messages, which has never been considered before. The existing approaches either detect only spam messages or smishing messages but our proposed approach is capable of detecting both spam and smishing messages. Moreover, this research work identified various new features for detection of spam and smishing messages. We have experimentally determined the importance of proposed features using the Information gain. Our scheme can also identify the zero-day smishing attack. Proposed method trains the neural network classifier and accurately filters spam and smishing messages. Our scheme is able to detect spam messages with an accuracy of 94.9% and smishing messages with an accuracy of 96%. The proposed work can be used as a mobile application, which blocks or shows a warning message whenever user receives the malicious messages.

5.2. Future Work

Several approaches have been proposed in order to protect users from SMS phishing attacks, but still, the threat is not elevated and demands more attention towards the improvement of defense solutions. Future work that can be done in the proposed mechanism includes –

- Analysis of the URL present in the message, so as to determine if this URL redirects the user to a malicious login page or leads to the download of some malicious application.
- The performance of our scheme may be improved by integrating various more features, which effectively detect spam and smishing messages, in turn, improving the accuracy of the scheme.
- Size of the dataset can be increased in order to enhance the richness of the scheme.

REFERENCES

Ali, S. S., & Maqsood, J. (2018). Net library for SMS spam detection using machine learning: A cross platform solution. In *Proceedings of the 2018 15th International Bhurban Conference on Applied Sciences and Technology (IBCAST)* (pp. 470-476). IEEE.

Almeida, T. A., Hidalgo, J. M., & Yamakami, A. (2011). Contributions to the study of SMS spam filtering: new collection and results. In *Proceedings of the 11th ACM symposium on Document engineering* (pp. 259-262). Mountain View, CA: ACM. 10.1145/2034691.2034742

Bottazzi, G., Casalicchio, E., Cingolani, D., Marturana, F., & Piu, M. (2015). MP-Shield: A Framework for Phishing Detection in Mobile Devices. In *Proceeding of IEEE International Conference on Computer and Information Technology; Ubiquitous Computing and Communications; Dependable, Autonomic and Secure Computing; Pervasive Intelligence and Computing*, Liverpool, UK (pp. 1977-1983). Academic Press. 10.1109/CIT/IUCC/DASC/PICOM.2015.293

Chan, P. P., Yang, C., Yeung, D. S., & Ng, W. W. (2015). Spam filtering for short messages in adversarial environment. *Neurocomputing*, *155*, 167–176. doi:10.1016/j.neucom.2014.12.034

Choudhary, N., & Jain, A. K. (2017). Comparative Analysis of Mobile Phishing Detection and Prevention Approaches. In *Proceedings of International Conference on Information and Communication Technology for Intelligent Systems,* Ahmedabad, India (pp. 349-356). Springer.

Delany, S. J., Buckley, M., & Greene, D. (2012). SMS spam filtering: Methods and data. *Expert Systems with Applications*, *39*(10), 9899–9908. doi:10.1016/j.eswa.2012.02.053

Diale, M., Celik, T., & Walt, C. V. (2019). Unsupervised feature learning for spam email filtering. *Computers & Electrical Engineering*, *74*, 89–104. doi:10.1016/j.compeleceng.2019.01.004

El-Alfy, E. S., & AlHasan, A. A. (2016). Spam filtering framework for multimodal mobile communication based on dendritic cell algorithm. *Future Generation Computer Systems*, *64*, 98–107. doi:10.1016/j.future.2016.02.018

Eshmawi, A., & Nai, S. (2013). Feature reduction for optimum sms spam filtering using domain knowledge. In *Proceedings of the International Conference on Security and Management (SAM)*. Academic Press.

Etaiwi, W., & Awajan, A. (2017). The Effects of Features Selection Methods on Spam Review Detection Performance. In *Proceedings of the 2017 International Conference on New Trends in Computing Sciences (ICTCS),* Amman (pp. 116-120). IEEE. 10.1109/ICTCS.2017.50

Goel, D., & Jain, A. K. (2017). Smishing-Classifier: A Novel Framework for Detection of Smishing Attack in Mobile Environment. In *Proceedings of the International Conference on Next Generation Computing Technologies,* Dehradun (pp. 502-512). Springer.

Goel, D., & Jain, A. K. (2018). Mobile phishing attacks and defence mechanisms: State of art and open research challenges. *Computers & Security*, *73*, 519–544. doi:10.1016/j.cose.2017.12.006

Hacked Hotel Phones Fueled Bank Phishing Scams. (2015, Feb 4). Krebson Security. Retrieved from https://krebsonsecurity.com/2015/02/hacked-hotel-phones-fueled-bank-phishing-scams/

Idris, I., Selama, A., Nguyen, N. T., Omatu, S., Krejcar, O., Kuca, K., & Penhaker, M. (2015). A combined negative selection algorithm–particle swarm optimization for an email spam detection system. *Engineering Applications of Artificial Intelligence*, *39*, 33–44. doi:10.1016/j.engappai.2014.11.001

Jain, A. K., & Gupta, B. B. (2018). Rule-Based Framework for Detection of Smishing Messages in Mobile Environment. *Procedia Computer Science*, *125*, 617–623. doi:10.1016/j.procs.2017.12.079

Jain, A. K., & Gupta, B. B. (2019). Feature Based Approach for Detection of Smishing Messages in the Mobile Environment. *Journal of Information Technology Research*, *12*(2), 17–35. doi:10.4018/JITR.2019040102

Joo, J. W., Moon, S. Y., Singh, S., & Park, J. H. (2017). S-Detector: An enhanced security model for detecting Smishing attack for mobile computing. *Telecommunication Systems*, *66*(1), 29–38. doi:10.100711235-016-0269-9

Ma, J., Zhang, Y., Liu, J., Yu, K., & Wang, X. (2016). Intelligent SMS Spam Filtering Using Topic Model. In *Proceedings of the 2016 International Conference on Intelligent Networking and Collaborative Systems (INCoS)m* Ostrawva, Czech Republic (pp. 380-383). IEEE. 10.1109/INCoS.2016.47

MOBILE THREAT REPORT. (2012, September). F Secure. Retrieved from https://www.f-secure.com/documents/996508/1030743/Mobile+Threat+Report+Q3+2012.pdf

Mr. Number - Caller ID & Spam Protection. (n.d.). Google Store. Retrieved from https://play.google.com/store/apps/details?id=com.mrnumber.blocker&hl=en

Silva, R. M., Almeida, T. A., & Yamakamia, A. (2017). MDLText: An efficient and lightweight text classifier. *Knowledge-Based Systems*, *118*, 152–164. doi:10.1016/j.knosys.2016.11.018

SMiShing & Vishing News. (2017). Tumblr Numbercop. Retrieved from https://numbercop.tumblr.com/post/119761088780/weekly-summary-54-517

Spam Blocker App. (n.d.). Google Store. Retrieved from https://play.google.com/store/apps/details?id=com.smsBlocker&hl=en

Why do phishing attacks work better on mobile phones? (2011, January 20). Retrieved May 23, 2019, from https://www.welivesecurity.com/2011/01/20/why-do-phishing-attacks-work-better-on-mobile-phones

Yadav, K., Kumaraguru, P., Goyal, A., Gupta, A., & Naik, V. (2011). SMSAssassin: crowdsourcing driven mobile-based system for SMS spam filtering. In *Proceedings of the 12th Workshop on Mobile Computing Systems and Applications,* Phoenix, AZ (pp. 1-6). ACM. 10.1145/2184489.2184491

This research was previously published in the International Journal of E-Services and Mobile Applications (IJESMA), 12(1); pages 21-38, copyright year 2020 by IGI Publishing (an imprint of IGI Global).

Chapter 15
Feature Based Approach for Detection of Smishing Messages in the Mobile Environment

Ankit Kumar Jain
https://orcid.org/0000-0002-9482-6991
National Institute of Technology Kurukhetra, Kurukshetra, India

B. B. Gupta
National Institute of Technology Kurukshtra, Kurukshetra, India

ABSTRACT

Smishing is a security attack that is performed by sending a fake message intending to steal personal credentials of mobile users. Nowadays, smishing attack becomes popular due to the massive growth of mobile users. The smishing message is very harmful since its target to financial benefits. In this article, the authors present a new feature-based approach to detect smishing messages in the mobile environment. This approach offers ten novel features that distinguish the fake messages from the ham messages. In this article, the authors have also identified the nineteen most suspicious keywords, which are used by the attacker to lure victims. This article has implemented these features on benchmarked dataset and applied numerous classification algorithms to judge the performance of the proposed approach. Experimental outcomes indicate that proposed approach can detect smishing messages with the 94.20% true positive rate and 98.74% overall accuracy. Furthermore, the proposed approach is very efficient for the detection of the zero hour attack.

1. INTRODUCTION

Mobile devices seem to be popular these days because of their small screen size, lower production cost, and portability (Kang, Lee, Kang, Barolli, & Park, 2014). Because of their popularity, these devices appear to be a perfect target of fatal malicious attacks like mobile phishing, SMS Spam, Smishing, ransomware, mobile multimedia application threat, etc. (Polla, Martinelli, & Sgandurra, 2013). Smishing

DOI: 10.4018/978-1-7998-8545-0.ch015

word is constructed by combining two words that are SMS and Phishing. Smishing is an SMS based online identity theft which steal sensitive personal information like username, password, and credit/debit card details by fooling the user to visit fake links, apps or webpages (Jain & Gupta, 2016; Gupta & Gupta, 2017; Jain & Gupta, 2017; Tewari, Jain, & Gupta, 2016). Sometimes the fake message also ask user to respond message with some personal details (Choudhary & Jain, 2017).

Short Message Service (SMS) is considered to be one of the widely used communication services. Some users prefer SMS messages over emails because it is simple and does not require the Internet connection. Moreover, the reduction in the cost of SMS services by telecom companies has led to the increased use of SMS and this rise attracted attackers to attack via SMS. Attackers can purchase any mobile number with any area code to send spam messages so that it becomes difficult to identify the attacker. Various mobile applications are also blocked spam messages, but people are not aware of these apps due to lack of knowledge (Zkik, Orhanou, & Hajji, 2017). Moreover, these applications cannot provide high accuracy because attackers continuously change their way of attacks. The bank customers are the traditional targets of phishers, and attacker send a significant number of smishing messages on behalf of telecom companies.

Smishing term was firstly used by David Rayhawk in a McAfee Avert Labs blog on August 25, 2006. In the Android's Google Play store, malicious apps have increased by approximately 388% from the year 2011 to 2013 (The Cybersecurity source report, 2014). In 2017, a global survey by dimensional research analysed various types of mobile device attacks and they found that smishing attack stands at 2nd position (The Growing Threat of Mobile Device Security Breaches, 2017). Table 1 presents the evolution of smishing attack from 2006-2017.

Table 1. Evolution of smishing attack

Year	Events
2006	Smishing term was used first time
2007	Smishing messages hits Canadian town
2008	Smishing messages target to Credit Unions
2009	Smishing attack targeted Buffalo Metropolitan Federal Credit Union customers in New York
2010	FBI's Internet Crime Complaint Center (IC3) warned consumers about smishing attack
2011	30% lookout uses clicks on malicious URL in text message
2012	Smishing phone scams proliferated in South Korea banks
2013	Serious Smishing vulnerability was reported in Samsung Galaxy S4
2014	IC3 reported 6495 users became victim of Smishing attacks
2015	Bank affiliations were used to send Smishing messages
2016	UK lost £2 million each day as a result of financial fraud
2017	Smishing attack targeted users in Czech Republic

Like emails, now mobile messaging also become common between enterprises and its users. Credence Research (Credence Research Statics, 2016) recently expected that organizations would send two trillion messages a year by 2017 and these huge numbers attract the attackers. Moreover, millions of people using messaging applications like WhatsApp and Facebook Messengers and VoIP, the attackers are also active there (Rasol, Kasasbeh, & Adwan, 2016; Bhardwaj & Goundar, 2017). According to MEF survey (MEF Global Consumer Trust Report 2016, 2016), 28% of mobile users receive an unwanted text message every day, and 58% of users receive at least once in a week. In the case of messaging apps, 26% of chat application users receive an unwanted text message every day, whereas 49% receive at least one a week. The research exposes that 33% of mobile users have received a text message of phishing type and it was estimated that out of 10, seven people do not take any action against unwanted messages.

Various existing approaches (Almomani, Gupta, Atawneh, Meulenberg, & Almomani, 2013) focus on email phishing as it is one of the oldest problems. In traditional e-mail phishing problem, blacklisting and content based solutions are often used to filter fake emails. Blacklist block the incoming e-mails from certain senders. Content-based solution filters the email based on textual content of message like keywords, header, images, etc (Gupta, Tewari, Jain, & Agrawal, 2017). However, the mobile phishing is hard to detect as compared to email because of the small size of the message. Moreover, the tokenization is also difficult because these messages usually contain symbols, emotions, and abbreviations.

This paper presents features based approach to detect the smishing messages. We have analysed the various characteristics of text messages in depth and then identified ten outstanding features that can efficiently filter smishing messages to the legitimate one. We then use popular classification algorithms to evaluate the performance of our proposed feature set on the dataset of ham and smishing messages. Our experiment results indicate that the proposed model can filter smishing messages with the accuracy of 98.20%. Followings are the major contributions of this paper:

- We proposed novel approach to detect smishing messages;
- We have identified ten outstanding features suitable for mobile smishing detection;
- Proposed features are extracted from the message content and they are not rely on message history or any third-party services;
- Proposed features are able to detect zero-day mobile smishing attacks with high accuracy;
- We have also conducted a sensitivity analysis to predict the most efficient features.

The rest of the paper is organized as follow. Section 2 presents the related work about the lifecycle of smishing attack, email and phone phishing problem, and categorisations of smishing attack. Section 3 describes the detailed discussion of our proposed approach. Section 4 presents the dataset detail and evaluation matrices. Section 5 discusses the experimental setup including various experimental results. Finally, Section 6 presents conclusion and future work.

2. BACKGROUND AND RELATED WORK

2.1. Lifecycle of Smishing Attack

The smishing attack mechanism is shown in Figure 1. An attacker steals the credential of the innocent mobile user by performing following steps:

Step 1 - Phishing website/application construction: In the first step attacker identifies the target as a well-known organization and create a fake copy of the website or malicious application. The fake application and website behave like genuine one. In this, the malicious application hosted on a fake site.

Step 2 - Sending fake link: Then attacker utilize Social engineering technique to spread the link of fake application or webpage. The attacker frequently used fake emails, Multimedia Messaging Service (MMS) with malicious attachment, SMS with embedded fake links. The benefit of social engineering technique is that they can reach a huge number of mobile users. In the case of spear phishing attack, an attacker sends the message to selected users.

Step 3 - Installing Malicious Application/Visiting fake website: The mobile user easily execute the malicious attachment or click those URL to open the fake webpage and download the fake software. Unknowingly user installs and executes the malicious application on his mobile device of which he is not at all aware. In the case of spear phishing attack, an attacker sends the message to selected users. In the case of the fake website, the website contains the bogus login form.

Step 4 - Identity Theft: This malicious application starts sending sensitive information to the attacker. The attacker uses the confidential information for evil purposes like sending some random requests and messages to different people using user's account. The attacker uses this credential of malicious purposes. For example, the attacker can purchase something by using credit card details of the user.

Figure 1. Lifecycle of smishing attack

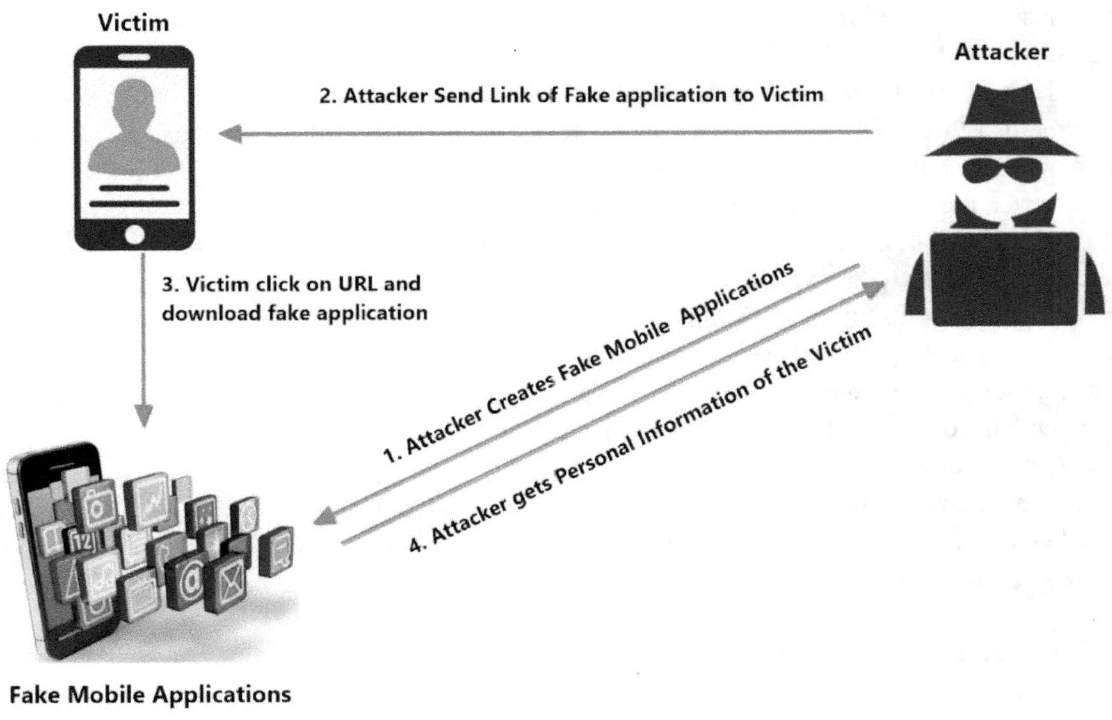

2.2. Mobile SMS and Phishing Detection Approaches

This section presents an overview of various mobile phishing detection methods proposed in the literature. Mobile phishing detection approaches categorised into the mobile webpage, malicious application and fake SMS based approached.

(Bottazzi, Casalicchio, Cingolani, Marturana, & Piu, 2015) proposed a method MP-Shield for detection of phishing attack in Android platform. MP-Shield is implemented on the top level of TCP/IP stack as a proxy service and inspect IP packets originated from the mobile application. It uses VPN service to check these IP packets. (Amrutkar, Kim, & Traynor, 2017) proposed a mechanism named KAYO which differentiates between the malicious and genuine mobile webpages. KAYO detects mobile malicious pages by measuring 44 mobile features. Out of these 44 features, 11 are newly identified mobile specific features. KAYO's 44 feature set is divided into four classes namely HTML, mobile specific, URL and JavaScript features. KAYO is considered to be the first method to identify mobile malicious webpages. KAYO provides 89% true positive rate and 90% detection accuracy.

(Wu, Du, & Wu, 2016) proposed an anti-phishing approach called as MobiFish which can protect against malicious mobile applications and web pages. This anti-phishing scheme is being tested on 100 phishing and legitimate URLs. MobiFish uses the OCR method to detect mobile phishing attacks. In OCR technique, the text is being extracted from the screenshot of the login page. The tool used for this process is Tesseract which considers being most efficient open source OCR engine.

An enhanced security model for detecting Smishing attack called S-Detector is developed by (Joo, Moon, Singh, & Park, 2017). S-Detector consists of SMS monitor, SMS analyser, SMS determinant, and Database. Through SMS monitoring, the logs and timestamps of the transmitted SMS is recorded. S-Detector also checks for the blacklist phone numbers and fake URLs. In this, the extracted words are learned by Naïve Bayes algorithm. After the learning, a weight value is given with the use of learning data. However, the author did not discuss the detection accuracy of their model.

(El-Alfy & AlHasan, 2016) proposed a spam filtering model for both email and SMS. Proposed technique used 11 features namely presence of URLs, likely spam words, emotion symbols, special characters, gappy words, message metadata, JavaScript code, function words, recipient address, subject field and spam domain. They have evaluated their proposed model on five email and SMS datasets. (Delany, Buckley, & Greene, 2012) discuss various SMS spam filtering approaches and problems associated with the dataset collection. This approach provides different SMS classification categories like dating, competitions, etc.

In the detection of smishing messages (Foozy, Ahmad, Abdollah, & Faizal, 2014) applied two rules namely 'winner announcement' and 'marketing advertisement'. This approach filters smishing messages from the spam messages while our approach distinguishes the smishing and ham messages.

2.3. Categorisation of Smishing Messages

We have analysed the dataset of 362 smishing messages from various sources and find that the smishing messages can be divided into six categories as shown in figure 2. This section presents the overview of all six type of smishing messages. Figure 3 gives the example of each category.

Figure 2. Categorisation of smishing messages

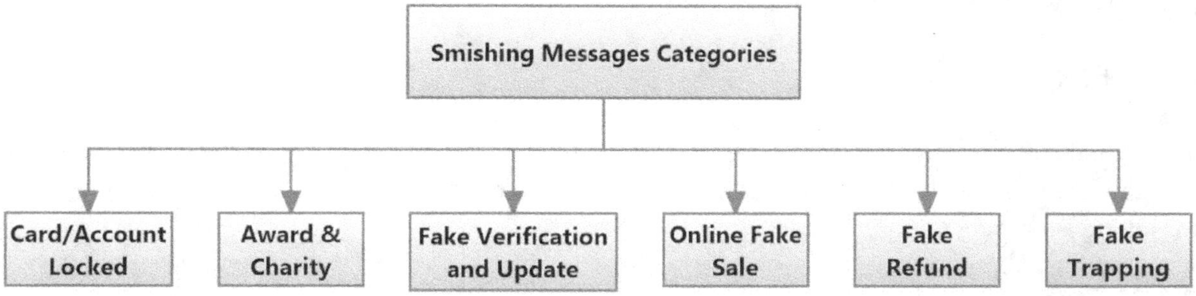

2.3.1. Card/Account Locked

Attackers lure victim by sending messages that look like coming from the leading bank and stating that user's account/ATM card has been locked and follow the link or call on the given phone number to unlock the account. When the user clicks on the link, he/she is being redirected to the phishing webpage and enters account login credentials. These credentials are stolen and further used by an attacker for some evil purposes of which user is not aware at all.

2.3.2. Award and Charity Type

In this type of smishing message, attacker notifies the user about winning a prize. The prize can be anything like money, Smartphone, laptop, holiday trip, etc. To claim the award, phishers asks the user to send money to their bank account that can vary from a few hundred dollars to several lakh dollars. As soon as the phishers received money from the victim, they disappear, and it is almost impossible to find them again. The attacker uses the name of legitimate lottery companies so that user get easily becomes the victim of smishing attack. An attacker can also pretend as a needy person and asks the user to send money for charity.

2.3.3. Fake Verification and Update

In these types of messages, attacker asks the user to verify his/her details or to update their software. These messages contain a link, and when the user follows the link, he is being asked to enter his details for verification purpose. These details are used by an attacker for malicious activities. This type of messages contains information like some suspicious activity detected in your account and click on the link to avoid account suspension.

2.3.4. Online Fake Sale

Attackers also lure users by providing some attractive offers in the name of sale which are actually fake. In this, the attacker creates a fake online shopping website and asks the user to visit that link. When the user clicks on the link, he is being redirected to a fake online shopping site and buys some product and make payment online. That amount is credited to attackers account, and the user is never delivered that product for which he made the payment.

2.3.5. Fake Refund

These kinds of smishing message lure user in terms of some refund and ask user to fill the form. The user thinks that he will get some refund, so he genuinely enters all his details which are being used by the attacker for malicious purposes.

2.3.6. Fake Trapping

In this type of message, Attack trap user by sending messages like their order has been approved and asks the user to confirm if they have placed an order or not. The user is easily trapped in this kind of messages as he gets confused whether he has actually placed an order or not. Therefore, he visits the link in the message and becomes a victim of the fake attack.

Figure 3. Examples of smishing messages

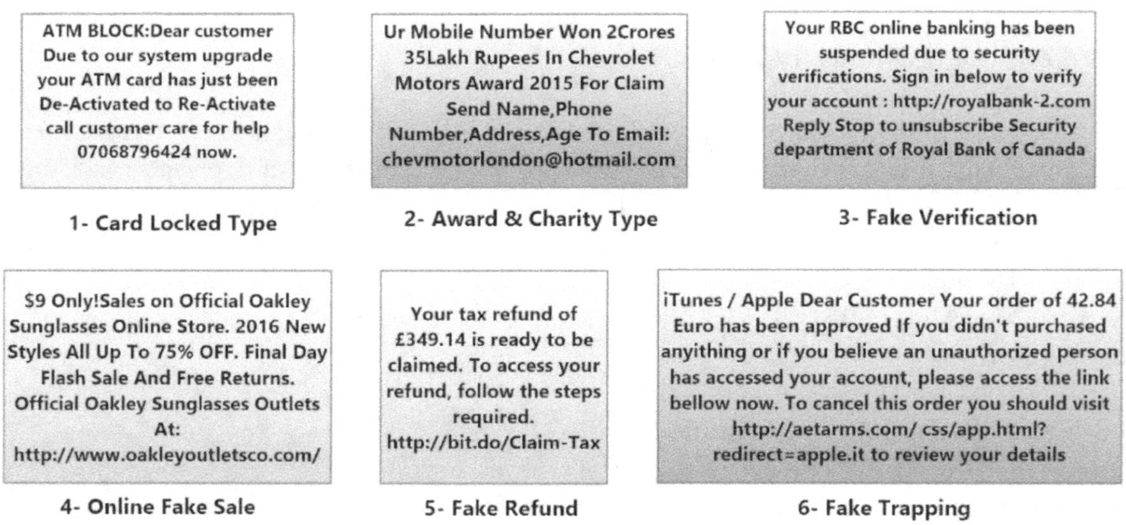

3. PROPOSED APPROACH

The primary objective of our approach is to identify the smishing messages received on the user's mobile device. Figure 4 presents the overall architecture of the proposed approach. In this, one preliminary filter is used before extraction of the features from the suspicious message. The filter contains a third party URL blacklist. The aim of this filter is used to reduce the false positive rate and speed up the overall system. The smishing messages generally contain the URL of the fake website. When the user clicks on the fake URL, it has been redirected to the phishing site. Therefore, our approach detects the fake URL in the message if it is present in the blacklist. We have used a Google safe browsing API (Google Safe browsing API, 2017) for the URL checking. Google safe browsing contains a list of fake URL, and it is regularity update. If the URL belongs to blacklist then it warns to the mobile user else, it goes in our next module. The next phase of our approach is the text pre-processing. Text pre-processing recognise boundary of sentences, removing duplicate instances, tokenization, auto word spacing, and lowercase text conversion. The pre-processing is used to normalised the data.

After pre-processing, a feature extractor fetches the features from the message. Based on the extracted features, a feature vector will be generated and used to a trained a binary classifier in training phase. In the testing phase, the trained classifier determines the category of the new message (ham and smishing). A classifier takes the decision based on the learning from the labelled dataset. When a user requested for a new message, the feature extractor generates the features values, and the binary classifier correctly identifies the message. We have tested our dataset on five widely used learning algorithms namely support vector machine (SVM), logistic regression, neural networks, Naive Bayes and random forest.

Figure 4. Architecture of proposed approach

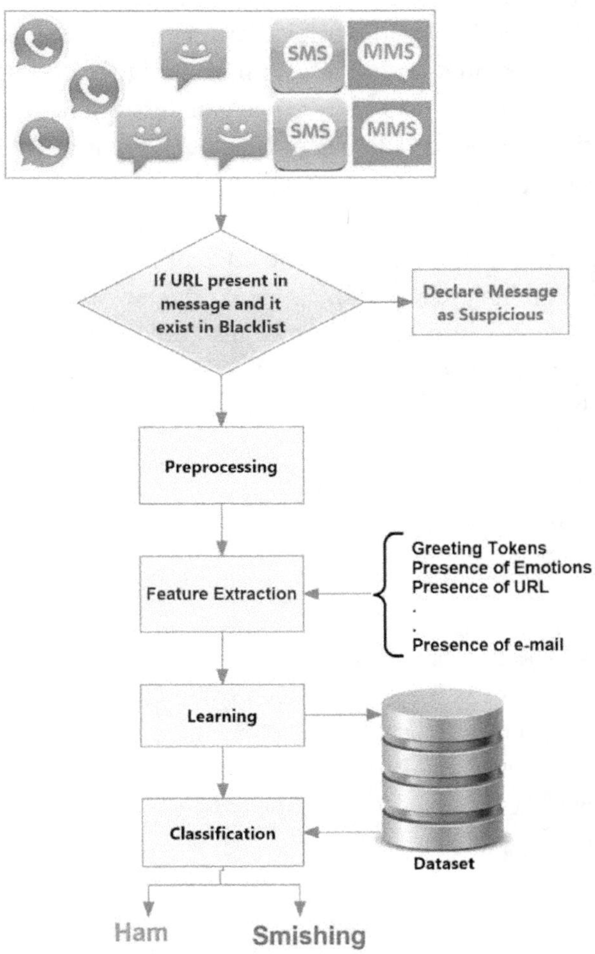

3.1. Google Safe Browsing API

Google provides a service for safe browsing (Google Safe browsing API, 2017) that allows the applications to verify the URLs using a list of suspicious domains which regularly updated by Google. It is an experimental API and used by most of the famous web browsers like Mozilla Firefox, Google Chrome, etc. The Safe Browsing Lookup API allows the clients to send the suspicious URLs to Safe Browsing service which tells whether the URL is legitimate or malicious. The client API sends the URLs with GET or POST requests, which are checked using the malware and phishing lists provided by Google. Approximately one billion people use this API. It protects the users when they try to access websites that would steal their secret credentials or install malicious software intended to take over their PCs.

3.2. Features Selection and Extraction

We have offered some powerful features that are not considered before and selected some useful features employed by existing approaches with some modification. These features are extracted from the text content of the message. Moreover, we have constructed the labelled dataset by using these features. In this, F = {F1,F2,F3. . ., F10} is defined as the feature vectors corresponding to each message. In this, two features (F1 and F2) are generally present in ham messages the good indicator for ham messages and. Moreover, remaining eight features (F3-F8) are normally present in smishing messages. Each feature produces the value in the form of 1 and 0, where 1 indicates for smishing and 0 show for ham. This section discusses selection and extraction of each feature:

F_1 - Greetings keywords: This feature checks the presence of greeting keywords like Good Morning, Good Afternoon, Good evening, Good night, hi, hello and hey. These keywords seem to be a good indicator of ham message because the legitimate user only uses these keywords while chatting. If a message contains greetings token, then it falls under the category of ham message. We have considered all the abbreviations of a token e.g. good night can be written as gud nyt, good ni8, good nyt, goodnight, etc.:

$$F_1 = \begin{cases} 0, & \textit{if message contains any greeting token} \\ 1, & \textit{otherwise} \end{cases} \tag{1}$$

F_2 - Presence of emotions: The presence of emotion symbols seems to be a good indicator for ham messages because a person usually uses emotions while chatting. Emotions can be of two types i.e. positive and negative. Positive emotion like love, nice, sweet, happy, etc. and negative emotion like sad, angry, hurt, nasty, etc. People use these emotions in the form of symbols to express his feelings or his views about the message. For example emotion ":)" is used for happy face and emotion ":(" is used for the sad face:

$$F_2 = \begin{cases} 0, & \textit{message contains positive or negative emotions} \\ 1, & \textit{otherwise} \end{cases} \tag{2}$$

F_3 - Presence of URL: This feature checks the whether the message contains any URL. We consider the presence of URL in the message as a smishing feature since phishers likely ask the user to click on a link to visit a website for a prize, fake update or to provide some personal information on the fake webpage. In this, if tiny URL present in the message, it is converted into long URL:

$$F_3 = \begin{cases} 1, & if\, URL\, present\, in\, message \\ 0, & Otherwise \end{cases} \tag{3}$$

F_4 - Presence of Mathematical Symbols: The presence of mathematical symbols usually refers to smishing messages. Mathematical symbols generally not used in genuine text messages because people use simple language to communicate rather than mathematical symbols. In this feature, we check whether the message contain +, %, -, /, ^ and % mathematical symbols. A regular expression is used to extract these symbols:

$$F_4 = \begin{cases} 1, & if\, message\, contains\, mathematical\, symbol \\ 0, & otherwise \end{cases} \tag{4}$$

F_5 - Message length: It includes the total length of the message including space, symbols, special characters, smileys, etc. The text limit of SMS message is 160 characters only. Chatting message is generally small as compared to the fake message. This feature results in 1 if the message length is greater than 150 characters:

$$F_5 = \begin{cases} 1, & if\, message\, length \geq 150 \\ 0, & otherwise \end{cases} \tag{5}$$

F_6 - Self Answering Message: The presence of self-answering message like asking the user to subscribe or unsubscribe any service falls under the category of smishing message. Sometimes this kind of messages asks the user to reply with 'yes' or 'no':

$$F_6 = \begin{cases} 1, & message\, is\, self\, answered\, type \\ 0, & otherwise \end{cases} \tag{6}$$

F_7 - Smishing symbol: This feature examines the presence of smishing symbols in Message. For example, specific symbol "$" is being used to represent money in the fake award messages. We have selected two symbols frequently present in the smishing message namely $ (Dollar) and £ (Pound):

$$F_7 = \begin{cases} 1, & if\, message\, contain\, any\, smishing\, symbol \\ 0, & Otherwise \end{cases} \tag{7}$$

F_8 - Smishing keywords: This feature examines the presence of suspicious words in Message. Attacker adds various suspicious keywords in the message to gain the trust on the message. We have identified nineteen keywords frequently present in the smishing message as shown in Table 2. If any of these keywords found in the message, this feature make it as smishing:

$$F_8 = \begin{cases} 1, & if\, message\, contain\, any\, smishing\, keyword \\ 0, & Otherwise \end{cases} \tag{8}$$

Table 2. List of smishing keywords

Smishing Keywords
award, congratulation, winner, alert, claim, activate, verify, attempts, gift voucher, blocked, suspend, unlock, won, prize, subscribe, activity, update, coupon, refund

F_9 - Presence of mobile number: This feature checks the presence of the mobile number in the message. Attackers use the mobile number in the message to get the personal information on the mobile number. This feature makes the message as smishing if a mobile number present in the suspicious message:

$$F_9 = \begin{cases} 1, & if\, message\, contain\, mobile\, number \\ 0, & Otherwise \end{cases} \tag{9}$$

F_{10} - Presence of email address: This feature checks the presence of email address in the message. The attacker also uses the email address in the message to get the personal information on the desired source. This feature makes the message as smishing if email address present in the suspicious message:

$$F_{10} = \begin{cases} 1, & if\, message\, contain\, email\, address \\ 0, & Otherwise \end{cases} \tag{10}$$

4. TRAINING DATASET AND EVOLUTION METRICS

4.1. Dataset Collection

We have collected SMS dataset v.1 from SMS Spam research work (Almeida, Hidalgo, & Yamakami, 2011). This dataset contains 5574 text messages in the English language labeled as ham and spam (4,827 ham and 747 spam). This dataset includes 425 spam messages taken from the Grumbletext website and 450 ham messages taken from Caroline Tag's Ph.D. Thesis. This corpus also contains 3,375 ham messages taken from NUS SMS Corpus (NSC). Moreover. A set of 1,002 ham and 322 spam messages included from the SMS Spam Corpus v.0.1 Big. Until now, benchmarked smishing dataset is not publically available. Smishing messages are the subset of the spam messages. A spam message is any unwanted message like free services, promotions, advertisements, etc. whereas Smishing message is bad spam message which intends to steal personal and financial information of the user. Therefore, we have manually filter smishing messages from the spam messages. The messages that do not fall under smishing marked as ham. Moreover, we have also collected 71 unique smishing message images from pinterest.com (Smishing message images, 2017). We have converted these images into text and added into the dataset. Moreover, we have also performed pre-processing which removes the messages having identical value. After pre-processing, our final dataset contains 5169 message of which 362 are smishing messages, and 4808 are the ham messages.

4.2. Evaluation Matrices

The evaluation measures set for our experiment are true positive rate (TPR), true negative rate (TNR), false positive rate (FPR), false negative rate (FNR), accuracy, and receiver operating characteristics (ROC) area. These are the slandered parameters to judge any classification model. The different evaluation metrics calculated using confusion matrix is described as follows:

- **True Positive Rate (TPR):** It is calculated as the number of true positive predictions divided by the total number of actual positive. It is also called as Recall:

$$\text{TPR} = \frac{TP}{TP + FN} \tag{11}$$

- **True Negative Rate (TNR):** It is calculated as the number of true negative predictions divided by the total number of actual negative:

$$\text{TNR} = \frac{TN}{FP + TN} \tag{12}$$

- **False Positive Rate (FPR):** It is calculated as the number of wrong negative predictions divided by the total number of actual negative:

$$\text{FPR} = \frac{FP}{TN + FP} \tag{13}$$

- **False Negative Rate (FNR):** It is calculated as the number of wrong positive predictions divided by the total number of actual positive:

$$\text{FNR} = \frac{FN}{TP + FN} \tag{14}$$

- **Accuracy:** It is calculated as the number of all correct predictions divided by the complete instances in dataset:

$$\text{Accuracy} = \frac{TP + TN}{TP + FN + TN + FP} \tag{15}$$

- **ROC:** In this, an area is plotted between True Positive Rate and False Positive Rate for different threshold values.

5. EXPERIMENT SETUP AND RESULTS OUTCOMES

5.1. Experiment Setup

The experiments were conducted on Pentium i5 computer with 2.4 GHz processor. We have used WEKA (Hall et al., 2009) to judge the performance of our proposed approach on various classifiers. Weka is the Java open source code which means "Waikato Environment for Knowledge Analysis." Numerous data mining and machine learning algorithms are implemented in WEKA. It contains the rich collection of modelling, clustering, classification, regression, and data pre-processing techniques. The experiments conducted on various traditional classification algorithms. Smishing detection is a binary classification problem. In this, either a message is belonging to smishing or not. We have evaluated our dataset with 10-fold cross-validation, which uses 90% of data for training purpose, and remaining 10% data for testing purpose.

5.2. Results of Proposed Approach

The classification results of our approach is shown in figure 5. This confusion matrix shows that our approach is pretty useful in identifying smishing messages (i.e. more than 94% smishing messages correctly identified), and very few number of ham messages misclassified as smishing (i.e. less than 1% false alarm). The detection accuracy (i.e. 98.74%) shows that our proposed features are useful in correctly distinguish between smishing and ham messages. We have tested our approach on traditional and widely accepted classifiers. Figure 6 presents the performance of our approach on different classifiers. It is noticed that random forest outperformed SVM, neural networks, logistic regression and naïve Bayes. Random forest performs best regarding highest TPR, TNR, and accuracy. The Naive Bayes also produce the same TPR, but its TNR is a low as compared to the random forest. The Neural Network delivers the lowest TPR of 88.67% followed by SVM and SMO. Moreover, SVM results in the same TNR, but its FPR is lower by 4.7% compared to the random forest. We have also explored the area under ROC (Receiver Operating Characteristic) curve to find a better metric of precision. In our experiment, the area under the ROC curve for phishing website is 98.79% for the Random forest as shown in figure 7, and it shows that our approach has the good detection accuracy.

Figure 5. Results of the proposed approach on random forest classifier

Confusion Matrix

	Smishing Messages	Ham Messages
Classified as Smishing	TP = 341	FP = 44
Classified as Ham	FN = 21	TN = 4764

Figure 6. Results of proposed approach on popular classifiers

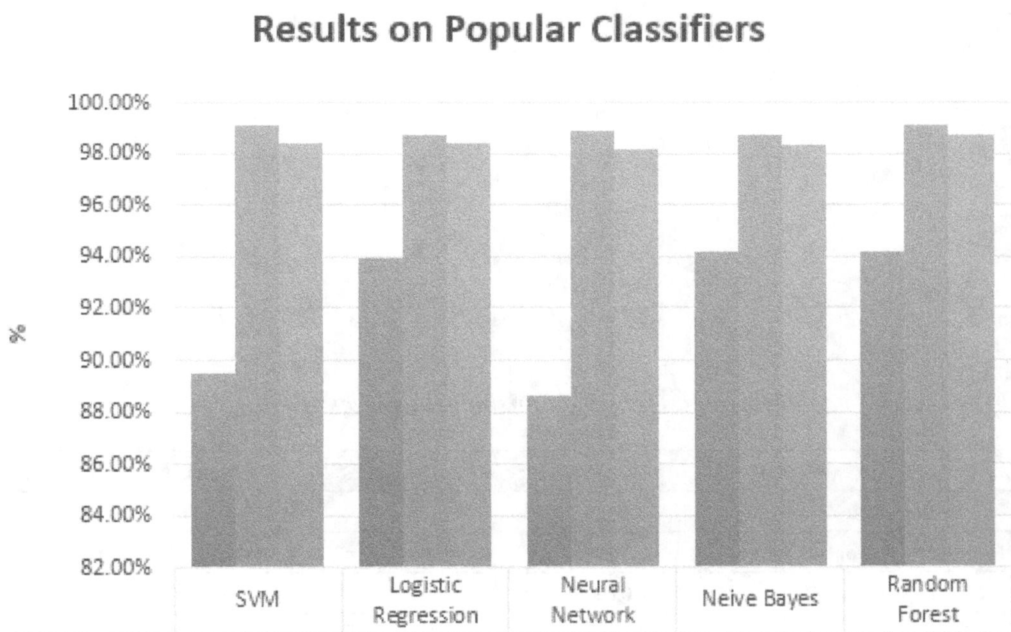

	SVM	Logistic Regression	Neural Network	Neive Bayes	Random Forest
TPR (%)	89.50%	93.92%	88.67%	94.20%	94.20%
TNR (%)	99.08%	98.73%	98.86%	98.69%	99.08%
Accuracy (%)	98.41%	98.39%	98.14%	98.37%	98.74%

5.3. Evaluation of Features

The proposed feature set is carefully projected to ensure the correct prediction of ham and smishing messages. The significance of proposed feature set is determined by calculating the frequencies for each feature in our datasets. Our feature set consists two ham (F1, F2) and eight smishing (F3-F10) oriented features. Table 3 presents the frequency of each feature in ham and smishing messages. The results indicate that the 'smishing keyword' feature is the most common in smishing messages and it is present in 87.85% of smishing messages, followed by 'smishing symbol' which present in 59.39% of smishing messages, the next attractive feature is 'message length,' with 55.25% frequency. The two features are commonly found in ham messages that are 'greeting token' and 'presence of emotions' with the frequency of 10.21% and 6.20% respectively. The lowest common feature in smishing messages is the 'presence of email address' which present in only nine smishing messages, followed by 'self-answering message' which founded in 25 smishing messages.

Figure 7. Area under ROC for random forest

Table 3. Frequency of each feature in Ham and Smishing messages

Sr. No.	Feature	Frequency in Ham Messages	Frequency in Smishing Messages
1	Greeting token	10.21%	1.10%
2	Presence of emotion symbols	6.20%	0.00%
3	Presence of URL	0.65%	28.45%
4	Presence of Mathematical symbol	8.57%	53.59%
5	Message Length	10.30%	55.25%
6	Self-Answering Message	0.42%	6.90%
7	Smishing Symbol	0.50%	59.39%
8	Smishing Keywords	1.08%	87.85%
9	Presence of Mobile number	0.73%	46.96%
10	Presence of email address	0.062%	2.49%

The detailed overview of greeting token (feature 1) is presented in Table 4. Table 4 shows that 'hey' greeting word is the most common in ham messages followed by 'hi' and 'good night'.

Table 4. Statics of greeting tokens (feature 1)

Greeting Token	Presence in Ham Messages	Presence in Smishing Messages
hi	90	2
hey	187	2
hello	30	0
Good evening	35	0
Good morning	48	0
Good night	74	0
Good afternoon	23	0

Table 5 presents the statics of emotion symbol. 285 ham messages contain the happy emotions while only 13 ham message contain sad emotions. Although, no smishing message include any emotion symbol. Table 6 presents the statics of mathematical symbol, and it shows that '/' mathematical symbol is frequently present in smishing message followed by '-' symbol. The URL contains '/' after http protocol. Therefore, it generally available in smishing messages. Most of the smishing messages are lottery type, and it includes either the pound or dollar sign. Table 7 shows that more than 57% of smishing message includes the £ (pound) sign. The frequency of smishing keywords is presented in Table 8, and it shows that 'claim' is the most frequent keyword present in the smishing message followed by 'price' and 'award'.

Table 5. Statics of emotion symbol (feature 2)

Emotion	Presence in Ham Messages	Presence in Smishing Messages
☺ (Positive)	285	0
☹ (Negative)	13	0

Table 6. Statics of mathematical symbol (feature 4)

Mathematical Symbol	Frequency in Ham Messages	Frequency in Smishing Messages
+	0.94%	11.04%
%	0.17%	1.10%
/	2.39%	33.97%
-	6.052%	25.14%

Table 7. Statics of smishing symbols (feature 7)

Symbol	Frequency in Smishing Message	Frequency in Ham Messages
$	1.934%	0.374%
£	57.459%	0.125%

Table 8. Frequency of smishing keywords (feature 8)

Keywords	Frequency in Smishing Message	In %	Frequency in Ham Message	In %
award	49	13.536%	3	0.062%
Congratulation	14	3.867%	2	0.042%
Winner	20	5.525%	1	0.021%
Alert	5	1.381%	1	0.021%
Claim	92	25.414%	12	0.250%
Activate	10	2.762%	0	0.000%
Verify	7	1.934%	2	0.042%
Attempt	18	4.972%	4	0.083%
Gift voucher	12	3.315%	0	0.000%
blocked	2	0.552%	1	0.021%
suspend	3	0.829%	0	0.000%
unlock	4	1.105%	0	0.000%
won	59	16.298%	2	0.042%
prize	78	21.547%	1	0.021%
subscribe	13	3.591%	8	0.166%
activity	4	1.105%	0	0.000%
update	13	3.591%	16	0.333%
coupon	2	0.552%	0	0.000%
refund	4	1.105%	2	0.042%

5.4. Importance of Features

We experimentally determine the importance of proposed features using the Information gain (IG). In this, we used ranker search algorithm to analyse the quality of each feature which helps in selecting the best features. Information gain is biased towards multivalued attributes. The information gain of an attribute tells you how much information on the classification target the attribute gives you. Therefore, the information gain depends on two things: how much information was available before knowing the attribute value, and how much was available after. The higher IG value indicates more dominant feature in classification result. Moreover, '0' IG value show that that feature is not contained any information. Figure 8 presents the plot of the IG value of each of the ten features of proposed approach with the label. From the figure, we analyses that IG values for all features are non-zero, represent that every feature in proposed approach is important. Moreover, the feature 8, 7, 9, 3, 4, and 5 had the high IG value and identified as the most useful features in our proposed feature set. However, these six features are not sufficient to detect all kind of smishing messages. Therefore, we have also used other features to improve the accuracy of the proposed approach.

Figure 8. Information gain values of proposed feature set

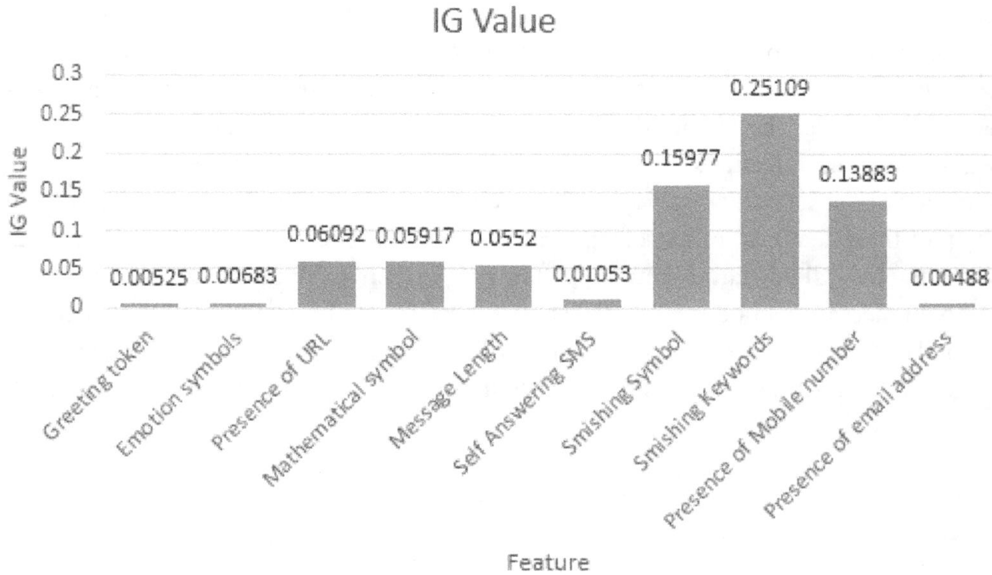

5.5. Discussion

With the rapid growth of e-commerce, e-banking, and social networking, the phishing attack is also growing day by days. Now a day, most of the Internet user prefer smartphone thus they become the perfect target of the attacker. It results in enormous amount financial losses to individual users and industries. There are various feature based approaches available in spam message detection, but these approached cannot filter smishing messages. Smishing and spam both are the different problems, and there is need of the effective solution to detect smishing attack which high accuracy. In this paper, our primary objective is to detect the phishing message which has high TPR. Results demonstrate that our approach correctly classified the phishing messages and achieved a decent TPR and detection accuracy. Our proposed approach incorporated various text content specific features into a model that have never been considered. Moreover, as per our knowledge and searching on journal publisher like IEEE, Springer, Science Direct, Wiley, Inderscience, Hindawi, etc. this is the first feature-based approach in the detection of smishing attack. Various anti-smishing approaches and mobile applications have been proposed recently. Most of these solutions detected messages based on URL features. After analysing dataset, we have found that only 28.45% of smishing message contains the URL. Moreover, our approach does not solely depend on a single feature, and it is more efficient compared to existing approaches.

However, there are certain limitations of our proposed approach. The feature set of our phishing detection approach entirely depends on the text content of the message, and it can detect the smishing message if it is written in English only.

6. CONCLUSION AND FUTURE WORK

In this paper, we have presented an efficient detector for the smishing messages with a pre-filter technique. The primary aim of our approach is to detect smishing messages accurately with high true positives. The pre-filter used in our approach reduces unnecessary computation to identify smishing messages that contain a fake URL. The feature extractor module provides various new features for identifying smishing messages. These features are extracted from the content of the message. We have used these features to train a classifier, which achieved high accuracy in detection of smishing and ham messages. With the help of these features, proposed approach proposed approach obtained TPR of 94.20%, TNR of 99.08% and the overall detection accuracy of 98.74%. Moreover, the existing SMS based approaches detect only spam messages, but our proposed approach is capable of detecting Smishing messages. Our approach can also detect the zero-day attack. In our future work, we will also introduce new Smishing features to get the better accuracy results. The accuracy of our approach may be improved by adding certain more features.

REFERENCES

Almeida, T. A., Hidalgo, J. M., & Yamakami, A. (2011). Contributions to the study of SMS spam filtering: new collection and results. In *Proceedings of the 11th ACM symposium on Document engineering* (pp. 259-262). Mountain View, CA: ACM. 10.1145/2034691.2034742

Almomani, A., Gupta, B. B., Atawneh, S., Meulenberg, A., & Almomani, E. (2013). A Survey of Phishing Email Filtering Techniques. *IEEE Communications Surveys and Tutorials*, *15*(4), 2070–2090. doi:10.1109/SURV.2013.030713.00020

Amrutkar, C., Kim, Y. S., & Traynor, P. (2017). Detecting Mobile Malicious Webpages in Real Time. *IEEE Transactions on Mobile Computing*, *16*(8), 2184–2197. doi:10.1109/TMC.2016.2575828

Bhardwaj, A., & Goundar, S. (2017). Comparing Single Tier and Three Tier Infrastructure Designs against DDoS Attacks. *International Journal of Cloud Applications and Computing*, *7*(3), 59–75. doi:10.4018/IJCAC.2017070103

Bottazzi, G., Casalicchio, E., Cingolani, D., Marturana, F., & Piu, M. (2015). MP-Shield: A Framework for Phishing Detection in Mobile Devices. In *Proceeding of IEEE International Conference on Computer and Information Technology; Ubiquitous Computing and Communications; Dependable, Autonomic and Secure Computing; Pervasive Intelligence and Computing*,). Liverpool, UK (pp. 1977-1983. 10.1109/CIT/IUCC/DASC/PICOM.2015.293

Choudhary, N., & Jain, A. K. (2017). Comparative Analysis of Mobile Phishing Detection and Prevention Approaches. In *Proceedings of International Conference on Information and Communication Technology for Intelligent Systems*, Ahmedabad, India (pp. 349-356). Springer.

Credence Research Statics. (2016). Retrieved October 26, 2017, from http://www.credenceresearch.com/

Delany, S. J., Buckley, M., & Greene, D. (2012). SMS spam filtering: Methods and data. *Expert Systems with Applications*, *39*(10), 9899–9908. doi:10.1016/j.eswa.2012.02.053

Dimensional Research. (2017, April). The Growing Threat of Mobile Device Security Breaches. Retrieved October 14, 2017, from https://blog.checkpoint.com/wp-content/uploads/2017/04/Dimensional_Enterprise-Mobile-Security-Survey.pdf

El-Alfy, E. S., & AlHasan, A. A. (2016). Spam filtering framework for multimodal mobile communication based on dendritic cell algorithm. *Future Generation Computer Systems*, *64*, 98–107. doi:10.1016/j.future.2016.02.018

Foozy, M., Ahmad, C. F., Abdollah, R., & Faizal, M. (2014). A practical rule based technique by splitting SMS phishing from SMS spam for better accuracy in mobile device. *International Review on Computers and Software*, *9*(10), 1776–1782.

Google Safe browsing API. (2017). Retrieved August 22, 2017, from https://developers.google.com/safebrowsing

Gupta, B. B., Tewari, A., Jain, A. K., & Agrawal, D. P. (2017). Fighting against phishing attacks: State of the art and future challenges. *Neural Computing & Applications*, *28*(12), 3629–3654. doi:10.100700521-016-2275-y

Gupta, S., & Gupta, B. B. (2017). Detection, Avoidance, and Attack Pattern Mechanisms in Modern Web Application Vulnerabilities: Present and Future Challenges. *International Journal of Cloud Applications and Computing*, *7*(3), 1–43. doi:10.4018/IJCAC.2017070101

Hall, M., Frank, E., Holmes, G., Pfahringer, B., Reutemann, P., & Witten, I. H. (2009). The WEKA data mining software: An update. *ACM SIGKDD Explorations Newsletter*, *11*(1), 10–18. doi:10.1145/1656274.1656278

Jain, A. K., & Gupta, B. B. (2016). A novel approach to protect against phishing attacks at client side using auto-updated white-list. *EURASIP Journal on Information Security,* (9).

Jain, A. K., & Gupta, B. B. (2017). Phishing Detection: Analysis of Visual Similarity Based Approaches. *Security and Communication Networks*, *2017*, 1–20. doi:10.1155/2017/5421046

Joo, J. W., Moon, S. Y., Singh, S., & Park, J. H. (2017). S-Detector: An enhanced security model for detecting Smishing attack for mobile computing. *Telecommunication Systems*, *66*(1), 29–38. doi:10.100711235-016-0269-9

Kang, A., Lee, J. D., Kang, W. M., Barolli, L., & Park, J. H. (2014). Security Considerations for Smart Phone Smishing Attacks. In Proceedings of Advances in Computer Science and its Applications (pp. 467-473).

MEF Global Consumer Trust Report 2016. (2016). Retrieved October 14, 2017, from https://mobileecosystemforum.com/programmes/consumer-trust/global-consumer-trust-report-2016/

Pinterest. (2017, October 22). Smishing message images. Retrieved from https://in.pinterest.com/seceduau/smishing-dataset/?lp=true

Polla, M. L., Martinelli, F., & Sgandurra, D. (2013). A survey on security for mobile devices. *IEEE Communications Surveys and Tutorials*, *15*(1), 446–471. doi:10.1109/SURV.2012.013012.00028

Rasol, M., Kasasbeh, B. A., & Adwan, F. A. (2016). An Improved Secure SIP Registration Mechanism to Avoid VoIP Threats. *International Journal of Cloud Applications and Computing*, *6*(2), 25–36. doi:10.4018/IJCAC.2016040103

Robinson, T. (2014). Report: Malicious apps in Google Play store grow 388 percent. *SC Magazine*. Retrieved October 26, 2017, from https://www.scmagazine.com/report-malicious-apps-in-google-play-store-grow-388-percent/article/538687/

Tewari, A., Jain, A. K., & Gupta, B. B. (2016). Recent survey of various defense mechanisms against phishing attacks. *Journal of Information Privacy and Security*, *12*(1), 3–13. doi:10.1080/15536548.2016.1139423

Wu, L., Du, X., & Wu, J. (2016). Effective Defense Schemes for Phishing Attacks on Mobile Computing Platforms. *IEEE Transactions on Vehicular Technology*, *65*(8), 6678–6691. doi:10.1109/TVT.2015.2472993

Zkik, K., Orhanou, G., & Hajji, S. E. (2017). Secure Mobile Multi Cloud Architecture for Authentication and Data Storage. *International Journal of Cloud Applications and Computing*, *7*(2), 62–76. doi:10.4018/IJCAC.2017040105

This research was previously published in the Journal of Information Technology Research (JITR), 12(2); pages 17-35, copyright year 2019 by IGI Publishing (an imprint of IGI Global).

Chapter 16
Securing Communication 2FA Using Post-Quantic Cryptosystem:
Case of QC-MDPC- Mceliece Cryptosystem

Kouraogo Yacouba
LabMiA-SI, Faculty of Sciences, Mohammed V University in Rabat, Rabat, Morocco

Orhanou Ghizlane
LabMiA-SI, Faculty of Sciences, Mohammed V University in Rabat, Rabat, Morocco

Elhajji Said
LabMiA-SI, Faculty of Sciences, Mohammed V University in Rabat, Rabat, Morocco

ABSTRACT

Many financial institutions interact with their customers via short message services (SMS), which is today one of the fastest and most powerful means of communicating information around the world. This information can sometimes be an access code such as the unique password (OTP) for two-factor authentication (2FA) or banking information and personal identities. All this data is confidential, and it is a major disadvantage to send them since an SMS service does not provide data encryption during network transmission and on mobile. Recently, OTPs via SMS have suffered from strong attacks that intercept messages. In order to avoid attacks and offer effective content security to 2FA credentials sent via SMS, the authors propose an SMS encryption mechanism using a post quantic cryptosystem quasi-cyclic MDPC and an electronic signature of the OTPs. Finally, this article performs an implementation and a security analysis of the proposal.

DOI: 10.4018/978-1-7998-8545-0.ch016

INTRODUCTION

Nowadays, many organizations and companies operate online. Commercial institutions, banks and many other parts of the society are taking advantage of the benefits of e-business by opening the access to networks and services to their employees, partners and customers. Thus, mobile devices are used today as an alternative to online transaction systems for authentication.

In this context, the traditional login / password authentication mechanism is not considered secure enough to be used in many security-critical applications, such as online banking. Two-factor authentication systems (2FA) promise a higher level of protection by extending the authentication factor to what the user owns (a hardware token or a smartphone) or who the user is (biometrics) (Dmitrienko, Liebchen, Rossow, & Sadeghi, 2014) However, today, it is very easy to carry out increasingly sophisticated attacks on two-factor authentication systems (2FA), which not only compromise a device (PC) but also take control of other devices (Mulliner, Borgaonkar, Stewin, & Seifert, 2013) On the other hand, biometric authentication is relatively expensive. Hardware tokens such as OTP generators (Schartner & Burger, 2011) are cheaper, but still generate additional costs for users. In this context, single passwords (OTPs) offer a promising alternative for 2FA systems. Thus, 2FA systems that use mobile devices (such as smartphones) to manage OTPs have become popular recently and have been adopted by many banks and major service providers. These 2FA mobile systems are considered to offer an appropriate compromise between security, ease of use and cost. An important example of 2FA mobile are SMS-based TAN (Transaction Authentication Number) systems, such as mTANs, smsTAN, mobileTANs.

Unfortunately, today, OTP SMS can not be considered secure for two different reasons:

1. First, the security of OTP SMS relies on the confidentiality of SMS messages that depends heavily on network security. Several attacks against GSM and even 3G networks have shown that the confidentiality of SMS messages is not offered by default;
2. Second, attackers have adjusted and created specialized trojans for mobile phones (Maslennikov, 2018; Klein, 2017) in order to recover the OTPs, since many service providers have adapted SMS OTP to secure transactions (Mulliner, Borgaonkar, Stewin, & Seifert, 2013).

Today, there are several mechanisms for encrypting SMS content using cryptography based on difficult mathematical problems like factorization and discrete logarithm. However, the advent of the quantum computer, which could have a much higher computational power than our conventional computers, would be able to break most conventional encrypted systems based on the discrete logarithm and the factorization problem. To solve and anticipate this security problem, the authors propose to use post-quantum cryptography that would be a priori resistant to the quantum computer. Thus, in this paper, they will use post-quantum cryptography to ensure confidentiality of the transmitted information such as SMS OTP on the unsecured public channel. They propose an SMS encryption mechanism using the Quasi-Cyclic MDPC version of the MC-Ellice cryptosystem and an electronic signature of OTPs based on elliptic crossovers ECDSA (Elliptic Curve Digital Signature Algorithm).

After presenting in section 2 the various existing research works, the authors will then describe briefly the cryptosystems which will be used in their proposal. In section 4, they will present their proposed SMS-based 2FA system using post quantum cryptosystems. Finally, the authors will present an implementation of their proposed system and a security issues discussion.

RELATED WORK

Many articles have been written about attacks and securing two-factor authentication methods. In this section, the authors will present the mobile application security related work, the existing two-factor authentication systems, then the SMS-based attacks on 2FA. Finally, some existing security proposals on 2FA based on SMS will be presented as well.

Mobile Application Security Related Work

- Kumawat et al. (2017) propose in an Android application allowing the user and the developer to easily test the application, whether it is benign or vulnerable. Thus, the authors have implemented the static and dynamic analysis of Android application for the detection of vulnerable and malicious applications;
- Kouraogo et al. (2017) discuss the security issues of android applications. Then, they perform the reverse engineering of banking applications and a DDOS attack on the bank server. Finally, they propose countermeasures to better secure these vulnerable applications;
- Vidyarthi et al. (2017) propose a complete approach combining static verification and dynamic analysis for the detection of malicious programs. Static analysis is used to check the specific characteristics of the code while dynamic analysis is used to analyze the runtime behavior of malware.

2FA Systems Related Work

- M. H. Eldefrawy et al. (2012) propose a novel 2FA scheme whereby multiple OTPs are being produced by utilizing an initial seed and two different nested hash chains: one dedicated to seed updating and the other used for OTP production;
- I. Švogor, et al. (2012) propose two-factor authentication using the EEG (Electro EncephaloGraphy) device which allows a ceaseless exploitation of the user. This system allows users to improve the password with the mental state of the user;
- J. Qu et al. (2014) propose a two-factor authentication scheme using smart cards in which the elliptic curves are implemented;
- M. Khamis et al. (2017) offer GTMmoPass, an authentication architecture that allows authentication of multi-factor users on public displays. This is usable and secure against opponents who can discover a user's password through the shoulder.

In the papers presented above, the researchers propose two-factor authentication systems based on different types of factors. The factors used such as EEG, smart card or the GTMmoPass system are not commonly used because they are either expensive or very difficult to use.

- In another work, F. Aloul et al. (2009) propose a method of implementing two-factor authentication system using a mobile phone. The proposed system involves the use of a mobile phone as a token generator which is OTP (One Time Password).

In the present work, the authors are also interested in studing 2FA systems using mobile phone and based on SMS communication channel sincemany financial institutions use this type of system and face a lot of security problems. In the next subsection, the authors will present the different types of SMS based 2FA system attacks.

Attacks on 2FA Based on SMS OTP

- M. Adham et al. (2013) propose an attack on banks with two-factor authentication systems that are connected to the internet by developing browser rootkit that performs an automated attack to retrieve confidential client information;
- C. Mulliner et al. (2013) analyze the attacks on OTPs sent by SMS and describe howsmartphone Trojanscan intercept TAN via SMS. They also describe countermeasures against these attacks, which remain insufficient today against increasingly sophisticated attacks;
- P. Schartner et al. (2011) present an attack against TAN solutions sent via SMS for the case of a single device considered as online banking service. The attack scenario presented is relatively simple as it eliminates challenges such as cross-platform infection;
- A. Dmitrienko et al. (2014) show that the current 2FA mobile has conceptual weaknesses. In this work, attackers can intercept the transmission of the OTP service or steal private key equipment for the production of the OTP.

As presented above, 2FA systemsbased on SMS communication suffer from numerous attacks that could occur atmanylevels. There are attacks which aimto access the confidential data such as mTAN or OTP stored on the mobile phone, and othersthat focus on attacking the SMS communication channel itself. That's why, several researchers have focused on SMS-based 2FA security by presenting differents solutions. Some of these solutions are presented below.

Security of 2FA Based on SMS OTP

- S. Rangarajan et al. (2013) propose an application called ECC SMS (Elliptic Curve Cryptography Short Message Service) to encrypt SMS using elliptic curves;
- J. Pan et al. (2012) propose the encryption of SMS using cryptography based on chaotic encryption;
- In the same perspective, other researchers such as S. Jha U. Dutta (2015) have ensured the confidentiality of SMS messages by implementingthe MNTRU algorithm;
- T. Arudchelvam et al. (2016) propose the encryption of SMS by combination César and Vigenere encryption algorithm;
- Another SMS security service has been studied by S. Karale et al. (2015), providing confidentiality and integrity of SMS data by applying AES and RC4 as a hybrid cryptography program to provide more powerful security. AES was used for encryption / decryption while RC4 was used for key generation;
- In the same way, the authors in Puyitha and Mallu (2013) propose another hybrid cryptography program known as Cipher-SMS. This protocol offers an end-to-end SMS encryption by combining the AES and MD5 algorithms.

From the studied literature, the authors noticed that many researchers have focused on the confidentiality of SMS by implementing cryptographic methods. With the advent of the quantum machine, these methods are not efficient and can not guarantee the confidentiality of the data. In addition, at the best of their knowledge, the security of mTAN and OTP in the mobile phone environnement has not been studied. For this purpose, the authors focused on the objective of ensuring the security of SMS and confidential data even if the quantum machine appears.

The following section describes the post-quantum cryptographic systems that will be used in their proposed authentication solution.

QC-MDPC MCELIECE CRYPTOSYSTEM AND ECDSA

In this section, the authors give a consise description of the QC-MDPC McEliece post-quantum cryptosystem algorithm and the used electronic signature ECDSA (Elliptic Curve Digital Signature Algorithm). QC-MDPC will enable encryption of SMS content and ECDSA will enable the system to authenticate the user.

The QC-MDPC McEliece Cryptosystem

In this sub-section, the authors describe the QC-MDPC version of McEliece cryptosystem presented in [21]. They present the key generation, the encryption and the decryption mechanisms.

Key generation: To generate a pair-key (t-error correcting (n, r, w)-QC-MDPC code where $n = n_0 r$) for the QC-MDPC version of McEliece cryptosystem, the authors follow these steps: h_i:

1. select the first rows $h_0, \cdots, hn_{0-1} \in F_2^r$ the n_0 parity-check matrix blocks:

$H_0, \cdots, Hn_{0-1} \in F_2^{rxr}$ with weight $\sum_{i=0}^{n0-1} wt(hi) \le w$. w is choosen randomly

2. Generate the parity-check matrix blocks $H_0, \cdots, Hn_{0-1}$ by r−1 quasi-cyclic shifts of $h_0, \cdots, hn_{0-1}$;
3. The parity-check matrix H, which is the private key, is formed by concatenating $H_0, \cdots, Hn_{0-1}$;
4. Generator matrix $G = [I_k|Q]$, which is the public key, is computed from H in row reduced echelon form by concatenating the identity matrix I_k with:

$$Q = \begin{pmatrix} \left(H_{n_0-1}^{-1} \cdot H_0\right)^T \\ \left(H_{n_0-1}^{-1} \cdot H_1\right)^T \\ \vdots \\ \left(H_{n_0-1}^{-1} \cdot H_{n_0-2}\right)^T \end{pmatrix}$$

Encryption: To encrypt a message $m \in F_2^k$, the authors follow just two steps:

1. Generate an error vector $e \in F_2^n$ with at most t set bits choosen randomly;
2. Compute $x = mG \oplus e$.

Decryption: To decrypt a ciphertextx $\in F_2^n$, compute $mG \leftarrow R_H(x)$ using a t-error correcting QC-MDPC decoder R_H. Since G has a systematic form, extract m from the first k positions of mG.

In this asymmetric cryptosystem, the public key is the generator matrix G and the private key is the control matrix H. This cryptosystem is very resistant to attacks because it relieson the resolution of very complex mathematical problem consisting in obtainingthe matrix H from the matrix G.

After the encryption $x = mG\oplus e$, the decryption of $x \in F_2^n$ depends on an R_H function.

The authors used this post-quantum cryptosystem because of its following advantages:

- Its security is based on the syndrome decoding problem which is resistant to quantum computer;
- It provides compact public key size since the authors simply need to stock the first line of their matrix generator;
- It is easy to implement and very fast since its implementation uses only bitwise and bit rotation operations.

ECDSA

In this sub-section, the authors present the Elliptic Curve Digital Signature Algorithm (ECDSA) and present its characteristics (see Table 1), as defined in the standard FIPS 186-4 (ECDSA) [20].

Table 1. ECDSA algorithm (Timothy & Hall, 2014)

Parameters	Finite field Fq elliptic curve E on F_q G point of E of order r, great prime number $a\in [1,r-1]$ $A = aG$	Public Publics Public Secret Public
Signature	**Alice** $k\in [1, r-1]$ $K = kG$ $t=X_K \bmod r$ $s = (m + at)k^{-1} \bmod r$	**bob**
Verification		$1 \leq t, s \leq r-1$? $P = ms^{-1} G + ts^{-1} A$? Xpmodr $\Leftrightarrow$ t

The authors use an electronic signature based on the elliptic curve E on F_q presented in table 1 above. Thus, $a \in [1, r - 1]$, is the private key and $A = aG$ is the public key of the electronic signature.

To sign a message m, Alice chooses the private key $k \in [1, r - 1]$, calculates the public key in the following way $K = kG$. Then, she calculates $t = X_K \bmod r$ of which X_K is the X coordinate of K modulo r.

The message is thus signed with the following formula: s = (m + at) k-1 mod r. After Alice's signature, the following data are sent to Bob for verification:

- m: Message
- s: Signature
- t: X coordinate of K modulo r

Received by Bob, the check is made by calculating P = ms^{-1} G + ts^{-1} A. Then, the X coordinate of P modulo r is compared with t. If the signature verification is successful, then the authentication step is completed, otherwise, the whole transaction is canceled.

The authors found it interesting to choose the electronic signature based on ECDSA elliptic curves because it is very fast and is suitable for small memory card objects, asin the mobile phone environnement. It also uses small keys.

In the following section, the authors will presenttheproposedpost quantic two factor authentication system.

ENHANCEMENT OF SMS-BASED 2FA SECURITY USING POST-QUANTUM CRYPTOSYSTEM

Advanced Post-Quantic Security 2FA Based on SMS

Two-factor authentication systems (2FA) are one of the methods that give companiesthe ability to secure access to their platforms, ensuring the reliability of any transaction and gaining the trust of their customers. Thus, the most widely used authentication method by financial institutions today is two-factor authentication based on SMS. Figure 1 shows an existing two-factor authentication scheme based on SMS that theauthors would try to improve later (Schartner & Burger, 2011).

Figure 1. Existing 2FA system (Schartner & Burger, 2011)

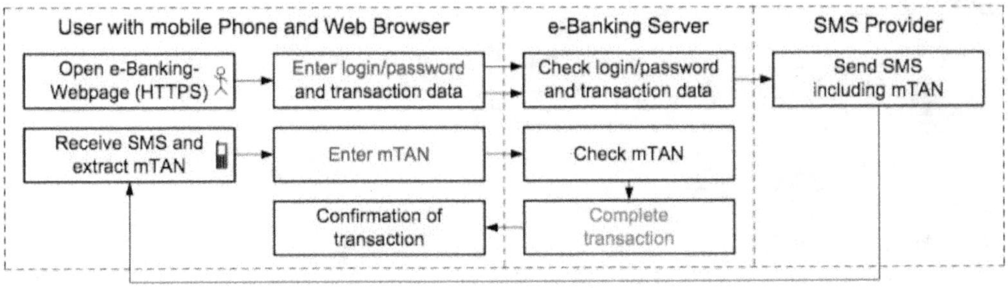

Figure 1 shows the operating principle of financial transactions authorized by the use of mTAN via SMS. This operation is described below:

- The user starts its web browser, opens its bank URL in an HTTP session and enters his username and password;
- This data is sent to the e-banking server for verification;
- In the next step, the user enters his transaction, which is sent again to the e-banking server;
- The user pressesa button labeled “Send mTAN”, then the e-banking server generates a random mTAN (usually 4 to 8 alphanumeric characters) and sends it by SMS to the user’s mobile phone;
- The user receives this short message, extracts the mTAN and enters the mTAN in the corresponding text field of the e-banking web page;

- If the verification of the mTAN on the e-banking server is successful, the transaction is completed. Otherwise, it is canceled. In both cases, the user is informed accordingly via the e-banking web page.

This two-factor authentication method seems to be secure, however it has many security vulnerabilities that allow attackers to conduct attacks against financial institutions. The most common and simple possible attacks are:

- Interception of SMS between the server and the phone;
- Brute force attack to generate valid mTAN;
- Retrieving SMS by malware like a trojan on the smartphone.

To overcome these security problems, the authors propose two-factor authentication method based on SMS which is an enhancement of the existing method. In this proposed 2FA system, they ensure the confidentiality of mTAN on the communication channel by using a QC-MDPC McEliece post-quantum cryptosystem, and then a client authentication by the bank's server using an ECDSA (Elliptic Curve Digital Signature Algorithm) electronic signature. Figure 2 below showsthe proposed method.

Figure 2. Enhancement of security 2FA based on SMS using post-quantum cryptosystem

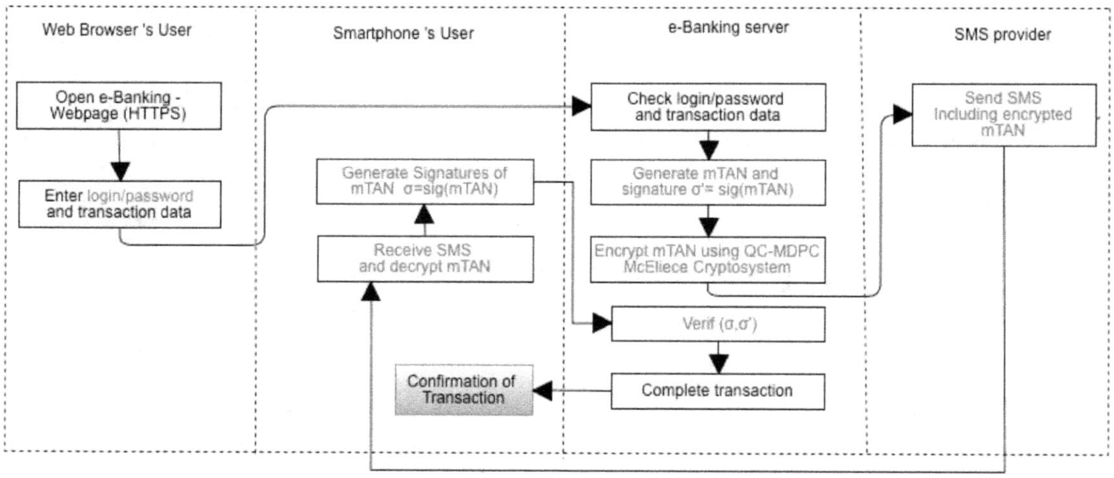

In the following, the authors describe the proposed authentication system operation:

Step 1: The customer of the financial institution launcheshis web browser, opens the URL of his bank in an https session and enters his login and password. Then, this data is sent to the e-Banking server, where they are checked.
Step 2: To complete the transaction, the customer clicks a button called "Send mTAN".
Step 3: The e-Banking server generates a random mTAN.
Step 4: The generated mTAN is encrypted using post-quantum cryptosystem MDPC Mc Eliece.
Step 5: An SMS containing the encrypted mTAN is sent to the customer's phone.

Step 6: Decryption is done by banking application to retrieve the mTAN.
Step 7: A notification is presented on phone banking application to confirm sending the mTAN to e-Banking server.
Step 8: After confirmation, a signature σ of the mTAN is generated by the phone banking application using ECDSA.
Step 9: The mTAN signature is sent to e-Banking server for verification.
Step 10: The server performs the signature verification by comparing σ of the customer with the signature generated by e-Banking server σ', using the Verif(σ,σ') function.
Step 11: If the signature verification on the e-banking server is successful, the transaction is completed. Otherwise, it is canceled.
Step 12: In both cases, a notification is sent to the customer's phone.

After describing their proposed post quantic SMS-based 2FA system, the authors will present below the results of its implementation and its security evaluation.

Implementation and Evaluation

In this subsection, the authors carry out an implementation and evaluation in order to demontrate the efficiency and the security of the proposed two-factor authentication system. Thus, they implemented a virtual banking server and then installed a mobile banking client application on the bank customer smartphone. The technologies used to implement the enhanced 2FA system are presented below:

- **Java Cryptography Extension (JCE):** An extension of Java Cryptography Architecture (JCA) which adds APIs for encryption and decryption, key generation and message authentication with Message Authentication Code (MAC) algorithms. In this work, the authors use this technology for the implementation of their virtual banking server;
- **Java for Android:** Android has a virtual machine named Android RunTime (ART) whichruns Java madeprograms. It is a virtual machine designed for mobile devices and their reduced resources. Java for Android is designed for programming applications that can run on this virtual machine. Thus, it was used for the implementation of client applications;
- **IDE Eclipse:** Eclipse is famous for the Java Integrated Development Environment (IDE). It allows to develop Java architecture applications (JCA), Java Cryptography Extension (JCE) and Java for Android. It was used as the development environment;
- **Android Virtual Device (AVD):** A configuration that defines the characteristics of an Android phone, tablet, Wear OS, or Android TV device that one wants to simulate in an Android Emulator. The AVD Manager is an interface that can be ran from Android Studio or from eclipse configured for the development of android applications.

These technologies were used to implement the secure 2FA system using thetwo cryptographic methods: The Quasi-Cyclic-MDPC post-quantum method for encryption and the ECDSA electronic signature.

For the implementation of the Quasi-Cyclic MDPC post-quantum encryption method to encrypt SMS content, the authors present in Table 2 a set of parameters with different key sizes, which depend on:

- The security level;

- The length n of the code C;
- The dimension r of the code C;
- The error correcting capacity t;
- The weight w of each line of the parity check matrix H;
- The number of blocs n_0 of the parity check matrix H i.e n= n_0.

Table 2. Recommended QC-MDPC parameters (Arnault, 2017)

n	r	W	t	n_0	Key Size	Security
9602	4801	90	84	2	4801	80
19714	9857	142	143	2	9857	128
65542	32771	274	264	2	327771	256

Since the authors aim to encrypt a message (Code mTAN) whose characters number is 4 to 8, they use the following parameters: $n_0 = 2$, n = 9602, r = 4801, w = 90, t = 84.

These settings are reasonable because the maximum character size of an SMS message is 160 (Brown, Shipman, & Vetter, 2007), which means that they will have no problem sending the encrypted content in a SMS over the network. In addition, it allows them to have a security level of 80 bits.

Moreover, in the proposed 2FA system, the authors perform an electronic signature of mTAN code using ECDSA on the customer's smartphone to authenticate with the bank's server. To choose the parameters values for the ECDSA signature, NIST recommends the key sizes listed in Table 3.

Table 3. NIST recommended key sizes (Olenski, 2015)

Symmetric Key Size (bits)	RSA and Diffie-Hellman Key Size (bits)	Elliptic Curve Key Size (bits)
80	1024	160
112	2048	224
128	3072	256
192	7680	384
256	15360	521

So, for the implementation of the ECDSA elliptic curves based electronic signature method, the authors choose 224 bit as key size which is equivalent to the RSA 2048 key size including a 112-bit security.

After presenting the cryptographicparameters used for setting up the proposed 2FA system, the authors willnow show a description of the system operation.

The 2FA two-factor authentication system is simulated on a virtual environment that was developed for this purpose. The server and the Android client use a communication channel established via sockets. The authentication begins by entering the login and the password on the bank web page. Theinformation is received by the bank server which generates the mTAN, as shown in Figure 3.

Figure 3. mTAN generation by the e-Banking server of the proposed 2FA system

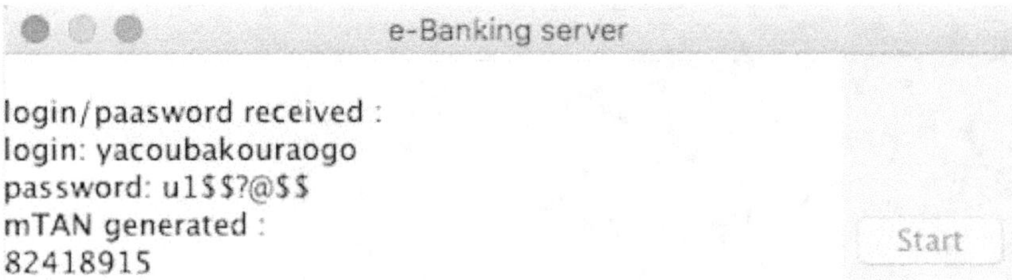

The generated mTAN is encrypted by the server and sent to the client. Upon reception of the mTAN-code, it is decrypted and displayed in a dialog box as shown in Figure 4.

Figure 4. mTAN reception by the proposed 2FA system client

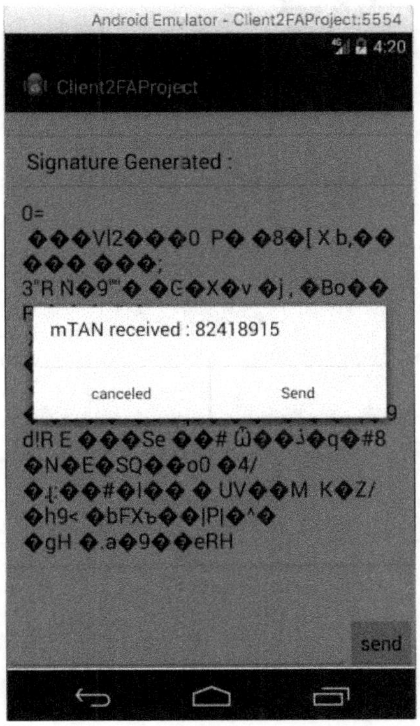

As soon as the client clicks on "send", the electronic signature of this mTAN is generated in the background and sent to the server for verification.

Evaluation and Comparison

Table 4 presents a comparison between the proposed system and existing ones. It's important to note that the proposed system addresses many two-factor authentication systems based on SMS communication security issues by adding, for exemple, the integrity feature that doesn't exist in the existing SMS-based 2FA systems, in addition to its post-quanctic property.

Figure 5. Authentication confirmation message

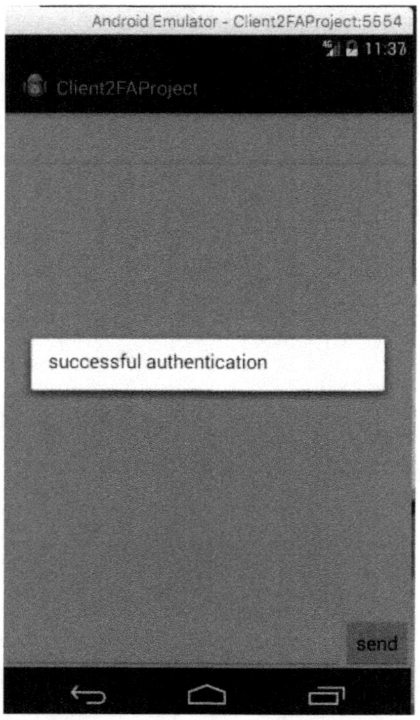

Table 4. Comparison of different 2FA security features

	Schartner et al. (2011)	Mulliner et al. (2013)	Aloul et al. (2017)	Our Proposed System
Authentification	Yes	Yes	Yes	Yes
Confidentiality	No	Yes	Yes	Yes
Integrity	No	no	no	Yes
Post-Quantic system	No	no	no	Yes

In the next subsection, the authors present a security analysis of their proposed system as a proof of concept of its security features.

Security Analysis

Client-Side Security

The strength of the proposed client-side system lies on the non-storage of confidential data, mTAN, in clear on the smartphone. Indeed, when the mTAN code is retreived and decrypted by the client-side banking application, this later automatically generates an electronic signature using the ECDSA method, whitout storing the mTAN itself. Thus, this mechanism allows to securethe system against the theft of confidential data by malicious programs on the client's smartphone and also allows the authentication of the client by the bank server. This means that a potential customer identity theft would be nearly impossible. In addition, on the proposed 2FA system, the brute force attack is almost impossible because the mTAN is not manipulated in clear but rather its signature, which presents a major benefit on the client side of this 2FA system.

Server-Side Security

In the proposed two-factor authentication system, server security is left to the financial institution. Nevertheless, it is important to know that the big advantage of the proposed server-side system is that the mTAN code is not manipulated in clear because it is processed either by its cipher or by its electronic signature. So, the recovery of mTAN as a clear text by malicious programs would be nearly impossible.

Security During Transmission

In client / server communication, security on the network is essential to mitigate certain types of attacks such as man-in-the-middle attack and network sniffing. Thus, many researchers have proposed different methods to ensure the confidentiality of SMS during its transmission. In the present system, the proposed method has a great advantage over the other approaches. Indeed, the authors provide security during transmission using quantum crypto post system Quasi-cyclical MDPC between the bank's server and the client smartphone. This method not only ensures confidentiality today but also during the quantum advent. The quantum revolution is very close with the success of the first inviolable communication on September 29, 2017, and the proposed system is designed to withstand the computer attacks of this revolution.

CONCLUSION

In this study, the authorsexamined existing two-factor authentication systems and their security issues. They also focused on SMS-based systems that are today used by most financial institutions. Starting from an existing system, they discovered numerous security problems such as the interception of SMS between the server and the phone, the generation of valid mTANs, the recovery of the SMS on the client phone by malicious programs. To overcome all these attacks, they have improved the existing system to better secure online banking by offering a secure 2FA system. The new authentication scheme implements cryptographic methods that ensure data privacy and authentication of the system users. Thus, for the confidentiality of data such as mTAN of the 2FA system, they used a post-quantum cryptosystem QC-MDPC McEliece. Then, for the client to authenticate the server, they used the secure HTTPS protocol. In addition, to avoid client identity theft, the authors implement the electronic signature using ECDSA based on elliptic curves. This will authenticate the client to the server.

Finally, it is important to note that the security analysis of the proposed system shows that this later is able to resist to attacks today and even with the arrival of the quantum machine.

REFERENCES

Adham, M., Azodi, A., Desmedt, Y., & Karaolis, I. (2013). How to attack two-factor authentication internet banking. *Proceedings of the International Conference on Financial Cryptography and Data Security*. Springer. 10.1007/978-3-642-39884-1_27

Aloul, F., Zahidi, S., & El-Hajj, W. (2009). Two factor authentication using mobile phones. *Proceedings of the International Conference Computer Systems and Applications* (pp. 641-644). IEEE.

Arnault, F. (n.d.). Theorie des nombres et cryptographie. Retrieved from http://www.unilim.fr/pages/_perso/francois.arnault/documents/polycrypto.ps

Arudchelvam, T., & Fernando, W. W. E. N. (2016). Maintaining User-Level Security in Short Message Service, *International Journal of Computer, Electrical, Automation. Control and Information Engineering*, *10*(6), 1121–1124.

Brown, J., Shipman, B., & Vetter, R. (2007). SMS: The short message service.

Dmitrienko, A., Liebchen, C., Rossow, C., & Sadeghi, A. R. (2014). On the (in)security of mobile two-factor authentication. *Proceedings of the International Conference on Financial Cryptography and Data Security*. Springer.

Eldefrawy, M. H., Khan, M. K., Alghathbar, K., Kim, T. H., & Elkamchouchi, H. (2012). Mobile one-time passwords: two-factor authentication using mobile phones. *Proceedings of the International Conference Security and Communication Networks* (pp. 508-516). Academic Press.

Hall, T. A., & Keller, S. S. (2010). The FIPS 186-4 Elliptic Curve Digital Signature Algorithm Validation System (ECDSA2VS).

Jha, S., & Dutta, U. (2015). Review on SMS Encryption using MNTRU Algorithms on Android. International Journal of Computer Science and Information Technologies. *International Journal of Computer Science and Information Technologies*, *6*(4).

Karale, S. N., Pendke, K., & Dahiwale, P. (2015, March). The survey of various techniques & algorithms for SMS security. *Proceedings of the 2015 International Conference on Innovations in Information, Embedded and Communication Systems (ICIIECS)* (pp. 1-6). IEEE.

Khamis, M., Hasholzner, R., Bulling, A., & Alt, F. (2017). GTmoPass: two-factor authentication on public displays using gaze-touch passwords and personal mobile devices. *Proceedings of the 6th ACM International Symposium on Pervasive Displays* (p. 8). ACM. 10.1145/3078810.3078815

Klein, A. (2012). The Song Remains the Same: Man in the Mobile Attacks Single out Android. Security Intelligence. Retrieved from https://securityintelligence.com/man-in-the-mobile-attacks-single-out-android

Kouraogo, Y., Zkik, K., Idrissi, N. E. J. E., & Orhanou, G. (2017). Security model on mobile banking application: Attack simulation and countermeasures. *International Journal of Intelligent Enterprise*, *4*(1-2), 155–167. doi:10.1504/IJIE.2017.087014

Kumawat, A., Sharma, A. K., & Kumawat, S. (2017). Identification of Cryptographic Vulnerability and Malware Detection in Android. *International Journal of Information Security and Privacy, 11*(3), 15–28. doi:10.4018/IJISP.2017070102

Maslennikov, D. (n.d.). ZeuS in the Mobile is back. Retrieved from https://securelist.com/zeus-in-the-mobile-is-back/29830/

Misoczki, R., Tillich, J. P., Sendrier, N., & Barreto, P. S. (2013, July). MDPC-McEliece: New McEliece variants from moderate density parity-check codes. *Proceedings of the 2013 IEEE international symposium on information theory* (pp. 2069-2073). IEEE.

Mulliner, C., Borgaonkar, R., Stewin, P., & Seifert, J. P. (2013). SMS-based one-time passwords: attacks and defense. *Proceedings of the International Conference on Detection of Intrusions and Malware, and Vulnerability Assessment*. Springer. 10.1007/978-3-642-39235-1_9

Olenski, J. (2015). ECC 101: What is ECC and why would I want to use it? GlobalSign GMO Internet Group. Retrieved from https://www.globalsign.com/en/blog/elliptic-curve-cryptography

Pan, J., Ding, Q., & Qi, N. (2012). The Research of Chaos-based SMS Encryption in Mobile Phone. *Proceedings of the Second International Conference on Instrumentation, Measurement, Computer, Communication and Control (IMCCC)*, (pp. 501-504). IEEE.

Pujitha, S., & Mallu, B. V. (2013). SMS based mobile banking. *International Journal of Engineering Trends and Technology*, *4*(4), 1211–1220.

Qu, J., & Tan, X. L. (2014). Two-factor user authentication with key agreement scheme based on elliptic curve cryptosystem. *Journal of Electrical and Computer Engineering*, 16. doi:10.1155/2014/423930

Rangarajan, S., Ram, N. S., & Krishna, N. V. (2013). Securing SMS using cryptography. *International Journal of Computer Science and Information Technologies*, *4*(2), 285–288.

Schartner, P., & Bürger, S. (2011). *Attacking mTAN-applications like e-banking and mobile signatures*. University of Klagenfurt.

Švogor, I., & Kišasondi, T. (2012). Two factor authentication using EEG augmented passwords. *Proceedings of the International Conference In Information Technology Interfaces (ITI)* (pp. 373-378). IEEE.

Vidyarthi, D., Choudhary, S. P., Rakshit, S., & Kumar, C. S. (2017). Malware Detection by Static Checking and Dynamic Analysis of Executables. *International Journal of Information Security and Privacy*, *11*(3), 29–41. doi:10.4018/IJISP.2017070103

This research was previously published in the International Journal of Information Security and Privacy (IJISP), 14(2); pages 102-115, copyright year 2020 by IGI Publishing (an imprint of IGI Global).

Section 2

Data Protection, Privacy, and User Risk Awareness

Chapter 17
Smartphone Confrontational Applications and Security Issues

Abhishek Kumar
https://orcid.org/0000-0003-4161-508X
Chitkara University Research and Innovation Network (CURIN), Chitkara University, Punjab, India

Jyotir Moy Chatterjee
https://orcid.org/0000-0003-2527-916X
Lord Buddha Education Foundation (Asia Pacific University of Technology & Innovation), Kathmandu, Nepal

Pramod Singh Rathore
ACERC, Delhi, India

ABSTRACT

In today's society, there is a high volume of smartphones, with Android being the most popular and most commonly used smartphones. In the last few years, the Android market has been booming, making lots of developers join the industry so as to create various mobile applications that are a benefit to people's lives. However, its over-popularity has brought many crime issues, including security. One of the major common incidents to mobile users is having their mobile phones lost or stolen. Since most mobile users want to find their lost phones, they are looking for the most reliable features that can help them locate their smartphones. Luckily, there are some developed applications and services that have been designed to track down and locate lost or stolen smartphones. In this work, the authors tried to identify a collection of these applications and the information they send to the user in aiding them to find their phone. Since some applications are able to send location information or a photo, this work will look at what metadata is usually sent with the message.

DOI: 10.4018/978-1-7998-8545-0.ch017

INTRODUCTION

In present days, the quantity of smart phones was expanding drastically because of the significance they have on individuals' day to day lives. Its ubiquity is developing quick every single day controlling a huge number of clients in excess of 200 countries (Zhang et al., 2017). Clients are introducing more recreations, apps and other advanced substance on their cell phones. Accordingly, it is expanding trustworthiness and dependability to numerous cell clients in noteworthy zones, for example, web-based banking, business, information stockpiling, and in all spaces of human services. In any case, practically all cell phone clients are terrified of losing significant information or records whenever the devices get lost or stolen. Insights were gathered from power from Cell towers demonstrates practically more than 12,000,000 devices were lost each academic year with 700,000 of devices got theft (Broida, 2011).

Therefore, this situation causes clients to endure a great deal financially, yet additionally they get the chance to lose significant information incorporating contact subtleties put away over the devices as well as memory card storage. Because of this, it ends up critical for a cell phone client to get back their smart cells. Nearly everybody may have encountered lost or stolen cell phones. Along these lines, it is principal to have an application in the gadget that can help distinguish a stolen phone. This paper highlights different apps with particular highlights, for example, discovery, call checking, area bringing through GPS, email notice, SMS notices and so on which helps in identifying the accurate location of the lost phone or theft phone.

Dominant part of cellphone clients for all time loses their smart cells as well as their own information. As a cure, the reason for the exposition is to feature a portion of the definite gathering of apps which helps cell phone client to identify the smart cells. The hugeness scenario of this concept is to reuse engineers in growing increasingly powerful and productive application frameworks for Android gadgets. Likewise, it will enable cellphone clients to be associated with the most recent innovation that can enable them to follow their stolen or lost phones. Which unmistakable highlights in an application helps in identifying the lost or stolen device? The general target of this thesis is to feature a portion of the apps that can help serve individuals viably whenever the smart cells are theft or lost. The apps are loaded up with different highlights and the objective of the thesis is to bring up different highlights on every application and how successful they can be on tracking gadgets and individual information. Android phones Operating System has caught over 75% of the all-out piece of the pie, leaving its rivals iOS, Windows Phone and Blackberry a long way behind. Despite the fact that cell phones were utilized in the earlier decade, dispatch of iOS and Android has changed the scene by creating a connectivity fascination globally among shoppers and engineers alike. Cell phones have turned out to be omnipresent because of wide scope of network alternatives, for example, GSM, CDMA, Wi-Fi, GPS, Bluetooth and NFC (G. Inc., 2013) demonstrated an expansion of 42.3% in cell phone deals from 2012.

Android does not prescribe, however permits establishment of outsider apps on gadget, which has worked up many territorial just as global application stores (Pandaapp, Baidu, Opera Cell App Store, AppChine, GetJar). Be that as it may, assurance and nature of apps accessible in outsider application stores involves concern (Zhou et al., 2012).

As portable apps are increasing expanding prominence among clients, the protection and security of cell phone clients become a worry. Because of the huge client base, keen gadgets are utilized to store touchy individual data more much of the time than PCs and work areas. As a result, a malignant outsider application cannot just take private data, for example, the contact list, instant messages, and area from its client, yet can likewise cause money related loss of the clients by making shrouded premium-rate phone calls and instant messages (Rothman). In the meantime, the quick development of the quantity of utilizations on Android markets makes it hard for application commercial centers, for example, Google App Store for instance, to altogether check if an application is genuine or malignant.

Use of cell phones has been summed up and use of it is not constrained to calling and messaging administrations. Because of improvement of cell phones, different jobs, for example, messages, web, digital book, Internet banking, and payments can be handled. Also, presently, the use of cell phones has turned into an everyday life, extending its region to all fields of life. Be that as it may, amidst reliably developing and growing new capacities, the security issues to ensure client data are not staying aware of the pace of advancement. Albeit different security innovations are looked into and created to develop reliably, there are as yet numerous dangers.

LITERATURE REVIEW

(Shah & Modi, 2019) gave a review on how an Android-based intelligent gadget can be safely utilized. To conquer the escape clauses in work done where significant work is done in the territory of intrusion detection and exceptionally less measure of work is done in the region of intrusion avoidance, they proposed dynamic preventive engineering for Android security. They demonstrated the outcomes for their proposed structure and contrast it and the comparable outcomes given by the accessible utility of Android gadgets. In their past work, they proposed a structure for their system and in continuation to that, here, they demonstrated the usage situation and the outcomes. The outcomes are tried for various sizes of apps from various authority/informal sources to get the affirmation of the proposed work. (Yoo et al., 2-019) introduced a visual analytics (VA) framework in cases of android privacy risks. They displayed risk factors dependent on the permission insights and figured three diverse risk scores, for example, access privileges problem analysis result, as well as category problem result. These problems resultant data were used in dissect individual privacy issues. So as effectively dissect the privacy information, we introduced a fixed privacy faults in VA along the privacy faults overall the life VA after some time period. Fixed perspectives were utilized in finding spammers were introduced at that point introduced over Android devices. The risk factor sees empowered cell phone clients to effortlessly perceive unintended security exercises. Their VA framework has been conveyed on Android cell phones from eight distinct clients and privacy related terms life log was gathered in very own gadgets. The overall information was not exchanged over any types of servers, along these lines, the protection was stored. They demonstrated regarding examination of the information gathered and demonstrated that our framework was valuable to comprehend the privacy terms over the Android gadgets. At long last, they have demonstrated a client study and criticism from clients. (Bhatt & Dutta, 2019) clarified assaults on various layers of Android in detail alongside their present arrangements, which will support scientists and programming engineers figure their methodologies for future malware detection and anticipation. They included forbidden clarifications of different dangers and detection approaches to give a superior understanding and similar examination. A portion of the significant discoveries incorporate the need of utilizing unique equipment items to keep the framework secure; the requirement for a powerful two-advance malware detection approach; utilizing both static and dynamic detection procedures, which can distinguish referred to just as obscure assaults; and making it mandatory for designers to pursue significant security conventions. Not at all like different overviews, they additionally give a top to bottom examination of different vulnerabilities in the Android framework misused by aggressors. The Android working framework will keep on developing after some time for various reasons thus will different apps running on it.

(Hussain et al., 2018) proposed mHealth Apps Security Framework (MASF) it is a type of framework helps to check the working of the application along with their customers' information. MASF results in diverse privacy along insurance risks of mHealth applications, for instance, information spillage, DMB strikes, benefits increasing attacks, and maltreatment of approvals. This framework offers parts to fine-grained get the chance to control, setting careful access control, and affirmation for the personal data by polluting the examination along information hiding. Additionally, MASF outfits customers by the capability in describing their own methodologies as demonstrated by their necessities to control application passageway to the framework resources subject to a couple of conditions which includes present setting of, for instance, the time and territory. The structure can renounce certain assets, repudiate access for framework peripherals, debilitate desires, and give shadow information. The proposed structure is executed over certifiable circumstance to overcome the certain properties of this architecture. As per the arrangement in MASF, it involves two rule components: The secure Module Layer (SML) as well as System Interface Layer (SIL). SML plays out overall basic privacy settings, privacy checking including portrays of all privacy plans which are actualized over applications at running time period, whenever SIL outfits SML within basic mediator for the Android internals. Everything considered, SML was an essential component, along the customer can simply speak by the help of this layer. Customers were able to describe the customized process in one of kind procedures along preserve them inside SML servers. SML contains a couple of parts and tackles top of the SIL. Due to the utilization, a stock Android structure was connected with a couple of catches along latest parts in realizing MASF.

Finally, we designed a structure that surveyed along separated similarly as reasonability and efficiency. Investigations prompted evaluate capability which includes examining the different types of attacks such as malware and spyware attacks. In the same manner by performing different test operations the performance shows the behavior on information de-duplicating, approval access privileges, weakening plans, along the behavior of engaging framework components. The test results of these examinations exhibited the MASF was convincing among overall the recorded attacking, along all the different attacks successfully verified mHealth apps as well as customer information on attempted circumstances. (Chang et al., 2018) displayed permission utilized security dangers and existing strategies for counteracting permission leak. To identify the malignant codes and improvement of authorization the executives and proposed a utilization and access control model and a powerful strategy for counteracting permission leakage dependent on ARM TrustZone security expansion. Rather than past work, the proposed security design gives a consent put together obligatory access control permission-based mandatory access control with respect to Android middleware, Linux bit, and equipment layers. The assessment results exhibit the adequacy in alleviating permission leakage vulnerabilities.

(Hoida 2018) identified Cerberus is the application among many applications which helps in identifying the lost or theft smart android devices. This application gives guarantee in finding the device in three different modes; remotely located, identifying the remotely located device by sending SMS from one device to our lost device which is in unknown place and the last method is getting alerts. Identifying by help of remote-control mode is a great challenging task which includes the following tasks such as location discovery, capturing events, getting alerts when it is in the tranquil mode. This application also helps in automatically talking of the text messages. This application also helps in taking the pictures of the person who is using the stolen mobile. This application has a feature of saving the history where the person moved from place to place. This application guarantees in capturing the stolen person data by removing any data either in phone or memory card data.

In such manner, it gives the alerts to the people who have done last calls from time to time. While performing such multiple operations by that application without any mobile device in our hand we can also do a trick like if our lost or theft device connected to a wifi or to a PC with the help of USB wire we can manage to extract the information of our device such as data modifications done on our smart phone, last calls used, last browser history. By all these clues we can catch our device very easily by the help of contagious Wi-Fi frameworks.

In a programmed method, Cerberus completes the processing where all the terms and conditions meet (Hoida, 2018). For example, sending the email alerts, sending the text message alerts, and sending the voice message alerts to the individual concerned person including the picture of the victim who uses application.

And mostly someone can establish the standards of the phone that is using this application in a customized manner this process called as "Auto-Task Section" which includes a alert called geo-enforcing where the owner of the phone gets a message that if the theft moves away from one geo-graphical location to another. (Hoida, 2018) depicted other highlights of the application that consolidate the fastest method of uninstallation confirmation that connects with Cerberus to endure through a plant reset and automatic empowering mechanism works with the help of GPS when one begins following the gadget. By temperance of the unknown location that draws in the app to perform with respect to that there is none other relationship from web in transferring messages and its data. On the off chance that there is a modification in the SIM slot on the smartphone, the SIM verifier who verifies highlight empowers by understanding the latest one that can build up an association on. Cerberus app has no impact on the data got and does not cripple the telephone when begun thusly making it an incredible client control for checking telephone or tablet. (Jonnalagadda, 2018) saw Google verification of smart phone feature that utilizes properties of top-level GPS navigation in identifying the gone or theft items. The accompanying element over this app work whenever the app is available over gadget along a person should authenticate sign over the site. This app which provides the region old data by drawing in over to see whether the gadget has been for the length of the day. One can in like way find the contraption by stamping on account wherever the sort "discover my telephone" over Google. Regardless on the task the configuration of the telephone envisions that was fought over web as well as zone changed as a manner that can connect with one access and discover the telephone. Best segment over the app is which it shocks gadget as well as makes data by means of going with the telephone number. That way, whoever has the contraption can just knob the "call me" jump over the device in this way knobbing the predefined phone number. The lock highlight is valuable when one has not protected the contraption with a mystery key or unique finger impression hence protecting one from getting to the home screen and just an acquaintance an arrival with number drawing in the individual helps in calling you. Right whenever the decree is no any alternate decision, this app can able to enable to play a sound whenever a person touches and press any button on the phone or performs and action like deleting telephone information. As appeared (Jonnalagadda, 2018), annihilate choice kills the majority of the information on the telephone long any type of data related with memory card whenever the mobile phone is isolated.

The fundamental unit of task when the contraption should be on Wi-Fi arrange or on cell information for the assistance of unequivocally identifying its accurate location. In any case, in context on the creator and variety of the Android sort out, the cutting-edge office reset outlines my recoup all of the data regardless of when the telephone is slaughtered. (Johnston, 2018) perceived McAfee which is a virus removing software that includes along with privacy app that shields on gadget by back to the front by providing a one week of unlimited usage preliminary by seeing the trial pack version, its features, working capacity and all if we are interested we can go to the complete version of software by paying the yearly packs of worth of $30 per academic year to cooperation. By performing in that limit, anyone can standardize the standard adversary of burglary affirmation, antivirus security, stronghold of information and reestablishes limits. The application checks any one personal information along updates presentation of smart devices. it provides safe way in looking at, Wi-Fi security, progress of showcases, fortress of SMS and contacts and different assurances. The features such as anti-theft and device privacy which consolidate into over the app which shocks our contraption by picking up a see over an individual holding the gadget. Crooks may be phishing as well as getting our own one of a kind data that can complete as long it have an enemy of different types of virus attacking is consolidate. By counter stolen highlight, this app can automatically remove the user data and store in a secure server and when the user logins they can be provided by their privacy information along that offers messages as well as user details strongholds. With guide an area benefits, the application finds, locate a present a notification alert that a system for following our theft or gone device. By the help of counter, the robbery victim catching process fuses, empowers on shielding cheats by removing the security patch over the wireless. Security consolidate having the app which helps in safely shocks delicate information along the call block by the help of SMS channel. Utilizing the bleeding edge antivirus programming cleans and remove toxic data by means of records, messages, memory card along web downloads. Secure web accessing and Wi-Fi services were squares openness to goals which were hazardous that gives security by any type of unbound structures. Improvement in execution over app supports energy in expanding its lifetime rack, help data size with help of stimulating one's contraption, and in following use of information. Highlights such as calling, blocking, along reading the text messages were not provided access permissions in contraptions which have the android alteration which is greater than 4.5. To different androids, the client must enable straightforwardness of the application to empower device, securing our app by different techniques.

(Hindy, 2018) worked out regarding AVG anti-virus software, it is a software helps to detect and remove the malware from the system. Even though we browse the internet it checks the server address is trustable one or not by the help of URL address. The same app is tested and working on the mobile devices in different applications and installed on smart held devices. With application lock, each sensitive datum and applications can be dashed up hence guaranteeing security and thriving of private data.

As appeared (Hindy, 2018), this app has a special feature of securing model by which it permits app in the picture which evaporates over the app plate in this manner persuading the chance to be unpretentious for criminals and hard to remember. the data which is in unknown location can eradicate complete device information such as phone call details, messages details, applications details, email records which makes difficult in cheat in getting whatever data over the phone. Making mobile restart remotely highlight which makes us comfort in restarting the device by shocks by the help of referencing a PIN number. The alert component which makes the device to ring continuously resoundingly to along whenever criminal undertakings quietness, the total amount of quantity gives out the best. By the help of navigating the device location from unknown place which includes, the phone can be masterminded through GPS, Wi-Fi or any PDA structure. Precisely whenever the sim slot is modified the app gets the alert message that to be Discovered bearing which can arise the (Lee et al., 2017) showed SEALANT, a strategy that unites static investigation of utilization code, which deduces helpless correspondence channels, with runtime seeing of between application correspondence through those channels, which foresees attacks. SEALANT's expansive evaluation demonstrates that (1) it recognizes and squares between application attacks with high accuracy in a corpus of in excess of 1,100 real worldwide applications, (2) it encounters less false cautions than existing frameworks in a couple of representative circumstances, (3) its show overhead is superfluous, and (4) end-customers do not imagine that it is hard to get. (Karthick & Binu, 2017) tells about the maltreatment of use approvals using Shared User ID, how two-factor affirmations fizzle due to wrong and not recommended utilization of use assents using spyware, information burglary in Android applications, security breaks or ambushes in Android and investigation of Android, iOS and Windows working framework as for its security.

(Pothitos, 2017) identified the lost device which identifies the mind-boggling app that overall segment high in quality that advances in agreeable along impact customers that they are able to screen, controlling as well as likewise locate their telephone whenever it is theft or stolen. Smart devices are ought to be presented as well as it is continued running until telephone been theft as it empowers by enlisted by Google account owner. Not in the slightest degree like diverse applications that are profitable concerning making a mystery expression, Lost Android will provide the basic when one person registers with a google account holder creation, so they did not need to ace complete private data. Whereas hugeness, gone devices engages in case of screen along controlling the smartphone by the following present region along by getting to text data along archives which by means of a potentially developed whenever the cell app along the server's data on the devices were merged. This app gives permissions to each and every user in scrutinize text data from unknown location and gives bearings by means of text messages. In like manner, it is immaculate by telephones that are having android of version of 2.3 and more consequently it might be used in for all intents and purposes all telephones. In spite of the way that it will, in general, be confined to build up telephones, the features of this application are for all intents and purposes that we can get on overall telephones. None repressions whenever using the app along these lines, it helps in a wide scope on customers in the help of engaging to use the functionalities such as zone following option else the impelled one can accessed from unknown location scrutinizing from the starting point record over the cell phone.

Zhang et.al. (2017) recognized mobile navigation application that generally used in a place. This stored data identifies over discovering theft or else gone telephone by face recognition area took by using covered rear camera as a result it provides the photo of a particular theft from starting at now by using telephone. Right whenever endorsed access is recognized as well as pictures are taken, a MMS message like image, audio, video and text was forwarded to the latest PDA number. Here user performs consequently without any problems were faced when using the application until an individual just requires consenting to acknowledge the record through the association. The app is monetarily smart since it cost insignificant in the wake of perceiving affirmed get to and we get phenomenal chances in having an undeniable photo of the theft person until app was just snaps an image consequently to distinguishing faces. Mobile number was set in the application with the link of customer to wherever a person needs getting reply by the presented app. From long back the app was starting at now presented in the telephone, the customer just requires pushing the application and setting the mobile number including particular country code along mystery state. Customer moreover helps in verifying along assert for checking whether the value was endorsed. If it is affirmed, the variables that arrange by the possible have mystery state. This app having composed examination fundamentally based with reasonable devices along limits which satisfies the basic conditions of the association. For example, the acquainted application uses different cases which traces with verify whether the customer was affirmed else unapproved. The app was made by the present instruments along advancements in following items which includes the Android Operating System, Java Programming and Platform and Software Development Kit (SDK). Telephone goes by Android OS-dependent savvy cells by extraordinary web bolster similarly as a not too bad objective of rear camera. Here customer should at have to look after mobile number along mystery expression that is given along the configuration settings. elective in order to customer get endorsed get to. The alert that encases shows up over the device in which aided over planning primary term by the help of mystery express along these lines choosing whether the individual is endorsed or unapproved. A basic component that goes by the help of app identifying the face acknowledgment here the camera snaps an image exactly whenever there is face recognizable proof. The application uses the face acknowledgment development in getting a sensible face of the person. exactly when the face is recognized, this app forwards Multimedia data by the help of message to the user telephone number by then makes the customer to track the individual knowing their device exact spot/place.

(Riley 2017), perfumed different operations and done many spamming, blockings, attackers and send many inappropriate mails to check the efficiency of the software. The post is secured with an extraordinary fragment “Find My smartPhone” that draws to find a theft or stolen device by using Google maps notwithstanding whenever the Global Positioning System of the device is butchered. By the commencement over resounding caution, part over the app can able to find the phone regardless of whenever the device is set over quiet position. Additionally, can teach somebody on the last recognized region in spite of when there are come up short on the power. This app guarantees over protected along privacy on connecting to Wi-Fi devices with help of certification from unknown Wi-Fi assaults by providing an affirmed assessment over quietness whenever interfacing with obscuring Wi-Fi structures. This app was a structure expert by checking the primary recognizing confirmation along guaranteeing working framework over on contraption was performing palatably. Post gives assurance in robbery from watching precious along money related data along it can notify to somebody on the off chance that something is found uncovered on the web. As per (Riley 2017), this app has a feature of SSN to observe the unquestionable setting of names, locations and different records related with one’s SSN and gives proactive notice for any type of change which flag false record advancement and burglary. In addition, it gives a watch by methods for electronic frameworks organization news by means of watching record along verifying weather the particular data was in danger. Overall this guarantees with the shielded by robberies which get to our electronic life user details whenever our device is gone or theft. This application has a mechanism of user identity as well as restoring and Insurance fuse the ideas for the duration of the day, reliably recreating undertakings help. If there should be an occasion of robbery, the application guarantees it re-establishes and recuperates lost a character like ensured ID. The part likewise which gives the discounts like charge less to our wallet recuperation in providing an infusion in accomplishing report guarantors so as they can drop and reorder charge cards comparatively as give perceiving proof substance. With robbery arranged part, Lookout which engage in getting the message by the help of photograph along district wherever the victim is. In the event that it winds up being difficult to get to the mobile against criminal, lock along wipe highlight over the app can access from unidentified places to dash up to contraption along deleting the information.

(Acar et al., 2016) shown a systematization of Android security and assurance look at with an accentuation on the appification of programming frameworks. To put Android security and assurance examination into the setting, they took a gander at the possibility of appification with normal working framework besides, programming environments. While the application has improved a couple of issues (e.g., grandstand get to and usability), it has in like manner exhibited a whole extent of new issues and annoyed a couple of issues of the old biological systems (e.g., coarse and ill-defined technique, poor programming headway practices). A part of our key disclosures is that contemporary research as regularly as conceivable stays on the by and large acknowledged approach as opposed to following unusual and as often as possible empowering new courses. Various security and assurance suggestion base by and large on the Android OS and do not abuse the intriguing features and performers of an appified environment, which could be used to take off new security frameworks less hazardously. Their work highlights regions that have gotten the greater ideas of thought, which aggressor models were kept an eye on, who is the goal, and who has the limits and inspirations to realize the countermeasures.

According to (Michael 2016), gone smartphone is a difficult task for cheats in recognizes that there is someone following along checking until the app was hidden by an individual note application. The web interface in like manner expects an imperative occupation in the application by engaging one to set up a discretionary telephone where one can impart something explicit headings to your own one of a kind phone. Moreover, the application can be for all intents and purposes worthless if it has not been continued running on any occasion once before losing the contraption. With everything considered, Lost Android is the best and most valuable resource which can give benefit with following theft or else stolen telephone.

(Cecchinato et al., 2016) orchestrated Kaspersky Cell privacy app that which can be used in checking telephones by threatening different viruses, yet furthermore by discovering theft or stolen android telephones. An adversary of burglary privacy in the app which is a licensed one in block else unblocking in finding our Smartphone, setting off alert, snapping an image by the phone, cleaning the device which includes singular data, for instance, contacts, calls, along texts. messages headings give the setting off alert, catapulting along finding contraption similarly as covering singular information. Not in the slightest degree like other security applications which do not recognize pre-arranged digits, Kaspersky tasks performs speedily with quality enduring digits to see that bearings have puzzle PIN pre-doled out from the customer by tolerant telephone. Security confirmation feature of the application helps sequestered from everything all exchanges with a picked contact of the customer. Regardless, this is not valuable to the customer of Android 4.5 and greater versions. voice and text type of communication is filtered by an imperative segment which goes for impeding specific person from the overall contacts. In any case, it needs an astonishing proportion of time for structure the contact records. Data attacking by anti-phishing helps by checking moving toward texts and in preventing those that include associations with dangerous destinations. This is in light of the fact that risky destinations may put your contraption in peril causing cash related and another personal information thievery. App locking function gives benefit in case of incorporate puzzle while getting to personal text data, pictures, along different types which are near and dear data which makes it difficult for individual for get too tricky information specially in case of contraption lands over unwanted persons (Cecchinato et al., 2016).

Discover of the telephone over the app that screens your telephone that gets theft or gone. The alarm that turns on automatically without any user interaction on the telephone along individual information like photos, works and chronicles are cleaned away. This app can in like manner which takes a lot of individual by and by handling all by contraption by making straightforward in discovering the telephone. The telephone having the option of setting the help in voice mode from the vary starting Find My telephone feature along these lines promising you are verified unrivaled. The app provides the telephone numbers, calls, as well as such any type of delicate data is abstained from prying eyes. By making filte mechanism standard can moreover it can be used in block as well as in boycotting any unknown, unrelated, unwanted calls or text messages.

As defined by (Minarovic 2016) the application uses a top tier GPS zone advancement in specifying the constant of whereabouts of those that you are associated within the circle from this time forward dividing the territory. By this manner this app gives benefit in providing confirming of relatives just as assistants in checking your telephone by finding its region when our device was theft or stolen. Driver Care which provides the aspect of the application goes along with one to a live specialist through telephone from now on engaging others to see where you are especially when in inconvenient conditions, for instance, vehicular effects. In like manner, a single push catch of the part approves that you are the one using the telephone.

(Thomas, Beresford & Rice, 2015) mapped the staggering arrangement of members on the Android environment who is having capable in mixing in giving revives, along build up that inaction by specific creators and framework directors infers various handsets are weak against fundamental vulnerabilities. They portrayed the FUM privacy function in ranking the best introduction of device producers and framework heads, in light of their course of action of updates along prologue to fundamental vulnerabilities. By the benefits of 20400 products, exhibited where tremendous anxiety over the advantageous movement of privacy invigorates transversely over different contraption makers along framework heads. This gives a relationship point to purchasers and controllers to make sense of which contraption makers and framework directors give security updates and which do not. They found that all things considered 88.8% of Android contraptions were displayed in any occasion to the essential accountability along, over biological system, all in all, consign a FUM security score of 2.87 out of 10.

(Area 2014) brought up on "Where's My Droid" app to find that there are no settings required for one to introduce Where's My Droid application and no impact on the data quality that is gotten. In any case, when the Android phone is stolen, the primary thing that is required is to verify the contraption in order to guarantee each close to home datum is protected. The contraption can be dashed by lock thought content which is sent trailed by a 4-6-digit stick like "WMD lock 1234" (Parish, 2014).

(Shull 2013) featured on Plan B app by which assistance telephones that starting at now are theft or lost. This app was exact that can introduce even though the device was vanished. Precisely when plan B begins, one thing is to set up to get mail messages from the app to the zone of the missing or stolen phone. Each district email involves an exactness rating in meters. Regardless of the path that now and again there is a need to send an SMS in beginning, it generally begins when the application is downloaded. The domain is sent to Gmail address which was starting at now paired with the device. Thus, making it difficult for others to get to the data sent to you by the application. The system stays dynamic for a couple of minutes and after relentless interims, it continues sending new messages with new organizes.

(Abhi 2012) explored the same type of app from the internet, affiliation and internet browser, which provides finding the goe or theft device. A basic as well as great fragment in the complete app was "Spy Camera" that aids review photos taken from the cam and by anyplace overall the world. Any progressions happened over the sim provides by a moment notice mail including subtleties of mobile number related to new card, if we want to send announcements, Antidroidtheft app should be in a working condition with connected to internet. As appeared (Abhi 2012), amazing information affiliation guarantees one gets a notice on any updates. Consequently, crippling telephone utilities rapidly may save some exertion for the cheat to a relationship with the web.

According to (Abhi 2012), highlights give data on the control zones where the cell gadget may utilize a guide. A privacy alarm which makes the user warns whenever the theft changes the sim to the device owner by email if the person links his/her mail to the app. GPS Geolocation gives facilitates that help in pinpointing precision thought about the guide. Prey Anti-Theft likewise has a wipe fuse that erases and shields one's confirmation furthermore it has a record recovery highlight that aides in recovering the document. Not only does the application offer security and protection in close to home, home and in independent company, yet it additionally has enormous needs as it discovers it use in various gigantic organizations. Along these lines, Prey Anti-robbery is a valuable application for all people and it does not inactivate the phone when introduced. Antidroidtheft is secret word verified and nobody can totally observe or utilize client data in the application. It utilizes consent when scrutinizing the contacts and reaches in putting away in the application and in showing picked contact number when one gets into this application. Additionally, it utilizes authorization when examining the condition of the phone, while getting the phone number of the cell phone, the present data on the cell compose and any rundown of phone records enlisted. This data is gotten in a customized mail and SMS.

In spite of the manner in which that approach B is a helpful application for discovering lost or stolen telephones, (Rehman 2011) expressed that the application is not noteworthy on Android OS 3.2 and OS 4.0 since it involves different changes that reason the application not to work appropriately at establishment or through a text that involves solicitations of areas. Right when the application is downloaded, a notice on the zone of the phone shows up in the notification bar. This can be an issue since the criminal is no doubt going to uninstall the application quickly, he/she understands somebody is endeavoring to tail them. Plan B application has no remote lock and wipes capacities which are basic for any stolen or lost phone. Plus, the application may not work respectably on the off chance that one has introduced an outcast SMS application in their gadgets. Likewise, the gadget must have a SIM card for Plan B to work appropriately. On the off chance that in the event that the hoodlum evacuates or changes the SIM card, it might make it difficult for one to find the contraption since there will be no Google email matched up to the phone, in this way no messages on district can be gotten. In the event that the phone is executed or it is in the rest mode and there is no framework affiliation, Plan B may not work. Additionally, if the application is introduced on a tablet, SMS zone may not work since this choice uses a phone number which most tablets are not outfitted with. Along these lines, it meddles with the possibility of data that is being gotten when following and finding the phone.

(Munoz 2011) raised on SeekDroid Lite application which has all the standard highlights like the remote lock and wipes, the foundation of the SMS, keep up a key separation from seeing mode, arranged ring despite when the phone is quiet, and drawing in the piece of the GPS. Advancing call logs can likewise be recovered and the application work outstandingly even without a SIM card related. SeekDroid Lite uses a history map in finding the developments of the stolen contraption in this manner drawing in one to know where the cell phone is going. With the application, one can likewise get protection for different telephones and get a notice at whatever point it leaves a particular territory. Some different highlights of the application that it offers are that of finding one's phone from their site, remotely dashing up the contraption, setting a caution with a custom message despite when the phone is quiet and cleaning the phone and the SD card at whatever point it is lost or stolen. Likewise, through a text, one is set up to find the gadget. Precisely when the application is not being utilized, there is no immediate in battery and the application can draw in one get data like the consistent calls, IMEI and the element of the battery. Breadcrumb zone is a valuable choice for the application that aides in ceaselessly following of the gadgets inside a multi-day time length.

(Broida, 2011) saw that SeekDroid Lite uses the authorization of the Device Administrator and takes the perfect position of its helpfulness inside Android. Contraption Administrator is a helpful handiness since it has a power lock fuse that bolts the gadget compellingly from its site or through SMS. Likewise, it has a reset-secret phrase highlight that is utilized in resetting the lock screen on the cell phone. Wipe-information join is utilized in eradicating or cleaning the contraption from its site, yet it ought to be done as the last choice for your information since when the gadget is rebooted, all data.

DISCUSSION

Cell following applications are of surprising use to various remote clients in such a way they shield the information of the client from being deleted, keep the cheat from physically getting to the information and of criticalness is the way wherein the application empowers remote access component for the proprietor to get to the contraption while still under physical control of the hoodlum. While there exist different particular applications, none of them is set up to fulfill these properties. A portion of the applications makes it difficult for the hoodlum to recognize whether he has stolen a conventional gadget or the one with an introduced cell following application. Research completed demonstrated that 34% of Android gadgets are not verified by a locking instrument (Groß, Tiwari, and Hammer, 2017). This makes it difficult for a criminal to recognize which has the application and which does not have. Some applications have a detectable application locking instrument that is at first utilized in guaranteeing client application information. The presence of a locking application suggests to the cheat non-presence of the application.

The lost or stolen cell gadget is an essential issue that is expanding every single day. The research demonstrated that more than 3 million gadgets are stolen or lost each year (Salim and Vargese, 2016). Along these lines, solutions for unraveling this have been established in order to straightforwardness hurt if there should be an occurrence of gadget misfortune or robbery. Android cell following applications together with their foe of theft functionalities accept a significant action in following and finding stolen or lost telephones. Most have been seen to utilize GPS and Wi-Fi for finding the region. Subordinate upon the SMS, subtleties of the lost or stolen phone can be watched. Undoubtedly, even change in SIM card, zone, picture and contact subtleties of the criminal can be recognized. Since all of this data and activities is running at the foundation and not appeared in the undertaking manager, it makes difficult for the cheat to remember he/she is being sought after. In any case, some applications have appeared at work when the contraption is turned off, there is a need for future applications to be built up that can thusly switch on at whatever point the phone is off for over 48 hours.

Having incorporated a portion of the telephone security applications, it is phenomenal to guarantee one uses different techniques for verifying their keen cells. One path is to keep subtleties by making a record of all data on the gadget and having it kept in a verified spot. Such data incorporates a security lock code, PIN no. phone number, IMEI number among others like concealing and appearances. A security etching can be consolidated by utilizing an Ultra Violet pen to print the house number and the postcode onto handset of the cell. This is to guarantee the phone is unquestionable in the event that it is stolen or lost. The ID of the email and contact number can likewise be added to guarantee the pioneer contacts the proprietor. It is extraordinary to enlist the phone in the framework overseer since it is through IMEI number one is set up to hinder the cell contraption and any delicate information (De, 2016). To jolt the phone, the security PIN highlight or lock code ought to be utilized. This is to make it less essential to the one holding it and give them expulsion when they attempt to get to private numbers that are put away on the SIM card of the gadget. A police report ought to be documented in a flash as it will make the remote provider progressively valuable particularly if protection is concerned. An essential, formal solicitation ought to be mentioned from the transporter in order to vanquish the transport from inceptions particularly when matters get stunning. Another technique one can likewise apply is of changing the secret phrase quickly it escapes locate. This is on the grounds that, when not changed, the phone stays marked in into one's Google records, Facebook and different stages. Likewise, it is fantastic too in every way that really matters change the secret key once in a while on each record marked in. Along these lines will make it difficult for the criminal to get to any close to home data from your gadget. Likewise, it is phenomenal to get a portion of Google's WearOS that can enable you to connect with your phone to your wrist all occasions and all over the place. The wears will keep you invigorated on whatever happens to your phone.

CONCLUSION

Appeared differently in relation to different keen cells, Android shrewd cells are the least expensive and their uses are expanding quickly. Reliably, its qualification is developing quick and due to its adaptable functionalities, remote clients are directly bending up continuously subject to the Android phone. A ton of information including contact subtleties and private information can be put away. Regardless, the rate of misfortune and burglary is expanding altogether making it difficult for the remote client to get to all data in the contraption. Along these lines, it has enabled some application designers to consider applications that can enable cell clients to return to their novel gadgets. Such applications have been structured with incredible highlights like face recognizing evidence, lock and wipe remotely, ring so anyone may hear caution and all altogether help in following and finding the gadgets. As included above, such cell security applications are of mind-boggling help to cell phone clients as they can enable them to arrange back their gadgets and catch the individual who was holding their phone.

REFERENCES

Abhi, T. (2012). 9 Android tracking apps to find lost Android smartphones. *Journal of Gadget explore*.

Broida, R. (2011). Recover a Missing Android Phone with Where's My Droid. *PCWorld*, *10*(125).

Cecchinato, M. E., Sellen, A., Shokouhi, M., & Smyth, G. 2016, May. Finding email in a multi-account, multi-device world. In *Proceedings of the 2016 CHI Conference on Human Factors in Computing Systems* (pp. 1200-1210). ACM. 10.1145/2858036.2858473

De, D. (2016). *Mobile cloud computing: architectures, algorithms, and applications*. Chapman and Hall/CRC. doi:10.1201/b19208

Gibson, J. M. (2001). U.S. Patent No. 6,198,431. Washington, DC: U.S. Patent and Trademark Office.

Groß, S., Tiwari, A., & Hammer, C. 2017, October. ThiefTrap–An Anti-theft Framework for Android. In *Proceedings of the International Conference on Security and Privacy in Communication Systems* (pp. 167-184). Springer.

Hill, S. (2018). Here's how to track an Android phone, tablet, or smartwatch. *Digital trends*, *47*(8).

Hindy, J. (2018). 5 best find my phone apps and other find my phone methods too! *Phone Spy*, *10*, A2.

Hoid, B. (2018). Use Cerberus to take control of Anyone's Android Phone. *Wonder How To: Null Byte*, *10*(4).

Johston, N. (2018). McAfee Total Protection (LiveSafe) Review. *Journal of Top 10 Reviews*, *2*(34). Retrieved from https://www.toptenreviews.com/software/security/best-internet-security-suites-software/mcafee-review/

Jonnalagadda, H. 2018. Find My Device: Everything you need to know. *Package tracking portal*, *23*(856). Retrieved from https://www.androidcentral.com/find-my-device

Michael, K., McNamee, A., & Michael, M. G. 2016. Lost Android for Android. GPS tracking and monitoring. Retrieved from https://download.cnet.com/Lost-Android/3000-20432_4-75623112.html

Mick, S., & Mcgraw, K. 2018. How to use Find My iPhone to rescue your iPhone, iPad, Mac, and AirPods. *iMore*, *12*(436). Retrieved from https://www.imore.com/find-my-iphone

Minarovic, J. T. (2016). U.S. Patent No. 6,198,431. Washington, DC: U.S. Patent and Trademark Office.

Munoz, J. (2011). Android App Review: SeekDroid Lite. *British Journal of Educational Technology*, *43*(1), E1–E5.

Parrish, S. (2014). Where's My Droid App Review: AndroidGuys. *Idea*, *31*, 297.

Pothitos, A. (2017). Lost Android app review- Control & find your lost phone remotely. For free. cell phone tracker. Retrieved from https://cellphonetracker.net/lost-android-app-review-find-control-your-lost-phone-remotely-for-free/

Rehman, A. (2011). Remotely install Plan B to track down your lost Android phone. Journal of Addictive Tips. Retrieved from https://www.addictivetips.com/mobile/remotely-install-plan-b-to-track-down-your-lost-android-phone/

Riley, S. (2017). Lookout Security & Antivirus: Taking Your Chances. Tom's Guide. Retrieved from https://www.tomsguide.com/us/lookout-mobile-security,review-2099.html

Salim, S., & Vargese, S. M. (2016). Monitoring system for detecting mobile theft. *International Journal of Computational Science and Information Technology*, *4*(2), 39–45. doi:10.5121/ijcsity.2016.4205

Shull, D. (2013). Plan B-Locating your phone AFTER it's lost or stolen. *Cell phone tracker*. Retrieved from https://cellphonetracker.net/plan-b-locating-your-phone-after-its-lost-or-stolen/

Zhang, M., Joshi, A., Kadmawala, R., Dantu, K., Poduri, S., & Sukhatme, G. S. (2009, October). Ocrdroid: A framework to digitize text using mobile phones. In Proceedings of the *International Conference on Mobile Computing, Applications, and Services* (pp. 273-292). Berlin: Springer.

This research was previously published in the International Journal of Risk and Contingency Management (IJRCM), 9(2); pages 1-18, copyright year 2020 by IGI Publishing (an imprint of IGI Global).

Chapter 18
Theft Preventive Measures for Interconnected Personal Computer Devices as Proactive Physical Security of Data

Ekaterina Pshehotskaya
Moscow Polytechnic University, Moscow, Russia

Oleg Mikhalsky
Moscow Polytechnic University, Moscow, Russia

ABSTRACT

This article is concerned with the arising problems and implications of physical security and privacy of personal and control data on portable computer devices, especially smartphones. The authors consider various classifications of portable computer devices, isolate smartphones as a most common device, and study types of user behavior regarding the involved security risks of unauthorized access to the data stored both locally and remotely with accent of physical data access via device theft. Based on provided categorization the researchers discuss the factors and criteria suitable to generalize user patterns and evaluate the corresponding vulnerability level against specified statistics. The considered statistical criteria can be formulated as a mathematical model of relative risks and implemented as a service or an application to be used for improving user awareness on current threats to his personal data and respective interconnected personal portable devices.

1. INTRODUCTION

The ongoing advance of computer technologies has already resulted in almost total civilization dependence on exponentially growing amounts volume of produced and stored data and the means of processing it. Modern hardware and software capabilities are generally both affordable and user-friendly to the general public. The simplicity of operation has already resulted in the fact that manufactured devices with WAN

DOI: 10.4018/978-1-7998-8545-0.ch018

access are outnumbering human population. The growing network of such standardized communication nodes is usually termed as "Internet of things" (IoT).

However, the simple and comfortable way of operating portable devices doesn't imply appropriately high level of security. On the contrary, the security measures are usually subdued to increase usability and achieve more cost reduction. Therefore, the remaining unresolved issue of most concern is the vulnerability of majority of devices, lacking any plausible means to provide reliable authentication of incoming connections. The range of the devices spans from network-accessible digital modules with MQTT-protocol for primitive sensors (e.g. house-hold Wi-Fi digital thermometers) to embedded single-board computers (e.g. Raspberry Pi) for more sophisticated consumer electronics like refrigerators, washing machines and TV-sets. Due to standardization of both machine architecture and communication protocols every node can serve as a breaching point for malicious cyber activity. Thus, the modern technologies gave rise to wide diversity of threats to device-stored data with multiple vectors of attack.

The overall problem of data security already gained a planetary scale. According to the recent Breach Level Index by Gemalto (Gemalto, 2017) the volume of the reported data breaches exceeded 1.9 billion of records only in the first half of 2017. The primary sources of the breaches where identified as malicious outsiders (74 percent), accidental loss (18 percent) and malicious insiders (8 percent). Narrow focused Bitglass report on financial services breaches (Bitglass, 2016) shows that the major proportions of breaches in the financial services sector where attributed to lost or stolen devices (25.3 percent) and hacking (19.2 percent), while malicious insiders where considered a source in 13.1 percent of cases. These facts indicate that device theft poses a serious threat to data security, and is a major factor contributing to large scale data breaches. The survey of Kensington IT Security Systems (Kensington, 2016) demonstrate that, despite such circumstances, more than a third (34 percent) of IT personnel lacks any physical security policy for corporate electronic assets, over half (54 percent) of survey participants do not consider physical locks for IT equipment, and 80 percent of survey respondents disregard use of locks for accompanying auxiliary equipment. Thus, 52 percent of the devices are stolen from the offices, and 24 percent from the conferences. The Recipero report on US Cell Phone Crime Statistics in 2016 (Recipero, 2016) addresses the issue of personal thefts, and reveals that 31 percent of devices reported stolen were NOT blocked by operators. Moreover, 7 percent of devices that are blocked when an insurance claim is made for loss or theft were unblocked within 30 days.

Regardless of possible data compromise, personal device theft results in financial losses for the owner even in case of empty smartphone. To counter such kind of crimes various measures were taken on nation-wise scale, including Cell Phone Theft Prevention Act of 2016, and introduction of regional-scale device check services like CTIA's Stolen Phone Checker (https://stolenphonechecker.org) and Recipero CheckMEND (https://www.checkmend.com/). The smartphone checking functionality of these services is built on blacklist IMEI databases taking granted the permanent nature of device identification number. However, this assumption is wrong, since the practice of replacing IMEI to render the smartphone untraceable became common and inexpensive (Pankhuri, 2017). The smartphone models built on MediaTek processors are most vulnerable to IMEI-change. Qualcomm Snapdragon processors are more resistant to flashing, but still yield if enough efforts are applied. Apple devices are immune to IMEI-replacement, however, they are usually dismantled to be sold as separate spare parts. Considering above mentioned facts, one can see that the theft mitigation is still greatly undervalued, and new proactive measures are required.

We aim is to reveal and estimate risks to the data based on categorizations and specifications of user devices and behavior. The provided categorizations serve to compute risks of smartphone theft and consequent compromise of personal information. Due to the statistical nature of the proposed mathematical model, only risks relative to a specified reference profile are computable. Our goal is to provide a mean to detect immediate changes of risk value based on user profile and current environment based on geolocation readings. Tracking of risk value changes can be implemented in the form of automated smartphone application, which will inform user and adjust smartphone security settings to the actual surroundings. This paper is a rework of conference paper (Mikhalsky & Pshehotskaya, 2017) presented at 20-th Conference of FRUCT Association. The paper is augmented with recent statistical data on cybersecurity in Russian Federation and updated data on smartphone thefts for London and Moscow. The mathematical model for smartphone theft risks has been reformulated to accept various reference user profiles. More transparent examples based on supplied data are provided. Possible fields of application, including software solutions are added at the conclusion section of the paper.

This article contributes the proposed model of risk estimation relative to reference profile as well as the revealing of possible deprecation factor impacting the theft-attractiveness of smartphone models.

The paper is organized as follows. The second section considers research of various aspects of statistics on personal data security dependence on user qualities and behavior as well as advances in software design. In the third section we study device data security from the points of threats and risks associated with type and manufacturer of device, as well as user experience and awareness of security threats. We argue that the majority of threats and risks, the user can take care of, are to smartphone segment of devices. Accounting for statistical surveys, we identify main threat to personal data as physical theft of the associated device. Also, we confirm qualitative agreement between independent data of smartphone thefts, and reveal deprecation factor impacting the attractiveness of theft-targeted smartphones. Fourth section deals with an approach to predict risk of personal device theft and compromising personal data with respect to the specified reference profile. The proposed model is based on the available statistics, and due to the scarcity of it, implements simple principles. However, if the regular flow of actual statistical data is provided, the proposed model can adequate give risk estimations for supplied user and reference profiles.

2. RELATED WORKS

The approaches to secure data within interconnected devices are considered in many surveys as well as in papers proposing implementations to achieve desired level of security and minimize threats to sensitive data. The initial exhaustive statistical research on loss of physical user control over owned device is given in paper (Mailley et al., 2008) concerned with phone thefts. This paper pioneered a theft index for mobile phones built in a similar fashion to widely acknowledged car theft index. The primary goal for a phone theft index was to incept phone industry with anti-theft designs to ensure decline of respective thefts. The similar goal for automobile industry was achieved nearly immediately after the publication of first car theft index in 1993. The paper proposed preliminary risk-based mobile phone theft indices for manufacturers and phone models.

The paper (Theoharidou, Mylonas & Gritzalis 2012) provides very detailed taxonomy of threats and risks to user data with respect to smartphone-targeted attacks. The proposed risk assessment method is based on risk matrix. However, estimation for the threats likelihood and risks severity is somewhat questionable, e.g. unauthorized physical access to the device is rated as having low thread likelihood and medium risk.

Recent studies of Metropolitan police (Home Office, 2014; Home Office, 2015) provide more refined data on factors that affect theft risks. The major contributing factors are categorized as smartphone model, type of surroundings and user age.

The paper (Mylonas, Kastania, & Gritzalis, 2013) studied the awareness of smartphone users on privacy intrusiveness of applications as well as their readiness to follow protective measures. An emphasis was given to the risky user behavior of ignoring security messages, using pirated software or excessively relying on trusted repositories, that exposes the device and data to the malware. Similar survey (Boyles, Smith, & Madden, 2012) was conducted concerning correlation of user age, education and other personal parameters with ability to securely manage smartphone applications. Other researchers provided software solutions to provide security to the sensitive data. The study (Ben-Asher et al., 2011) reveals the need of users to graded security measures like biometrical authorization via fingerprints or voice prints. (Vincent et al., 2011) considered implementation of special software firewall to prevent processing undesirable data requests from local applications and remote services. However, these related works address only cases where the portable interconnected device is physically controlled by the owning user. The provided solutions are mostly irrelevant if the device is occasionally lost or deliberately stolen. Once the device is unreachable by the owner, the whole data content can be duplicated and its compromise is only the matter of time and computational resources.

The conference paper (Mikhalsky & Pshehotskaya, 2017) precedes current work and deliver the overview of recent facts and trends in both data vulnerabilities and data protection. The outlined mathematical approach to estimate risks was based only on single set of then-available statistics of Metropolitan police.

We consider various cases of device loss or thefts and outline the simple mathematical means to assess such risks and make user aware of them. To the best of our knowledge, the existing papers do not provide any similar mathematical models at all.

3. ESTIMATION OF DEVICE DATA SECURITY

3.1. Threats to User and System Data

According to the intended purpose, the data can be categorized into two distinct classes, namely:

- User data
- System data

This classification is generalized from comprehensive smartphone data taxonomy provided in paper (Theoharidou, Mylonas & Gritzalis 2012). Our category of system data constitutes both authentication and connection service data types, whereas user data consists of personal, business, government and financial data types. If the taxonomy is built on data source property, system data includes device data, application data, sensor data and user input data. At the same time, our category of user data mainly represents messaging data.

The user data literally represent information of the user in his everyday living, including personal identification, credentials for financial accounts, health records and other less sensitive mundane data like favorite pieces of entertainment such as photos and music. In contrast, the system data constitute of command and control information for technological systems varying from toy radio-controlled model to nuclear plant. The consequences of compromising the control circuits can also range from the nuisance to the world-scaled catastrophe. However, considering the balance between complexity and profitability, the researchers assume that most of the attacks are directed toward the mostly used existing and emerging devices along with corresponding data types and formats. There are following device categories considered to be the most frequent targets for either real or anticipated data breaching:

- Portable personal devices
- Network hospital devices
- Implanted medical device
- Financial embedded devices
- Smart-house embedded devices
- Vehicle embedded device

Usually, the hospital, medical and financial devices are out of reach for private persons. Thus, we assume that private users only have maximum control over portable personal devices, and partial control over smart-house and vehicle embedded devices. Therefore, we only consider the data security for personal portable devices.

The considered problem of device data security is a complex one. However, we decreased the complexity by reducing the diversity of portable devices to the single category of smartphone devices. In our opinion, this reduction is justified, since for the bulk of commercial devices the underlying hardware architecture shares many identical components like CPU, RAM and SSD. The vendors are few in number and their product list is also relatively short. Non-PC devices usually based on hardware CPU solutions of the following manufacturers: Broadcom, Qualcomm, Apple, Intel, NVidia, Samsung, MediaTek and HiSilicon, with Qualcomm, MediaTek and Samsung holding about 90 percent of the android smartphones market. Moreover, the overwhelming majority of the solutions are produced under the licenses of only two corporations, namely Arm Holdings and Imagination Technologies. Similar situation can be observed in RAM and SSD segments.

There is a remaining category of portable devices, based upon single-board computers like Raspberry Pi or Arduino. These architectures differ from smartphone hardware, and share much more similarity with traditional PC-systems. However, these portable devices are not truly personal, since their dedicated purposes (like GNSS-receiver or laser rangefinder) do not assume storing private data. Therefore, we assume that largest user-dependent risks for personal data are related to smartphones. This consideration can be further strengthened by the smartphone build-in universality and the trend to use smartphones in multitude of roles like Wi-Fi routers and Wi-Fi/Bluetooth remote controllers for digital photo cameras, copters, hi-end toy-robotic constructors and other end-point devices, as well as embedded core units, running main control programs.

There are several ways for unauthorized persons potentially to get access to the user and/or system data. These ways vary in the scale and complexity as well as in the associated expenses and frequencies of occurrence. The common types of unauthorized data access are duplication to remote media and/or modification in situ. Accordingly, the vectors for data access of such kind are:

- Theft of physical device with user/system data.
- Duplication/modification of user/system data via malware applications installed locally on the device.

- Duplication/modification of user/system data via interception of communication channels.
- Duplication/modification of user/system data replicas via breach in the remote storage.

These vectors of data access correspond to the risks considered in paper (Theoharidou, Mylonas & Gritzalis 2012). Thus, physical theft directly relates to threat "T7 – loss, theft, disposal or damage" and implicitly relates to threat "T10 – unauthorized device (physical) access" which, in turn can lead to more general threat "T11 – unauthorized access" resulting in compromise of not only locally stored user data, but also user data remotely stored in synchronized cloud resources. Duplication of user data via malware corresponds to threats "T17 – sensitive information disclosure (SID) via spyware", "T18 – corrupting or modifying private content", "T19 – disabling applications or the device", "T20 – client-side injection/ malware". Duplication of user data via duplication of communication channels corresponds to threats "T1 – spoofing, T2 – scanning, T5 – eavesdropping, T15 – electronic tracking/surveillance/exposure of physical location".

It is well accepted fact, that unauthorized access to the locally and remotely stored confidential data is greatly facilitated by the physical access to the associated device. Moreover, theft is the most common and identifiable crime affecting portable electronic devices. The studies of Consumer Reports indicate that only in USA in 2013 about 3.1 million consumers fell victims to smart phone theft (Tapellini, 2014) nearly doubling the same annual indicators for 2012. In 2014 the number of reported thefts dropped to 2.1 million (Deitrick, 2015) due to deterring measures like locks and kill-switches introduced by some but not all manufacturers. Despite the trend changes to decline the overall number of device thefts all over the world remains high and can be roughly estimated in tens of millions per year. According to Lookout Mobile Security (cited by Eadicicco, 2014), 44 percent of all lost mobile devices are left in public places, while 14 percent are taken from a house or car, and 11 percent are pick-pocketed. Moreover, the major aspect of the device theft shifts from data loss to data compromise, since it much easier to recover data from a backup than to reliably purge data from a media before it undergoes unauthorized replication.

Research results, presented annually by Russian Analytic center Infowatch, show that Russian Federation holds second place in data leak distribution across the states, significantly lagging USA by the number of data compromise incidents. Thus, the report of InfoWatch (InfoWatch, 2017) states that there were 213 incidents in Russia during 2016, nearly doubling the number of incidents in 2015, while being four times less the respective values in USA (838 data leaks). Also, the total channel of leaks, consisting of stolen or missing hardware, leaks from portable data media, and leaks from mobile devices, equaled 9.3 percent, being slightly less than the leaks via paper documents. Additionally, the major factor of data compromise is revealed to be a network channel, generalizing leaks via local browser, malware of cloud services.

3.2. The Estimation of Risks to User and System Data

3.2.1. Device Classification

At the current age of worldwide communications, the user and system data is nearly always exposed to risks by various degree, depending on the user device type, user applications and services as well as the qualities of data itself. This part of the paper mainly considers the following qualities categorizing personal electronic devices:

- Cost.
- Portability.
- Diversity.
- Controllability.
- Hardware security measures.

This categorization is built upon more generic commonly accepted categorization of theft-preferred goods, having CRAVED- properties (Concealable, Removable, Available, Valuable, Enjoyable, and Disposable) as described in e.g. paper (Johnson, 2010). For example, category of portability includes properties of concealability, removability and disposability, whereas controllability and hardware security measures contribute to enjoyability and partially to valuableness. These relations are reasonable, since disabled (bricked) phone neither deliver joy nor possess any sound commercial value.

The device cost is one of the top properties contributing to the risk of theft. The study on smartphones thefts conducted by the Behavioural Insights Team of Home Office (Home Office, 2014) reveals that various manufacturer brands and product models differ greatly in likeliness of being stolen. The likeliness ratio is presented in the Table 1. The values on these figures are the ratio of measures T_m and R_m. First measure is a fraction of target thefts for the considered model m among target thefts for all models (the phones have been snatched). Second measure is a ratio of untargeted thefts for the model (the phone has been stolen in a burglary) to the share of untargeted thefts for all models. The statistics also indicates status of a certain smartphone model, since the iPhones are stolen more deliberately that other brands, while Apple having lesser market share than the Samsung brand. The statistics varies with time and should be updated at regular time intervals of about quarter of a year, or even monthly, if circumstances permit. For example, the table also contains updated data from Behavioural Insights Team of Home Office (Home Office, 2015). One can see that the most stolen model shifted from Apple iPhone 5 in 2014 to HTC M8 in 2015.

Two statistical sets allow one to assess theft dynamics for particular models. For example, Apple iPhone 5 ceased to be the most targeted smartphone dropping slightly below the averaged theft likeness threshold, due to the release of a more advanced model Apple iPhone 6. The same pattern can be seen for Samsung Galaxy S4 and Apple iPhone models 4S and 4. The likeness of thefts decreased in a lesser degree for Apple models 5C and 5S. The pattern for Samsung Galaxy Mini is different, due to its likeness of being stolen remained nearly the same. The most interesting case, however, is for Samsung Galaxy Note and HTC One, since these models actually rose in the rating, becoming more attractive targets. We assume that these patterns reveal different deprecation rates for different models within respective consumer strata. For example, the Galaxy Note model becomes obsolete at a lesser rate for its users than iPhone 5 model for its core users. Therefore, the demand for the first model can even rise, while the demand for the second one will drop significantly. Another explaining factor can be the strength of theft countermeasures, rendering easy-to-crack models more attractive to steal.

According to the statistical research of portable device thefts in Russia (AlfaStrakhovanie, 2014), smartphones are stolen the most, accounting for 65 percent of total thefts. The mobile phones are at the second position, rating for 18 percent, and tablets are at the third place, constituting 10 percent of all cases. These positions are followed by e-books (5 percent) and notebooks (2 percent). The rating of the most stealable smartphone models was calculated as the ratio of the number of stolen smartphones to the total amount of insured smartphones. The rating of the most stealable models is as follows: Apple iPhone 5 (2.1 percent), Samsung Galaxy S4 mini (1.8 percent), Apple iPhone 4s (1.7 percent), HTC

Table 1. Reference for targeted thefts, (Home Office, 2014; Home Office, 2015)

Device Model, 2014	Likeliness of thefts	Device Model, 2015	Likeliness of thefts
Apple iPhone 5	220%	HTC M8	267%
Apple iPhone 5C	190%	Samsung Galaxy S6 Edge	252%
Apple iPhone 5S	185%	Apple iPhone 6 Plus	250%
Apple iPhone 4S	179%	Apple iPhone 6	242%
BlackBerry 9790	172%	Apple iPhone 6S	206%
Samsung Galaxy S4	156%	Samsung Galaxy S6	193%
Samsung Galaxy S3	151%	Samsung Galaxy S5	162%
BlackBerry 8520	149%	Samsung Galaxy Note	157%
BlackBerry Z10	147%	Apple iPhone 5C	143%
HTC One X	147%	HTC One	137%
Apple iPhone 4	145%	Apple iPhone 5S	130%
BlackBerry 3G	140%	Averaged Theft Likeliness	100%
HTC One	130%	Nokia 635	98%
BlackBerry 9900	127%	Samsung Galaxy S4	97%
Samsung Galaxy Note	122%	Apple iPhone 5	92%
HTC One S	119%	Samsung Galaxy Mini	91%
Samsung Galaxy S2	111%	HTC Desire	82%
Averaged Theft Likeliness	100%	Samsung Galaxy	79%
Samsung Galaxy S	97%	Apple iPhone 4S	72%
Samsung Galaxy Mini	93%	Apple iPhone 4	71%
BlackBerry 9780	91%	Apple iPhone 3	70%

Desire 500 Dual Sim/Huawei Ascend P6 (1.6 percent), Apple iPhone 5s (1.5 percent), Samsung Galaxy S III (1.3 percent), HTC One X (1.1 percent), Samsung Note 3 (0.9 percent), Nokia Lumia 800/Sony Experia ZL (0.7 percent).

At first glance, the expensive top-tier state-of-art devices attract more unhealthy attention that aging low and medium cost counterparts. However, this is not always true, as provided data (RBC, 2017) indicate. The corresponding statistics of Moscow police shows the following iPhone-model theft ratings (see Table 2).

One can see, that proportion of cases (e.g. cases for 6-th vs 5-th series or 6-th vs 4-th series) are in good agreement with data of Behavioural Insights Team of Home Office. This can be considered as a sign of correctness of provided mutually independent statistics.

The portability essentially means the size-weight-operating time characteristics of the device. It is obvious that small devices, like smartphones and tablets or portable media storages are easier both to lose and steal than large notebooks, not mentioning tabletop PCs and server racks. However, relatively large devices are usually having modular structure that can be dismantled on site for retrieval of data sensitive components like HDD and SSD units or DRAM modules in case of Cold Boot attack.

Table 2. Number of iPhone theft cases, (RBC, 2017)

iPhone model series	6	7	5	4	8
Number of cases	46	35	20	12	1

The diversity is indicating how rich the product line is. In other words, it shows the variability of the product architecture in aspects of implemented hardware, firmware and software. If the product line is extremely small, then the devices are more often aimed by hardware- and firmware-specific attacks. The products with the same single architecture can be more resistant to exploits, but once breached they expose to the threat all users of the whole product line. The products with varying and diverse elements of architecture can be less resilient to attacks, but one breached product can compromise only the respective part of users. Moreover, users can select another product of the line without any changes in usability. Some additional resilience to attacks can be possessed by discontinued models like BlackBerry Leap running BlackBerry 10 OS. However, some architecture flaws can be common for a global-wise range of device models, regardless of their manufacturers. For example, discovered "Quadrooter" set of vulnerabilities in Qualcomm chipsets (Khandelwal, 2016) affect more than 900 million of android operated smartphones and tablets gaining the root-level access. The recently revealed widest vulnerabilities (Meltdown and Spectre) affect entire families of Intel and AMD CPU designs that date back for over a decade.

The controllability indicates how complete the user control over the device is. For example, is an operating system allows user by default to operate with a file system, or require a custom jail-break, effectively terminating the warranty on the device. Usually, reduced controllability is considered to improve device-resistance to hacking as not all malware is properly designed to bypass installation and data access OS-policies.

The hardware security measures are comprised of biometrical sensors as well as a circuitry protection against physical intrusion. The rapid innovations for biometrical sensors are praised as a new tool of authentication. The identification and authorization methods of everyday life rely on mobile biometric technologies. The most widespread technology of static biometrics is fingerprint recognition, successfully implemented in iPhone devices. Other common technologies are the voice, face and iris recognition. However, these approaches to authentication are harder to apply for personal mobile devices with robust performance. For example, recent Apple-X smartphone models have already revealed issues with false positive face recognition.

3.2.2. Types of User Behavior

The authors also consider user behavior, especially Internet-activity, since it also contributes to security risks. Per 2014 study of Consumer Reports, only 36 percent of smart-phone users set a four-digit PIN screen lock, 14 percent install antivirus application, 11 percent implement multidigit PIN or unlock pattern, 8 percent install data purge software and 7 percent utilize encryption. In contrast, 34 percent of users did not apply any of these measures. However, the proper and secure smart-phone handling is only a part of risk mitigation policies, since it reduces only two threats, namely: device theft and malware data theft. Two remaining risks essentially depend on user Internet-activity. Thus, the user behavior can be divided into following categories:

- Local device handling.
- Network handling.
- Remote services handling.

The network handling category indicates, how strict or relaxed are the user choices to connect to unknown communication nodes. Currently the most common are Wi-Fi and Bluetooth nodes. The recent Kaspersky Lab report (Jackson, November 28, 2016) shows that among 31 million of analyzed Wi-Fi hotspots almost 22 percent lack any security and additional 2.7 percent has the obsolete WEP-protection, which is easy to defeat. Another research (Jackson, June 29, 2016), conducted by Norton (Symantec) reveals that 61 percent of 9135 respondents completely disregard risks of public Wi-Fi networks, considering their information to be safe. This study indicates that exaggerated trust to hotspots is somewhat age-related, since Millennials trust public Wi-Fi more (68 percent) that users over age of 55 (55 percent). Nevertheless, the recent survey (Jackson, October 25, 2016) of 1.516 respondents conducted by ISPreview.co.uk discovered that users prefer mobile broadband 3G/4G over public Wi-Fi hotspots with respective percentage of 72 and nearly 21. However, even mobile broadband communications can be compromised via the use of fake base stations delivering a man-in-the-middle (MITM) attacks (Borgaonkar et al., 2015).

There are many habits, which indirectly impact the risks for user data. The Great Britain Home Office report on the mobile phone theft ratio (Home Office, 2014; Home Office, 2015) reveals dependence of theft frequency on location (see Table 3). Thus, the most of thefts in London occur in public transport (37 percent), while the least thefts happen in the street (6 percent). The Lookout report (Lookout, 2014) basically agrees to the ratios, except for the public transport, not accounting for public places (see Table 3). So, the habit of visiting specific locations implicitly affects the overall risks to data security.

Table 3. Percentage of London mobile phone thefts by location, (Home Office, 2014; Home Office, 2015; Lookout, 2014)

Area	Percent of thefts, London, 2014	Percent of thefts, London, 2015	Percent of thefts, USA, 2014
Street	6%	17%	5.0%
Public transport or Public place	37%	35%	6.0%
Bar/Pub/Club + Restaurant	30%	22%	11% + 16%
Place of work	19%	15%	11%
Elsewhere	8%	11%	—

The likeness of theft is also highly dependent on victim age. According to the reports (Home Office, 2014; Home Office, 2015) the age theft ratio for London has following distribution functions provided in Table 4.

The frequency of smartphone and mobile phone thefts exhibit not only spatial but also seasonal variations. The research of AlfaStrakhovanie (AlfaStrakhovanie, 2014) reveals that in June the thefts overshoot the half-year-averaged value by 40 percent, and thefts in July exceed averaged value by 27 percent. The thefts frequency averaged over the year yielded the value of 0.25 percent, the June and July

Table 4. Percentage of London mobile phone theft ratio by age, (Home Office, 2014; Home Office, 2015)

Theft ratio, 2015, [%]	15.4	49.6	149.5	229.0	183.2	147.6	125.4	99.1	89.4	82.5	62.6	41.3	37.6	14.5
Theft ratio, 2014, [%]	—	84.2	217.2	274.3	196.0	136.6	100.1	80.1	70.6	55.3	40.6	27.1	19.4	24.7
Age	<10	10–14	15–19	20–24	25–29	30–34	35–39	40–44	45–49	50–54	55–59	60–64	65–69	70+

frequencies equaled 0.30 and 0.27 percent respectively. In was revealed that the most calm months are January (0.15 percent) and February (0.17 percent).

The measured risk of theft exposure against user age has weak dependence on location. This is confirmed by highlights on 2016-year statistics of Irish National Crime Prevention Unit (Moonan, 2017) indicating 75 percent of phone-theft victims are aged between 18 and 39. The statistics also states that 40% of all smartphone-related thefts and robberies occurred on a street, road or footpath with a further 20% took place in a licensed premises or restaurant-type location. The measured percent of occurred thefts for different locations are provided in Figure 1.

Figure 1. Percent of occurred thefts for different locations

4. MATHEMATICAL MODEL OF RISK ESTIMATION

The major goal of threats and risks estimation is to proactively reduce their realization for the arbitrary smartphone user. Therefore, we aim to provide an approach to assess risks of smartphone theft and consequent data compromise relative to the specified baseline reference user profile. Proposed model accepts parameters of theft statistics and return amplification factor to the specified benchmark. If the factor is less than 1, then the current risk estimation is relatively low. Correspondingly, the amplification factor above 1 indicates heightened risk and acts as a signal to take theft-preventive measures.

As we established in the previous section, the risks depend on user hardware and software preferences, user digital footprint, overall network and social behavior as well as the types of device-stored data. Moreover, the risks also depend on user experience, especially on user age.

To the best of our efforts, we failed to discover any model of risk estimation for personal data or even device theft. This issue may arise due to the sole fact that it is impossible to estimate absolute probabilities of risks, since there is no underlying detailed raw personal statistics for such prediction.

To resolve this issue, we propose to introduce a baseline representing a profile of a generic average user and estimate risks relative to it. We should notice that even a baseline is subjected to shifts, depending on freshly provided statistics. The summary relative risk R_{Σ} can be defined as a product of supplied amplification factors a_i corresponding to i-th risk category (country, environment, device model, user age, etc.) with respect to reference amplification factors a_i^{ref}. The baseline amplification factors a_i^{base} are equal to one by definition.

$$R_{\pounds} = \prod_{i} \left(a_i \times \left(a_i^{ref} \right)^{-1} \right) \tag{1}$$

The value R_{Σ} indicates how great the risk for specified user profile is, in comparison with reference risk profile. For example, during the 2014 the smartphone user of 16 years old, living in London and owning Apple iPhone 5, was 3.68 times likely to experience a theft than a reference Londoner at an age 37 possessing HTC One. On the next 2015 year the same user of age 17, still living in same place and owning the same smartphone model, was 0.8 times likely to experience a theft than a same reference user at an age 38 possessing same smartphone and living at the same city.

The same approach can be applied to statistics on data breaches and records compromising. Revisiting the above considered example, we can estimate the relative risk of data compromise if the theft occurred for both teenager user and a reference Londoner. The paper (Home Office, 2015) provides age-wise proportions of users who implement various security features on their smartphones. For the sake of brevity we assume that any implemented feature renders data on stolen phone unusable. So, the only required proportions are for the users, who implement none of the features (see Table 5).

Table 5. Age-wise ratio of users, who implement none of the security features (Home Office, 2015)

Age bin	16-24	25-34	35-44	45-54	44-64	65-74	75+
User ratio	14%	18%	19%	29%	43%	60%	78%

Considering this statistics, the teenager would be 0.74 times likely to encounter data compromise of his stolen smartphone than the reference Londoner of the previous example.

Aggregating these two cases we can estimate the unconditional risk ratio for data compromise by multiplying ratio of theft risk with conditional risk ratio of data being unprotected. Thus, the estimated unconditional risk ratio (teenager/reference Londoner) would be 2.57 times for the 2014 and 0.6 for the 2015.

The outlined mathematical model for risk estimation of data compromise via smartphone thefts can be used in various practical fields. The most obvious usage is to implement the model as an online service or as a smartphone application to inform ordinary users on current risk level of their device to be stolen. For example, one can fill online form, specifying smartphone model as well as frequently visited locations, and receive risk level relative to an average reference factor. Moreover, the smartphone application can perform similar actions in automated manner, reading such parameters as geolocation datum including social surroundings, current date and time, device model and other relevant parameters. If the model returns increased risk of theft, the application can signal device owner and advise him to be cautious, as well as provide suggestions to mitigate risks. The model can be also used by commercial organizations like insurance companies. For example, the underwriter can consider such detailization of risks, and provide more precise insurance fee to the client. If the client stores sensible data on the most stealable smartphone without encryption, while visiting the most theft-related places, it is obvious that the fee should be greater than in the case with all security requirements are met. However, to succeed in these applications, the proposed methodology should be supplemented with detailed statistical data on the above aspects of data compromise. To the knowledge of the authors, the similar research and model implementation is only done to provide estimations of website security (https://sitesecure.ru/). The proposed methodology can be also implemented in this form.

The management of risk mitigation is always divided among user and third parties that provide online services and smartphone applications. To increase data safety, the user can implement recommendations considered above, e.g. enable encryption on the smartphone for the data stored locally and remotely, as well as to avoid uncertified open Wi-Fi hotspots.

CONCLUSION

This paper was concerned with the problem of physical thefts of the portable personal devices, especially smartphones. The classifications of various theft-related threats and risks to the smartphone stored data were described. The authors considered various classifications of computer devices and their software services as well as types of user behavior regarding the involved security risks of unauthorized access to the data stored both locally and remotely. The provided classifications proved to be consistent with existing ones described in various papers. It was shown, that location- and age-related theft risks only slightly depend on specified geographical regions, thus allowing one to consider and use gathered statistical data as invariant. The researchers outlined the simple approach for estimation of user relative risks. The proposed mathematical model can be successfully used if supplemented by detailed statistical data on above mentioned aspects of data compromise. The examples of practical use where considered, namely smartphone-insurance underwriting and environment-aware smartphone application to inform owner on dynamics of current theft-risks relative to baseline reference profile.

REFERENCES

AlfaStrakhovanie. (2014, June 09). iPhone 5 and Samsung Galaxy S4 Mini are the most stealable smartphones, the thefts peak at June [Press-release]. Retrieved from https://www.alfastrah.ru/news/1860355/

Ben-Asher, N. (2011, August). On the need for different security methods on mobile phones. In *Proceedings of the 13th International Conference on Human Computer Interaction with Mobile Devices and Services* (pp. 465-473). ACM. 10.1145/2037373.2037442

Bitglass(2016). Financial services breach report. Retrieved from https://pages.bitglass.com/rs/418-ZAL-815/images/Financial-Services-Breach-Report-2016.pdf

Borgaonkar, R., & Shaik, A. (2015). LTE and IMSI catcher myths. *Blackhat*. https://www.blackhat.com/docs/eu-15/materials/eu-15-Borgaonkar-LTE-And-IMSI-Catcher-Myths.pdf

Boyles, J. L., Smith, A., & Madden, M. (2012). Privacy and Data Management on Mobile Devices.

Deitrick, C. (2015). Smartphone thefts drop as kill switch usage grows but Android users are still waiting for the technology. *Consumer Reports*. Retrieved from http://www.consumerreports.org/cro/news/2015/06/smartphone-thefts-on-the-decline/index.htm

Eadicicco, L. (2014, May 7). People Are Willing To Go To Extreme Lengths To Retrieve Their Stolen Smartphones. *Business Insider*. Retrieved from http://www.businessinsider.com/smartphone-theft-statistics-2014-5

Gemalto(2017). Poor Internal Security Practices Take a Toll. http://breachlevelindex.com/assets/Breach-Level-Index-Report-H1-2017-Gemalto.pdf

InfoWatch. (2017, March 23). Investigation of confidential data leaks in 2016. (in Russian). Retrieved from https://www.infowatch.ru/report2016

Jackson, M. (2016, November 28). Kaspersky Lab Reports 25% of WiFi Internet Hotspots are Unsecured. *ISP News*. Retrieved from http://www.ispreview.co.uk/index.php/2016/11/kaspersky-lab-reports-25-wifi-internet-hotspots-unsecured.html

Jackson, M. (2016, June 29). Norton. Only 42% of People Can Tell if a Wi-Fi Network is Secure or Not. *ISP News*. Retrieved from http://www.ispreview.co.uk/index.php/2016/06/norton-42-people-can-tell-wi-fi-network-secure-not.html

Jackson, M. (2016, October 25). Study -- Mobile Broadband Still More Popular than Risky Public Wi-Fi. *ISP News*. Retrieved from http://www.ispreview.co.uk/index.php/2016/10/study-mobile-broadband-still-popular-risky-public-wi-fi.html

Johnson, S. D. (2010). Theft of Customers' Personal Property in Cafés and Bars. US Department of Community Oriented Policing Services. Washington, DC. Retrieved from http://www.popcenter.org/problems/pdfs/cafe_bar_theft.pdf

Kensington. (2016, August). Survey: IT Security & Laptop Theft. https://www.kensington.com/a/312684

Khandelwal, S. (2016, August 07). Warning! Over 900 Million Android Phones Vulnerable to New "QuadRooter" Attack. *The Hacker News*. Retrieved from http://thehackernews.com/2016/08/hack-android-phone.html

Mailley, J., Garcia, R., Whitehead, S., & Farrell, G. (2008). Phone theft index. *Security Journal*, *21*(3), 212–227. doi:10.1057/palgrave.sj.8350055

Mikhalsky, O., & Pshehotskaya, E. (2017). Mobile Device Security, Management of Personal and Business Privacy. In *20th conference of FRUCT association*, FRUCT Oy, Finland (pp. 643-649).

Moonan, N. (2017, March 1). Fifteen mobile 'phones stolen every day in Ireland as Gardai launch 'Streetwise' blitz to crack down on thefts. *Irish Mirror*. Retrieved from https://www.irishmirror.ie/news/irish-news/fifteen-mobile-phones-stolen-every-9939024

Mylonas, A., Kastania, A., & Gritzalis, D. (2013). Delegate the smartphone user? Security awareness in smartphone platforms. *Computers & Security*, *34*, 47–66. doi:10.1016/j.cose.2012.11.004

Pankhuri, Y. (2017, August 28). Why tracking your stolen mobile phone may be tougher than ever. *The Times of India*. Retrieved from https://timesofindia.indiatimes.com/city/delhi/why-tracking-your-stolen-mobile-phone-may-be-tougher-than-ever/articleshow/60250036.cms

FCC, U.S. Government(2014). Phone Theft in America: Breaking down the phone theft epidemic. *Lookout*. Retrieved from https://transition.fcc.gov/cgb/events/Lookout-phone-theft-in-america.pdf

RBC. (2017, November 3). Presentation of new smartphones provokes thefts of deprecated ones. (in Russian) Retrieved from https://www.rbc.ru/society/03/11/2017/59fb24489a794730c4143feb

Tapellini, D. (2014, May 28). Smart phone thefts rose to 3.1 million in 2013. Industry solution falls short, while legislative efforts to curb theft continue. *Consumer Reports*. Retrieved from http://www.consumerreports.org/cro/news/2014/04/smart-phone-thefts-rose-to-3-1-million-last-year/index.htm

The Behavioural Insights Team. (2014). Reducing Mobile Phone Theft and Improving Security. *Home Office*. Retrieved from https://www.gov.uk/government/uploads/system/uploads/attachment_data/file/390901/HO_Mobile_theft_paper_Dec_14_WEB.PDF

The Behavioural Insights Team. (2015). Reducing Mobile Phone Theft and Improving Security Paper 2. *Home Office*. https://www.gov.uk/government/uploads/system/uploads/attachment_data/file/509846/WEB-6-1946-HO-Reducing_Mobile_Phone_Theft_and_Improving_Security_March_2....pdf

Theoharidou, M., Mylonas, A., & Gritzalis, D. (2012). A risk assessment method for smartphones. In *Information security and privacy research* (pp. 443-456). Retrieved from https://hal.inria.fr/hal-01518232/document

SCMagazine(2016). US Cell Phone Crime Statistics in 2016. Retrieved from https://media.scmagazine.com/documents/253/financial-services-breach-repo_63173.pdf

Vincent, J. (2011, June). Privacy Protection for Smartphones: An Ontology-Based Firewall. In *5th Workshop on Information Security Theory and Practices (WISTP)*, Heraklion, Crete, Greece, June, LNCS (Vol. 6633, pp. 371-380). Springer.

This research was previously published in the International Journal of Embedded and Real-Time Communication Systems (IJERTCS), 9(2); pages 97-110, copyright year 2018 by IGI Publishing (an imprint of IGI Global).

Chapter 19
Mobile Cyber Information Leakage:
Users' Attitudes and Information Protection Strategies

Hongbo Lv
Zhejiang Wanli University, China

Zhiying Zhou
Zhejiang Wanli University, China

ABSTRACT

With the wide use of smartphones, mobile cyber has become an indispensable part of our lives. While smartphones are used almost every aspect of our lives, awareness of personal information protection is underdeveloped, and information leak has become one particular problem on mobile cyber. At the same time, personal information resources have become more valuable than ever. This chapter investigates users' attitudes toward cyber information leakage and methods to protect personal information. The software SPSS 19 was used to analysis the relationship among cyber environment, applicability and practicability of protection methods, economic cost, and overall evaluation (satisfaction). A suggestion that may promote personal information awareness and promote cyber security from technical, legal, and social aspects is also provided.

1. INTRODUCTION

With the rapid development of mobile communication technology, information interactions between individuals are now more frequent than ever. Smart phones are the preferred device for web browsing, emailing, using social media and making purchases. Personal information is also transferring from PC (personal computer) to smart phones. The global smartphone users are 2.1 billion in 2016, with smartphone penetration rates still increasing. China, the most populous country in the world, leads the smartphone industry. The number of smartphone users in China is around 563 million in 2016. It is

DOI: 10.4018/978-1-7998-8545-0.ch019

projected that approximately half of the Chinese population will use a smartphone by 2020 (Statista, 2016). Information security has become a major concern and the security of mobile cyber information is also facing with unprecedented challenges. Many lawless merchants and enterprises found methods to exploit personal information into their offense, especially through personal information value research by the social research institutions. There is a rapid expansion of information leakage in mobile network.

Most mobile apps require personal information to demonstrate authenticity and authority. However, improper authority will lead to a direct access to information by smartphone apps, which may cause personal information to leak occasionally or deliberately. In 2016, China's Android mobile phone privacy security report pointed out that,58% of Android apps leak users' private information. They provided users' data to developers, advertisers and third parties of unidentified identities. There are a lot of leakage problems in apps. The user information is hard to protect. Wang et al. (2015) found that smartphone users in China were vulnerable to malware than anywhere else through multiple data comparisons. In addition to spam messages and harassing phones, malicious software such as decoy fraud, system destruction, malicious deductions, privacy theft and others have been constantly evolving.

This chapter will discuss the growing concern about mobile information security. A strategy to use questionnaire is discussed for investigating and analysis the information leakage state, as well as for information cognition and protection. Possible reasons for the mobile cyber information leakage are explored, and suggestions for preventing information leakage are provided.

2. CATEGORIES OF MOBILE INFORMATION LEAKAGE

Smartphone has become an indispensable part of modern life. With the growing maturity of mobile cyber technology, the use of mobile network is ever increasing. A series of O2O business apps, such as Baidu Takeout[1], Didi Kuaidi[2], and Uber allow phone users to do what they want to do without leaving their homes. Feng (2016) mentioned that social apps, such as QQ and WeChat[3] are must-have SNS(Social Networking Services) apps in almost every smartphone in China. Watching videos on phones is also getting easier. The Statistical Report on Internet Development in China (2017) showed that the utilization rates of mobile network were 67.1% and 70.8% for watching video and listening to music online, respectively.

The mobile network makes life better and smoother. It provides convenience support for model life and travels. However, at the same time risks of information leakage increased. The risks can mainly dividend into two categories. One category is that data stolen by direct intrusion, which can implement by viruses or irregular apps. Tan (2014) envisioned that the smartphones have become a new virus disaster area. Ma (2016) mentioned that because of diversity and openness of the smartphones, malicious attackers, as part of software developers, can also get the users' app authority. They use users' network, access to users' communication record and SMS, charge users, and provide "service" for them. All these greatly increased the security risks of mobile cyber users. Another category is the leakage caused by the users' information sharing by themselves. Such leaks usually occur during online account registration, online shopping, commercial app downloading and uses SNS apps for self sharing. The popularization of real name network makes users use their real information during websites and game id registering. Commercial apps need use real information to fulfill its accurate service. It makes accurate personal information transmitted into the network and promotes the risk of information leakage. Personal information is in an open state as users are using social apps while filling their information. This is one of the easiest ways for information to be stolen. Vorobyov et al. (2015) demonstrated that the more frequently use of information, the higher risk of information leakage. Although the current network is well developed, it is not completely safe.

3. INVESTIGATION AND ANALYSIS OF MOBILE CYBER INFORMATION LEAKAGE

3.1 Questionnaire Design

In order to further analyze the status of information leakage, we designed a questionnaire from four aspects.

1. Personal information. This includes user's age, gender, occupation, and annual income, etc. This information will allow researchers assess factors causing information leakage among different users.
2. Usage of mobile network. This part will collect basic information about usage on mobile network. We then analyze users' awareness level of non-invasive leakage in the mobile network and clarify users' main use of mobile network. We will further acquire users' mobile operation steps and possible causes of information leakage.
3. Status of information leakage. We will check the extent of information leakage among mobile users, explore users' awareness of information leaked, search users feedback method, find out connections between information leakage with economic benefits, and collect information of how much users pay attention to their personal data.
4. Satisfaction about mobile cyber information security. This information will help us to conduct further research and generate constructive suggestions to protect personal information and enhance cyber security.

The questionnaires were distributed through both online and offline at the same time. Online questionnaire was issued to the public living forums, bulletin board systems (BBS), and opening public platforms. Offline questionnaire was distributed in business centers, living areas, and public spaces. After 5 days, we received 432 questionnaires, among them 418 were valid, resulting in a valid rate of 96.8%. The questionnaire sample size is sufficient for the analysis of this chapter.

3.2 Mobile Cyber Information Leakage Status Analysis Based on Questionnaire

3.2.1 Main Usage of Mobile Network

The distribution of mobile cyber usage is shown in Figure 1. It shows that SNS apps rank the highest among smartphones usage, with a proportion of 97.1%. Smartphones use various kinds of SNS apps, which collect our private data. SNS apps mostly have hidden information security risk. Their credential management would allow an attacker to view personal information stored on the server (2016). Private data leakage, as a potential hazard, is caused by the current design flaws in the app industry, which is increasing in size but with various level of personal information protection security. Other usages of mobile network, such as online entertainment and online shopping, are also have large proportion, as shown in Figure 1. Entertainment information, taking online game as an example, needs real name register

Figure 1. Distribution of mobile cyber usage

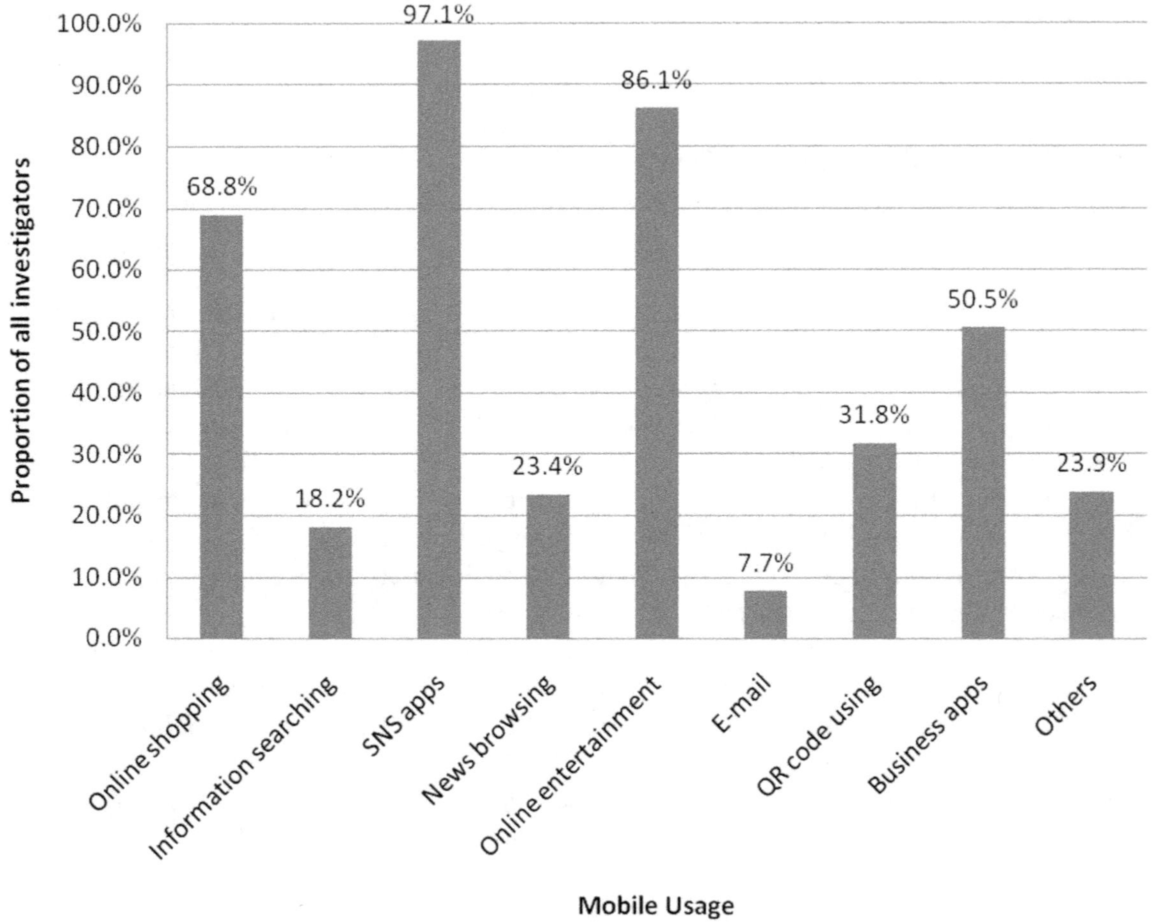

information supervised by official. Information leakage will occur if the personal information is spread out or sold to profiteers. Online shopping apps information can also be obtained by sellers, and then be used as target of propaganda or promotion. Other remainder usage proportions are relatively low. They are not main channels of information leakage.

3.2.2 Attitude Toward Information Leakage

The problem of information leakage is quite common among questionnaire responders. According to questionnaire data, 87.1% of the responders believe that their information have been leaked, and only 12.9% users believe that their data have never been leaked. As for the attitude about information leakage, as shown in Table 1, 35 of the responder are seriously concerned about personal information, 52 are not much concerned, while the rest are at a normal concern level.

Table 1 also shows that only 2.11% male responders are seriously concerned about information leakage, and 33.8% feel indifferent to it. In contrast, female responders are more seriously about personal information. A fraction of 11.47% female responders are seriously concerned about information leakage,

Table 1. Attitude toward personal information leakage

	Seriously Concerned		**Normal**		**Indifferent**	
	Number	**Percent**	**Number**	**Percent**	**Number**	**Percent**
Male	3	2.11%	91	64.08%	48	33.80%
Female	32	11.47%	243	87.10%	4	1.43%
Total	35	8.38%	334	79.90%	52	12.44%

and only 1.43% responders feel indifferently. It shows that females pay more attention to information leakage than males.

3.2.3 Type and Frequency of Information Leakage

Based on the questionnaire, there is no large proportion of direct benefits lost about information leakage, only 14 among 418 (3.3%) questionnaire responders suffered economic lost. The percentages of responders suffered from phone harassment and spam message harassment are relatively high (Table 2). In particular, all responders received spam messages at various frequencies, among whom 46.7% often received spam harassment. Unwanted advertising message caused by the leakage of phone numbers is one of the cheapest advertising methods. Currently, there is no relevant law that regulates this type of activity. Because it is not deemed illegal and its risks are lower than those of internet account embezzlement or fraudulent phone, profiteers use this channel to advertise. This is the reason for such high occurrence of spam message harassment.

Table 2. Statistics of type and frequency of information leakage

	Often		**occasionally**		**Never**	
	Number	**Percent**	**Number**	**Percent**	**Number**	**Percent**
Stolen of account password	2	0.5%	121	29.0%	295	70.6%
Phone harassment	74	17.7%	269	64.4%	75	17.9%
Spam message harassment	195	46.7%	223	53.4%	0	0.0%
Telephone fraud	31	7.4%	119	28.5%	268	64.1%
Others	97	23.2%	141	33.7%	180	43.1%

3.2.4 Methods to Prevent Information Leakage

In the questionnaire, we designed four choices, i.e., *software protection, download app from authority, not fill in real information*, and *no protections* to investigate users' preference to protect personal information. According to the collected data, most users (91.2%) prefer protective software, followed by *download app from authority* (38.5%). There are 99 users among the 418 responders do not take any protective measures.

As shown in Table 2, 100% questionnaire responders suffered from spam message harassment at least occasionally. We investigated the methods taken by people to prevent such harassment and found that 69.1% of them chose to block information from the harassment source. Table 2 also shows that 29.0% users suffered from account number being stolen. Among these victims, 84.5% chose to modify their password, indicating that not all of them chose to modify password even they found that the password was stolen. Only less than 1% victims chose to use law to defend their own interest. This is likely caused by the perception of high legal cost and long time to settle the case leading the victim to think not worthwhile pursuing the legal route.

The questionnaire also collected information about the responder's occupation, whether or not related to network. Among all of those who provided valid responses, the occupations of 77 responders were irrelevant to network while those of the remaining 341 responders were relevant. Table 3 summarizes main causes for information leakage. It shows that information *Left over* and *Be sold* were the main reasons, independent of whether their occupation is computer relevant or not. The proportion was all over 70%. A total of 55.3% of responders think their information was leaked because of *friend list sharing*. Only 29.2% of the leakage was caused by *virus*, and that by *others reason* was 15.6%.

Table 3. Statistics of main causes for information leakage

Way of Information Leakage	Irrelevant(77)		Relevant(341)		Total(418)	
	Number	Percent	Number	Percent	Number	Percent
Left over	67	87.0%	244	71.5%	311	74.4%
Virus	40	51.9%	82	24.0%	122	29.2%
Friend list sharing	24	31.2%	207	60.7%	231	55.3%
Be sold	61	79.2%	237	69.5%	298	71.3%
others	11	14.3%	56	16.4%	67	15.6%

3.2.5 Satisfaction About the Information Security

Figure 2 shows questionnaire responders' perception for causes of personal information leak. Most responders believed that the main reason for information leakage is that sellers think that personnel information profitable. Therefore, sellers will collect all kinds of buyers' information. The under-developed legal system regarding personal information protection is the second cause. An improved legal system will significantly reduce cases of personal information leak. Another 31.6% responders believe that the information leakage is caused by the ineffective network supervision. Only 14.6% responders think that the society does not pay enough attention to the information leakage problem.

3.2.6 Regression Analysis

To further mine the 418 valid questionnaire responses, we used SPSS19.0 to analyze the reliability and validity of designed variables and the correlation among five variables: *cyber environment, easy-to-use of methods, practicability of protection methods, economic cost and overall evaluation(satisfaction).*

Figure 2. Causes for personal information leakage

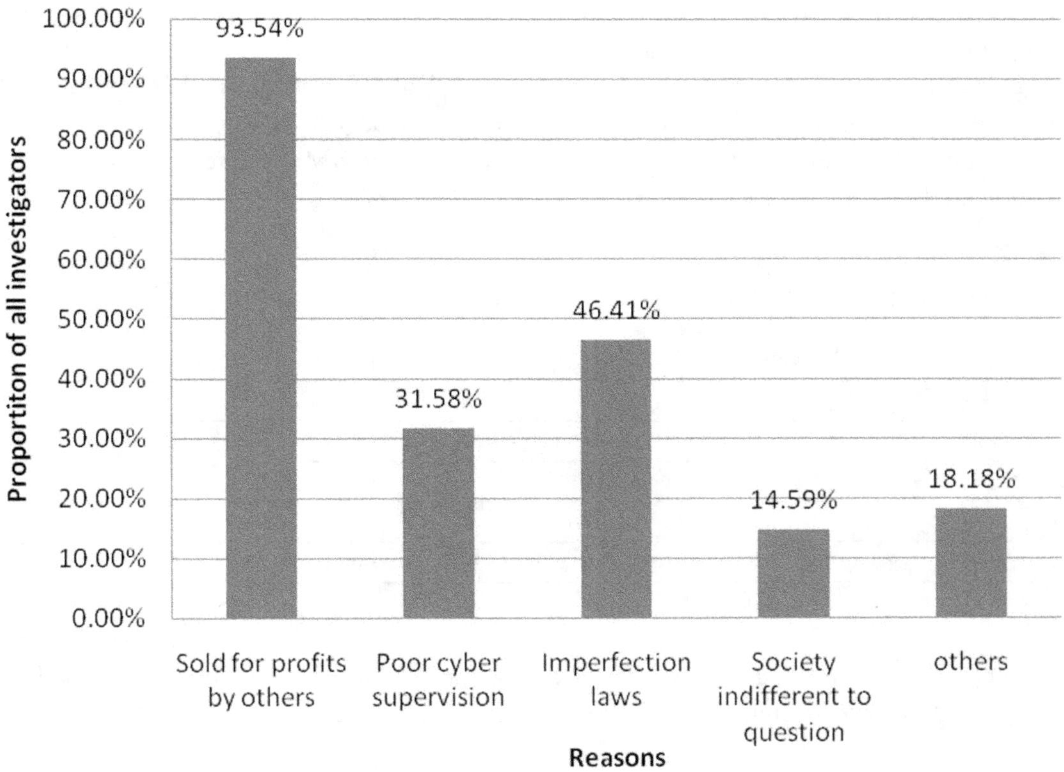

1. Reliability analysis.

Reliability analysis is to study the effectiveness of the designed variables and to determine the reliability of the questionnaire. Cronbach's alpha is usually used as an indicator for reliability analysis. Its value is between 0 and 1. 0, with 0 indicating not reliable while 1 indicating absolutely reliable. Analysis results are shown in *Table 4*.

Table 4. Cronbach's alpha of each variables

Variable	Cronbach's Alpha
Cyber environment	0.837
Easy-to-use of methods	0.792
Practicability of protection methods	0.931
Economic cost	0.916
Overall evaluation(Satisfaction)	0.802

As seen from *Table 4*, all Cronbach's alpha values are larger than 0.79, indicating that they have good reliability for further analysis.

2. Validity analysis.

Validity refers to the validity or accuracy of the measurement. It is closely connected with reliability analysis. Factor analysis is usually used to assess validity. Only high values of KMO (Kaiser-Meyer-Olkin) can be used for factor analysis. Its value is between 0 and 1. 0, with 0 indicating not validate while 1 indicating absolutely suit for further validity analysis. The KMO values calculated by SPSS19 for this study are listed in *Table 5*:

Table 5. KMO values for factor analysis

Variable	KMO Value
Cyber environment	0.733
Easy-to-use of methods	0.862
Practicability of protection methods	0.801
Economic cost	0.794
Overall evaluation(Satisfaction)	0.753

The KMO values are all larger than 0.7, indicating that the chosen variables have a good internal consistency and that they are suitable for factor analysis. Further results of factor analysis are shown in *Table 6.*

Table 6. Factor analysis of variables

Factor	Questions	Factor Loading	Variance (%)
Cyber environment	1	0.865	71.654
	2	0.853	
	3	0.846	
Easy-to-use of methods	4	0.952	86.384
	5	0.924	
	6	0.912	
Practicability of protection methods	7	0.864	70.796
	8	0.848	
	9	0.836	
Economic cost	10	0.869	80.021
	11	0.923	
	12	0.884	
Overall evaluation(Satisfaction)	13	0.926	83.956
	14	0.895	
	15	0.926	

All values of *Cyber environment*'s *factor loading* are higher than 0.8 with a total variance of 71.654%, indicating that the variable *Cyber Environment* has a good construction validity. Similarly, the others four variables all have the good construction validity, indicating that a good validity analysis has been achieved and correlations between variables can be calculated.

3. Correlation analysis.

The correlation analysis checks the internal relationship among variables. The correlation coefficients (r) are summarized in *Table 7*. *Cyber Environment, Easy-to-use of methods* and *Practicability of protection methods* all have positive impact on *Overall evaluation(Satisfaction)* of modern cyber information security. *Cyber Environment* has the highest degree of influence with a r value of 0.881; *Practicability of protection methods* ranks the second with a r value of 0.761; and the third is the *Easy-to-use of methods* with a r value of 0.632. Yet *Economics Cost* has a negative influence on *Overall evaluation(Satisfaction)* with a r value of -0.532.

Table 7. Correlation coefficient between variables

	Cyber Environment	**Easy-to-use of Methods**	**Practicability of Protection Methods**	**Economic Cost**	**Overall Evaluation (Satisfaction)**
Cyber Environment	1	0.543	0.532	-0.765	0.811
Easy-to-use of methods	0.543	1	0.891	-0.673	0.632
Practicability of protection methods	0.532	0.891	1	-0.611	0.761
Economic cost	-0.765	-0.673	-0.611	1	-0.532
Overall evaluation(Satisfaction)	0.811	0.632	0.761	-0.532	1

4. Regression analysis.

The above correlation analysis shows that *Cyber Environment* is best correlated with overall evaluation. Here we take *Cyber Environment* as the predictor variable and *Overall evaluation(Satisfaction)* as the dependent viable for the regression analysis (further analyze shows that the other three factors are all have significant probability big than 0.1, not fit for further regression analysis). The results are shown in *Tables 8* and *9*. According to the analysis results in *Table 9*, the statistical value of the model is 61.541 and the significant probability is 0.000 (<0.05), indicating that the overall regression effect of the model is statistically significant.

Table 8. Model Summary[b] and Anova[b]

Model	**R**	**R Square**	**Adjusted R Square**	**Std. Error of the Estimate**	**F**	**Sig**
1	0.793[a]	0.629	0.614	0.71334	61.541	0.000 [a]

a. **Predictors:** (Constant), Cyber Environment
b. **Dependent Variable:** Overall Evaluation (satisfaction)

As shown in Table 9, the regression coefficient of *Cyber Environment* on *Overall Evaluation (satisfaction)* is 0.793, and the significant probability of *t* value is less than 0.05, indicating that *Cyber Environment* can be used to explain *Overall Evaluation (satisfaction)*. Furthermore, its impact on *Overall Evaluation(satisfaction)* is positive, i.e., the better the cyber environment, the more users will trust and be satisfied with the mobile network. The standard regression equation from this analysis is: *Overall evaluation (satisfaction) = 2.694 + 0.793* Cyber environment.*

Table 9. Coefficient

Model		Unstandardized Coefficients		Standardized Coefficients	t	Sig
		B	Std. Error	Beta		
1	Constant	2.694	0.511		5.482	0.000
	Cyber environment	0.604	0.085	0.427	7.070	0.000

According to the analysis of the results of the questionnaire investigation, we make the following observations and preliminary conclusions:

1. In mobile devices, a high proportion of users use SNS apps. The possibility of information leakage through SNS apps is high.
2. Users, independent of gender, do not pay enough attention to their personal information.
3. The mainly risk of personal information leakage lies in the little legal consequence.
4. Information harassment is very serious, with all responders being a victim at least once.
5. The majority of users use security software to protect personal information. However, there are still a large proportion of users who do not take any measures to protect their personal information.
6. The advantages and disadvantages of the whole cyber environment are the most direct and fundamental factors of whether the users perceiving the mobile network is safe.
7. If the methods of protection and feedback channels are too complicated, users will choose to give up taking actions.

4. STRATEGIES FOR MOBILE CYBER INFORMATION PROTECTION

4.1 Ways to Reduce the Information Leakage Risk

1. Using security software.

Xu and Gong (2016) pointed out that in facing of the current mobile cyber information leakage, security software can effectively reduce the urgent problem. Base on the questionnaire analysis, 69.6% of the responders chose at least one security software to protect information leakage. And within those taking protection methods, the percent of using software protection is 91.2%. The high percent use of security software also proves that it can provide effective protection. Nowadays security software can protect the user's information from many aspects. Firstly, security software can effectively resist the invasion of the virus. A total of 29.1% of the users thought their information leakage was caused by virus, indicating that virus has a certain degree of threat to information. Security software has been an effective first line of defense against virus. Strengthening the usage of security software can effectively reduce the occurrence of information leakage caused by the virus. Secondly, some malwares will automatically be detected when they apply for mobile authority permission. In that case, users can intuitively judge whether authority should be given and prevent direct access to mobile information effectively. Third, part of the security software has the function of blocking advertising information. Our data analysis shows that the proportion of SMS harassment is up to 100%. That is to say, after the information is leaked, it will most likely lead to the harassment of SMS. And some of the security software has the function of advertisement filtering, which can effectively reduce the harassment of SMS.

2. Avoid installing unknown software and browsing unknown sites.

There are lots of free cyber software and a wide range of websites online now. However, except a few of high-profile websites, the majority of the websites' privacy statements are lack of standardization. There is a lack of relevant regulation that requires the privacy statement to include which part of information. Therefore, the content, collecting methods, security requirements for information usages, corresponding secrecy measures, duration of information retention about personal data are not clearly defined. Website managers can easily find loopholes in them, sell users' personal information or make use of it for profits. If there's an unknown website requires user registration, one should avoid filling in the real and effective information for protection from information leakage. Now most of apps need specific privileges before installation. If the user does not provide permissions to those apps at will, it will not let them any chance to access or steal information from smartphones. Then it won't cause any economic loss, either. In facing this situation, users must know well about the content of mobile access management, decide whether to give permission or not. Then they can effectively prevent malwares from stealing information.

3. Prudently sharing of personal information.

The results of the questionnaire show that 97% of the investigators transmitted their personal or family information on mobile network. If one does not know the importance of information and the danger that might cause, the transmitted information might be in an unknown environment. The result of the questionnaire also shows that SNS apps were the most popular in smartphones apps. The proportion is up to 97.1%. Such a high ratio of usage, information security must be paid attention to. There are two main reasons for the leakage of personal information through SNS apps. The first reason is sharing information within friends list. One should consider whether the information will cause personal information leak before releasing the information within the friends list. After being shared, the information is in an open state. It could be randomly shared by friends and friends' friends and it can be obtained by strangers after two or more times of re-sharing. If the information contains important message it would be exploited by others. Therefore, before information release one should consider the risk of re-sharing. The second reason is to fill in the personal information of an account. Now even well-known dating websites have reports of information leakage occasionally. So one should be careful when logging into these kinds of websites. At least one should hide part of the important personal information during registration.

4.2 Strategies for Preventing Information Leakage

1. Technical aspect.

Nowadays people pay more attention to the protection of personal information on mobile network. However, modern technologies in changing this situation still have a long way to go. Our questionnaire results show the frequency of SMS harassment and phone harassment are high. The proportion of SMS harassment is up to 100% and the phone harassment is 82.06% among all responders. The cause of harassment information mainly due to personal information leak. Hiding part of the contact information using technical methods would reduce the occurrence of information leak. Today's network phone can partly conceal information and effectively realize this technique. Two kinds of methods can be used.

The first method is to make temporary web phone exist for a certain time. The questionnaire results show that about half (211 of the 418) responders used O2O type of business apps, such as Baidu Takeout and Didi Kuaidi. These kinds of apps require sharing personal contact information for the sellers to offer service. Take Didi Kuaidi as an example. When Didi Kuaidi was first launched, it transferred uses' real names and the real phone numbers to drivers (service provider). After users and drivers reported harassment information during the first stage, the company switched the real data to a web phone. This switch shields both the users and the drivers from obtaining the real phone numbers, thereby protecting user's information. However, this method has its drawbacks. The web phones are unable to persist for a long time after the phone hung up. It makes it unable to re-contact when the users use a temporary account. The web phone is an approach for personal information protection, but it needs more technical development to resolve the current shortcomings.

The second method is to use QR code to generate web phone contacts. The results of the questionnaire shows that 64.8% of the responders used mobile shopping online. At same time, up to 100% investigators have been harassed by business ads caused by cell phone numbers breach. Therefore, it is necessary to hide the contact information. Yet in the face of online shopping users must provide effective contact information. To solve this dilemma, users can use some apps to transfer personal information into a QR code on both customers and sellers sides. If couriers need to communicate buyers, they scan the QR code to generate a temporary web phone contact again to reach the customer. Once the transaction completes, the user can choose to cancel that QR code at any time. This can be a good way to avoid harassment SMS.

2. Legal aspect.

According to the questionnaire data, illegal activities involving fraud and account embezzlement that are prohibited by laws have lower occurrences than those not explicitly prohibited by a law. The problems caused by information leakage usually do not cause direct economic losses or personal injury, but in a form of advertising which is beyond severe legal punishment. Our data also shows that only 1% of the responders will take legal actions to defend their interests due to the high cost and long duration of the legal process. In our opinion, the legal system should be improved from the following three aspects to for personal information protection.

First, make clear what kind of advertisement is illegal. Our questionnaire data show that 3.3% of the responders incurred direct economic losses because of information leakage. These low-risk and high-frequency behaviors are spread among businesses. It is necessary to develop regulations to restrict unsolicited anonymous calls and spam messages, clearly states the legal provisions, and forbid spreading advertisements randomly or targeted to others smartphones without the users' consent. Once such a legal system is in place, the problem of information leakage can alleviate.

Second, clearly legislate accountability the infringement of personal information. For the reasons of information leakage, 46.4% of the responders believed that the current law was imperfect. Many of the low danger of information leakage activities are beyond the current laws. It is necessary to determine the infringement responsibility of personal information. Guo (2012) called for improving legislation to meet the challenges in the current mobile cyber world. This need is ever increasing with the increasing number of cases of personal information infringement in the big data age. Therefore, a state needs to increase the punishment of personal information theft and transaction, strengthen the protection of personal information rights, and make full use of various forms of responsibility.

Third, simplify legal complaints channels. In the survey of users' satisfaction to current network, the questionnaire set up questions for ways of simplicity feedback and attributed it to *Usability of protection methods*. From the overall analysis of data, the correlation coefficient of *Usability of protection methods* and *Overall evaluation (satisfaction)* is 0.632. The simplicity of feedback channels is an issue that we deeply concern. The reason is that the cost of using the law as a means of feedback is too high and the steps are too time-consuming. It is necessary to establish a relatively convenient way of feedback. One approach might be setting up a special complaint web page on the websites, where all users need to do is only to provide the illegal advertising information and business information online. This way more feedback can be collected.

3. Social aspect.

According to our data analysis, the impact of *Cyber environment* on *Overall evaluation (satisfaction)* is up to 0.811. It shows that users now have a high level of concern about the cyber environment. The embodiment of cyber environment is mainly on the protection of network and the quality of cyber merchants. A total of 31.5% responders believed that the reasons for information leakage were attributed to poor supervision. It also proved that the current cyber regulation is not in place. The improvement of cyber merchants' quality can put an end to unhealthy competition. A total of 93.5% responders believed that the fundamental reason for the information leakage was that the business was making a profit from it. This means that sellers can effectively prevent information from leakage once unhealthy competition is removed. All aspects of society need to give publicity to promote the importance of personal information. Only 8.37% of responders are seriously concerned about their information. This situation is more prominent among male responders, and only 2.11% of the males surveyed are very concerned with personal information. This is partly because males do not pay enough attention to personal information. It is necessary to make the public understand the importance of personal information.

The information leakage can't be solved by law and technology alone. Take the virus as an example, the protection of security software is undoubtedly a very favorable barrier. But new viruses will ceaseless appear. What security software can do is to reduce the probability of the problem and reduce the losses. Technology is only a tool for an inanimate network, and only people's self-discipline can eradicate the nature of virus. There will be a qualitative change in the improvement of business autonomy. The mobile network is just like a society: if the social atmosphere become well, the surrounding crime rate will decline. Technologies and laws are necessary in dealing with information leakage, but it is impossible to solve the problem of information leakage by them only. Information leakage is not a single aspect that can be completely eradicated, and the whole society needs to work together.

CONCLUSION

The rapid development of mobile network gives us unprecedented convenience. Our lives have also undergone a dramatic change. However, the hidden problems behind the prosperity are increasingly sharpened. Information leakage leads to economic losses, injury of mind, and harass of life.

This chapter starts from the cognition and experience of information security in modern mobile network. In the form of questionnaire survey, we collected users' view on the security of modern mobile network. The data was analyzed to gain further insights into the extent, causes, and perception of personal information leak. We found that the cyber security is not so optimistic. A total of 87.1% of the users discovered information leakage by themselves. However, the loss of the economy after information leakage is not as serious as we imagined. Only 3.5% responders have had substantial economic losses. Most leakage causes SMS harassment; all questionnaire responders experienced SMS harassment. It showed that the result of information leakage harassment occurs mostly by the way of this kind of smaller form of illegal responsibility. The satisfaction of responders on mobile cyber security is basically in the middle of a mediocre attitude, neither perfect nor worthless. They all think they are under safe protection, but it should be strengthened. Combined with the state cyber environment, we put forward feasible suggestions from three aspects of technology, law and society. Suggestions to achieve the better development of the mobile cyber information security are also provided.

REFERENCES

Chen, P., & Huang, Z. (2012). *Ibm SPSS 19 statistical software application guide*. Beijing, China: People's Health Press.

Choi, J., Na, G., & Yi, J. H. (2016). Hardware-assisted credential management scheme for preventing private data analysis from cloning attacks. *Multimedia Tools and Applications*, *75*(22), 1–16. doi:10.100711042-016-3271-z

CNNIC. (2017). *Statistical Report on Internet Development in China*. Retrieved from http://www.cnnic.net.cn/

Dai, W., Qiu, M., Qiu, L., Chen, L., & Wu, A. (2017). Who moved my data? privacy protection in smartphones. *IEEE Communications Magazine*, *55*(1), 20–25. doi:10.1109/MCOM.2017.1600349CM

Fan, X. (2016). *Last year, the national 4G user reached 386 million permeability to 29.6*. Beijing, China: Beijing Daily.

Feng, Z. (2016). *Research on network privacy in the big data age: from rights protection to personal choice* (Unpublished doctoral dissertation). Jilin University, Changchun, China.

Guo, M. (2012). *Protection of the tort law of personal information rights*. Beijing, China: Legal Publishing House.

Ma, L. (2015). On privacy leaks in mobile Internet social networks. *News Knowledge,* (6), 3-5.

Statista. (2016). *Number of smartphone users worldwide from 2014 to 2020 (in billions)*. Retrieved from https://www.statista.com/statistics/330695/number-of-smartphone-users-worldwide/

Tan, J. (2014). *Research and system implementation of mobile phone virus network detection and system implementation* (Unpublished doctoral dissertation). South China University of Technology, Guangzhou, China.

Vorobyov, K., Krishnan, P., & Stocks, P. (2012). A Low-Overhead, Value-Tracking Approach to Information Flow Security. *International Conference on Software Engineering and Formal Methods*, (73), 367-381. 10.1007/978-3-642-33826-7_26

Wang, Y., Liao, Z. & Xia, C. (2015). Discussion on the information security of smart phones. *Technology Square*, (11), 50-60.

Xing, K. (2014). *College students mobile phone: mobile market share accounted for about 70%*. Retrieved from http://news.youth.cn/gn/201404/

Xu, H., & Gong D. (2016). Research on the way, hidden danger and prevention strategy of personal information leakage. *Legal System*, (13), 176-177.

This research was previously published in Global Information Diffusion and Management in Contemporary Society; pages 212-232, copyright year 2019 by Information Science Reference (an imprint of IGI Global).

APPENDIX

Mobile Cyber Information Leakage Questionnaire

Dear Mr. / Miss:

To better understand the encounter and understanding of the information leakage of the mobile network. We launched this questionnaire. This survey is used only for academic research. Your answer is very important to us. Thank you very much for your help.

1. Your gender ()
   A.Male B. Female
2. Is your occupation related to the computer ()
   A.Y B. N
3. How much is your monthly cost of living?()
   A.lower than 1000 B.1000-1499 C.1500-1999 D. more than2000
4. Do you often get, transfer, and share information through mobile networks (mobile phones, tablet and other mobile terminals)? ()
   A. Often B. Occasionally C. Never
5. What is the main use of your mobile brower ()(Multi-choice)
   A. Online shopping
   B. Finding out information
   C. SNS apps
   D. Browse news
   E. Entertainment (games, movies)
   F. Message passing (E-mail)
   G. Use of QR code
   H. Others
6. Is there any disclosure of information that has been leaked? ()
   A. Yes B. No
7. What do you think of the cause of your information leak? ()(Multi-choice)
   A. Sold by bad merchants after account register.
   B. Necessary cyber information left over (Online shopping, Takeouts, Didi Kuaidi)
   C. Virus
   D. Friend list sharing
   E. Others

8. What are the usual hazards after the information leak? How about the frequency? (Table 10)

Table 10. Frequency of information leakage

	Often	Occasionally	Never
Account password is stolen			
phone harassment			
Spam message harassment			
telephone fraud			
Other method			

9. Is there any economic loss after information leakage ()
   A. Yes B. No
10. Do you care about the leakage of personal information ()
    A. Yes B. Don't like but don't dig into it ether C. Indifferent to me
11. What action did you take if information leaked ()(multi-choice)
    A. Shielded harassment source
    B. Modify account password
    C. Reflect to the operator
    D. Take no steps
    E. Take legal actions
    F. Others
12. What do you think is the reason for the disclosure of information()
    A. For profit by website administer
    B. Poor cyber supervision
    C. The relevant law is not perfect enough
    D. The society does not pay enough attention to the problem
    E. Others
13. What do you do to protect your information ()(multi-choice)
    A. Software protection
    B. Download App from authority
    C. Not fill in real information
    D. No any measures be taken
    E. Others
14. Your attitude with the current mobile cyber information security

Table 11. Attitude of cyber information security

		1 Totally Disagree	2	3	4	5 Totally Agree
Cyber Environment	**1. I feel safe about the network software**					
	2. Sellers won't hurt my interests					
	3. Few seller steal information					
Usability of protection methods.	**4. Protection software are easy to use**					
	5. The way of feedback is simple and convenient					
	6. The method of reducing and avoiding harm is simple					
Practicability of protection methods	**7. Protection software puts my mind in peace**					
	8. There will be no loss of information by downloading regular software					
	9. Reporting to operators can be a good way to curb information leaks					
economic cost	**10. The cost of pay ware is reasonable**					
	11. Protection software is free					
	12. The lower the software sales standard, the more user					
Overall evaluation (Satisfaction)	**13. Protection software can be good for protecting information**					
	14. Normal software does not steal user information					
	15. My current security assessment for mobile networks is good					

1An online food ordering-and-delivery company.

2A major ride-sharing company, providing transportation services for more than 450 million users across over 400 cities in China.

3 QQ and Wechat are Chinese social media (instant messaging, commerce and payment services) mobile application software developed by Tencent.

Chapter 20
Protection of the Communication and Digital Security of Mobile Phones as a Fundamental Right to Privacy:
A Study of the Jurisprudence of the Brazilian Superior Court of Justice

Nestor Eduardo Araruna Santiago
https://orcid.org/0000-0002-2479-7937
Universidade de Fortaleza, Brazil

Ana Monica Anselmo Amorim
Universidade de Fortaleza, Brazil

ABSTRACT

The central idea of this chapter is to discuss the notion of privacy arising from the use and exchange of messages through the use of mobile phones and whether police officers in charge of patrolling and public security may have access to the messages contained therein without proper judicial authorization. The authors focused on the position of the Brazilian Superior Court of Justice (STJ) on the subject, based on the decision rendered in the Appeal in Habeas Corpus (RHC) n. 51.531/RO. At this trial, protection was given to the new forms of communication and data contained in mobile phones under the constitutional terms of protection of privacy and non-unlawful obtaining of evidence, so that it will be necessary to issue a court order to get any informational elements found in messaging applications installed on a smartphone.

DOI: 10.4018/978-1-7998-8545-0.ch020

INTRODUCTION

The rights to privacy and intimacy (article 5, item X, of the Brazilian Federal Constitution of 1988 - CF) are protected as fundamental rights of individuals, so they can ensure the most intimate forum of the human being. The confidentiality of communications and data (article 5, item XII, CF) are aimed at preserving the fundamental rights referred to above, especially when there is state criminal prosecution activity targeted to obtain information contained in mobile phones or smartphones seized in situations of characterization of criminal offense.

In such cases, it is common for police officers to search text messages found on the device seized even without a court order, with the intention of verifying any information that may lead to the elucidation of crimes. It is important to say that the owners do not allow with the police practice, and in most cases they are "slightly threatened" to cooperate with law enforcement, providing the password to access the device.

It is discussed whether such behavior can be characterized as unlawful evidence, with the consequent breaking of the chain of custody. In order to do so, it is necessary to analyze the judgment of Appeal in Habeas Corpus (RHC) 51.531/RO from the Brazilian Superior Court of Justice (Superior Tribunal de Justiça, 2018), which considered *unlawful* to inspect the messages and data stored on the mobile phone of the person arrested without prior judicial authorization.

BACKGROUND

It should be emphasized that the present work is not intended to exhaust the topic of privacy and intimacy. The notions brought in are necessary to the aspiration of the research, considering that it is essential to conceptualize the right to privacy and intimacy and how it is related to the use of mobile phones. Or, being more specific, how technology is related to privacy and should be protected, taking for granted that this everyday relation brings some concerns, and the answers given by law and lawmakers not always are satisfactory.

Although the day-to-day use of smartphones rises, it seems important to accentuate it does not imply that the owners of this devices should give up the right to privacy. Protecting privacy should not be confused with hiding illicit behavior, once privacy and intimacy (as a narrower approach of privacy) are strictly linked to human dignity, which is protected by the Constitution (Vianna, 2004). The paradox is inevitable: the more we share our life experiences through the internet, the more we need to protect privacy, as it is a way to demonstrate freedom. Technological advances cannot condition the proper protection of freedom (Badaró, 2014; Schreiber, 2016). It is not wrong or exaggerated to say that this is one of humanity's greatest dilemmas today.

We can say that human dignity, treated as a fundamental principle of the federal constitution, can only be considered in its entirety if privacy is assured. The right to be with oneself, to have a right to an intangible sphere, far from the public eye (even to celebrities and public people) is revealed as one of the facets of personality. Of course, the understanding of privacy and intimacy extends also to all the belongings and objects of the human being, since these are often extensions of the personality.

It is well-known that in the chapter devoted to the description of fundamental individual rights in the CF there is the need to protect the confidentiality of correspondence, telegraphic communications, data and telephone communications. The same Constitution considered such inviolable secrets as a means of exercising the right to privacy among private individuals (article 5, item X), it also relativized this protection, allowing the "breach" of confidentiality of telephone communications by means of authorization issued by a judicial authority, in cases and in accordance with the procedure legally established only for purposes of criminal investigation or for criminal procedural instruction. Even if the seized telephone device does not have a password, judicial authorization for its analysis becomes indispensable (Lemos, Caldeira, Quintiere, 2016).

Thus, unlike the texts of the Brazilian Constitutions of 1946 and 1967, which had a categorical form on the inviolability of the secrecy of communications, oral or written, generating doubt about the validity of the Brazilian Telecommunications Code (Law n. 4177/1962, derogated by Law n. 9472/1997 and subsequent updates) regarding the possibility of violation of telecommunication by means of judicial authorization (Greco Filho, 2005), the current Constitutional rule relegated to the ordinary federal legislator the competence to discipline the subject.

The Brazilian Federal Supreme Court (STF – Supremo Tribunal Federal) was called to position itself regarding the reception of article 57 of the Brazilian Telecommunications Code according to CF. STF decided that, as long as there was no authorizing law, any evidence obtained through telephone interception, would be considered unlawful even if judicially granted.

While the Brazilian Federal Constitution of year 1988 protects the right to free expression of thought, regardless of the path chosen by the citizen, as a true exteriorization of the principle of the dignity of the human person (article 1, item III), on the other hand, defends the interest of the society not only for the search for factual and/or authoritative proof of the occurrence of an offense, but also - and especially - stipulates limits in the form of obtaining the same evidence, in the same line of protection of human dignity. What is interesting to note is that the CF, even though it was granted before the commercial emergence of the Internet and cutting-edge technological innovations in smartphone communication, has already provided for a general privacy protection clause for distance communications.

There is yet another point to consider, which somehow conflicts with the right to privacy. In Brazil, criminal organizations, as well as criminals in general, constantly use the message applications (apps) on cell phones to exchange information about illicit conduct planning. Among all apps, the most used in Brazil is *WhatsApp*, with about 120 million active users – one and a half billion worldwide. With an end-to-end encryption system, there is no way for a third-party to intercept the content of messages exchanged between two or more users. So, if someone wants to have access to the messages, a person must have the desired mobile phone to read or hear what was written or said.

Brazil has been in the news for some time due to their involvement in judicial decisions regarding the use of WhatsApp; forcing the company to deliver information to the Judiciaries that allegedly contained messages that pertained to criminal investigations. Of course, court decisions have failed to understand how information is sent in the application, because the encryption used not only prevents the company from having access to the content, but also prevents data from being stored on the server (Sydow, 2017).

So, it is possible for someone to say that the way information was obtained by the Judiciary was not modified by technology in the exchange of messages. Despite this, it is still common for police officers in charge of street surveillance to conduct personal searches in case of suspected crimes, immediately seizing the suspects' cell phones. They then ask the owner to enter the password so that they can read any messages exchanged via WhatsApp that can serve as evidence of criminal conduct.

THE LEADING CASE

The Brazilian Superior Court of Justice (Superior Tribunal de Justiça - STJ) ruling originated from Habeas Corpus judged by the Court of Justice of Rondônia (TJRO), a Northern Brazilian State, filed in favor of Leri Souza e Silva, accused of drug trafficking (Article 33 of Law n. 11343/ 2006). On March 18, 2014, Military Police of the State of Rondônia received information about unlawful conduct, possibly drug trafficking. As a result, they waited in a certain place until the arrest of the suspect and the seizure of his cell phone.

In the *writ* of habeas corpus, the thesis of nullity of the search promoted by the police authority in the mobile phone device seized during the caught-in-the-act arrest was supported, since the information contained in the apparatus has the same constitutional protection inherent to telephone secrecy. Finally, it was argued that access to data contained in the telephone could only be affected by judicial decision, and it was then an unlawful evidence, as prescribed by art. 157 of the Criminal Procedure Code (Código de Processo Penal – CPP).

The court said that the phone device was seized with the accused during his arrest, pointing out that the forensic examination performed in the device is based on article 6, items II, III and VII of the CPP, which determines to the police authority, as soon as it becomes aware of the criminal practice: a) to apprehend the objects related to the fact, after being released by the criminal forensic experts; b) collect all the evidence that serves to clarify the fact and its circumstances; and c) to determine, as the case may be, the examination of a body of evidence and any other forensic examination. It also stated that access to data contained in the device does not encounter the same impediment of telephone interception and that the police authority acted strictly for law enforcement.

TJRO denied the order of habeas corpus, citing precedent of the Brazilian Federal Supreme Court (Supremo Tribunal Federal, 2018), in which protection conferred by the CF would be inherent only to the communication of data and not to the data in themselves, although stored in computers or similar. Due to this decision, the defense appealed to STJ, supporting the same thesis already presented to the TJRO.

In April 2016, in the judgment of Appeal in Habeas Corpus (RHC) 51.531/RO (Superior Tribunal de Justiça, 2018), STJ considered *unlawful* to inspect messages and data stored on the mobile phone of the person arrested without prior judicial authorization, since it is a violation of the privacy and privacy of the person being investigated. After the seizure of the smartphone, the police authority should have filed a petition for the breach of confidentiality of the stored data. Lastly, the unlawfulness of the evidence produced were recognized, as well as those derived from it (Article 157 of the CPP), which should be taken from the case file.

ISSUES, CONTROVERSIES, AND PROBLEMS

The purpose of this chapter is to investigate the position of the Brazilian Superior Court of Justice on the subject, based on the decision rendered in the Appeal in Habeas Corpus n. 51.531/RO, of the rapporteur of the Justice Nefi Cordeiro (Superior Tribunal de Justiça, 2018). At this trial, protection was given to new forms of communication and data contained in mobile telephones under the constitutional terms of protection of privacy and non-unlawful obtaining of evidence, so it will always be necessary to issue a court order to get any informational elements found in messaging applications installed on one's smartphone.

In addition, other Justices of STJ also examined to determine whether there is a distinction in treatment arising from facts other than caught in the act arrest; for example, the issuance of a search warrant and seizure, or delivery of a mobile device by to the police.

Problems and research hypotheses are aimed at examining the inviolability of privacy and intimacy, the inviolability of data, and prohibiting the use of illegal evidence and its derivatives as fundamental rights, with an emphasis on protecting the confidentiality of data and telephone messages, and their subjection to judicial authorization for evidence in criminal proceedings, without mitigating the protection of fundamental rights as a result of a significant increase in crime.

THE JURISPRUDENCE OF STJ ON ACCESS TO SMARTPHONE DATA

In the global village, there is an era of communications dynamism, and smartphones are a palpable record of this. Modern mobile handsets have a multifunctionality that houses a significant portion of records relating to privacy and the privacy of its owner. They store not only messages or links, but also bank passwords, credit card data, photographs, appointment schedules, so constitutional protection must adapt and extend to new technologies. Not because they are new, but because they refer to a facet of the privacy and intimacy of their user.

This was a point made by Justice Rogério Schietti Cruz when speaking at the judgment of the Appeal in Habeas Corpus n. 51.531/RO (Superior Tribunal de Justiça, 2018), referring to the third-generation evidentiary law, which refers to "highly technological, invasive tests that allow the achievement of knowledge and results unattainable by senses and the traditional techniques adopted so far" (Knijnik, 2016).

It is important to emphasize the *distinguishing* made by Justice Schietti Cruz in the RHC 51.531/RO judgment regarding a 2004 positioning of Justice Gilmar Mendes in the judgment of Habeas Corpus (HC) 91.867/PA of the STF, which also deals with seizure of cell phone in the moment of arrest and access to data, but without access to the conversations contained in the device (Supremo Tribunal Federal, 2018). It is reminded that at the time, mobile phones were not connected to broadband Internet, which reveals the less intrusive character of the police check. It also points out that the messaging, photos and documents applications store the files in the phone memory, and it should be noted that most companies that offer such features do not save the files on their servers. What is deduced from the argument is that if the proof that were wanted to be stored on the servers that run the applications, it would be essential to submit a court decision determining the data routing so that the evidence would be considered lawful.

Regarding the seizure of mobile phones in situation of flagrant arrest and access to the content of messages and data contained therein, the STJ has solidified understanding on the subject. In a database search using the expression "WhatsApp prox10 conversation", 13 occurrences (judgments) were found, the oldest being judged in April 2016 – exactly the case that is being examined in this chapter - and the most recent in February 2018. And about the content of the judgments, there are distinctions of a qualitative nature as to the form of access to text messages.

In Habeas Corpus 372.762/MG, it should be noted that there were no arrests in the act of seizing the mobile phones, which were seized in compliance with the judicial order that authorized the search and seizure in the addresses linked to the patient and the other e-mails (Superior Tribunal de Justiça, 2018). Thus, new judicial authorization would be dispensable for analysis and use of stored data, since there was search and apprehension of the physical basis of the cellular telephone handsets.

Concerning the judgment by majority of votes of Appeal in Habeas Corpus 86.076/MT, it was decided by lawfulness in the investigation of mobile telephone equipment in the hypothesis in which the victim is dead, and that telephone was delivered to the police authority by his wife, interested in clarifying the facts that held him, since there was no more secrecy to protect the holder of that right (Superior Tribunal de Justiça, 2018). In this trial, Justice Sebastião Reis Júnior proposed a divergent vote, due to the fact that he understood that judicial authorization was essential for access to WhatsApp application data from the victim's telephone, even if the telephone been delivered by the wife of the victim to elucidate the crime committed against him, since it cannot be lost sight that the conversation maintained through this application is a form of written communication and immediate between the interlocutors, and, if accessed without judicial authorization, represents unauthorized interception of communications, making possible to break the chain of custody (Prado, 2014).

It can be said that the Brazilian Superior Court of Justice acted as a guarantor by requiring the issuance of a court order to access the data contained in the cellular device; distancing itself, however, from the activist form as it has been conducting the theme of telephone interceptions (Santiago, Costa, 2016), a matter that is related to the access to mobile data, and which, for consistency, deserves the same treatment, which shows that the STJ adopts a double behavior in relation to evidence obtained through telephone and data interception. In this sense, the STJ has been failing in its constitutional commitment to unify national jurisprudence.

FUTURE RESEARCH DIRECTIONS

In the year 2018 the Brazilian Data Protection Law was approved (Law n. 13708). The text guarantees greater control of citizens about their personal information, requiring explicit consent for data collection and use and requires the provision of options for the user to view, correct and delete such data.

Amongst many subjects, the law expressly excludes from the scope of protection the processing of personal data for public security purposes. The law also says that in this case, there will be special legislation, to be elaborated observing proportional measures and necessary to the service of the public interest, according to due process of law. It also prohibits companies from accessing this data.

In a society that is increasingly interconnected by digital means, and even access to the Internet is considered a fundamental right, it is important that only public power has access to data that can be used to the detriment of citizens. But even in these cases, it is important that there be legally established limits for state action.

Considering this, it is important that law researchers are concerned with establishing what the state boundaries would be, since the parameters established by law are too vague, and may give rise to relativizations based on the good faith of state agents.

CONCLUSION

Even in the case of *distinguishing* in relation to Appeal of Habeas Corpus 51.531/RO, there was consistency with the contents of protection of intimacy and privacy foreseen in the CF. Since there is a judicial warrant for search and seizure, there is no need to speak about issuing a new order to scan the contents of a smartphone. In the same way, the voluntary delivery of the device to the authorities eludes the unlawfulness in obtaining the proof, even if made by the wife of the deceased victim, in genuine exercise of renouncing the right of privacy. Although it is objected the necessity of a judicial warrant for access to the data of the telephone of the deceased for the validation of the informative elements contained in the evidentiary context, its delivery by the wife made the dispatch of the warrant unnecessary.

In any case, it is crucial to note that the position of the STJ in the decisions about non-invasion of smartphones at the moment of police approach is an important achievement aimed at the protection of intimacy and of privacy, despite a punitive discourse and violating fundamental rights and guarantees that has as its background the fight against crime.

ACKNOWLEDGMENT

We thank the University of Fortaleza (UNIFOR) for the financial and professional support regarding the development and presentation of this work.

REFERENCES

Badaró, G. (2014). *Processo penal*. Rio de Janeiro: Elsevier.

Brasil, Superior Tribunal de Justiça. (2018). *Habeas Corpus n. 372.762/MG*. 5ª. Retrieved from https://ww2.stj.jus.br/processo/revista/documento/mediado/?componente=ITA&sequencial=1631821&num_registro=201602540301&data=20171016&formato=PDF

Brasil, Superior Tribunal de Justiça. (2018). *Recurso Ordinário em Habeas Corpus n. 51.531/RO*. Retrieved from https://ww2.stj.jus.br/processo/revista/documento/mediado/?componente=ITA&sequencial=1497056&num_registro=201402323677&data=20160509&formato=PDF

Brasil, Supremo Tribunal Federal. (2018). *Habeas Corpus n. 91.867/PA*. 2ª. Retrieved from http://redir.stf.jus.br/paginadorpub/paginador.jsp?docTP=TP&docID=2792328

Brasil, Supremo Tribunal Federal. (2018). *Recurso Extraordinário n. 418.416/SC*. Retrieved from http://www.stf.jus.br/portal/jurisprudencia/listarJurisprudencia.asp?s1=%28418416%2ENUME%2E+OU+418416%2EACMS%2E%29&base=baseAcordaos&url=http://tinyurl.com/hews2gu

Knijnik, D. (2016). A trilogia Olmstead-Katz-Kyllo: o art. 5º da Constituição Federal do século XXI. *Revista da EMAGIS - Revista da Escola da Magistratura do TRF da 4ª Região*, 2(4), 77-90.

Lemos, B., Caldeira, F., & Quintiere, V. (2016). Sobre as (i)legalidades no processo penal: breve reflexão a respeito do Whatsapp a partir da Lei n. 9.296/96 – um estudo de caso. In Interceptação Telefônica: os 20 anos da Lei n. 9.296/96 (pp. 45-62). Belo Horizonte: D'Plácido.

Prado, G. (2014). *Prova penal e sistema de controles epistêmicos: a quebra da cadeia de custódia das provas obtidas por métodos ocultos*. São Paulo: Marcial Pons.

Santiago, N. E. A., & Costa, D. K. A. (2016). Garantismo, encontro fortuito decorrente de interceptações telefônicas e a jurisprudência do Supremo Tribunal Federal e do Superior Tribunal de Justiça: rumo à ilicitude das provas. In Interceptação Telefônica: os 20 anos da Lei n. 9.296/96 (pp. 319-346). Belo Horizonte: D'Plácido.

Schreiber, S. (2016). Algumas reflexões sobre a medida cautelar de interceptação telefônica, 20 anos depois. In Interceptação Telefônica: os 20 anos da Lei n. 9.296/96 (pp. 445-463). Belo Horizonte: D'Plácido.

Sydow, S. T. (2017). Criptografia e ciência penal informática. *Revista dos Tribunais*, *985*, 343-374.

Vianna, C. S. M. (2004). Da privacidade como direito fundamental da pessoa humana. *Revista de Direito Privado*, *17*, 102–115.

ADDITIONAL READING

Dutra, L. (2007). *Busca e apreensão penal: da legalidade às ilegalidades cotidianas*. Florianópolis: Conceito Editorial.

Fayet, F. A., & Carvalho, A. V. (2018). *Whatsapp*, sigilo de dados e prova ilícita: Para dizer o óbvio. *Revista Brasileira de Ciências Criminais*, *26*(140), 297–322.

Santiago, N. E. A. (Ed.). (2013). *Proibições probatórias no processo penal: análise do direito brasileiro, do direito estrangeiro e do direito internacional*. Brasília: Gazeta Jurídica.

Santoro, A., & Maduro, F. (Eds.). (2016). Interceptação Telefônica: os 20 anos da Lei n. 9.296/96. Belo Horizonte: D'Plácido.

Teixeira, T., Sabo, I., & Sabo, P. (2017). Whatsapp e a criptografia ponto-a-ponto: Tendência jurídica e o conflito privacidade vs. interesse público. *Revista da Faculdade de Direito da UFMG*, *71*, 607–638.

Tucci, R. (2006). Ordem judicial de busca e apreensão e ilicitude da prova dela extrapolante. *Revista dos Tribunais*, 95 (848), 457-470.

KEY TERMS AND DEFINITIONS

Appeal of Habeas Corpus: It is a judicial measure that can only be used by the citizen because of an unfavorable decision rendered in the judgment of a writ of habeas corpus decided by the state courts or by the federal courts.

Brazilian Superior Court of Justice (STJ): According to Brazilian Federal Constitution, STJ is the constitutional court, empowered to resolve issues that directly concern constitutional violations.

Brazilian Supreme Court of Justice (STF): According to Brazilian Federal Constitution, the STJ is responsible for the uniformization of national jurisprudence, as well as ensuring the uniform application of federal legislation.

Caught in the Act Arrest (Flagrant): Arrest made by public agents when there is visual certainty of the infraction, or when the suspect has just practiced the alleged offense.

Habeas Corpus: It is a constitutional writ that seeks to protect the citizen against abuse of state power that curtails his freedom.

This research was previously published in Constitutional Knowledge and Its Impact on Citizenship Exercise in a Networked Society; pages 231-241, copyright year 2019 by Information Science Reference (an imprint of IGI Global).

Chapter 21
Achieving Secure and Privacy-Preserving in Mobile Social Networks

Mohamed Amine Ferrag
https://orcid.org/0000-0002-0632-3172
Guelma University, Algeria

Abdelaziz Amara korba
Badji Mokhtar-Annaba University, Algeria

ABSTRACT

This chapter proposes a security framework for achieving secure and privacy-preserving mobile social networks named ASPP. Based on the cooperative neighbor, reactive routing protocol, and short signatures technique, the proposed scheme can not only detect and avoid but also can preserve the message privacy against elemental attacks and compound attacks. In addition, ASPP is robust against eavesdropping attack, wormhole attack, packet analysis attack, packet tracing attack, and replay attack. Extensive analyses and experiments are given to demonstrate its high detection rate and practicability under various scenarios.

INTRODUCTION

A mobile social network fueled with heterogeneous wireless infrastructures (e.g., cellular/ WiFi) and mobile devices (e.g., smartphones, tablets), which can facilitate multimedia services by providing ubiquitous connections between service providers and users in a mobile environment (Zhang et al., 2014). Mobile social networks still face many security and privacy challenges, including private information leakage, cheating detection, Sybil attacks, DDoS attacks and so on (Liang et al., 2014). Based on spoofed identities, pseudonyms, locations, and profiles, an adversary can launch active or passive attacks (deliberately delays, drops, corrupts, or modifies messages) in order to steal the social data as well as to damage P2P communications (Ferrag et al., 2016). According to the work (Ferrag et al., 2017), the privacy preserva-

DOI: 10.4018/978-1-7998-8545-0.ch021

tion models for mobile social networks can be divided into location privacy, identity privacy, anonymity, traceability, interest privacy, backward privacy, and content-oriented privacy.

In this chapter, to address both security and performance challenges in ad hoc social networks, we propose a security framework for achieving secure and privacy preserving in mobile social networks, called ASPP, for ad hoc social communications. With the proposed ASPP scheme, each node user can be privacy-preserving authenticated before joining other nodes using routing protocol. The contributions of this chapter are fourfold.

- First, we formalize the system model where we consider the social characteristics, i.e., human mobility, human group and preferences in a typical MANET which consists of trusted authority (TA), some stationary social unit (SU) deployed at the social space, and a large number of mobile equipped with WiFi technology moving on a social space. Next, we improve the AODV routing protocol basing on some concepts of social theory to be suitable for ad hoc social communications, i.e., degree centrality, closeness centrality, and betweenness centrality.
- Second, we propose an efficient certificate scheme, where the TA issues the private key SK_{n_i} and certificate $Cert_{TA,n_i}$ using the Schnorr signature algorithm (Schnorr, 1991). The node n_i can verify the certificate $Cert_{TA,n_i}$ by the procedure *S.check* and cannot use these certificates directly in ad hoc social communication. Based on the proxy re-signature cryptography technology (Toshiyuki et al., 2013), the node request N_y resignature key from S_x and then re-signs the certificates issued by the TA to be the same as those issued by S_x itself. With this method of key distribution, the proposed ASPP scheme guarantees the node identity confidentiality.
- Third, we provide conditional privacy preservation to the nodes with demand response. Although the SU act as certificate issuers in ASPP, they do not know what certificates are held by a node. Therefore, the adversaries cannot trace the interested nodes although they had compromised all SU. Then, to detect and verify the attack against routing protocol, based on three control messages {*Detectreq, Detectrep, Notifreq*}, each node initiates a response request and send *Detectreq* signed to nodes in its routing table at 1-hop, and waits the response of his request for runs the notification phase. After receiving a request response, the node checks its signature. If valid and $TDetect_{n_i} < 0$, it considers that the link with the node is proved. Otherwise, it returns suspicious and starts the notification phase.
- Finally, to validate the efficiency and effectiveness of the proposed ASPP, we integrate in the AODV implementation, being the modified protocol designated AODV-ASPP. Extensive simulation results in the first scenario show that the proposed ASPP scheme can detect the black hole attack more in the configuration where attack is launched on a number of more hops. Thus, in the second scenario, we focus on the transmission delay of ASPP at the node with extensive performance evaluation, which further convinces its practicality.

This work extends our previous work that introduced a new security mechanism for ad hoc on-demand distance vector in ad hoc social network (Ferrag et al., 2013a). In our previous work, we have provided the taxonomy of attacks and their influence on the security property using the AODV as routing protocol. Then, we have presented a security mechanism based on use of two messages *Detectreq* and digital signatures *Champ_sig* to detect malicious links. In this chapter, we focus on investigating the privacy of node under the routing attacks.

The remainder of this chapter is organized as follows: Section 2 surveys some related research work. Section 3 presents the system model, the routing improvements, and the research objectives. Section 4 gives some preliminaries, including the secure hash function, bilinear pairings, and the short signatures technique. Then, Section 5 presents the proposed ASPP scheme, followed by the security analysis and performance evaluation in Sections 6 and 7, respectively. Finally, Section 8 draws our conclusion.

RELATED RESEARCH WORK

In this section, we review some of these existing works (Deng, 2003; Deng, 2006; Jian, 2007; Kamat, 2005; Ozturk, 2004; Yipin, 2010; Lu, 2012a, 2012c; Gupta, 2016; Chaudhary, 2017; Stergiou, 2018; Li, 2018;Ferrag, 2018). Generally, to achieve contextual privacy, the existing approaches can be categorized into two types: one is by source location privacy, and the other is by destination location privacy.

- By source location privacy, Kamat et al. (2005) describes two techniques for location privacy: 1) fake packet generation technique in which a destination creates fake sources whenever a sender notifies the destination that it has real data to send; 2) the phantom single-path routing, which achieves location privacy by making every packet generated by a source walk a random path before being delivered to the destination. Another technique to protect the source location privacy is presented by Ozturk et al. (2004) where it requires a source node to send out each packet through numerous paths to a destination to make it difficult for an adversary to trace the source.
- By destination location privacy, Deng et al. (2003) describes a technique to protect the locations of destinations from a local eavesdropper by hashing the identification fields in packet headers, then in (Deng et al., 2006) presents four techniques to protect the location privacy of destination from a local eavesdropper who is capable of carrying out time correlation and rate monitoring. Another technique to protect the destination location privacy is presented in (Jian et al., 2007) proposed the location privacy routing protocol (LPR) for destination location privacy. The LPR algorithm provides privacy to the destination with help of redundant hops and fake packets when data is sent to the destination.

The privacy-preservation scheme has been improved for different types of network. Yipin et al (2010) propose an efficient pseudonymous authentication scheme with strong privacy preservation (PASS), for vehicular communications. Unlike traditional pseudonymous authentication schemes, the size of the certificate revocation list (CRL) in PASS is linear with the number of revoked vehicles and unrelated to how many pseudonymous certificates are held by the revoked vehicles. For achieving vehicle user's privacy preservation while improving key update efficiency of location based services in vehicular ad hoc networks, Lu et al. (2012c) propose a dynamic privacy-preserving key management scheme, called DIKE. Another privacy-related study but for smart grid communications, where Lu et al. (2012a) propose an efficient and privacy-preserving aggregation scheme, named EPPA. EPPA uses a super increasing sequence to structure multidimensional data and encrypt the structured data by the homomorphic paillier cryptosystem technique, and to reduce authentication cost, it adopts the batch verification technique.

Our proposed ASPP is a privacy-preservation scheme with response requested and demand response. Compared with previously reported privacy-preservation schemes, cannot only satisfy the security and privacy requirements of ad hoc but also can detect, prevent and notify the elemental attacks and composed attacks. More importantly, ASPP is the first study on privacy preservation against the routing attacks, i.e., the elemental attacks and composed attacks in this chapter.

SYSTEM MODEL, ROUTING IMPROVEMENTS, AND RESEARCH OBJECTIVES

In this section, we formalize the system model, including routing improvements, and research objectives. We provide an overview of the architecture, and highlight several interesting properties.

System Model

Mobile Ad hoc NETworks (MANETs) have a wide range of applications. In fact, they are robust, inexpensive and adaptable to both urban as rural areas. Among the applications, we include: military applications, operations of rescue, commercial applications, vehicular networks (VANET) or Social Network (VSN) (Lu, 2012). In our work, we consider the social characteristics (human mobility, human group and preferences) in a typical MANET which consists of trusted authority (TA), some stationary social unit (SU) deployed at the social space, and a large number of mobile equipped with WiFi technology moving on a social space, as shown in Figure 1. Furthermore, we use the abbreviation ADSocial in this chapter for ad hoc social network (Ferrag et al., 2013b).

Figure 1. System model

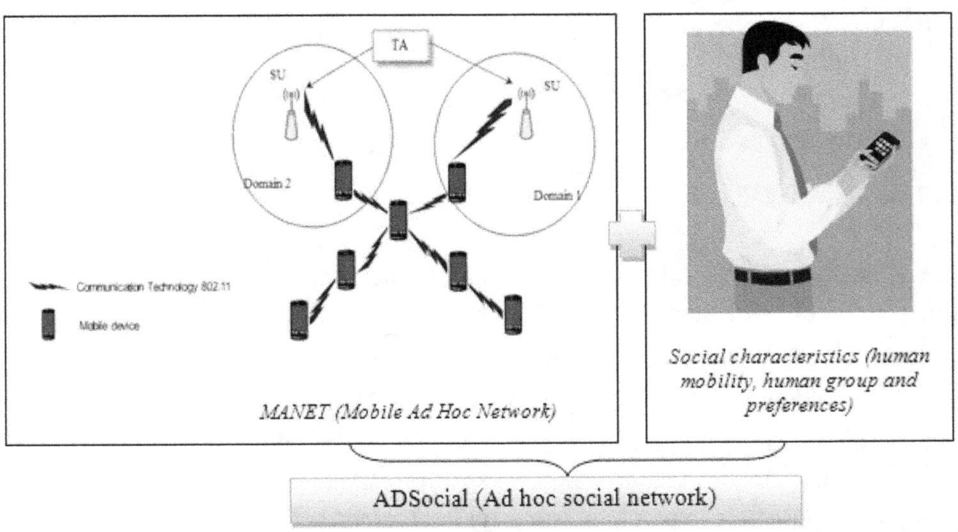

- The TA is fully trusted by all parties in the system and in charge of the registration of SUs and mobile device. The TA can divide its huge precinct into several domains and deploy SUs at the boundary between these domains. The domain information is available to all entities. In addition, the TA is assumed powered with sufficient storage capability and is infeasible for any adversary to compromise (Calandriello et al, 2007).

- The SU connect with the TA by wired links in the system. They provide service for information dissemination and certificate updating. The certificates issued by a SU can only be used in the domain where the SU is located.

Figure 2. ADSocial under consideration $\mathcal{G} = (\mathcal{V}, \varepsilon)$

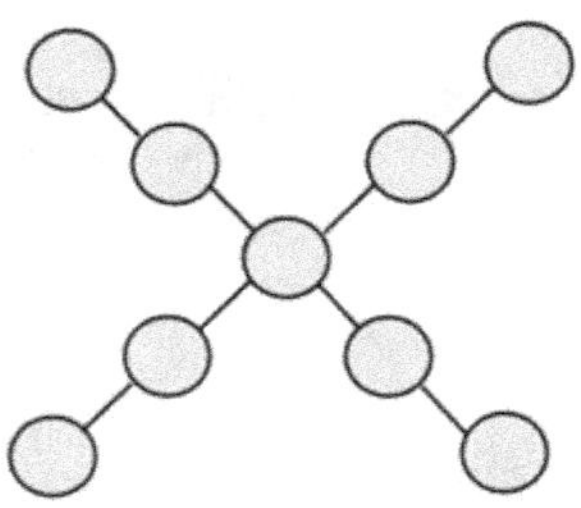

- Formally, such a wireless ad hoc network, as shown in Figure 2, ADSocial can be represented as an undirected graph $\mathcal{G} = (\mathcal{V}, \varepsilon)$, where $\mathcal{V} = \{v_1, v_2, \ldots\}$ is the set of all node $\mathcal{N} = \{n_1, n_2, \ldots\}$, and $\varepsilon = \{(v_i, v_j) \mid v_i, v_j \in \mathcal{V}\}$ is the set of edges. Let $d(v_i, v_j)$ denote as the distance between v_i and v_j, the each e_{ij}, which indicates whether there exists a communication edge between two nodes v_i and v_j or not, is defined as

$$e_{ij} = \begin{cases} 1, d(v_i, v_j) \leq R; \\ 0, d(v_i, v_j) > R; \end{cases} \tag{1}$$

ROUTING IMPROVEMENTS

In our work, we use the AODV (Ad hoc On demand Distance Vector) (Perkins et al., 2003) which is a reactive routing protocol by distance vector. We consider a typical wireless ad hoc network which consists of links created by AODV and large number of mobile nodes $\mathcal{N} = \{n_1, n_2, \ldots\}$ randomly deployed at a certain interest region of social with the area S. In ADSocial with AODV, when a source node $n_i \in \mathcal{N}$ needs a route to a certain destination, and no path is available (the path may be non-existent, have expired or be faulty). The node source broadcasts a route request message RREQ (Route REQuest). Then it records the identifier of RREQ ([RREQ_ID, @ SRC]) in its history (buffer) and associates it with a timer which will count down his life beyond which this entry will be deleted. However, when an intermediate node has no valid path to the destination, it receives the RREQ and then adds or updates the neighbor from which the packet was received. The node finds that the RREQ is already covered; he abandons and does not rebroadcast. Otherwise, it updates its routing table with the information contained in the application in order to reconstruct the reverse path to the source later. To ensure the absence of routing loops, AODV uses the principle of sequence numbers. Upon receipt of a RREQ packet, the destination adds or updates its routing table in a path to the neighboring node from which it received the packet and a path to the source. The destination then generates a route reply RREP and it sends in unicast to next hop toward the source. When the route reply reaches the source, a bidirectional path is established between the source and the destination and the transmission of data packets can begin.

We improve the AODV routing protocol basing on some concepts of social theory presented in (Batallas and Yassine, 2006; Kamat, 2005; Opsahl, 2010). We consider the following three measures for AODV:

- **Degree Centrality:** In ADSocial, a central node is one that relates to a large quantity of other nodes in the network. Therefore, we propose that groups are formed around the nodes with the densest neighborhood; the node that has the largest number of neighbor's symmetrical hop is considered the center of group. In this way, we are sure that the center of the group is the node that covers the largest number of nodes in group. Thus, the degree of node n_i is equal to sum of RREQ messages received, the centrality of node n_i is calculated by:

$$C_D(n_i) = d\left(n_i\right) = \sum_{\forall i \neq j} e_{ij} \tag{2}$$

where $C_D(n_i)$ is the degree centrality of node i; $d(n_i)$ is the degree of node i; e_{ij} is 1 if a route request message RREQ sent by j to node i; and 0 otherwise. However, we distinguish three states:

- **State (1):** $C_D(n_i)=0$ node n_i has happened, or he just left his group and has no neighbor in its neighborhood.
- **State (2):** $C_D(n_i){}^{1}0$ and with low density, node n_i part of group members.
- **State (3):** $C_D(n_i){}^{1}0$ and with highest density, node n_i is the center of a group.

- **Closeness Centrality:** A node which is the center of a group can reach the other nodes or groups using paths available in ADSocial. Therefore, the classical definition of centrality proximity is modeling the dissemination of information through the use of shortest paths. Assuming that the network is fully connected, the proximity of a center node n_i can be measured by:

$$C_c(n_i) = \left[\sum_{\forall i \neq j} dr\left(n_i, n_j\right)\right]^{-1} \tag{3}$$

where $C_c(n_i)$ is closeness centrality of node n_i and $dr(n_i,n_j)$ is the path between node n_i and node n_j. In AODV routing protocol, $dr(n_i,n_j)$ is calculated by the information in routing table.

- **Betweenness Centrality:** This measure was introduced by Freeman (1978) as a measure to quantify the control of a human being in communication between human beings. In ADSocial, betweenness is focusing on nodes that lie in the path between other nodes, is another one to measure the centrality. The node communicates in AODV with another node uses its routing table to send messages on the shortest path using routing table. Hence, we suggest that if other paths are available in the AODV routing protocol, nodes that have a high probability to occur on a shortest path selected at random between two nodes chosen at random have a high betweenness. More precisely, the formula betweenness centrality of a node can be defined by:

$$C_b\left(n_i\right) = \sum_{i \neq j \neq k \in G} \frac{g_{jk}\left(n_i\right)}{g_{jk}} \tag{4}$$

where $C_b(n_i)$ is the betweenness centrality of node n_i, $g_{jk}(n_i)$ is total number of shortest paths from node n_j to node n_k and g_{jk} is the number of those paths that pass through n_i.

Research Objectives

Since the ad hoc social is a large-scale wireless network scenario for public service, it faces serious security and privacy challenges. In the ASPP scheme, we aim to achieve the following security and privacy objectives.

- **Availability:** This property implies that the services or resources requested are available in a timely manner, even if there is a problem or a malfunction in the system. In our context, it is necessary to ensure access to the routing service at any time.
- **Authentication:** This includes user authentication and message integrity. All accepted messages should be from legal members and delivered unaltered.
- **Content Oriented Privacy:** Ensure this property is achieve the following three properties:
    - **Immutability**: This property says that it must be impossible to change a part of the message designated as unmodifiable by the signer, while maintaining a valid signature.
    - **Transparency (Privacy):** It should be infeasible to decide whether a message has been sanitized or not. This may be desirable in applications where one should not be able to discriminate against messages produced by the sanitizer. This property implies privacy. If the adversary against privacy exists, it is able to recover the original message from messages changed. So, it will be possible to use this adversary for construct an adversary against transparency.
    - **Accountability:** It should be impossible to produce, for a person other than the signer, an original signature or a modified signature. So, in case of problems on the origin of a signature, it must be able to properly find that is the origin of the signature.

PRELIMINARIES

In this section, we introduce some preliminaries, including secure hash chains (Mao, 2003), bilinear pairing technique (Grewalet et al., 2013), and short signatures technique (Boneh and Boyen, 2008), which are the bases of our proposed ASPP scheme. In addition, the notations used throughout this chapter are given in Table 1.

Table 1. Notations

Symbol	Notation
ADSocial	Ad hoc Social Network
n_i	The node *i* in ADSocial
Detectreq	The message of response requested
Detectrep	The message of demand response
Notifyreq	The message of notification
T	The current timestamp
$TNormal_{\text{Detectreq}}$	Time to send the message *Detectreq*
K	The temporary key
H_1,H_2	A hash function
S.Init	Generates an environment bilinear
S.GenK	Generates the secret key
$S.Sign\left(K,SK_{n_i},Detectreq\right)$	Signing the message *Detectreq* with the temporary key *K* and secret key SK_{n_i}
$S.Check\left(Detectreq,\sigma_{n_i,Detectreq},K\right)$	Verifying the signature $\sigma_{n_i,Detectreq}$ of the message *Detectreq* with the temporary key *K*
$Enc_s(.)$	A secure symmetric encryption algorithm with the secret key *s*
$\|\|$	The concatenation
Send_Request (C)	Send the message C
PID_{n_i}	The pseudo identity of n_i
SK_{n_i}	The secret key of n_i
PK_{n_i}	The public key of n_i
σ_{TA,n_i}	A signature signed by TA for n_i
$Cert_{TA,n_i}$	A certificate of n_i issued by TA
ΔT	The privacy requirement on the validity period length of a certificate
S_x	The x-th SU
$TDetect_{n_i}$	The attack detection time of n_i
R	The maximum transmission
SD	The speed of light
SIMS	Short Inter-Message Space

Hash Chains

A hash function is a one-way function in particular allowing to reduce a bit string of any size {0,1} in a digest of fixed size $\{0,1\}^{\lambda}$ with λ a security parameter.

Definition 1 (Cryptographically Secure Hash Function $\mathcal{H}$): *A hashing function* $\mathcal{H}:\{0,1\}^{*}\rightarrow\{0,1\}^{\lambda}$, *with* λ *a security parameter, is cryptographically secure if it satisfies the following three security properties.*

1) Preimage-resistance, given $y\in\{0,1\}^{\lambda}$, regardless the adversary, the probability of finding x as $\mathcal{H}(x)=y$ is negligible. This corresponds to the definition of a one-way function;
2) Second pre-image resistance, given $x\in\{0,1\}^*$, regardless the adversary, the probability of finding $x'\neq x$ as $\mathcal{H}(x)=\mathcal{H}(x')$ is negligible.

3) Collision-resistance, regardless the adversary, the probability of finding a couple (x, x') as $\mathcal{H}(x) = \mathcal{H}(x')$ and $x' \neq x$ is negligible.

Supposing that $\mathcal{H}^i(x) = \mathcal{H}(\mathcal{H}^{i-1}(x))$, a hash chain of length L, i.e., $\{S_i\}$, is constructed by recursively applying $\mathcal{H}$ to an initial seed value SD, where $S_i = \mathcal{H}^i(SD), i \in [1, L]$. Obviously, given S_i, it is easy to compute $S_j = \mathcal{H}^{j-i}(S_i)(j > i)$ but infeasible to obtain S_{i-1}.

BILINEAR PAIRING TECHNIQUE

Let $\mathbb{G}_1, \mathbb{G}_2, \mathbb{G}_T$ be three cyclic groups of prime order q. Coupling is an application $e : \mathbb{G}_1 \times \mathbb{G}_2 \rightarrow \mathbb{G}_T$ with the following three properties. 1) Calculable, for all $(x,y) \in \mathbb{G}_1 \times \mathbb{G}_2$, there is an efficient algorithm to compute $e(x,y)$; 2) Bilinear: for all $(x,y) \in \mathbb{G}_1 \times \mathbb{G}_2$, and all $(a, b) \in \mathbb{Z}_q^2, e(x^a, y^b) = e(x, y)^{ab}$; 3) Non-degenerate: there exists g_1 generator of $\mathbb{G}_1$ and g_2 generator of $\mathbb{G}_2$ such as $e(g_1, g_2) \neq 1_{\mathbb{G}_T}$.

Definition 2 (The Strong Diffie-Hellman Problem (SDH)): *Let $(q, \mathbb{G}_1, \mathbb{G}_2, \mathbb{G}_T, e, g_1, g_2)$ an environment bilinear of prime order q. Given $\left(g_1, g_2, g_2^x, g_2^{x^2}, \ldots, g_2^{x^q}\right)$ with x randomly chosen in $\mathbb{Z}_q$, the strong Diffie-Hellman problem is to find a couple $\left(c, g_1^{\frac{1}{x-c}}\right)$ with $c \in \mathbb{Z}_q$.*

SHORT SIGNATURES TECHNIQUE

The technical short signature is defined by the following four algorithms:

- ***S.Init***(1^λ)**:** This algorithm generates an environment bilinear $(q, \mathbb{G}_1, \mathbb{G}_2, \mathbb{G}_T, g_1, g_2, e, \psi)$. $\mathbb{G}_1, \mathbb{G}_2$ two groups of prime order q with g_1 a generator of $\mathbb{G}_1$ and g_2 a generator of $\mathbb{G}_2$. ψ is an isomorphism such as $g_2 = \psi(g_1)$. We note *S.param*=$(\lambda, q, \mathbb{G}_1, \mathbb{G}_2, \mathbb{G}_T, g_1, g_2, e, \psi)$.
- ***S.GenK(S.param)*:** This procedure randomly selects r and r' in $\mathbb{Z}_q^*$ and computes $u = g_1^r \in \mathbb{G}_1$ and $v = g_2^{r'} \in \mathbb{G}_2$. We note $z = e(g_1, g_2) \in \mathbb{G}_T$. The secret key of the signer is:

$$SK = (S.param, u, v, z) \tag{5}$$

- ***S.Sign(PK,SK,m)*:** Is the signature algorithm. It takes as input the signing key pair (*PK,SK*) and a message *m*. The signer randomly chooses $r \in \mathbb{Z}_q^*$ such as $x+m+yr^{10}\text{mod}q$ and computes:

$$A = g_1^{\frac{1}{x+m+yr}} \tag{6}$$

The signer obtains signature σ=(A,r) of message m.

- ***S.check(m,σ,PK)*:** With this procedure, the verifier ensures that σ=(A,r) is a signature of the signer, designated by the public key PK=(S.*param, u, v*, z), of message m checking the following equality:

$$e\left(A, u \cdot g_2^m \cdot v^r\right) = z \tag{7}$$

If equality (7) is valid, it returns true, otherwise it returns false.

OUR PROPOSED ASPP SCHEME

In this section, we will present our ASPP scheme, which mainly consists of six phases: 1) system initialization; 2) pseudo identity, private key and certificate issued by the TA; 3) node certificate updating; 4) message signature and verification; 5) response requested; and 6) demand response.

System Initialization

Assume that there exists a TA in the system, which initializes the whole system. For side routing, the AODV initializes three control messages {*Detectreq, Detectrep, Notifreq*}, which have the same message format of a route request message RREQ and includes three parts, namely, message signing cost, verification cost, and communication overhead, (including the certificate and the signature, as shown in Table 2). The TA initializes the system by running the following steps.

1. The TA initializes the security settings by running $S.Init\left(1^{\lambda}\right)$ to generate an environment bilinear $\left(q, \mathbb{G}_1, \mathbb{G}_2, g_1, g_2, e, \psi\right)$, where $\left(\lambda, l_1\right)$ is the security parameters.
2. The TA selects r and r' in $\mathbb{Z}_q^*$ and computes $u = g_1^r \in \mathbb{G}_1$ and $v = g_2^{r'} \in \mathbb{G}_2$, and picks up a random number $s \in \mathbb{Z}_q^*$ as the master-key. TA also chooses two secure cryptographic hash functions H_1, H_2, where $H_1: \{0,1\}^* \rightarrow \mathbb{G}_1$, $H_2: \{0,1\}^* \times \mathbb{G}_2 \rightarrow \{0,1\}^{l_1}$, and a secure symmetric encryption algorithm $Enc_s(.)$.
3. The TA chooses ΔT according to the privacy requirements of most nodes and sets the validity period of the certificate equal to ΔT. Then, the TA estimates the number of certificates that a node has to update from a SU once according to the SU density in each domain (Lu, 2012).
4. The TA publishes in ADSocial the public system parameters

$$\left(q, \mathbb{G}_1, \mathbb{G}_2, g_1, g_2, e, \psi, H_1, H_2, Enc_s\left(.\right), \Delta T\right)$$

and keeps the master-key s secretly.

Table 2. Format of the signed message Detectreq

<table>
<tr><td>0</td><td></td><td></td><td></td><td></td><td></td><td></td><td></td><td></td><td></td><td>1</td><td></td><td></td><td></td><td></td><td></td><td></td><td></td><td></td><td></td><td>2</td><td></td><td></td><td></td><td></td><td></td><td></td><td></td><td></td><td></td><td></td><td></td></tr>
<tr><td>0</td><td>1</td><td>2</td><td>3</td><td>4</td><td>5</td><td>6</td><td>7</td><td>8</td><td>9</td><td>0</td><td>1</td><td>2</td><td>3</td><td>4</td><td>5</td><td>6</td><td>7</td><td>8</td><td>9</td><td>0</td><td>1</td><td>2</td><td>3</td><td>4</td><td>5</td><td>6</td><td>7</td></tr>
<tr><td colspan="8">Type</td><td>J</td><td>R</td><td>G</td><td>D</td><td>U</td><td colspan="11">Reserved</td><td colspan="4">Hop Count</td></tr>
<tr><td colspan="28">Detectreq ID</td></tr>
<tr><td colspan="28">Destination IP Address</td></tr>
<tr><td colspan="28">Destination Sequence Number</td></tr>
<tr><td colspan="28">Originator IP Address</td></tr>
<tr><td colspan="28">Originator Sequence Number</td></tr>
<tr><td colspan="13">Certificate</td><td colspan="15">Signature</td></tr>
</table>

PSEUDO IDENTITY, PRIVATE KEY, AND CERTIFICATE ISSUED BY THE TA

When a new node n_i desire to communicate with other nodes, for an SU S_x in the domain D_y, the TA issues the private key SK_{n_i} and certificate $Cert_{TA,n_i}$ as shown in the list that follows.

1. The TA starts to compute the pseudo identity $PID_{n_i} = Enc_s\left(ID_{n_i}\right)$ with the master-key s.
2. The TA chooses a random number $r \in \mathbb{Z}_q^*$, sets the secret key $SK_{n_i} = H_1\left(PID_{n_i}\right)$ and the public key $PK_{n_i} = rg_1$.
3. The TA generates the signature σ_{TA,n_i} using the Schnorr signature algorithm (Schnorr, 1991), where $\sigma_{TA,n_i} = S.Sign\left(s, PK_{n_i}\right)$.
4. The TA securely delivers PID_{n_i}, SK_{n_i} and $Cert_{TA,n_i}$ to n_i where $Cert_{TA,n_i}\left(PK_{n_i}, \sigma_{TA,n_i}\right)$. Then, it stores the mapping between the real ID_{n_i} and $Cert_{TA,n_i}$.

The node n_i can verify the certificate $Cert_{TA,n_i}$ by the procedure $S.check\left(\sigma_{TA,n_i}, g_{1pub}, PK_{n_i}\right)$. With this method of key distribution, the ASPP guarantees the node identity confidentiality.

Node Certificate Updating

ASPP adopts a restore strategy, wherein each node can obtain a large set of certificates from the TA. But it cannot use these certificates directly in ad hoc social communication. In the domain D_y, only the certificates issued by an SU S_x belonging to this domain are valid. However, ASPP adopts the proxy re-signature cryptography technology (Toshiyukiet al, 2013), it only needs to request N_y resignature key from S_x and then re-signs the certificates issued by the TA to be the same as those issued by S_x itself. Given that the current time window is TW_k, the node n_i can submit the signing certificates *Scert* to request the corresponding re-signature keys from the S_x.

Figure 3. The flowchart for certificate updating

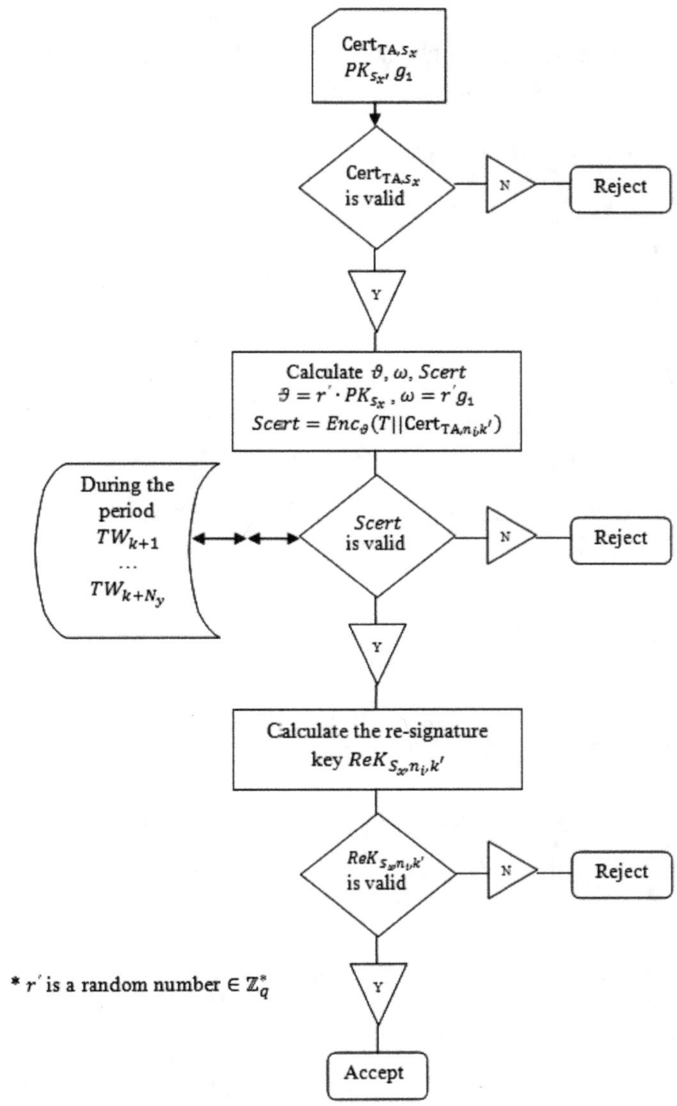

The whole process of certificate updating is shown in figure 3 and the list that follows.

1. The S_x broadcasts its certificate $Cert_{TA,S_x}$ periodically, e.g., every 3 s.
2. The node n_i verify the certificate $Cert_{TA,S_x}$. If $Cert_{TA,S_x}$ is valid, the node n_i selects a random number $r' \in \mathbb{Z}_q^*$ and calculates the shared secret key $\vartheta = r' \cdot PK_{S_x}$ and the hint $\omega = r'g_1$.
3. The node n_i calculates the signing certificates

$$Scert = Enc_\vartheta\left(T \,||\, Cert_{TA,n_i,k'}\right)$$

where $k' \in \left[k+1, k+N_y\right]$ and *T* is the time stamp, then sends the request message (*ω*, *Scert)* to S*x*.

4. To decrypt the request message and checks whether T is fresh, the social unit S_x calculates the shared secret key $\vartheta' = \omega \cdot SK_{S_x}$, then verify the signing certificates *Scert* during the period from TW_{k+1} to TW_{k+N_y}. If the verification is valid, S_x calculates the re-signature key

$$ReK_{S_x,n_i,k'} = \left(\frac{1}{SK_{S_x}}\right) \cdot PK_{TA,n_i,k'}$$

for each $Cert_{TA,n_i,k'} \in Scert$.

5. The social unit S_x sends $\{ReK_{S_x,n_i,k'}\}$ back to the node n_i.
6. The node n_i verifies each re-signature key in $\{ReK_{S_x,n_i,k'}\}$ by checking that

$$\left(ReK_{S_x,n_i,k'}, PK_{S_x}\right) = e\left(\left(\frac{1}{SK_{S_x}}\right) \cdot PK_{TA,n_i,k'}, PK_{S_x} \cdot g_1\right).$$

Message Signature and Verification

To sign the three messages {*Detectreq, Dectectrep, Notifreq*}, the node n_i should use Schnorr signature algorithm, i.e., $S.Sign\left(K, SK_{n_i}, Detectreq\right) \| \sigma_{n_i,Detectreq}$ where $K = H2\left(T \cdot Cert_{S_x,n_i}\right)$ is the temporary key, $Cert_{TA,n_i}$ is the certificate of n_i issued by the TA, and SK_{n_i} is the secret key of n_i. After receiving the message *Detectreq* from n_i, the other nodes first verify whether $Cert_{S_x,n_i}$ is valid and then accept the message if $S.Check\left(Detectreq, \sigma_{n_i,Detectreq}, K\right)$ is true.

Based on the transmit power, the time of sending the message *Detectreq*, and environmental conditions, the attack detection time $TDetect_{n_i}$ of n_i can be then defined as follows:

$$TDetect_{n_i} = TNormal_{Detectreq} - \frac{R}{SD} \tag{8}$$

where *R* the maximum transmission of node transmitter, *SD* is the speed of light, and $TNormal_{Detectreq}$ is time to send the message *Detectreq*. $TDetect_{n_i}$ must be calculated carefully to avoid erroneous decisions. If $TDetect_{n_i} < 0$, there's probably an attack. (See algorithm 1, line 10)

Response Requested

The adversary can delete, modify or copy the control messages to send false messages. However, to detect and verify the attack, each node $n_i \in$ ADSocial initiates a response request (runs algorithm 1) to nodes n_j in its routing table at 1-hop, and waits the response of his request for running the notification phase.

Algorithm 1. Response_requested

Input: The control message $Detectreq$
Output: The link is *proved* or *suspicious*
Begin
  Obtain the current timestamp;
  Compute the temporary key $K = H2\left(T \cdot Cert_{S_x,n_i}\right)$;
  Compute $C = S.Sign\left(K, SK_{n_i}, Detectreq\right) \| \sigma_{n_i,Detectreq}$;
  Obtain the short time $SIMS$ (Short Inter-Message Space);
  Send_Request (C);
  When node receive the encrypted control message $Detectrep$, recover $Detectrep$ from D ;
  Compute $\gamma = S.Check\left(Detectrep, \sigma_{n_j,Detectrep}, K\right)$;
  if $\gamma = true$ then begin
    Compute $TDetect_{n_i} = TNormal_{Detectreq} - \frac{R}{SD}$;
    if $TDetect_{n_i} < 0$ then begin
      return *suspicious*;
    end;
    else return *proved*;
  end;
  else Begin
    return *suspicious*;
  end;
End;

Algorithm 2. Notify

Input: The control message $Notifyreq$, the adversary node n_x
Output: R ready for transmission
Begin
  Compute $R = S.Sign\left(K, SK_{n_i}, Notifyreq\right) \| \sigma_{n_i,Detectreq}$;

```
3:      if  n_x ∈ (1 − hop of n_i) then
4:                begin
5:                    remove n_x in the routing table of n_i at 1-hop;
6:                    return;
7:                end;
8:      else return;
9:  End;
```

This phase of response request is summarized in the following steps:

1. The node n_i first signs the message with the private key SK_{n_i} and the temporary key $K = H2\left(T{\cdot}Cert_{S_x,n_i}\right)$. To avoid collisions, he waits the exhaustion of *SIMS* time. Then, it sends to all nodes at 1-hop in its routing table and continues to the next step. (See algorithm 1, line 1 to 6)
2. After receiving a request response sent by n_j, the node n_i retrieves from *R* the encrypted control message *Detectreq*. Then, it checks its signature. If valid and $TDetect_{n_i} < 0$, it considers that the link with the node n_j is proved. Otherwise, it returns suspicious and starts the notification phase. (See algorithm 1, line 7 to 19)

Figure 4. Response requested and demand response

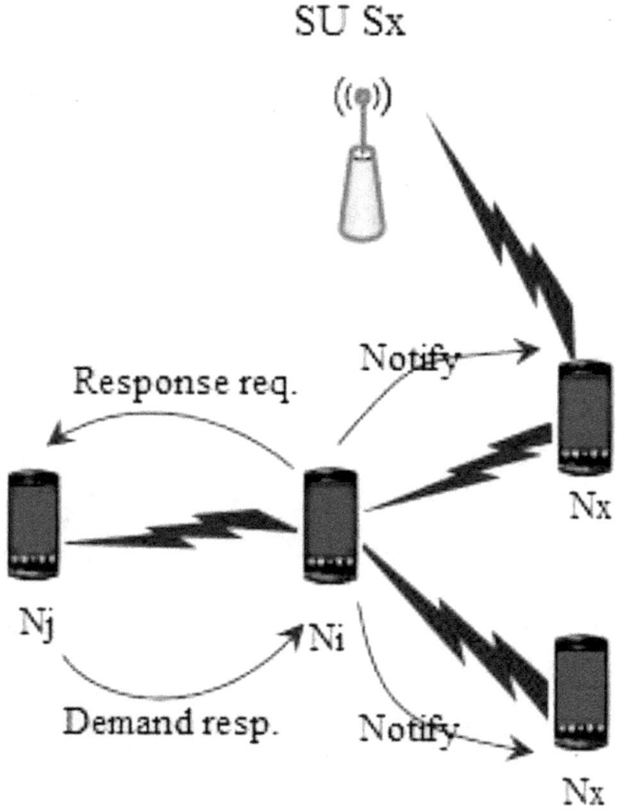

3. If the node n_i determines that the link with the node n_j is suspicious. The notification phase (See algorithm 2) is summarized in the following steps:
    a. The node n_i prepares a notification message *Notifyreq*. Then, it removes the adversary node n_j in its routing table 1-hop.
    b. The node n_i sends a notification message to its neighbors. When the neighbor node receives the notification request, it removes the node adversary in its routing table. In the end, the neighbor node broadcasts the notification message to all nodes in its routing table. (See Figure 4)

Remarks*:*

- In ASPP, a node takes a large number of certificates, but each certificate validates in different time slots. It can restrict the credential misuse.
- With the adopted re-signature cryptography technology (Toshiyuki et al, 2013), the node cannot generate a correct signature for the forged certificate.
- The notification phase (Algorithm 2) is executed after the verification of response message *Detectrep*.

Demand Response

When the node n_j receives request detection, it executes Demand_Reponse () (Algorithm 3).

Algorithm 3. Demand_Reponse

```
    Input: C the encrypted control message Detectreq
    Output: D the encrypted control message Detectrep
Begin
    Obtain the current timestamp ;
    Compute the temporary key K = H2(T · Cert_{S_x,n_j}) ;
    Compute D = S.Sign(K, SK_{n_j}, Detectrep) || σ_{n_j,Detectreq} ;
    When node receive the encrypted control message Detectreq,
          recover Detectreq from C ;
     Compute γ = S.Check(Detectreq, σ_{n_i,Detectreq}, K);
     if γ = true then Send_reponse (D, n_i) ;
                        Else Response_requested () ;
  End;
```

This phase of demand response is summarized in the following steps:

1. When the node n_j receives the encrypted control message *Detectreq*, it retrieves *C* and checks its signature $S.Check\left(Detectrep, \sigma_{n_i,Detectreq}, K\right)$.

2. If the signature is valid, the node n_j signs the response $S.Sign\left(K, SK_{n_j}, Detectrep\right) \| \sigma_{n_j,Detectreq}$ back to n_i. Otherwise, the node n_j sends response requested to the node n_i.

SECURITY ANALYSIS

In this section, we first describe the oracles, and then we prove that the proposed ASPP has the content oriented privacy, and conditional privacy for certificate. Finally, we discuss the robustness of the ASPP against elemental attacks and compound attacks.

The Oracles

The semantic security of *S.Sign(PK,SK,m)* in ASPP is defined using a game between a challenger and an adversary. Let a polynomial adversary $\mathcal{A}$ trying to gain experience against challenger $\mathcal{C}$. In ASPP, the adversary can use the following four oracles:

- $\mathcal{O}.\boldsymbol{Sign}\left(\boldsymbol{m}, \boldsymbol{PK}, \boldsymbol{VAB}\right)$ allows the adversary $\mathcal{A}$ to obtain an original signature on a message *m* of his choice with the signer's public key *PK* and the variable *VAB* that the adversary has set.
- $\mathcal{O}.\boldsymbol{Modify}\left(\boldsymbol{m}, \boldsymbol{\sigma}, \boldsymbol{ALT}\right)$ allows the adversary $\mathcal{A}$ to modify a pair of message-signing (m,σ) choice with the changes A*LT* that it wishes. The oracle returns a modified signing valid if changes are eligible and $\perp$ when error.
- $\mathcal{O}.\boldsymbol{Prove}\left(\boldsymbol{m}, \boldsymbol{\sigma}, \boldsymbol{DB}\right)$ allows the adversary $\mathcal{A}$ to obtain proof of the origin of a pair message-signing (m,σ) according a database $DB = \left\{\left(m_k, \sigma_k\right)\right\}_{k \in [1,q]}$.
- $\mathcal{O}.\boldsymbol{Sign} \,/\, \boldsymbol{Modify}_b\left(\boldsymbol{m}, \boldsymbol{VAB}, \boldsymbol{MOD}\right)$ allows the adversary $\mathcal{A}$ to take as input a message *m*, *VAB* and *MOD*. The oracle returns $\perp$ if *MOD* does not match with *VAB*. If $b=0$, it returns an original signature of signatory on m' (m' is the corresponding message to the message *m* modified by *MOD*). And if $b=1$, it performs an original signature of signatory on the message *m* by *VAB* and modify it by *MOD*.

Content Oriented Privacy

In our proposed ASPP, the content-oriented privacy can be guaranteed by the security of *S.Sign(PK,SK,m)*. If the ciphertext *S.Sign(PK,SK,m)* is provably secure, so does the content oriented privacy in ASPP. Therefore, we will prove the semantic security property of *S.Sign(PK,SK,m)* by using the techniques from provable security proved by Brzuska et al. (2009).

Theorem 1: Let $\left(q, \mathbb{G}_1, \mathbb{G}_2, \mathbb{G}_T, e, g_1, g_2\right)$ *be an environment bilinear of prime order,* λ *a security parameter and* $\mathcal{A}$ *be a polynomial adversary in ASPP seeking to gain experience against challenger* $\mathcal{C}$*. This adversary can access to the four oracles:*

$\mathcal{O}.Sign\left(m, PK, VAB\right)$,

$\mathcal{O}.Modify\left(m, \sigma, ALT\right)$,

$\mathcal{O}.Prove\left(m, \sigma, DB\right)$

$\mathcal{O}.Sign\ /\ Modify_b\left(m, VAB, MOD\right)$.

The ASPP scheme is secure, i.e., it is immutable, transparent, and accountable.

Proof: We stepwise go through the properties. Most times we describe each property for ASPP.

Immutability

Definition 3 (ASPP Immutability): *Let* λ *be a security parameter,* $S.Init\left(1^{\lambda}\right)$ *an initialization algorithm and S.GenK a key generation algorithm of signer (PK,SK). Let* $\mathcal{A}$ *an adversary with access to the signing oracle* $\mathcal{O}.Sign\left(m, PK, VAB\right)$ *and the oracle of prove* $\mathcal{O}\ .Prove\left(m, \sigma, DB\right)$. *We consider the following random experiment:*

$Experiment\ EXP_{\mathrm{Im}\,m,\mathcal{A}}^{ASPP}\left(\lambda\right)$

$S.param \leftarrow S.Init\left(1^{\lambda}\right)$

$\left(PK, SK\right) \leftarrow S.GenK\left(S.param\right)$

$\left(m^{*}, \sigma^{*}, PK^{*}\right) \leftarrow \mathcal{A}^{Sign,Prove}\left(S.param, PK\right)$

if $S.check\left(m^{*}, \sigma^{*}, PK^{*}\right) =$ true then $b \rightarrow 1$ else $b \rightarrow 0$

return b

We define the success of the adversary polynomial $\mathcal{A}$ in the experiment $EXP_{\mathrm{Im}\,m,\mathcal{A}}^{ASPP}\left(\lambda\right)$ via:

$$Succ_{\mathrm{Im}\,m,\mathcal{A}}^{ASPP}\left(\lambda\right) = \Pr\left[1 \leftarrow EXP_{\mathrm{Im}\,m,\mathcal{A}}^{ASPP}\left(\lambda\right)\right] \quad (9)$$

The ASPP is said to be $\left(\lambda, t, \epsilon\right)$ immutable secure, if no adversary $\mathcal{A}$ running in time *t* has a success $Succ_{\mathrm{Im}\,m,\mathcal{A}}^{ASPP}\left(\lambda\right) < \epsilon$, with ϵ negligible.

Transparency (Privacy)

Definition 4 (ASPP Transparency): *Let* λ *be a security parameter,* $S.Init\left(1^{\lambda}\right)$ *an initialization algorithm and S.GenK a key generation algorithm of signer (PK,SK) and b one bit chooses randomly. Let* $\mathcal{A}$ *an adversary with access to the four oracles:* $\mathcal{O}.Sign\left(m, PK, VAB\right), \mathcal{O}.Modify\left(m, \sigma, ALT\right), \mathcal{O}.Prove\left(m, \sigma, DB\right)$ *and* $\mathcal{O}.Signe\,/\,Modify_{b}\left(m, VAB, MOD\right)$ *. If b=0, the oracle of challenge returns a signature ; or a signature changed if b=*1*. We consider the following random experiment:*

$$Experiment\ EXP_{Transp,\mathcal{A}}^{ASPP}\left(\lambda\right)$$

$$S.param \leftarrow S.Init\left(1^{\lambda}\right)$$

$(PK,SK) \rightarrow S.GenK(S.param)$

$b \rightarrow \{0,1\}$

$$b^{*} \leftarrow \mathcal{A}^{Signe,Modify,Prove,Signe/Modify_{b}}\left(S.param, PK\right)$$

if $b^{*}=b$ then $a \rightarrow 1$ else $a \rightarrow 0$
return a

We define the advantage of the adversary polynomial $\mathcal{A}$ in the experiment $EXP_{Transp,\mathcal{A}}^{ASPP}\left(\lambda\right)$ via:

$$Adv_{Transp,\mathcal{A}}^{ASPP}\left(\lambda\right) = \left|\Pr\left[1 \leftarrow EXP_{Transp,\mathcal{A}}^{ASPP}\left(\lambda\right)\right] - \frac{1}{2}\right. \qquad (10)$$

The ASPP is said to be $\left(\lambda, t, \epsilon\right)$ transparent secure, if no adversary $\mathcal{A}$ running in time *t* has an advantage $Adv_{Transp,\mathcal{A}}^{ASPP}\left(\lambda\right) < \epsilon$, with ϵ negligible.

Accountability

Definition 6.3 (ASPP Accountability): *Let* λ *be a security parameter,* $S.Init\left(1^{\lambda}\right)$ *an initialization algorithm and S.GenK a key generation algorithm of signer (PK,SK). Let* $\mathcal{A}$ *an adversary with* *accesstotheoraclofchange* $\mathcal{O}.Modify\left(m,\Sigma,ALT\right)$ *andreturnsaquadruplet* $\left(SK^*,\pi^*_{or/ALT},m^*,\sigma^*\right)$.

We consider the following random experiment:

$Experiment\ EXP^{ASPP}_{Acc,\mathcal{A}}\left(\lambda\right)$

$S.param \leftarrow S.Init\left(1^{\lambda}\right)$

$(PK,SK) \rightarrow S.GenK(S.param)$

$\left(PK^*,\pi^*_{or/ALT},m^*,\Sigma^*\right) \leftarrow \mathcal{A}^{Modify}\left(S.param,PK\right)$

Let $\left(m'_k,\sigma'_k\right)_{k\in[1,n]}$ the response of the oracle modified
if $S.check\left(m^*,\sigma^*,PK^*\right) = \text{tru}$
and $(PK^*,m^*) \neq (PK^*_k,m^*_k)$ where $k\in[1,n]$
then $a\rightarrow 1$ else $a\rightarrow 0$
return a

We define the success of the adversary polynomial $\mathcal{A}$ in the experiment $EXP^{ASPP}_{Acc,\mathcal{A}}\left(\lambda\right)$ via:

$$Succ^{ASPP}_{Acc,\mathcal{A}}\left(\lambda\right) = \Pr\left[1 \leftarrow EXP^{ECPDR}_{Acc,\mathcal{A}}\left(\lambda\right)\right] \tag{10}$$

The ASPP is said to be $\left(\lambda,t,\epsilon\right)$ accountable secure, if no adversary $\mathcal{A}$ running in time *t* has a success $Succ^{ASPP}_{Acc,\mathcal{A}}\left(\lambda\right) < \epsilon$, with ∈ negligible

This completes the proof.

Conditional Privacy for Certificate

In ASPP, conditional privacy for certificate is preserved by re-signature technology where SU S_x acts as the certificate issuer for the node n_i and only the certificates issued by an SU S_x belonging to this domain are valid. With adopts the proxy re-signature cryptography technology (Toshiyukiet al, 2013), SU S_x has no idea about infringing on n_i's privacy.

Robustness

In this subsection, we discuss the robustness of ASPP. Specifically, we show how the ASPP prevents known attacks against the routing protocols, as elementary attacks and compound attacks. Both types of attacks affect the privacy of messages and the performance of routing protocols in ad hoc social communications.

- **Resistance Against Elemental Attacks:** An adversary could make the following four basic actions on messages:
    - Deleting a message, the adversary does not participate in routing, it's like he is not part of the social network.
    - Modification of a message, the adversary can play the sequence number of the destination and / or the number of hops in control messages by increasing the first and decreasing the second.
    - Manufacture of a message, the adversary can be made operational without being picked on the path. Like making a route reply even if no valid path to the destination.
    - Reduce the time, the adversary can reduce the processing time for control messages and retransmit faster so they reach the destination more quickly. This will ensure for the adversary a place on the path.

These attacks do not affect ASPP proposed because the node checked the validity of received message with *S.Check*. If message is not valid, it will be directly eliminated.

- **Resistance Against Compound Attacks**: An adversary can combine elemental attacks to perform compound attacks potential to achieve more advanced goals. As insertion into an already established route or not yet established, the creation of a routing loop or tunnel. However, the ASPP can detect and check the compound attacks and then notify other nodes, by use of control messages(*Detectreq*, *Detectrep*, *Notifyreq*) and authentication of this messages based on pseudo identity, the private key and certificate issued by the TA have been integrated in the proposed ASPP.

PERFORMANCE EVALUATION

To evaluate the proposed ASPP, we conducted simulations in two different scenarios. Our proposals were integrated in the AODV implementation, being the modified protocol designated AODV-ASPP. In this section, we first introduce the used simulation environment and afterwards present our simulation results for each scenario.

Scenario 1

In the first scenario, we evaluate the effectiveness of ASPP against black hole attack (Kamatchi et al., 2013) on reactive routing AODV using the simulator Ns-2 (Information Sciences Institute, 2007) configured with the standard IEEE 802.11 (11 Mbps and 2 Mbps were used to transmit unicast and broadcast traffic, respectively). We generated a number of random topologies with N nodes on a square field, where N is between 30 to 80 square field size varies of 600x600m to 1500x1500m depending on the size of ADSocial. The maximum transmission range is $\pi R2$ and traffic model is CBR.

Let $P_{k.neigh.adversary}$ denote the probability that there are at least k neighbors in the transmission range $\pi R2$ of an adversary node with the area S, then

$$P_{k.neigh.adversary} = P(N \geq k \mid \pi R^2)$$

$$= 1 - P\left(N < k \mid \pi R^2\right)$$

$$= 1 - \sum_{i=0}^{k-1} P\left(N = i \mid \pi R^2\right)$$

$$= 1 - \sum_{i=0}^{k-1} \binom{|\mathcal{V}|}{i} \left(\frac{\pi R^2}{S}\right)^i \cdot \left(1 - \frac{\pi R^2}{S}\right)^{|\mathcal{V}|-i} \tag{11}$$

And the black hole detection rate D_r

$$D_r = \frac{1}{T_{detect}} \tag{12}$$

where T_{detect} is the detection time of black attack.

Figure 5 shows the probability of k neighbors of an adversary $P_{k.neigh.adversary}$ in a parameterized wireless ad hoc social network with different k, ($1 \geq k \geq 20$). It can be seen the expected high probability of black hole attack can be achieved when choosing a proper k, i.e., $k \geq 5$.

Figure 6 (a,b,c) shows the black hole detection rate D_r varies with the tunnel length which is the number of hops between the adversary nodes, where k=5 and R=10m,15m,20m. As seen in Figure 6, the black hole detection rate D_r rises with the increase of R on the whole. For the same R, the black hole attack is detected more in the configuration where attack is launched on a number of more hops. That is obvious fact, as through the link of hole, packets are encapsulated. In addition, from Figure 6, we can also roughly observe the relation between D_r and the number of node N, i.e., with the increase of N, the black hole detection rate will also increase.

To further discuss on the black hole detection rate, Figure 7 (a,b,c,d) shows the black hole detection rate D_r varies for different HELLO emission T_{Hello} interval and different the black hole attack duration, where k=5 and R=10m,15m,20m,25m. As seen in Figure 7, if the black attack duration is shorter than black hole detection rate D_r becomes poor (i.e., less than 0.55). This is due to the fact that there are some nodes that do not perform the algorithm 1 and algorithm 2. As seen in Figure 6, also in Figure 7, the black hole detection rate D_r rises with the increase of R on the whole. This result demonstrates the impact of T_{Hello} on the detection time. If T_{Hello} emission interval is long enough, then it takes more time to detect. Therefore, the ECDPR scheme needs to use small message emission intervals. In addition, the earlier the black hole attacks are detected, the more energy can be saved in the whole network.

Figure 5. Probability of k neighbors of an adversary $P_{k.neigh.adversary}$ with S=200×200m², R=5m,10m,15m,20m, $|\mathcal{V}| = 100$, and 1≥k≥20

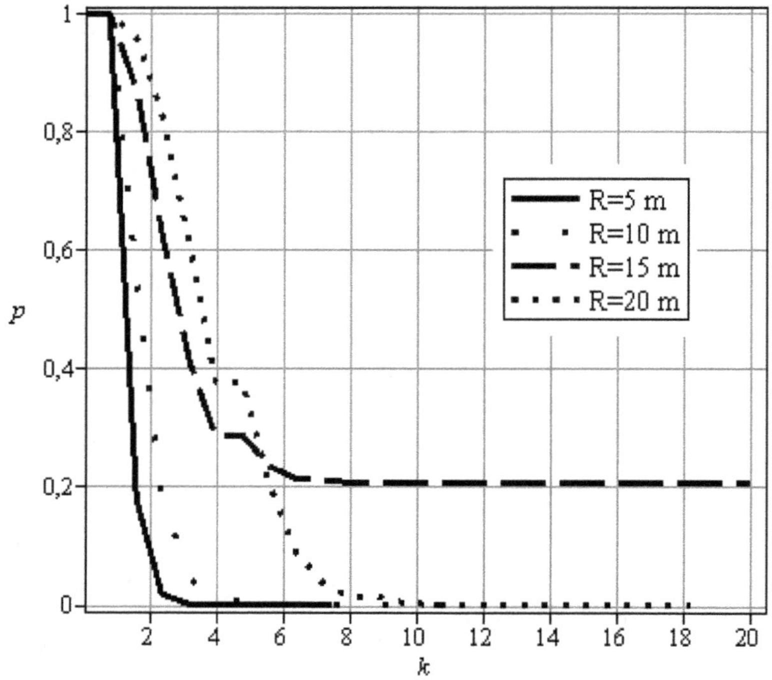

Scenario 2

An important performance metric in ADSocial systems is how long it takes for a node sends the message, i.e., *Detectreq* to reach its neighboring nodes. Thus, in the second scenario, we focus on the transmission delay of ASPP at the node. Roughly, the computation costs of ASPP include, detection, verification and notify of attack, which mainly involve the following cryptographic operations: authentication, genk, encrypt, decrypt, multiplication in Z_q^* and hash operations. We implement the Tate pairing with

Figure 6. The black hole detection rate D_r varies with the tunnel length

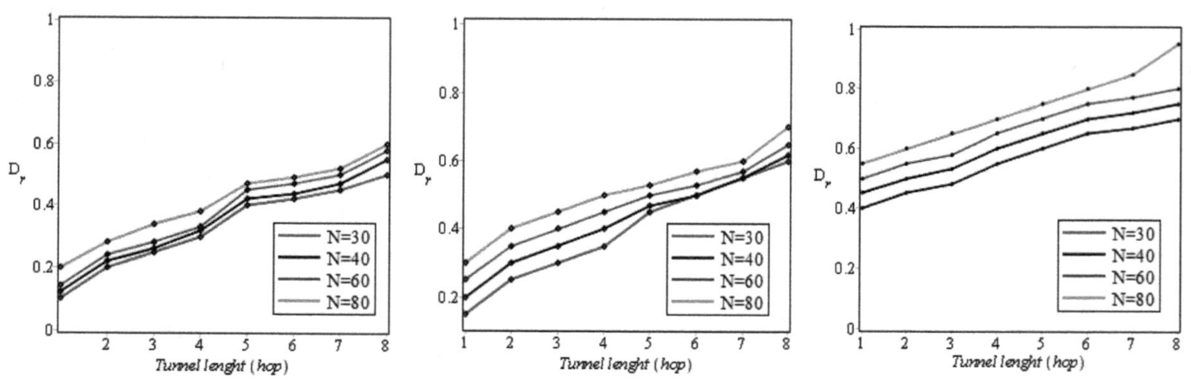

(a) D_r v.s. *Tunnel lenght* with $k = 5, R = 10\,m$ (b) D_r v.s. *Tunnel lenght* with $k = 5, R = 15\,m$ (c) D_r v.s. *Tunnel lenght* with $k = 5, R = 20\,m$

Figure 7. The black hole detection rate D_r varies for different HELLO emission interval T_{Hello} (s) and different the black hole attack duration

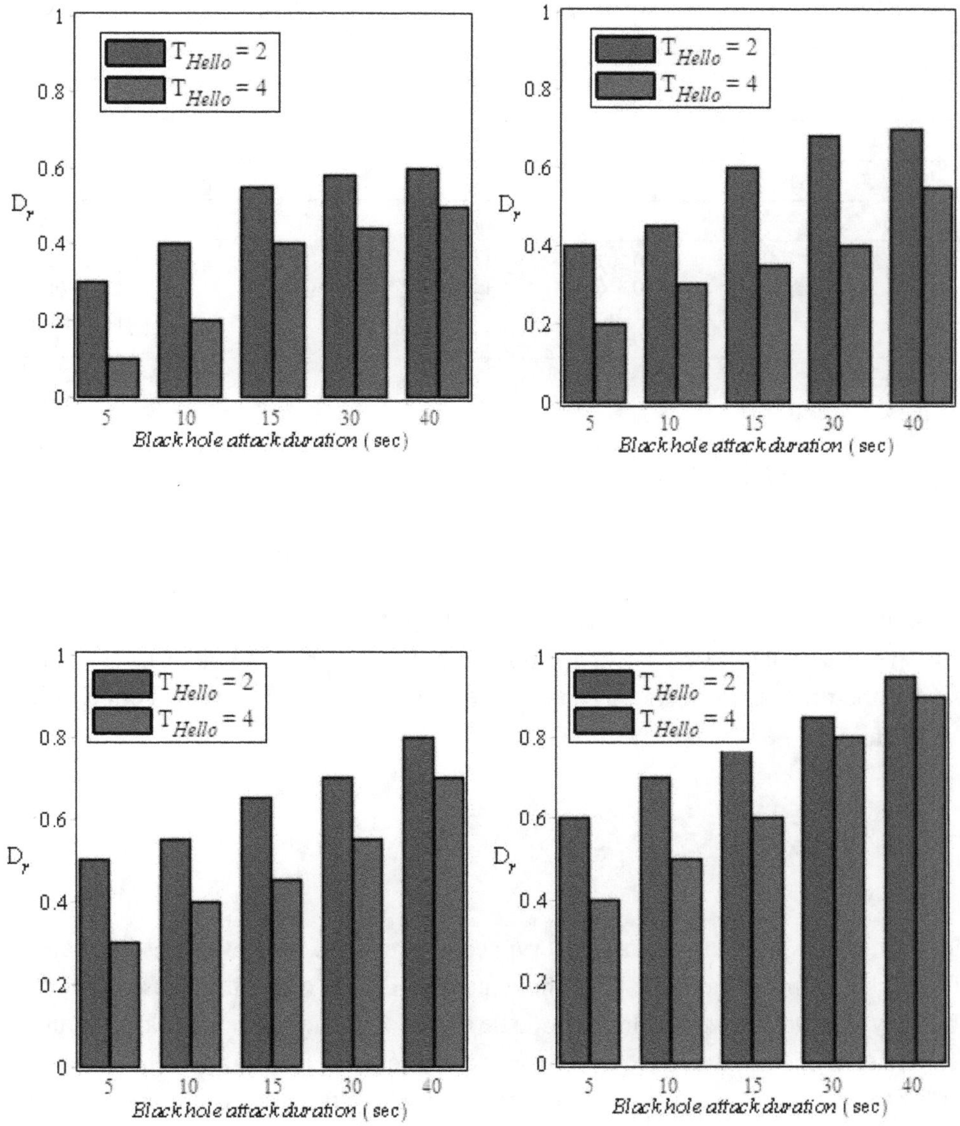

an embedding degree k = 2 Cocks-Pinch (CP-80) curve. CP-80 is over $\mathbb{F}_p$ with 512 bit prime p . Then, we implement the Ate paring with an embedding degree k = 6 Miyaji-Nakabayashi-Takano curve (MNT-80). MNT-80 is $\mathbb{F}_{p^3}$ with 160 bit prime p . Benchmarks for the selected pairing were running on a modern workstation, where the processor is 64-bit Intel i5 520M, clocked at 2.4GHz. The measured results are given in Table 3. Based on these benchmark numbers and the adopted Inner-Product Predicate Encryption (IPE) (Park, 2011), we can estimate the computation costs in ASPP, and the relevant results are given in Table 4 (Scott, 2011).

Then, we have

Table 3. Time costs of required operations

Curve	CP-80		MNT-80	
G_2 type	$\mathbb{F}_p$		$\mathbb{F}_{p^3}$	
k	2		6	
Modulus (bits)	512		160	
Paring	Tate		Ate	
with/without precomp.	w	w/o	w	w/o
GenK Encrypt Decrypt 2 Decrypt *	0.207ms 0.366ms 1.213ms 0.834ms	1.020ms 1.695ms 2.360ms 1.991ms	0.663ms 0.194ms 1.392ms 1.043ms	2.239ms 0.767ms 3.788ms 3.383ms

2: 2 pairings / *: multi-pairing

$$\mu = \begin{cases} 169.3 / s, w / o \ pairing \ precomputation; \\ 377.3 / s, with \ pairing \ precomputation. \end{cases}$$

Next we evaluate the transmission delay of ASPP. Based on the M/D/1 process (Donald, 2008), we consider the average arrival of *Detectreq* at the node is a Poisson process with arrival Rate λ, departure rate μ, and move the process from state i to i+1. The average delay time of Detectreq before being put into the node buffer is t_v,

$$t_v = \frac{1}{\mu} \cdot \frac{2-\rho}{2-2\rho}, \text{ where } \rho = \frac{\lambda}{\mu} \tag{13}$$

By broadcasting *Detectreq*, encrypt and decrypt can be reduced. However, this mechanism will incur the transmission delay. In addition, the black hole attack will also cause the transmission delay. Let the invalid probability of a *Detectreq* arriving at node be *p* due to the black hole attack. We study the average delay time in buffer of the node as follows. We first consider how long it takes the *i*-th *Detectreq* in the node to wait for the arrival of the next *i*+ 1-th Detectreq. Since the invalid probability of a *Deectreq* is *p*, when a valid *Detectreq* is put into buffer of the node, the number of *Detectreq* authentications at the node is a geometrically distributed random variable:

Table 4. Time costs of required operations in ASPP

Curve	CP-80		MNT-80	
With/Without Precomp.	w	w/o	w	w/o
Detectreq authentication Encrypt (t_e) Decrypt (t_d)	0.207ms 0.366ms 1.213ms	1.020ms 1.695ms 2.360ms	0.663ms 0.194ms 1.392ms	2.239ms 0.767ms 3.788ms

$$P(\text{number of authentication} = k) = p^{k-1}\left(1-p\right) \quad (14)$$

where =1,2,⋯. We define $t_{i(i+1)}$ to be the average waiting time,

$$t_{i(i+1)} = \sum_{k=1}^{\infty} \frac{k}{\mu} \cdot p^{k-1}\left(1-p\right) = \frac{1}{\mu\left(1-p\right)} \quad (15)$$

For i=1,2,…,n–1. Also, for the trivial case i=n, t_{ii}=t_{nn}=0. Thus, before *Detectreq* is sent, the waiting time for each *Detectreq* in buffer of the node is

$$T_i = \begin{cases} \frac{n-i}{\mu\left(1-p\right)}, i = 1,2,\cdots,n-1; \\ 0, i = n. \end{cases} \quad (16)$$

And the average waiting time is

$$t_w = \sum_{i=1}^{n} \frac{1}{n} T_i = \frac{1}{n} \cdot \frac{1}{\mu\left(1-p\right)} \cdot \left(1+2+\cdots+\left(n-1\right)\right)$$

$$= \frac{1}{n} \cdot \frac{1}{\mu\left(1-p\right)} \cdot \frac{n\left(n-1\right)}{2}$$

$$= \frac{1}{n} \cdot \frac{n\left(n-1\right)}{2\mu\left(1-p\right)}$$

and the transmission delay t_r of ASPP at the node in status the receiver node is

$$t_r = t_v + t_w + t_d = \frac{2-\rho}{2\mu\left(1-\rho\right)} + \frac{n-1}{2\mu\left(1-p\right)} + t_d, \; \rho = \frac{\lambda}{\mu} < 1 \quad (18)$$

Fixing the parameters n and p, Figure 8 (a,b,c,d,e,f) shows the transmission delay t_r varies with the average arrival rate λ, where $1 \geq \lambda \geq 100$. As seen in Figure 8, the transmission delay t_r rises with the increase of λ on the whole. Also, the transmission delay t_r with Cocks-Pinch curve (CP-80) is less than that Miyaji-Nakabayashi-Takano curve (MNT-80) .This result indicates that CP-80 the transmission delay could be reduced when the performance of the mobile device (the node) is improved. In addition, from Figure 8, we can also roughly observe the relation between t_r and p,n, i.e., with the increase of p and n, the transmission delay t_r will also increase.

Figure 8. Average transmission delay t_r varies with the average arrival rate λ, where $1 \geq \lambda \geq 100$

(a) t_r v.s. λ with $n = 1, p = 0$

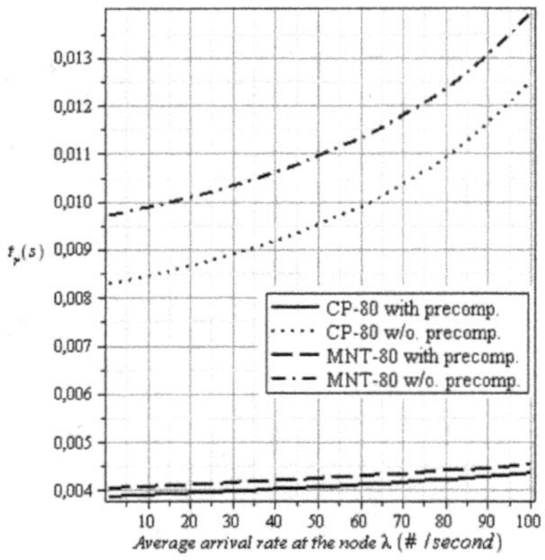

(b) t_r v.s. λ with $n = 10, p = 10\%$

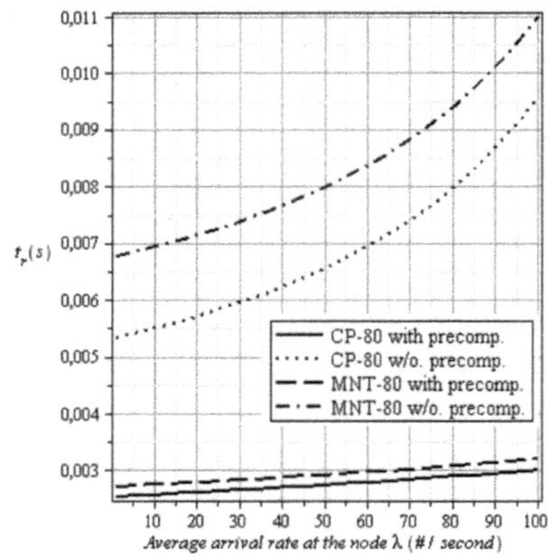

(c) t_r v.s. λ with $n = 20, p = 15\%$

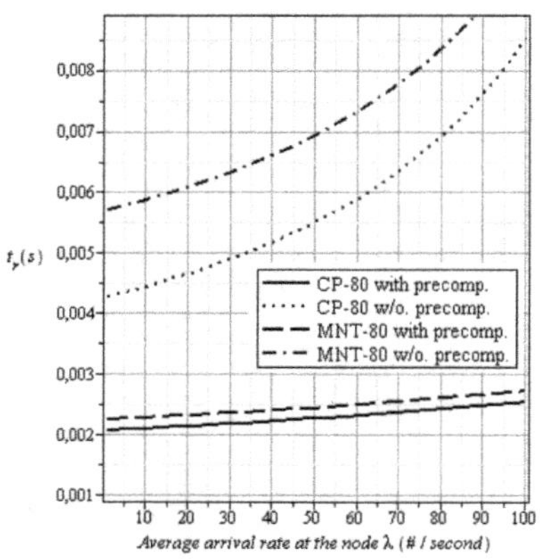

(d) t_r v.s. λ with $n = 35, p = 25\%$

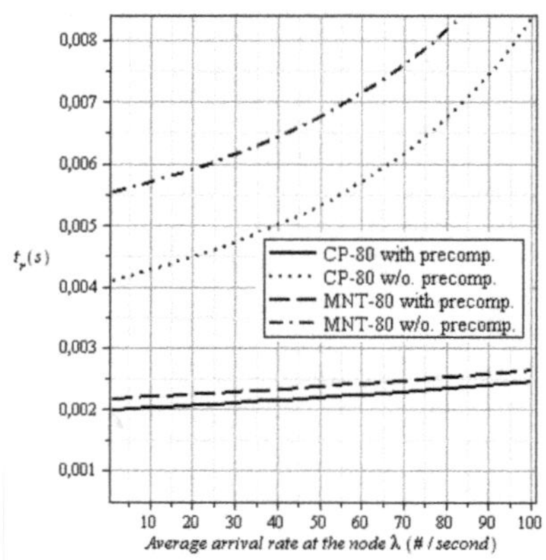

(e) t_r v.s. λ with $n = 50, p = 30\%$

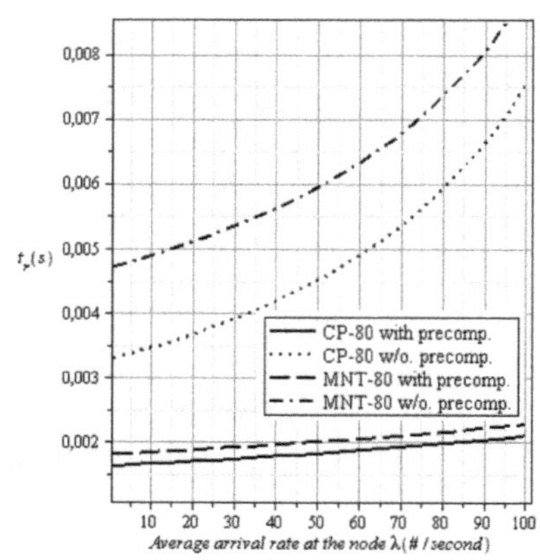

(f) t_r v.s. λ with $n = 80, p = 55\%$

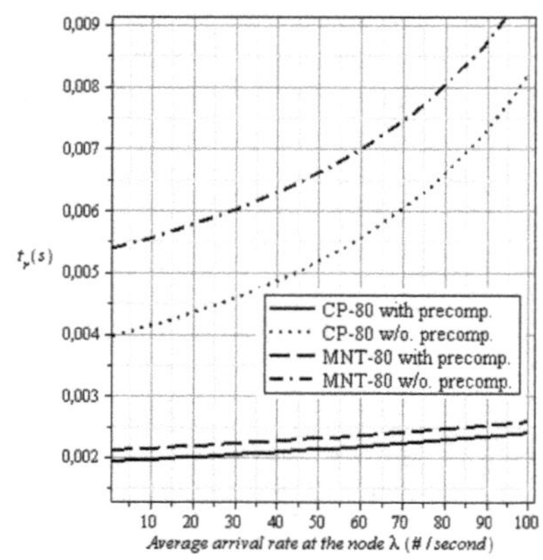

Figure 9. Average transmission delay t_r varies with p and n, where $1\% \geq p \geq 80\%$ and $1 \geq n \geq 100$

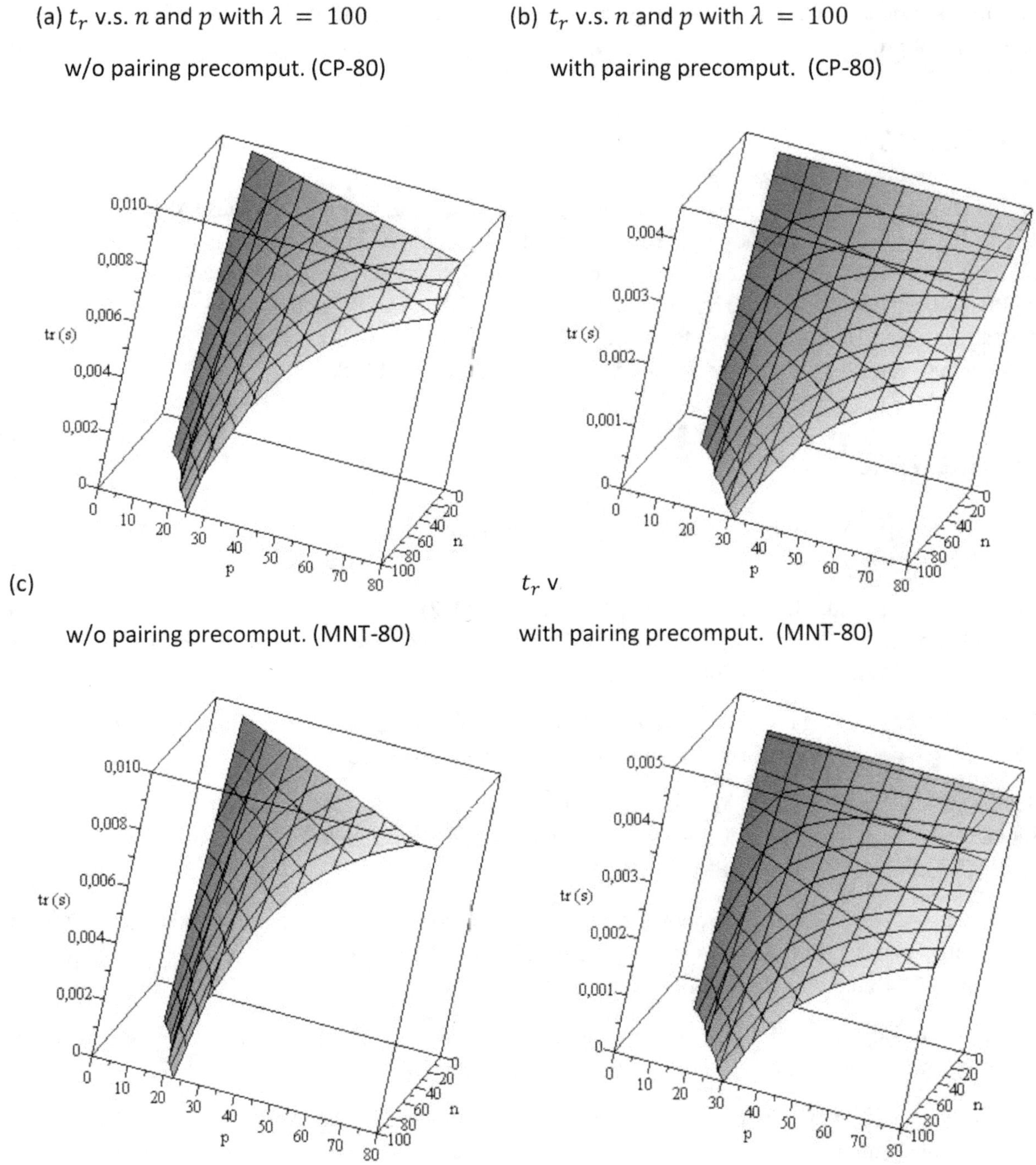

To further discuss the relation subtly, we plot t_r varies with p and n in Figure 9, where λ is fixed as 100. It can be seen that for small p, the transmission delay increases very slowly with n. However, for large p, the transmission delay increases quickly. This indicates that the parameter p due to the black hole attack is the dominant factor for the transmission delay. Therefore, for small, the proposed scheme can gain a good performance in terms of transmission delay. Also, the difference between Figure 9(a),

Figure 9(b), Figure 9(c) and Figure 9(d) demonstrates that the precomputation of CP-80 can reduce the transmission delay. In addition, the implementation of cryptographic methods used by ASPP on another type of processor, such as ARM-M series and Intel Atom, is underway.

CONCLUSION

In this chapter, we have proposed ASPP for secure ad hoc social communications. ASPP cannot only satisfy the security and privacy requirements of ad hoc but also can detect, prevent and notify the elemental attacks and composed attacks. In addition, through the extensive performance evaluation, the ASPP has been demonstrated to be disinfected and effective for AODV routing protocol against the black hole attack and efficient in terms of transmission delay. Our future work will focus on investigating the relation between user mobility and privacy under the black hole attack.

REFERENCES

Ad, I., Hubaux, J., & Edward, K. (2008). Impact of denial of service attacks on Ad Hoc networks. *IEEE/ACM Transactions on Networking*, *16*(1), 791–802. doi:10.1109/TNET.2007.904002

Awerbuch, B., David, H., Cristina, N., & Herbert, R. (2002). An on-demand secure routing protocol resilient to byzantine failures. In *Proceedings of the 1st ACM workshop on Wireless security* (pp. 21-30). Atlanta, GA: ACM. 10.1145/570681.570684

Azarderakhsh, R., Longa, P., Hu, S., & Jao, D. (2013). Efficient implementation of bilinear pairings on ARM processors. *Springer Selected Areas in Cryptography*, 149-165.

Batallas, D., & Yassine, A. (2006). Information leaders in product development organizational networks: Social network analysis of the design structure matrix. *IEEE Transactions on Engineering Management*, *53*(4), 570–582. doi:10.1109/TEM.2006.883706

Bender, A., Adam, B., Neil, S., Bobby, B., & Daniel, S. (2009). Persona: an online social network with user-defined privacy. In *Proceedings of the ACM SIGCOMM 2009 conference on Data communication* (pp. 135-146). Barcelona, Spain: ACM.

Boneh, D., & Boyen, X. (2008). Short signatures without random oracles and the SDH assumption in bilinear groups. *Journal of Cryptology*, *21*(2), 149–177. doi:10.100700145-007-9005-7

Brzuska, C., Fischlin, M., Freudenreich, T., Lehmann, A., Page, M., Schelbert, J., ... (2009). Security of sanitizable signatures revisited. In *Proceeding of the 12th International Conference on Practice and Theory in Public Key Cryptography* (pp. 317-336). Springer-Verlag Berlin.

Calandriello, G., Papadimitratos, P., Hubaux, J. P., & Lioy, A. (2007). Efficient and robust pseudonymous authentication in VANET. In *Proceeding of 4th ACM International Workshop VANET* (pp. 19–28). Montréal, QC, Canada: ACM. 10.1145/1287748.1287752

Chaudhary, P., & Gupta, B. B. (2017). A novel framework to alleviate dissemination of XSS worms in online social network (OSN) using view segregation. *Neural Network World, 27*(1), 5–25. doi:10.14311/NNW.2017.27.001

Clausen, T., & Jacquet, P. (2003). *Optimized link state routing protocol. RFC 3626*. IETF.

Deng, J., Han, R., & Mishra, S. (2003). *Enhancing base station security in wireless sensor networks. CU-CS-951-03: University of Colorado*. Department of Computer Science Technical Report.

Deng, J., Richard, H., & Shivakant, M. (2006). Decorrelating wireless sensor network traffic to inhibit traffic analysis attacks. *Elsevier Pervasive and Mobile Computing Journal, 2*(2), 159–186. doi:10.1016/j.pmcj.2005.12.003

Donald, G., John, F. S., James, M. T., & Carl, M. H. (2008). *Fundamentals of queueing theory* (4th ed.). Wiley.

Emre, S., Riva, O., Stuedi, P., & Alonso, G. (2009). Enabling social networking in ad hoc networks of mobile phones. *Proceedings of the VLDB Endowment International Conference on Very Large Data Bases, 2*(2), 1634–1637. doi:10.14778/1687553.1687611

Ferrag, M. A. (2012a). *Study of attacks in ad hoc networks*. LAP Lambert Academic Publishing.

Ferrag, M. A., Derdour, M., Mukherjee, M., Derhab, A., Maglaras, L., & Janicke, H. (2018). *Blockchain Technologies for the Internet of Things: Research Issues and Challenges. IEEE Internet of Things Journal.*

Ferrag, M. A., Maglaras, L., & Ahmim, A. (2017). Privacy-preserving schemes for ad hoc social networks: A survey. *IEEE Communications Surveys and Tutorials, 19*(4), 3015–3045. doi:10.1109/COMST.2017.2718178

Ferrag, M. A., Nafa, M., & Ghanemi, S. (2012b). OlsrBOOK: a privacy-preserving mobile social network leveraging on securing the olsr routing protocol. In *Proceeding of The 8 th International Scientific Conference eLearning and Software for Education* (pp. 133-139). Bucharest, Romania: Editura Universitara.

Ferrag, M. A., Nafa, M., & Ghanemi, S. (2013a). A new security mechanism for ad-hoc on-demand distance vector in mobile ad hoc social networks. In *Proceedings of the 7th Workshop on Wireless and Mobile Ad-Hoc Networks (WMAN 2013) in Conjunction with the Conference on Networked Systems NetSys/KIVS.* Stuttgart, Germany: WMAN.

Ferrag, M. A., Nafa, M., & Ghanemi, S. (2013b). Security and privacy in mobile ad hoc social networks. In D. B. Rawat, B. B. Bista, & G. Yan (Eds.), *Security, Privacy, Trust, and Resource Management in Mobile and Wireless Communications* (pp. 223–244). IGI Global.

Ferrag, M. A., Nafa, M., & Ghanemi, S. (2016). EPSA: An efficient and privacy-preserving scheme against wormhole attack on reactive routing for mobile ad hoc social networks. *International Journal of Security and Networks, 11*(3), 107–125. doi:10.1504/IJSN.2016.078390

Freeman, L. C. (1978). Centrality in social networks: Conceptual clarification. *Elsevier Social Networks, 1*(3), 215–239. doi:10.1016/0378-8733(78)90021-7

Guha, S., Tang, K., & Francis, P. (2008). NOYB: privacy in online social networks. In *Proceedings of the first workshop on Online social networks* (pp. 49–54). Seattle, WA: ACM. 10.1145/1397735.1397747

Gupta, B., Agrawal, D. P., & Yamaguchi, S. (Eds.). (2016). *Handbook of research on modern cryptographic solutions for computer and cyber security*. IGI Global. doi:10.4018/978-1-5225-0105-3

IETF MANET group. (n.d.). Retrieved June 01, 2013, from http://datatracker.ietf.org/wg/manet/

Information Sciences Institute. (2008). Retrieved June 01, 2013, from http://www.isi.edu/nsnam/ns/

Jian, Y., Shigang, C., Zhan, Z., & Liang, Z. (2007). Protecting receiver-location privacy in wireless sensor networks. In *Proceedings of 26th IEEE International Conference on Computer Communications* (pp. 1955-1963). Anchorage, AK: IEEE Computer Society. 10.1109/INFCOM.2007.227

Kamatchi, V., Rajeswari, M., & Raja, K. (2013). Securing data from black hole attack using aodv routing for mobile ad hoc networks. In *Proceedings of the Second International Conference on Advances in Computing and Information Technology* (pp. 365-373). Chennai, India: Springer Berlin Heidelberg. 10.1007/978-3-642-31552-7_38

Lee, J., & Hong, C. S. (2011). A mechanism for building Ad-hoc social network based on users interest. In *Proceedings of the 13th Asia-pacific Network Operatens and Management Symposium* (pp. 1-4). Taipei, Taiwan: IEEE Computer Society.

Li, H., Bok, K., & Yoo, J. (2013). Mobile P2P social network using location and profile. In Ubiquitous Information Technologies and Applications (Vol. 214, pp. 333-339). Springer Netherlands. doi:10.1007/978-94-007-5857-5_36

Li, T., Gupta, B. B., & Metere, R. (2018). Socially-conforming cooperative computation in cloud networks. *Journal of Parallel and Distributed Computing*, *117*, 274–280. doi:10.1016/j.jpdc.2017.06.006

Liang, X., Li, X., Luan, T., Lu, R., Lin, X., & Shen, X. (2012). Morality-driven data forwarding with privacy preservation in mobile social networks. *IEEE Transactions on Vehicular Technology*, *61*(7), 3209–3221. doi:10.1109/TVT.2012.2202932

Liang, X., Zhang, K., Shen, X., & Lin, X. (2014). Security and privacy in mobile social networks: Challenges and solutions. *IEEE Wireless Communications*, *21*(1), 33–41. doi:10.1109/MWC.2014.6757895

Lin, X., Lu, R., Shen, X., Nemoto, Y., & Kato, N. (2009). SAGE: A strong privacy-preserving scheme against global eavesdropping for ehealth systems. *IEEE Journal on Selected Areas in Communications*, *27*(4), 365–378. doi:10.1109/JSAC.2009.090502

Lu, R. (2012). *Security and privacy preservation in vehicular social networks*. University of Waterloo.

Lu, R., Liang, X., Li, X., Lin, X., & Shen, X. (2012a). EPPA: An efficient and privacy-preserving aggregation scheme for secure smart grid communications. *IEEE Transactions on Parallel and Distributed Systems*, *23*(9), 1621–1631. doi:10.1109/TPDS.2012.86

Lu, R., Lin, X., Liang, X., & Shen, X. (2012c). A dynamic privacy-preserving key management scheme for location based services in VANETs. *IEEE Transactions on Intelligent Transportation Systems*, *13*(1), 127–139. doi:10.1109/TITS.2011.2164068

Lu, R., Lin, X., Zhu, H., Liang, X., & Shen, X. (2012b). BECAN: A bandwidth-efficient cooperative authentication scheme for filtering injected false data in wireless sensor networks. *IEEE Transactions on Parallel and Distributed Systems*, *23*(1), 32–43. doi:10.1109/TPDS.2011.95

Mao, W. (2003). *Modern Cryptography: Theory and Practice*. Prentice Hall.

Nait-Abdesselam, F., Bensaou, T., & Taleb, T. (2008). Detecting and avoiding wormhole attacks in wireless ad Hoc networks. *IEEE Communications Magazine*, *46*(4), 127–133. doi:10.1109/MCOM.2008.4481351

Opsahl, T., Agneessensb, F., & Skvoretzc, J. (2010). Node centrality in weighted networks: Generalizing degree and shortest path. *Elsevier Social Networks*, *32*(2), 245–251. doi:10.1016/j.socnet.2010.03.006

Ozturk, C., & Zhang, Y. (2004). Source-location privacy in energy-constrained sensor network routing. In *Proceedings of the 2nd ACM workshop on Security of ad hoc and sensor networks* (pp. 88-93). Washington, DC: ACM. 10.1145/1029102.1029117

Park, J. H. (2011). Inner-product encryption under standard assumptions. *Springer Designs. Codes and Cryptography*, *58*(3), 235–257. doi:10.100710623-010-9405-9

Perkins, C., Belding-Royer, E., & Das, S. (2003). *Ad hoc On-Demand Distance Vector Routing*. Retrieved June 01, 2013, from http://tools.ietf.org/html/rfc3561

Sanguankotchakorn, T., Shrestha, S., & Sugino, N. (2012). Effective Ad Hoc Social Networking on OLSR MANET Using Similarity of Interest Approach. In *Proceedings of 5th International Conference IDCS*, (pp. 15-28). Wuyishan, Fujian, China: Springer Berlin Heidelberg. 10.1007/978-3-642-34883-9_2

Sanzgiri, K., Dahill, B., Levine, B. N., Shields, C., & Belding-Royer, E. M. (2002). A secure routing protocol for ad hoc networks. In *Proceedings of 10th IEEE International Conference on Network Protocols* (pp. 78-87). Paris, France: IEEE Computer Society. 10.1109/ICNP.2002.1181388

Schnorr, C. (1991). Efficient signature generation by smart cards. *Journal of Cryptology*, *4*(3), 161–174. doi:10.1007/BF00196725

Scott, M. (2011). On efficient implementation of pairing-based protocols. In *Proceedings of 13th IMA International Conference IMACC* (pp. 296-308). Oxford, UK: Springer Berlin Heidelberg.

Shirey, R. (2007). *Internet security glossary*. Retrieved June 01, 2013, from http://tools.ietf.org/html/rfc4949

Stergiou, C., Psannis, K. E., Kim, B. G., & Gupta, B. (2018). Secure integration of IoT and cloud computing. *Future Generation Computer Systems*, *78*, 964–975. doi:10.1016/j.future.2016.11.031

Toshiyuki, I., Nguyen, M., & Tanaka, K. (2013). Proxy re-encryption in a stronger security model extended from CT-RSA2012. In *Proceedings of The Cryptographers' Track at the RSA Conference* (pp. 277-292). San Francisco, CA: Springer Berlin Heidelberg.

Yanyong, Z., & Celal, O. (2005). Enhancing source-location privacy in sensor network routing. In *Proceedings of the 25th IEEE International Conference on Distributed Computing Systems* (pp. 599-608). Genova, Italy: IEEE Computer Society.

Yeung, C. A., Liccardi, I., Lu, K., Seneviratne, O., & Berners-lee, T. (2009). Decentralization: The future of online social networking. *W3C Workshop on the Future of Social Networking Position Papers.*

Yipin, S., Lu, R., Lin, X., & Shen, X. (2010). An efficient pseudonymous authentication scheme with strong privacy preservation for vehicular communications. *IEEE Transactions on Vehicular Technology, 59*(7), 3589–3603. doi:10.1109/TVT.2010.2051468

Zhang, K., Liang, X., Shen, X., & Lu, R. (2014). Exploiting multimedia services in mobile social networks from security and privacy perspectives. *IEEE Communications Magazine, 52*(3), 58–65. doi:10.1109/MCOM.2014.6766086

This research was previously published in the Handbook of Research on Cloud Computing and Big Data Applications in IoT; pages 94-126, copyright year 2019 by Engineering Science Reference (an imprint of IGI Global).

Index

E

F

H

I

J

K

L

M

N

O

P

Q

R

S

T

U

V

W

Y

www.ingramcontent.com/pod-product-compliance
Lightning Source LLC
LaVergne TN
LVHW080924210926
848485LV00004B/724

* 9 7 8 1 6 6 8 4 3 3 3 7 9 *